Synthese Library

Studies in Epistemology, Logic, Methodology, and Philosophy of Science

Volume 407

The aim of *Synthese Library* is to provide a forum for the best current work in the methodology and philosophy of science and in epistemology. A wide variety of different approaches have traditionally been represented in the Library, and every effort is made to maintain this variety, not for its own sake, but because we believe that there are many fruitful and illuminating approaches to the philosophy of science and related disciplines.

Special attention is paid to methodological studies which illustrate the interplay of empirical and philosophical viewpoints and to contributions to the formal (logical, set-theoretical, mathematical, information-theoretical, decision-theoretical, etc.) methodology of empirical sciences. Likewise, the applications of logical methods to epistemology as well as philosophically and methodologically relevant studies in logic are strongly encouraged. The emphasis on logic will be tempered by interest in the psychological, historical, and sociological aspects of science.

Besides monographs *Synthese Library* publishes thematically unified anthologies and edited volumes with a well-defined topical focus inside the aim and scope of the book series. The contributions in the volumes are expected to be focused and structurally organized in accordance with the central theme(s), and should be tied together by an extensive editorial introduction or set of introductions if the volume is divided into parts. An extensive bibliography and index are mandatory.

More information about this series at http://www.springer.com/series/6607

Stefania Centrone • Deborah Kant • Deniz Sarikaya
Editors

Reflections on the Foundations of Mathematics

Univalent Foundations, Set Theory and General Thoughts

Editors
Stefania Centrone
Technical University of Berlin
Berlin, Germany

Deborah Kant
University of Konstanz
Konstanz, Germany

Deniz Sarikaya
University of Hamburg
Hamburg, Germany

Synthese Library
ISBN 978-3-030-15654-1 ISBN 978-3-030-15655-8 (eBook)
https://doi.org/10.1007/978-3-030-15655-8

This Springer imprint is published by the registered company Springer Nature Switzerland AG.
The registered company address is: Gewerbestrasse 11, 6330 Cham, Switzerland

Introduction

Stefania Centrone, Deborah Kant, and Deniz Sarikaya

The present volume originates from the conference *Foundations of Mathematics: Univalent Foundations and Set Theory (FOMUS)*, which was held at the Center for Interdisciplinary Research of Bielefeld University from the 18th to the 23rd of July 2016. Within this framework approximately 80 graduate students, junior researchers and leading experts gathered to investigate and discuss suitable foundations for mathematics and their qualifying criteria, with an emphasis on homotopy type theory (HoTT) and univalent foundations (UF) as well as set theory. This interdisciplinary workshop, conceived of as a hybrid between summer school and research conference, was aimed at students and researchers from the fields of mathematics, computer science and philosophy.

A collected volume represents, it goes without saying, an excellent opportunity to pursuing and deepening the lively discussions of a conference. This volume, however, is not a conference proceedings in the narrow sense since it contains also contributions from authors who were not present at FOMUS. Specifically, 6 from the 19 contributions have been developed from presentations at the conference and only 9 from 24 authors were present at FOMUS.

As to the conference, the concomitant consideration of different foundational theories for mathematics is an ambitious goal. This volume integrates both univalent foundations and set theory and aims to bring some novelty in the discussion on the foundations of mathematics. Indeed, a comparative study of foundational

S. Centrone
Technical University of Berlin, Berlin, Germany
e-mail: stefania.centrone@tu-berlin.de

D. Kant
University of Konstanz, Konstanz, Germany
e-mail: kantdebo@gmail.com

D. Sarikaya
University of Hamburg, Hamburg, Germany
e-mail: Deniz.Sarikaya@uni-hamburg.de

frameworks with an eye to the current needs of mathematical practice is, even to this day, a *desideratum*.

The FOMUS conference was organized with the generous support of the Association for Symbolic Logic (ASL), the German Mathematical Society (DMV), the Berlin Mathematical School (BMS), the Center of Interdisciplinary Research (ZiF), the Deutsche Vereinigung für Mathematische Logik und für Grundlagenforschung der Exakten Wissenschaften (DVMLG), the German Academic Scholarship Foundation (Stipendiaten machen Programm), the Fachbereich Grundlagen der Informatik of the German Informatics Society (GI) and the German Society for Analytic Philosophy (GAP).

The editors received funding from the Claussen-Simon-Stiftung, Studienstiftung des deutschen Volkes, Heinrich-Böll-Stiftung, Hamburger Stiftung zur Förderung von Wissenschaft und Kultur, the Schotstek Network and the German Research Foundation (Temporary Positions for Principal Investigators, Heisenberg Programme).

Very special thanks go to the above-mentioned associations for the support of the conference as well as of the editors. Without this support the volume as it stands would not have been possible. The opinions in the volume do not necessarily match with those of the agencies.

The editors warmly thank Lukas Kühne and Balthasar Grabmayr for their help in organizing the conference. The encouragement at a decisive moment and the friendly advice from *Synthese Library*'s editor-in-chief, Otávio Bueno and from Springer's project coordinator, Palani Murugesan, were truly invaluable. A very special thanks goes to the authors of the contributions and to all anonymous referees who reviewed each single contribution.

The Topic

Set theory is widely assumed to serve as a suitable framework for foundational issues in mathematics. However, an increasing number of researchers are currently investigating Univalent Foundations as an alternative framework for foundational issues. This relatively young approach is based on HoTT. It links Martin-Löf's intuitionistic type theory and homotopy theory from topology. Such developments show the necessity of a novel discussion on the foundation of mathematics, or so we believe. The volume pursues two complementary goals:

1. To provide a systematic framework for an interdisciplinary discussion among philosophers, computer scientists and mathematicians
2. To encourage systematic thought on criteria for a suitable foundation

General criteria for foundations of mathematics can be drawn from the single contributions. Some candidates thereof are:

- Naturalness with regard to mathematical practice

- Applicability in mathematical practice
- Expressive power
- Possibility of extending the theory by justified new axioms
- Possibility of implementing the theory into formal proof systems
- Interpretability of non-classical (e.g. constructive) approaches
- Plausibility of the ontological implications

As far as set theory is concerned, the research literature is rich in perspectives and argumentations on foundational criteria. Roughly, set theory is seen as a theory with much expressive power, in which almost every other mathematical theory can be interpreted and which also may very well serve as ontological foundational framework for mathematics. However, it is not applicable in all mathematical areas and is not easily implemented in formal proof systems.

As far as HoTT is concerned, discussion on specific foundational criteria has not yet been properly carried out. Many scholars defend the thesis that HoTT is a very well applicable theory, easily implemented in formal proof systems (such as Coq and Agda) that can serve as a good foundational framework by dwelling rather on intensional than on extensional features of mathematical objects, at a variance with set theory.

Not least, the question about alternatives to the standard set-theoretic foundation of mathematics seems to be relevant also in view of the last developments of formal mathematics that appears to develop more and more in the direction of a mathematical practice focusing on the use of automated theorem provers.

Historical Background

Foundational disputes are not new in the history of mathematics. The *Grundlagen-streit* at the beginning of the twentieth century is but one ultimate example of a controversy, sometimes acrimonious, between various schools of thought opposed to each other.

Among the questions that mathematics poses to philosophical reflection on mathematics, those concerning the nature of mathematical knowledge and the ontological status of mathematical objects are central, when it comes to the foundations of mathematics. Is mathematics a *science* with an own content or is it a *language*, or rather, a *language-schema* that admits different interpretations? Does mathematical activity describe objects that are there, or does it constitute them? How do we explain our knowledge of mathematical objects?

If we believe that mathematics is a science with an own content, we have to say which objects mathematics is talking about. If these objects are mind-independent, how do we have access to them? If we constitute them, how do we explain the fact that the *same* mind-dependent objects are grasped by different subjects?

Philosophy of mathematics generally distinguishes three different views as to the foundations of mathematics: Logicism, Formalism and Intuitionism.

Logicism. The *descriptive view* at the turn of the century was represented by the *Logicism* of Gottlob Frege (1848–1925). Even the word "descriptive" hints at the fact that we are confronted with a version of the traditional standpoint of *platonism*. Numbers and numerical relations are abstract logical objects. Number systems are well-determined mathematical realities. The task of the knowing subject is to discover and to describe such realities that subsist independently of him *via* true propositions about such objects. The latter, organized systematically, make up the theory of that mathematical reality. Platonism is most often associated with an eminently *non-epistemic* conception of truth: the truth value of a proposition is independent of its being known.

The programme of logicism was to ground finite arithmetic on logic: the basic concepts of arithmetic (natural number, successor, order relation, etc.) had to be defined in purely logical terms, and arithmetical true propositions had to be *ideographically* derived from logical principles.

The arithmetization of analysis initiated by Karl Weierstrass (1815–1897) had concluded with the simultaneous publication in 1872 of the foundations of the system of real numbers by Richard Dedekind (1831–1916)[1] and Georg Cantor (1845–1918)[2]. Since, before then, it was well known how to define rational numbers in terms of integers and the latter in terms of natural numbers, the last question to be answered was how to lead back natural numbers to logic. To answer such question, Frege formulated in his *Grundgesetze der Arithmetik* (1901-1903) a system of principles from which the axioms of finite arithmetic should have been derived. Just as the second volume of the *Grundgesetze* was getting into print, Frege received a letter from Bertrand Russell (1872–1970), who called his attention on an antinomy arising by an indiscriminate use of the principle of unlimited comprehension, that is, the assumption that *each* concept has an extension. Frege could only recognize the mistake in a postscript.

Russell's Antinomy. Russell's antinomy, we recall it, says that class of all classes that are not elements of themselves is and is not element of itself. Indeed, if it is, it is not, since it is the class of all classes that do not have themselves as elements. If it is not, it is, for the same reason. The antinomy turns out to be relevant associated with Basic Law V of Frege's *Grundgesetze* that applies to functions and their course of values as well as to concepts (which for Frege are a special kind of functions) and their extensions. The gap is caused by the fact that Frege takes *courses of values* as well as *extensions of concepts* to be objects and takes Basic Law V as an *identity criterium* for such objects. Basic Law V stipulates that the extensions of two concepts (more generally the courses of values of two functions) are the same, if and only if the concepts apply exactly to the same objects (or the two functions have the same input-output behaviour).

[1] Hereto see Dedekind 1872.

[2] Hereto see Cantor 1872.

Basic Law V for concepts runs as follows:

$$Ext(P) = Ext(Q) \leftrightarrow \forall x(P(x) \leftrightarrow Q(x))$$

while its more general version for functions runs as follows:

$$\varepsilon' f(\varepsilon) = \varepsilon' g\,(\varepsilon) \leftrightarrow \forall x(f(x) = g(x))\,.$$

Russell's antinomy can be presented in Frege's system thus:
Let "R(x)" stand for the predicate "x is Russellian":

$$(*)\ \ R(x) \leftrightarrow \exists Y(\ x = Ext(Y) \wedge \neg Y(x))\,.$$

That is, x is Russellian iff x is the extension of a concept that does not apply to x.
Let "*r*" be short for "*Ext(R)*", the extension of the concept *R*. Then the contradiction

$$R(r) \leftrightarrow \neg R(r).$$

is easily derived as follows.

(1) ¬R(r) → R(r). Assume ¬R(r); then, according to the definition of r:

$$r = Ext(R) \text{ and } \neg R(r).$$

By existential quantifier introduction it follows thereof:

$$(**)\,\exists Y.\ r = Ext(Y) \wedge \neg Y(r),$$

and so, by (*), we get $R(r)$.

(2) $R(r) \rightarrow\ \neg R(r)$. Assume $R(r)$; then by (*)

$$\exists Y(\ r = Ext(Y) \wedge \neg Y(r))\,.$$

Let then (by ekthesis) Y be such that $r = Ext(Y)$ and $\neg Y(r)$.
So $r = Ext(R)$ and $r = Ext(Y)$, hence $Ext(Y) = Ext(R)$. At this point Basic Law V comes into play, yielding

$$\forall z(Y(z) \leftrightarrow R(z))\,.$$

Therefore from $\neg Y(r)$ we get $\neg R(r)$.
Thus, in conclusion, r is at once *Russellian* and *not-Russellian*, and Frege's system is, at least without amendments, inconsistent.

Formalism: Hilbert's Program. At the turn of the century Hilbert was trying to establish in Germany an interdisciplinary area of research for mathematicians, logicians and philosophers modelled on the kind of cooperation that was in bloom at Cambridge around Bertrand Russell. Hilbert held Russell's and Whitehead's work in high esteem, and he was "convinced that the combination of mathematics, philosophy and logic ... should play a greater role in science".[3] Hilbert's Program undergoes many phases. The initial phase of his reflections on foundations spans from 1898 to ca. 1901. Hilbert aimed at an axiomatic foundation not only of mathematics but also of physics and other sciences through the formal-axiomatic method. As to mathematics he took to be possible to reduce the axiomatic foundation of all mathematics to that of the arithmetic of real numbers and set theory. More specifically, Hilbert took the complete formalization of concrete mathematics, i.e. of the theory of natural numbers as well as analysis and set theory, to be possible. He took Peano-Arithmetik (**PA**), Peano-Arithmetik at the second order ($\mathbf{PA_2}$) and Zermelo-Fraenkel set theory (**ZF**) to be formal counterparts of the concrete mathematical theories. Formalization, however, was only the first step of Hilbert's Program. For, once the concrete theories were formalized, the proper task was to prove the consistency of the formalized theories. In this way, Hilbert thought to avoid antinomies like the Russellian one.

Hilbert addressed the problem of a proof of consistency for arithmetic in a conference, entitled "Mathematical Problems (*Mathematische Probleme*)", held to the second International Congress of Mathematicians in Paris in 1900. As he put it: "But above all I wish to designate the following as the most important among the numerous questions which can be asked with regard to the axioms: To prove that they are not contradictory, that is, that a definite number of logical steps based upon them can never lead to contradictory results."

A kind of criticism is often raised against formalism, namely, that the latter reduces mathematics to a meaningless symbolic game by investigating the logical consequences of axiom systems set up arbitrarily. However, formalization was not the primary goal of formalism, but was conceived of as a necessary condition for proving the consistency of mathematics. Only then, mathematics would have been secure.

A necessary condition for the formalization was to find for each concrete theory a formal counterpart able to capture *all* its truths. In particular, all arithmetical truths should have been provable in **PA**, all truths of analysis should have been provable in (a theory equivalent to) $\mathbf{PA_2}$ and all set-theoretical truths should have been provable in **ZF**. However, like the logicistic programme also the formalistic programme was doomed to fail. In 1931 Kurt Gödel (1906–1978) proved his famous incompleteness theorems. They can both be conceived as limits put to the power of formalisms. The first theorem says that arithmetic is syntactically incomplete. More specifically, the theorem says that each theory that is (**i**) consistent, (**ii**) effective and (**iii**) contains some arithmetic (precisely, as much as Robinson's arithmetic **Q**, a weak subsystem

[3]Cp. Hilbert et al.'s "Minoritätsgutachten" of 1917.

of **PA**) is syntactically incomplete: there is, at least, one sentence of the language of the theory that is neither provable nor refutable within the theory. So, **PA** is not able to capture formally all arithmetical truths. All the more so, as to the formalized counterparts of analysis (equivalent to $\mathbf{PA_2}$) and set theory (**ZF**).

The second step of Hilbert's Program was to make mathematics secure. In particular, Hilbert required the use of infinitary parts of mathematics to be justified. It must be said that Hilbert started to talk about "ideal elements" in a later phase of the foundational research (1920–1924), in which "Hilbert's Program" takes its proper shape. In a paper presented in Leipzig in 1922 entitled "*The Logical Foundations of Mathematics* (*Die logischen Grundlagen der Mathematik*)" (Hilbert 1923), Hilbert indicates *finitary mathematics* as that part of mathematics that "has a concrete content", that is, that it operates concretely with symbols and does not use infinitary procedures and principles. It is in this context that he starts talking about "finitary logic" (the logic of finitary procedures) and of "ideal elements". Hilbert presents his paper "*On Infinite* (*Über das Unendliche*)" (Hilbert 1926) in Münster in 1925, where he explicitly speaks of "ideal elements". Here *finitary mathematics* is said to be that part of mathematics that can be rightfully considered as "secure". It does not need a justification but must itself serve as justification for *infinitary mathematics*, which is that part of mathematics that deals with actual infinity, lacks a concrete content and is moreover a possible source of contradictions. Infinitary instruments are acknowledged as "useful": they are used to prove *real* propositions. However, they must be justified, that is one has to demonstrate that their use does not lead to contradictions.

To this aim, the proof of consistency for the formalized theories **PA**, $\mathbf{PA_2}$ and **ZF** turned out to be essential. But, to serve as a justification, the proof should have used means that did not need themselves a justification. It should have used only *finitary mathematics*.

However, while the first of Gödel's theorems had shown the impossibility of completely formalizing the concrete mathematical theories, the second theorem showed that it was not possible to prove the consistency of an arithmetical theory using only means formalizable in that very same theory. More exactly, Gödel's second incompleteness theorem says that each theory, that is (**i**) consistent, (**ii**) effective and (**iii**) contains as much arithmetic as **PA** (actually, a weaker system like, e.g. **PRA**, primitive recursive arithmetic, suffices), cannot prove (the statement formalizing) its own consistency. Thus, if one assumes that *finitary mathematics* is part of **PA**, is it not possible to prove the consistency of **PA** with *finitary* means.

Thus, even the formalistic approach was denied the likelihood of success.

Intuitionism. Intuitionism was first brought forward by Luitzen E. J. Brouwer (1881–1966). Brouwer had completed his academic studies in 1907 at the University of Amsterdam with a dissertation on the foundations of mathematics (*Over de grondslagen des wiskunde*). Already at this time he had taken up a position in direct opposition to formalism and logicism, many years before he would engage in the attempt to construct intuitionistic mathematics. What was at issue was, once again, to build mathematics on an absolutely secure ground with no risk to run into antinomies. Intuitionism is in its origin a particular way of conceiving mathematics. Mathematics is the exact part of the thinking activity of an idealized knowledge-subject. Such activity grounds on our *Ur-intuition* of the flow of time. Kant's conception of time explicitly works at the background. In his doctoral dissertation of 1907 Brouwer writes: "*Mathematics can deal with no other matter than that which it has itself constructed.*"[4] At a variance with Platonism, mathematical objects are *not* objects *that are there* and *stand in certain relations*, but rather mental constructions of an idealized knowledge-subject. At a variance with formalism, the role of the language is only peripheral, it is useful to communicate the results of the thinking activity of the idealized subject as well as *aide-mémoire*, but mathematics as a mental construction has not to be confused with its linguistic expression.[5]

The basic phenomenon of intuitionism, the *Ur-intuition* of time, consists, according to Brouwer, in the perception of the split-up of one moment of life into two different things which differ qualitatively. Therein consists the "first act of intuitionism" that is attributed, by Brouwer, to the mind and not to sensitiveness.

> Mathematics arises when the subject of two-ness, which results from the passage of time, is abstracted from all special occurrences. The remaining empty form [the relation of n to n+1] of the common content of all these two-nesses becomes the original intuition of mathematics and repeated unlimitedly creates new mathematical subjects.[6]

According to Brouwer, a mathematical proof is a mental construction. It is essential, for the intuitionistic view, to conceive mathematical objects as results of finite mental construction processes. This conception goes hand in hand with a strongly epistemic view about truth: the truth of a proposition depends, in an essential way, on the knowledge of the subject.

Another characteristic feature of intuitionism is the refusal of the actual infinite. Only finite reasoning can be justified. Measurable infinities are admitted in the sense that the step from n to n+1 can be repeated unlimitedly and is understood as principle of formation of the sequence of natural numbers. As Dummett puts it: "[T]he thesis that there is no completed infinity means, simply, that to grasp an infinite structure is to grasp the process which generates it, that to refer to such a structure is to refer to that process, and that to recognize the structure as being infinite is to recognize that the process will not terminate."[7] This view contrasts markedly with classical mathematics, which deals with the infinite process as a whole: "It is, however, integral to classical mathematics to treat infinite structures as if they could be completed and then surveyed in their totality."[8] To top it off, "an infinite process is spoken of as if it were merely a particularly long finite one".[9]

---

[4]Brouwer 1975, 51.

[5]Cp. Heyting's Introduction to Brouwer 1975, xiv.

[6]Cit. in Kline 1972, 1199–2000.

[7]Dummett 1977, 56.

[8]loc.cit., 56.

[9]loc.cit., 57.

Intuitionistic Logic. Arendt Heyting (1881–1996) makes intuitionistic logic systematic in his work *Die Formalen Regeln der intuitionistischen Logik* (1930). While classical logic makes use of non-epistemic concepts of truth and falsehood (the truth of a statement is independent of its being known), intuitionistic logic grounds on a basically epistemic conception of them. Truth and falsehood are not properties that a statement has independently of its being known. That a statement is true means that the knowledge-subject has direct evidence for it or can exhibit a proof, i.e. a suitable mental construction, for it. Similarly, that a statement is false means the knowledge-subject knows that no matter how his knowledge will develop, he will never be able to exhibit a proof for it. Thus, while intuitionism reads the sentence "that p, is true" as "the epistemic subject is in possession of a proof for the proposition, that p", the assertion "that p, is false", in symbols: "$\neg p$", is saying something much stronger than "the epistemic subject is not in possession of a proof for p". "$\neg p$" means "the epistemic subject is in possession of a proof of the impossibility of p", i.e. he knows he will never have evidence for p, no matter how his cognitive process develops further. On the basis of these reflections, intuitionism does not include, among others, the law of the excluded middle in the tautologies of logic. There are statements, such as Goldbach's Conjecture (every even, positive integer greater than two is the sum of two primes), that are at the present undecided. Intuitionism emphasizes the temporal aspect of knowledge. The fact that someday, it will be perhaps possible to determine whether Goldbach's Conjecture is true or false does not change the current indecision. Potential truth values cannot replace actual ones. Only if a proper mental construction lays before the epistemic subject, he or she knows that p, or, respectively, that $\neg p$. The excluded middle ($p \vee \neg p$) as valid logical principle is dropped.

After this argumentation, it becomes clear that intuitionistic logic cannot admit the formal counterpart of indirect reasoning:

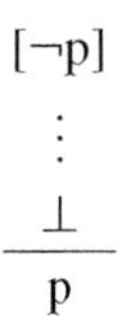

Another point of disagreement between classical and intuitionistic (predicate) logic is constituted by the interpretation of the quantifiers. Classical logic deals with quantified statements *as if* the multiplicity of objects the quantifier ranges over would be finite. To say it roughly: when the classical logician writes '$\forall x\alpha(x)$' he or she takes for granted that it is in principle possible to verify, for each single element of the domain in question, that it has the property of being α. But this is possible only if the quantifier ranges over a collection of only finitely many entities.

Current Foundations

Set Theory

Set theory is rather well known to be a quite efficient framework for foundational issues: mathematical as well as metamathematical objects can easily be represented within it, it gives clear answers to old philosophical questions on basic assumptions, allowed inferences and gapless proofs. Set-theoretic accounts, however, have to face one main challenge: the independence phenomenon, or syntactic incompleteness of set theory.

Let us recall that **ZFC** originated, in the first place, as an answer to the problems arisen within the context and set of problems of Cantor's *naïve* set theory. As is well known, Zermelo suggested a first axiomatization in 1908,[10] which was further developed to **ZFC**. **ZFC** as foundational framework helps to give an answer to the following fundamental questions:

(i) *Which are the basic assumptions of mathematics?*
(ii) *What is an allowed inference?*
(iii) *What counts as a proof without gaps?*

Indeed, one takes (**i**) the basic assumptions to be the axioms of **ZFC**, (**ii**) an allowed inference to be such that it only employs inference rules whose premises and conclusions are written in the language of first order logic, and (**iii**) a proof without gaps to be a derivation of a sentence from the **ZFC**-axioms through the use of allowed inferences.

Set theory can represent most *mathematical* objects from any mathematical area. There are standard representations of simple mathematical objects such as the following of natural numbers:

$$0 = \varnothing, 1 = \{\varnothing\}, 2 = \{\varnothing, \{\varnothing\}\}, \ldots, \mathrm{n} + 1 = \mathrm{n} \cup \{\mathrm{n}\}, \ldots$$

Functions are represented as sets of ordered pairs, groups $(G, \cdot)$ as the ordered pairs of a set G and an operation $\cdot$, and so on.

Set theory can represent *metamathematical* objects as well, for instance, through Gödel's coding.

It is worth mentioning that we do not need the full power of **ZFC** to formalize a number of metamathematical concepts. In particular, for that part of mathematics needed to set up **ZFC**, only some recursion theory is actually needed.

It is the representing property that makes it possible to take set theory as foundational framework for mathematics, metamathematics and yet, as some argue,

[10] Zermelo 1908.

for all mathematical reasoning. *De facto* "foundational framework" and "set theory" are most often used as tantamount.[11]

Nonetheless, the *independence phenomenon* constitutes no minor problem for any set-theoretic account. For, if set theory cannot *decide* every sentence of its own language, how can it work as a suitable framework for all mathematical reasoning? We recall that a theory **T** is syntactically complete when it decides every sentence **A** of its language. Sentences that are neither provable nor refutable in a theory are called "independent sentences". Set theory happens to be more sensible to the independence phenomenon than other mathematical areas, and it is often the use of a set-theoretic framework that makes the independence problem come to the fore in different mathematical fields, such as in operator algebra.[12]

A number of set theorists appear to be quite indifferent to the philosophical implications of the *independence phenomenon*. Others consider independence to be avoidable and search for stronger methods to decide whether an independent sentence is true or false. Two main questions arise in this context:

(i) Are *all* set-theoretic sentences either true or false?
(ii) If **ZFC** is extended to a stronger theory which axioms should be added?

Question **i.** is usually referred to as "realism/pluralism debate". "Realism" denotes that particular position in the philosophy of mathematics that takes *every* set-theoretic sentence to be either true or false, "pluralism", at variance, denotes the view that there are set-theoretic sentences that are true in one mathematical reality and false in another.[13] Ontological talk is often used to formulate question **i.**. Under the heading "universism" one usually understands the view that set theory has a unique model (as argued, for instance, by W. Hugh Woodin), under "multiversism" the view that there is a multiverse of various different universes of sets (as argued, for instance, by Joel D. Hamkins.)[14]

Answers to question **ii.** are often articulated in proposing criteria for new axioms. Already Kurt Gödel had suggested that new axioms could be justified either *intrinsically* or *extrinsically*.[15] We can render the idea of an *intrinsically* justified axiom by saying that it flows from the very nature of set or that it roots in the concept of set. The *extrinsic* justification plays, nowadays, a major role in Penelope Maddy's approach. Roughly, an axiom is justified if it meets the practical goals of set theorists. A further criterion welcomed by all parties (even those who are critical towards both intrinsicality and extrinsicality) is *maximality*. Shortly, *there exist as many sets as there can possibly exist*. A problem here is constituted by the fact that there are different ways to formally capture this property and that, depending

---

[11]See, for instance, Caicedo et al. (eds.) 2017.

[12]Hereto cp. Farah 2011.

[13]See, among others, the symposia on Set-theoretic Pluralism (https://sites.google.com/site/pluralset).

[14]See Hamkins 2012.

[15]Gödel 1947.

on this formal rendering, different existence axioms are introduced, to wit, large cardinal axioms or forcing axioms.[16]

The philosophy of set theory is today an active research field, in which new ideas are regularly brought into discussion, available concepts are constantly refined and farsighted perspectives are developed. The first chapter of the present volume gives more detailed insights into debated topics in current philosophy of set theory.

Homotopy Type Theory/Univalent Foundations

Part II of the present volume pursues two complementary goals, namely, **(i)** to explain the basic ideas of HoTT and **(ii)** to present special properties of it.

Classic set theory, as we saw, can represent all mathematical objects *via* sets, i.e. by *one kind* of objects. Type theory, on the contrary, works with a hierarchy of different *kinds* of objects, called "*Types*". The idea of a type-sensitive foundation has become more and more important along with the progressive development of *data-types*-talk in computer science. Moreover, actual mathematical practice seems to be aware of type differences even when single mathematicians plead for untyped frameworks: nobody would try to differentiate a vector space!

In type theory each term has a type.

Let "*nat*" be the type of natural numbers. Thus, "5 is a natural number" will be written as follows:

$$\mathbf{5} : \boldsymbol{nat}.$$

Operations are type-sensitive too. The addition on natural numbers, e.g., takes two objects of the kind *nat* as input and gives one object of the kind *nat* as output:

$$+ : \boldsymbol{nat} \times \boldsymbol{nat} \rightarrow \boldsymbol{nat}.$$

As already said, the type-hierarchy of mathematical objects and operations makes it possible to ascribe to each term its own type. Under "rewrite systems" is usually meant the reduction of complex terms to their *normal forms*. For instance, "5 + 2" and "7" are names of the very same object. The latter has a well-determined place in the natural number series. We write:

$$\mathbf{5 + 2} \twoheadrightarrow \mathbf{7},$$

to indicate the *reduction* of the complex term "5 + 2" to its normal form "7".

[16]Cp. Incurvati 2017, 162 for Large Cardinal Axioms, and 178ff. for Forcing Axioms.

Type theory was introduced by Bertrand Russell as a way out from the difficulties come into being within the context and set of problems of *naive* set theory. At a variance with Russell's, current type theory has the additional problem of reducing terms to their normal form (**rewrite problem**). This makes it necessary to look for a formal counterpart for the operation represented by "$\twoheadrightarrow$".

This operation is to be found formally implemented, for the first time, in combinatory logic,[17] a logic which was introduced by Schönfinkel with the aim to drop quantifiers from the language of predicate logic. Combinatory logic is usually seen as the forerunner of the λ-calculus. In this regard it is worth mentioning that A. Church elaborated, around 1930, a framework for a paradox-free reconstruction of mathematics based on the notion of a type-free function-in-intension. The original theory was proved inconsistent around 1933 by S. Kleene and J.B. Rosser. Subsequently, its sound fragment, the so-called untyped λ-calculus, was isolated and employed in the context of the theory of computability. In this theory the abstraction operator "λ" plays a fundamental role. Such approach is based on the procedural conception of function, i.e. its conception as rule, prescription, algorithm in a general sense.

Indeed, two natural ways of thinking of functions are often appealed to. The first way is the relational way: it takes a function to be a type of relation. The second way takes a function to be a rule, or algorithm in a general sense, as we said above. Under the extensional reading, the concept of function and the concept of set are interdefinable. We reduce functions to sets by means of the notion of "graph of a function" and, conversely, we reduce sets to functions by means of the notion of "characteristic function of a set". *This interdefinability no longer obtains when we consider functions as rules*, since the latter notion always implies the idea of a procedure that is more or less effective. Note that an implicit type-distinction is present in the set-theoretical account of functions: functions, on the one hand, and arguments and values, on the other hand, live at different levels of the cumulative hierarchy. Thus, from a set-theoretical perspective the application of a function to itself makes no sense: from $f : A \to B$ it follows, in standard set theory with the axiom of foundation, that $f \notin A$. By contrast, in the λ-calculus it does make sense to apply an operation to itself: intuitively, operations can be thought of as finite lists of instructions which, therefore, do not differ in principle from the arguments on which they act: both are, loosely speaking, finite pieces of data. The active role (behaving as an operation) and the passive role (behaving as an argument) of certain elements F, G of this untyped universe is only position-determined. When we write $F(G)$, F is a rule that is applied to the argument G; but this does not prevent us from considering also the application $G(F)$ of G to F, where now G is the rule.

Martin-Löf's type theory provides an alternative foundational framework inspired by constructive mathematics. In particular, the requirements we recalled above as to intuitionistic logic are taken to hold; to wit, the exhibition of a witness for any existential assumption, the existence of a proof for one of the

[17] Cp. Schönfinkel 1924 and Church 1932.

horns of a disjunction, if the disjunction is taken to be true, or the exhibition of the proof that a sentence **A** leads to contradiction if this sentence is taken to be false. The implementation of such requirements within the theory is referred to as *internalization* of the **B**rouwer–**H**eyting–**K**olmogorov interpretation (**BHK**-interpretation) of the logical operators. In addition, constructive type theory presents the particular feature of admitting dependent types.

HoTT is a form of type theory based on constructive type theory[18] endowed with some elements from topology that can be used to construct models for logical systems.[19] One important reading of HoTT sees types as spaces and terms as points. Dependent types correspond to fibrations in topology, and the identity between terms is rendered by the concept of *path* between points. At a variance with set theory there is here no need for a logic next to or on top of it: constructive types already act as a formal calculus for deduction. HoTT (in the mentioned interpretation) along with the corresponding models has properties that naturally bring to the fore candidates for new axioms, as for instance, the univalence axiom.[20] Voevodsky's univalent foundations programme that emerged from these investigations is but one good example for a foundational framework, which, moreover, turns out to be well suited as a background theory for the implementation of modern mathematics in formal proof systems such as Coq.[21]

HoTT is a very promising approach. Even the Institute of Advanced Study in Princeton has dedicated a *Special Year on Univalent Foundations of Mathematics* (2012/2013), during which topologists, computer scientists, category theorists and mathematical logicians have collaborated to jointly promote HoTT.

The Contributions

Let us now give a short overview of all contributions.

[18]Cp. Martin-Löf 1975.

[19]Hofmann and Streicher 1998 constructed a model of Martin-Löf's type theory in the category of groupoids. Moerdijk 2011 conjectured that there exists a general link between Quillen models (models that contain, in particular, classical homotopy of spaces and simplicial sets) and type theory. Awodey, Warren and Voevodsky proved fundamental connections between homotopy theory and type theory.

[20]This approach interprets types as spaces or homotopy types. This makes it possible to directly work with spaces, without having to define a set-theoretic topology. The univalence axiom additionally holds in several possible models. Hereto see Voevodsky 2009.

[21]In the univalent perspective one uses a modified type theory, which is extended by certain axioms, such as the univalence axiom. This approach can very well be implemented as theoretical background of formal proof system, such as Agda and Coq.

Part I: Current Challenges for the Set-Theoretic Foundations

Mirna Džamonja and **Deborah Kant** attempt in their chapter "Interview with a Set Theorist" to give insights into current set-theoretic practice with a focus on independence and forcing. After giving some technical remarks, Džamonja describes the introduction and adoption of the forcing method in set theory and they present important forcing results. In a next section, they discuss the meaning of the word "axiom", differing between ZFC-axioms, large cardinal axioms and forcing axioms, and mentioning the questions of existence and truth in set theory. Then, some experience-based statements on set theory are revealed in a conversation on surprising events. Džamonja and Kant argue that such statements are part of set-theoretic knowledge. They give three hypotheses and encourage further research on set-theoretic practice.

Laura Fontanella provides in her chapter "How to Choose New Axioms for Set Theory" a mathematically highly informed overview on specific arguments for and against new axiom candidates for set theory. First, she discusses common justification strategies and focuses on Gödel's and Maddy's ideas. Second, she considers in detail the axiom of constructibility, large cardinal axioms (small ones and large ones), determinacy hypotheses, Ultimate-L and forcing axioms.

Claudio Ternullo considers in his chapter "Maddy on the Multiverse" Maddy's objections to multiverse approaches in set theory and suggests counter-arguments to her concerns. He elaborates in particular on the role of set theory as a generous arena, and as fulfilling the goal of conceptual elucidation, and on the issues of the metaphysics and possible axiomatization of the multiverse. In particular, Ternullo identifies two forms of multiversism – instrumental and ontological multiversism – and while he agrees with Maddy that all metaphysical considerations relating to the status and prospects of the multiverse should be disregarded, he argues that a multiverse theory can fulfil set theory's foundational goals, in particular generous arena and conceptual elucidation, and that, ultimately, it might also be axiomatized.

Penelope Maddy clarifies her view on multiversism in her "Reply to Ternullo on the Multiverse". She points out that extrinsic reasons for new axioms cannot simply be rejected by evaluating them as practical, since multiverse views also use extrinsic considerations as theoretical. Categorizing important forms of multiversism, as Ternullo did, she considers theory multiversism as an explicit form of instrumental multiversism, and heuristic multiversism. The main problem for theory multiversism is in her view that there is only one axiomatic multiverse theory (given by John Steel), but this theory is not intended to replace ZFC. Maddy rejects the claim that a multiverse theory is a better foundational theory than ZFC. However, she embraces the heuristic value of multiverse thinking for set theory.

Philip Welch presents in his chapter "Proving Theorems from Reflection" some open problems of analysis which are independent of ZFC. For their solution, he suggests a global reflection principle. First, he describes important relations between large cardinal axioms, determinacy principles and regularity properties of projective sets of reals. Second, he explains how reflection principles can justify large cardinal axioms. Welch considers a family $\mathcal{C}$ of the mereological parts of V and formulates the global reflection principle which states that the structure (V, $\in$, $\mathcal{C}$) be reflected to some (V_α, $\in$, $V_{\alpha+1}$). This principle justifies the existence of certain large cardinals, which implies that all projective sets of reals have the regularity properties. This answers the problems presented at the beginning.

Part II: What are Homotopy Type Theory and the Univalent Foundations?

Thorsten Altenkirch motivates and explains in his chapter "Naive Type Theory" the use of homotopy type theory. He starts his presentation from a clear position by arguing that HoTT is a better foundation of mathematics than set theory. In the first half of the chapter, he revises important basic notions of type theory such as judgements, propositions, the Curry-Howard equivalence, functions, induction and recursion. Throughout the chapter, his explanations are supplemented by many exercises which invite the reader to use the introduced definitions. The second half is dedicated to HoTT. While referring to set-theoretic concepts for clarification, he literally denotes some classical principles as forms of lying and elaborates on several type-theoretic versions of AC. Altenkirch concludes with the topic of higher inductive types and gives an extended example of the definition of the integers.

Benedikt Ahrens and **Paige Randall North** focus in their chapter "Univalent Foundations and the Equivalence Principle" on the notion of equivalence of mathematical objects, which is one fundamental notion of the univalent foundations. Their aim is to prove the equivalence principle for different domains D, and defining equivalence through reference to D-properties and D-structures such that it coheres with mathematical practice. They review the univalence principle which holds in Voevodsky's simplicial set model and prove the equivalence principle for propositions, sets and monoids. Ahrens and North close their presentation with categories and show that the equivalence principle does not hold for arbitrary categories but for univalent categories.

Ulrik Buchholtz presents the construction of higher structures in his chapter "Higher Structures in Homotopy Type Theory". He distinguishes between two aims of HoTT: On the one hand its use as a foundation of mathematics in the univalent foundations, and on the other hand, the study of structures that are needed in mathematics. He introduces many technicalities to show in which specific aspects the constructions are complicated. Elementary HoTT is defined as MLTT with univalence, pushouts and propositional resizing, and can already be most needed. There remain, among others, the problems of constructing $(\infty,1)$-categories and developing a meta theory for HoTT/UF. Here, Buchholtz offers detailed and optimistic considerations on the challenges and possible means that could solve these problems.

In his chapter "Univalent Foundations and the UniMath Library: The Architecture of Mathematics" **Anthony Bordg** first explains the univalent foundations of mathematics, the Univalence Axiom and the so-called homotopy levels. Then, he gives insights into the UniMath library of formalized mathematics based on the univalent foundations, underlying in the process the compatibility of this library with classical reasoning supported by the Law of Excluded Middle and the Axiom of Choice. Second, he analyses some challenges for large-scale libraries of formalized mathematics. Bordg devotes his third part to an investigation of parallels between ideas from the architect Christopher Alexander and some organic features of mathematics. He argues that these features need to be preserved, hence they should become desirable properties of libraries of formalized mathematics if one wants these libraries to be scalable and a sustainable way of doing mathematics.

Andrei Rodin's article "Models of HoTT and the constructive view of theories" can be read in two ways: As the development of new formal tools to do philosophy of science and as a case study of the usefulness of the new developments around the Univalent Foundation program. On the philosophical side of things, he develops the constructive view of (scientific) theories in contrast to axiomatic approaches and those based on set theory. A crucial distinction here is between rules and axioms in their role as first principles. These general thoughts find their way to precise formal counterparts and similar to the part before both axiomatic deduction system and model theoretic tools based on set theory are contrasted with proof-theoretic alternatives benefitting from the tools of Martin-Löf Type Theory and Homotopy Type Theory. While still being a first study, it offers a very detailed account of the technical apparatus.

Part III: Comparing Set theory, Category Theory, and Type Theory

Neil Barton and **Sy Friedman** deal with the problem of different foundational perspectives – categorial and set-theoretic – in their chapter "Set theory and Structures" by presenting a modification of set theory to better respect structural discourse. In the first part, they respond to various concerns about both category-theoretic and set-theoretic foundations such as the problem of a determinate subject matter for category theory and too little isomorphism invariance in set theory. In a second part, they present a class theory with structures (NBGS), in which material set theory is combined with structures in such a way that a higher degree of isomorphism invariance is provable. Barton and Friedman advocate methodological pluralism for the foundations of mathematics and encourage further research on combining foundational theories.

Mirna Džamonja faces in her chapter "A New Foundational Crisis in Mathematics, Is It Really Happening?" the question of a potential rivalry between different foundational theories in mathematics. Her main aim is to describe the current situation of the foundations of mathematics and to resolve some worries. In order to illustrate some differences, she describes type theory, elaborates on identity and univalence and explains the topology behind the univalent foundations. She concludes by claiming that there does not have to be a unique foundation of mathematics. Džamonja argues for pluralism and that the univalent foundations and set theory can well complement each other.

Ansten Klev elucidates in his chapter "A Comparison of Type Theory with Set Theory" fundamental conceptual differences between Martin-Löf's type theory and set theory. He argues that type theory is better suited to clarify mathematical notions. In a first part, he describes types to be kinds or sorts of objects. In contrast, he explains sets as pluralities which can contain objects of arbitrary different types. In a second part, he considers in detail the syntax of type theory and presents types, propositions, terms, judgements and contexts (for hypothetical judgements). In his third part on functions, the advantage of having functions as primitive notions in type theory is highlighted. The final part is dedicated to identity, in which Klev elaborates on the difference between propositional and judgemental identity in type theory.

Penelope Maddy addresses in her chapter "What Do We Want a Foundation to Do?: Comparing Set-Theoretic, Category-Theoretic, and Univalent Approaches" the question of foundational theories by isolating rigorously the foundational jobs that are done by a suggested theory. First, she argues that set theory does the jobs risk assessment, generous arena, shared standard, and metamathematical corral. Second, she analyses that the foundational job that category theory was argued to be better suited for is essential guidance. Third, she introduces a new foundational job, proof checking, that is done by the univalent foundations. With her new terminology, Maddy provides a theoretic framework which is very well suited for a thorough comparison and discussion of foundational theories in mathematics.

Part IV: Philosophical Thoughts on the Foundations of Mathematics

In "Formal and Natural Proof: A Phenomenological Approach" **Merlin Carl** elaborates on the relation between formal proofs in the sense of a derivation and proofs in the sense we can find them in mathematical discourse. He motivates a positive connection *via* Gödel's Completeness Theorem and success of the generation of deduction with formal mathematics.

A crucial problem is the nature of formal and natural proofs. A classical dispute is the one between Rav and Azzouni, while the second explains the strong agreement of mathematicians and the statements they prove with a link to underlying derivation, the first claims that a derivation presupposes a sufficient understanding of the natural proof, which makes it absolute.

This is a challenge for the connection that the author answers in two ways. *Via* the success of formal mathematics and especially *via* the Naproche system, which tries to bridge the gap with linguistic methods including a historic case study, where a formalization attempt would have helped to find a flaw in a proof attempt. And *via* phenomenological methods, namely by enlarging Husserl's distinction between distinctiveness and clarity of judgements, to the mathematical proofs.

Michèle Friend provides with "Varieties of Pluralism and Objectivity in Mathematics" a revised version of an article published in *Mathematical Pluralism*, Special Issue of the *Journal of the Indian Council of Philosophical Research*, Mihir Chakraborty and Michèle Friend (Guest Editors) Springer. JICPR Vol. 34.2 pp. 425 – 442. DOI 10.1007/s40961-061-0085-3. ISSN: 0970-7794. pp. 425 – 442.

She offers an alternative to the realist and traditional accounts of objectivity in mathematics by elaborating the pluralistic position. To be more precise, there are several ways in which a philosophy of mathematics can be pluralistic, namely, epistemology, foundations, methodology, ontology and truth. Friend gives an overview on these different approaches, delivers thereby an interesting typography and analyses how each variation can explain objectivity in mathematics.

Graham Priest's chapter "From the Foundations of Mathematics to Mathematical Pluralism" tells the story of different foundational endeavours pinned at historically important actors, namely, Frege, Russell, Zermelo, Brouwer and Hilbert. These more historical aspects yield to the recent history of category theory. In § 9 and 10 the author introduces paraconsistent accounts of mathematics, motivated by impossible manifolds, which can be formalized by paraconsistent tools. This roundtrip to the study of such different mathematical structures motivates the final vote towards a more pluralistic philosophy of mathematics.

Roy Wagner's chapter "Does Mathematics Need Foundations?" is an important step back from our endeavour of the comparison of different foundational theories. It starts with a revisit of the classical foundational enterprises: Russell, Brouwer and Hilbert but also critical accounts like Poincaré's and Wittenstein or the maverick tradition. This offers insights into a variety of schools of thoughts in a broad narrative of the problems and possibilities of foundational thinking. The chapter ends with a very interesting case study from anti-foundationalist mathematics, namely, Kerala mathematics in the fourteenth to sixteenth century.

The chapter offers an account which preserves a lot of positive aspects often connected to the foundations of mathematics, like shared standards and risk assessment, and shows that this is not only a metaphysical possibility but historical fact.

Part V: Foundations in Mathematical Practice

Nathan Bowler's chapter "Foundations for the Working Mathematician, and for Their Computer" discusses the aspects of the candidates discussed in this volume which are reflected in the mathematical practice. The chapter focuses mainly on

ZFC but considers some advantages of categorical or type-theoretic foundations. A key distinction is in how far we differentiate between human-friendly and machine-friendly aspects of the foundations.

Bernhard Fisseni, Martin Schmitt, Bernhard Schröder and Deniz Sarikaya's Article "How to frame a mathematician: Modelling the cognitive background of proofs" uses the tool of frame semantics, developed in linguistics and artificial intelligence, as a basis for describing and understanding proofs. The key idea of frames is that concepts offer some roles (also: features, attributes): When talking about "selling" something, we know, for example, that the transfer involves a seller, a buyer, a sold item, money, etc., and, if not given explicitly, we assign default entities from the discourse universe to (some of) those roles. The general argument is that we proceed similarly when comprehending proofs. As an illustration, the authors develop a frame representation for induction proofs. They use a formalism based on feature structures, a general data structure prominent in formal linguistics.

Lawrence C. Paulson gives in his chapter "Formalising Mathematics in Simple Type Theory" an intriguing insight into the formalization of mathematics. Starting from the conviction that there will always be different formalisms that are better suited than others for specific purposes, he carefully delineates the possibilities of simple type theory. He shares his personal perspective, which is based on long experience with different automated proof systems and presents the original code of a theorem on stereographic projections in HOL Light and its translation into Isabelle/HOL. Paulson shows here how translations from one proof system to another are accomplished, and he emphasizes the importance of such translational work for the future use of proof systems in mathematics.

In "Dynamics in Foundations: What Does It Mean in the Practice of Mathematics?" **Giovanni Sambin** gives insight in the programme of *dynamic constructivism*. The first part gives a philosophical motivation and explains the interplay between mathematical practice and the philosophy of mathematics. We learn (without too many technical details) about a possible shared arena for classical and constructive mathematics, the minimalist foundation (developed jointly by the Author and M. E. Maietti). The second part offers a proof of concept for dynamic constructivism, we learn about point free topology and other theories developed by the author in this new paradigm. This is especially important given the motivation in the first section asking from a foundation to foster new and creative pieces of mathematics. The last part reflects on the benefits of the adaption to this position and draws a parallel to fruitful developments in the sciences *via* the vocabulary from philosophy of science, namely by elaborating how the adaption of this position yields a helpful new perspective - a new paradigm.

The Editors

Stefania Centrone holds an Heisenberg Stelle at the Technische Universität Berlin. She is author, among others, of *Logic and Philosophy of Mathematics in the Early Husserl* (Springer, Synthese Library 2010) and *Studien zu Bolzano* (Academia Verlag 2015). She is editor of the collected volumes *Versuche über Husserl* (Meiner 2013) and *Essays on Husserl's Logic and Philosophy of Mathematics* (Springer, Synthese Library 2017) and co-editor, together with Sara Negri, Deniz Sarikaya and Peter Schuster of the volume *Mathesis Universalis, Computability and Proof* (*Synthese Library*, 2019).

Deborah Kant is a PhD candidate in philosophy at the University of Konstanz with a project on independence and naturalness in set theory. As a member of the Forcing Project, led by Carolin Antos, she investigates the set-theoretic independence problem from a practical perspective. In 2019, she visited the Department for Logic and Philosophy of Science in Irvine, and, in 2017, the Institute for History and Philosophy of Sciences and Technology in Paris. Before her PhD, she studied mathematics in Berlin.

Deniz Sarikaya is a PhD candidate in philosophy at the University of Hamburg. Before his PhD, he studied philosophy (BA: 2012, MA: 2016) and mathematics (BSc: 2015, MSc: 2019) at the University of Hamburg with experience abroad at the Universiteit van Amsterdam, Universitat de Barcelona, and the University of California, Berkeley, and as a research intern at the University of British Columbia. His project was augmented through a stay at the EHT Zurich. He is mainly focusing on philosophy of mathematics and logic.

Literature

Brouwer, L. E. J. (1907). *Over de Grondslagen der Wiskunde*. Maas & van Suchtelen.

Caicedo, A., Cummings, J., Koellner, P., & Larson, P. (Eds.). (2017). *Foundations of Mathematics: Logic at Harvard: Essays in honor of Hugh Woodin's 60th birthday, March 27–29, 2015, Harvard University, Cambridge, MA* (Contemporary mathematics, v. 690). Providence, Rhode Island.

Cantor, G. (1872). Ueber die Ausdehnung eines Satzes aus der Theorie der trigonometrischen Reihen. *Mathematische Annalen, 5*(1), 123–132.

Church, A. (1932). A Set of Postulates for the Foundation of Logic. *Annals of Mathematics, 33*(2), 346–366.

Church, A. (1940). A formulation of the simple theory of types. *Journal of Symbolic Logic*, 5, 56–68.

Dedekind, R. (1872). *Stetigkeit und irrationale Zahlen*. F. Vieweg und Sohn.

Dummett, M. (1977). *Elements of Intuitionism*. Oxford University Press.

Farah, I. (2011): All automorphisms of the Calkin algebra are inner. *Annals of Mathematics, 173*(2011), 619–661.

Frege, G. (1893/1903) *Grundgesetze der Arithmetik: Begriffsschriftlich Abgeleitet* I and II. Julius. Jena: Springer.

Gödel, K. (1947). What is Cantor's Continuum problem? *The American Mathematical Monthly, 54*(9), 515–525.

Hamkins, J. D. (2012). The set-theoretic multiverse. *The Review of Symbolic Logic*, *5*(3), 416–449.

Heyting, A. (1930). *Die formalen Regeln der intuitionistischen Logik*. Deutsche Akademie der Wissenschaften zu Berlin.

Heyting, A. (1975). Intuitionism in Mathematics. *Journal of Symbolic Logic* 40(3), 472–472.

Hilbert, D. (1917). Minoritätsgutachten. In: Peckhaus, V. (1990). *Hilbertprogramm und Kritische Philosophie. Das Göttinger Modell interdisziplinärer Zusammenarbeit zwischen Mathematik und Philosophie*. Vandenhoeck & Ruprecht, Göttingen.

Hilbert, D. (1935). Die logischen Grundlagen der Mathematik. In *Dritter Band: Analysis· Grundlagen der Mathematik· Physik Verschiedenes* (178–191). Berlin/Heidelberg: Springer.

Hilbert, D. (1926). Über das Unendliche. *Mathematische Annalen, 95*(1), 161–190.

Hofmann, M. & Streicher, T. (1998). The groupoid interpretation of type theory. *Twenty-five years of constructive type theory (Venice, 1995)*, 36, 83–111.

Incurvati, L. (2017). Maximality principles in set theory. *Philosophia Mathematica*, 25(2), 159–193.

Kleene, S., & Rosser, J. (1935). The inconsistency of certain formal logics. *Annals of Mathematics, 36*(3), 630–636.

Kline, M (1990). *Mathematical thought from ancient to modern times* (Vol. 3). New York: Oxford University Press.

Martin-Löf, P. (1975). An intuitionistic theory of types: Predicative part. In *Studies in logic and the foundations of mathematics* (Vol. 80, pp. 73–118). Amsterdam: Elsevier.

Moerdijk, I. (2011), Fiber bundles and univalence, Based on a talk given at the conference: Algorithms and Proofs 2011, see: https://www.andrew.cmu.edu/user/awodey/hott/papers/moerdijk_univalence.pdf.

Russell, B. & Whitehead, A. N. (1910–1927). *Principia Mathematica* (3 Volumes). Cambridge: Cambridge University Press.

Schönfinkel, M. (1924). Über die Bausteine der mathematischen Logik. *Mathematische Annalen, 92*(3), 305–316.

Voevodsky, V. (2009). *Notes on type systems*, Unpublished notes.

Zermelo, E (1908). Untersuchungen über die Grundlagen der Mengenlehre. I *Mathematische Annalen*, 261–281.

Contents

Part I Current Challenges for the Set-Theoretic Foundations

1 Interview With a Set Theorist ... 3
Mirna Džamonja and Deborah Kant

2 How to Choose New Axioms for Set Theory? ... 27
Laura Fontanella

3 Maddy On The Multiverse ... 43
Claudio Ternullo

Reply to Ternullo on the Multiverse ... 69
Penelope Maddy

4 Proving Theorems from Reflection ... 79
Philip D. Welch

Part II What Are Homotopy Type Theory and the Univalent Foundations?

5 Naïve Type Theory ... 101
Thorsten Altenkirch

6 Univalent Foundations and the Equivalence Principle ... 137
Benedikt Ahrens and Paige Randall North

7 Higher Structures in Homotopy Type Theory ... 151
Ulrik Buchholtz

8 Univalent Foundations and the UniMath Library ... 173
Anthony Bordg

9 Models of HoTT and the Constructive View of Theories ... 191
Andrei Rodin

Part III Comparing Set Theory, Category Theory, and Type Theory

10 Set Theory and Structures ... 223
Neil Barton and Sy-David Friedman

11 A New Foundational Crisis in Mathematics, Is It Really Happening? ... 255
Mirna Džamonja

12 A Comparison of Type Theory with Set Theory ... 271
Ansten Klev

13 What Do We Want a Foundation to Do? ... 293
Penelope Maddy

Part IV Philosophical Thoughts on the Foundations of Mathematics

14 Formal and Natural Proof: A Phenomenological Approach ... 315
Merlin Carl

15 Varieties of Pluralism and Objectivity in Mathematics ... 345
Michèle Friend

16 From the Foundations of Mathematics to Mathematical Pluralism .. 363
Graham Priest

17 Does Mathematics Need Foundations? ... 381
Roy Wagner

Part V Foundations in Mathematical Practice

18 Foundations for the Working Mathematician, and for Their Computer ... 399
Nathan Bowler

19 How to Frame a Mathematician ... 417
Bernhard Fisseni, Deniz Sarikaya, Martin Schmitt, and Bernhard Schröder

20 Formalising Mathematics in Simple Type Theory ... 437
Lawrence C. Paulson

21 Dynamics in Foundations: What Does It Mean in the Practice of Mathematics? ... 455
Giovanni Sambin

Part I
Current Challenges for the Set-Theoretic Foundations

Part I
Current Challenges for the Set-Theoretic Foundations

Chapter 1
Interview With a Set Theorist

Mirna Džamonja and Deborah Kant

Abstract The status of independent statements is the main problem in the philosophy of set theory. We address this problem by presenting the perspective of a practising set theorist. We thus give an authentic insight in the current state of thinking in set-theoretic practice, which is to a large extent determined by independence results. During several meetings, the second author asked the first author about the development of forcing, the use of new axioms and set-theoretic intuition on independence. Parts of these conversations are directly presented in this article. They are supplemented by important mathematical results as well as discussion sections. Finally, we present three hypotheses about set-theoretic practice: First, that most set theorists were surprised by the introduction of the forcing method. Second, that most set theorists think that forcing is a natural part of contemporary set theory. Third, that most set theorists prefer an answer to a problem with the help of a new axiom of lowest possible consistency strength, and that for most set theorists, a difference in consistency strength weighs much more than the difference between Forcing Axiom and Large Cardinal Axiom.

1.1 Introduction

The current situation in set theory is an exciting one. In the 1960s, set theory was challenged by the introduction of the forcing technique, in reaction to which some researchers might have turned their back on set theory, because it gave rise to a vast range of independence results. Today, the independence results constitute a large part of set-theoretic research. But how do set theorists think about them? How do mathematicians think about provably independent statements?

M. Džamonja (✉)
School of Mathematics, University of East Anglia, Norwich, UK
e-mail: m.dzamonja@uea.ac.uk

D. Kant
Institute for Philosophy, University of Konstanz, Konstanz, Germany
e-mail: kantdebo@gmail.com

S. Centrone et al. (eds.), *Reflections on the Foundations of Mathematics*,
Synthese Library 407, https://doi.org/10.1007/978-3-030-15655-8_1

An answer to this question can be attempted through a detailed description of the current set-theoretic practice through the eyes of set-theorists themselves, which is an enterprise to be realised. The present article provides a step towards that program by giving a description of some important aspects of set-theoretic practice, formulated and observed from a joint mathematical and philosophical perspective. During several meetings in Paris,[1] the second author (PhD candidate in philosophy of set theory) has been talking to the first author (a logician specialising in set theory and a professor of mathematics) in order to gain insights into the current situation in set theory, and to understand how set theorists think about their work and their subject matter. Parts of these conversations are directly presented in this article. They are supplemented by descriptive paragraphs of related (mathematical) facts as well as comments and discussion sections.

The article is structured as follows. First, we argue for the relevance of this article and our method. The second section contains facts of set theory and logic that will be relevant in the following sections. In the third section, we present some important forcing results, which includes mathematical details but is self-contained. We also added many (historical and mathematical) references. We then elaborate on classical, philosophical concepts that can be found in set-theoretic practice, for instance platonism. The fifth section presents some general observations about independence: Set theorists have developed some intuition which problems might turn out independent and which ones might be solvable in ZFC, and they can organise and differ between set-theoretic areas in this respect. Finally, we present three hypotheses about set-theoretic practice: First, that most set theorists were surprised by the introduction of the forcing method. Second, that most set theorists think that forcing is a natural part of contemporary set theory. Third, that most set theorists prefer an answer to a problem with the help of a new axiom of lowest possible consistency strength, and that for most set theorists, a difference in consistency strength weighs much more than the difference between Forcing Axiom and Large Cardinal Axiom.

The intended audience of this article includes both set theorists and philosophers of mathematics. Set theorists can skip Sect. 1.3, where some central mathematical concepts are introduced. For philosophers of mathematics who do not focus on set theory in their work, this section is intended to facilitate an understanding of the following text. They can also scan Sect. 1.4 without a loss of understanding for the subsequent sections. The aim of Sect. 1.4 is to show a variety of applications of the forcing method, and to support the view that forcing is an integral part of set-theoretic research.

[1]The first author is an Associate Member at the IHPST (Institut d'histoire et de philosophie des sciences et des techniques) of the Université Panthéon-Sorbonne, Paris, France. The second author is grateful for the support of the DAAD (Deutscher Akademischer Austauschdienst) and the IHPST.

1.2 Methodological Background

We are interested in describing and analysing what set theorists do. We adopt Rouse's view that "a central concern of both the philosophy and the sociology of science [is] to make sense of the various activities that constitute scientific inquiry",[2] and apply it to set theory with a special focus on independence. In order to take a first step in the right direction, we present one specific view in current set-theoretic practice: the first author's view on her discipline. We cannot generalise this view to a description of set-theoretic practice, because other set theorists might have different views. Thus, we see the present article as an attempt to start a discussion on set-theoretic practice, which should lead to a more general, rigorous analysis.

1.2.1 How to Describe and Analyse Set-Theoretic Practice?

This question certainly deserves much more attention than we can devote to it in this article.[3] We briefly state our main points.

Our methods consist of a sociological method combined with philosophical considerations. On the sociological side, we can choose between two main methods: surveys and interviews. Surveys have the advantage of an easy evaluation of the information. They are suitable to check clearly formulated hypotheses. But there are two problems concerning hypotheses on set-theoretic practice: A missing theory and the question of a suitable language.

The study of set-theoretic practice is a rather new research area. In the literature, one can find specific views about set-theoretic practice,[4] or investigations of specific parts of set-theoretic practice,[5] but there is no analysis of current set-theoretic practice in general. This means that we do not have a theory at hand which could be tested in a survey. But we can use interviews to find reasonable hypotheses about set-theoretic practice.

Secondly, not only the theory has to be developed but also a suitable language for the communication between set theorists and philosophers has to be found. There are sometimes huge differences between the language which philosophers use and the language which set theorists use to talk and think about set theory. Therefore, communication between both disciplines can be difficult. There is a greater risk of a misunderstanding in a survey formulated by a philosopher and answered by a mathematician than in a question asked in an interview in which the possibility to immediately clear up misunderstandings is given. Hence, based on these aims of

---

[2]Rouse (1995, p. 2).

[3]But it will be considered more attentively in the PhD project of the second author.

[4]Maddy (2011) and Hamkins (2012).

[5]Rittberg (2015).

finding reasonable hypotheses and of a successful communication, we decided to conduct interviews.

On the philosophical side, a mathematical perspective is brought together with philosophical conceptions of mathematics. Furthermore, a mathematical perspective is communicated to philosophers, i.e. presented in a way that makes it comprehensible to other philosophers. The elaboration on our methodology in this section is part of the philosophical side of the methodology.

1.2.2 Why Describe and Analyse Set-Theoretic Practice?

We argue here that set-theoretic knowledge is not completely captured by gathering all theorems, lemmas, definitions, and the mathematical motivations and explications that mathematicians give to present their research. For example, when set theorists agree that cardinal invariants are mostly independent of each other, then we argue that this judgement, which is based on experience, is part of set-theoretic knowledge as well. Judgements of this kind are ubiquitous in set-theoretic practice. We extend our focus to set-theoretic practice in general. The notion of set-theoretic practice is taken here very generously as including all mathematical activities performed by set theorists as well as their thoughts and beliefs about mathematics (where the latter also includes definitions and theorems because we assume that set theorists believe what is an established definition and has been proven). In order to learn more about set-theoretic knowledge, we think that it is valuable to present the practices of the discipline, the similarities and differences between the views held by set theorists, and to formulate general ideas about the current situation in set theory.

The reasons for this are at least three, as we describe.

Visions Reflecting on the practices, the historical developments, the importance or role of specific objects and methods etc. gives rise to explicitly formulated visions for future set-theoretic research. What do set theorists wish to find out in the next ten years?[6]

Availability Set-theoretic knowledge is not easily accessible to other mathematical areas, the sciences in general and the philosophy of mathematics. With important exceptions, such as the philosopher Alain Badiou[7] or the musician François Nicolas[8] and of course many mathematicians and philosophers of mathematics, the independence phenomenon is not sufficiently known outside

[6]Large research programs have already been described by some set theorists. Consider, for example, the research programs by Hugh W. Woodin or Sy-David Friedman. For Woodin's program *see* Woodin (2017), and for Friedman's Hyperuniverse Program, *see* Arrigoni and Friedman (2013). Not everybody describes his/her program so specifically and we wish to discover more about these unspecified programs.

[7]Badiou (2005) bases ontology on set-theoretic axioms and also considers forcing.

[8]www.entretemps.asso.fr/Nicolas/

of the community directly studying set theory and logic. The formulation of set-theoretic ideas in a general and simple language could make set-theoretic knowledge available to other researchers who are interested in set theory.

Reasonable premises in philosophy In the philosophy of set theory, there is a debate on the independence problem and new axioms. Typical questions are *Is every set-theoretic statement true or false?* or *Which criteria can justify the acceptance of a new axiom?*[9] In some important philosophical approaches, set-theoretic practice plays a major role. For example, in Maddy's view, philosophical questions can only be answered adequately when considering in detail what set theorists are doing.[10] Hamkins' multiverse view is also strongly motivated by the current situation in set-theoretic research.[11] Therefore, some existing philosophical ideas require supplementation by an analysis of set-theoretic practice,[12] and such an analysis seems in general a promising starting point for future philosophical research.

1.3 Preliminary Facts

We summarise basic facts of logic and set theory such as the twofold use of the concept of set, the incompleteness of an axiomatic theory, independence proofs, the Continuum Hypothesis, forcing, and some new axioms. This section is intended to facilitate the understanding of the following interview parts, and it contains the necessary background for philosophical questions on set theory.

Set theory is the study of sets, and sets are determined by their elements. We can take the union of two sets, we can take their intersection, we can form ordered pairs and sequences. And we can consider infinite sets, like the set of all natural numbers or the set of all real numbers. Numbers can themselves be interpreted as certain sets,[13] and so can functions and many other mathematical objects. But also a sentence of a formal language can be interpreted as a certain set: every symbol of the formal language is interpreted as a set and then a sentence is just a finite sequence of these sets.[14] Therefore, also a formal theory—a set of axioms and all formal sentences that can be derived from these axioms—can be interpreted as a certain set.

[9] *See* the article by Laura Fontanella in this volume.

[10] *See* for example Maddy (2007, 2011).

[11] Hamkins (2012).

[12] Thanks to Carolin Antos for emphasising this fact.

[13] $0 = \emptyset$, $n + 1 = n \cup \{n\}$, $\mathbb{N} = \{n : n < \omega\}$, $\mathbb{R} = \mathscr{P}(\mathbb{N})$, and so on. (These equations should not be understood as the claim that numbers *are* sets.)

[14] For example, take the coding '$\exists$'=8, '$\forall$'=9, '$\neg$'=5, '('=0, ')'=1, 'v_0'=(2,0), 'v_1'=(2,1), then the statement '$\exists v_0 \forall v_1 \neg(v_1 \in v_0)$' can be coded as the sequence $\langle 8, (2, 0), 9, (2, 1), 5, 0, (2, 1), 4, (2, 0), 1 \rangle$.

In set theory, the most commonly accepted formal theory is the Zermelo-Fraenkel set theory with the Axiom of Choice (ZFC).

One could say that the concept of set in mathematics and in logic has an extremely wide scope of application. Set theorists study sets, and at the same time, they study the different models of ZFC (starting from the assumption that there is a model of ZFC). They can use their own sophisticated, set-theoretic methods on the models. (By contrast, when considering models of number theory, we cannot use number theory itself to manipulate the models but instead must resort to set theory.) Because of this twofold application of the concept of set, these two levels, which are sometimes distinguished as mathematics and metamathematics, are closely intertwined in today's set-theoretic practice.

The study of different models of ZFC is also the study of the independence phenomenon in set theory. The starting assumption to study a model is always that there exists a model of ZFC (by Gödel's Completeness Theorem for first order logic, this is equivalent to the assumption that ZFC is consistent).[15] This assumption itself is not part of ZFC (it could not be because of Gödel's second Incompleteness Theorem), but it is a natural assumption in practice. Thus, importantly, the set-theoretic method of building models is not a constructive method because of this starting assumption. If then, for a sentence φ, set theorists can build a model of the theory ZFC+φ, and they can also build a model of the theory ZFC+$\neg\varphi$, then φ is an independent sentence.

Let us give one example. In set theory, two sets have the same size if there is a one-to-one onto correspondence between them. So, the set of the natural numbers has the same size as the set of the even natural numbers, because $f : n \mapsto 2n$ is a one-to-one onto correspondence between them. All members of the first set can be completely paired up with the members of the second set. But if we take the real numbers as the second set, they cannot be completely paired up with the natural numbers (this is Cantor's Theorem). This gives rise to different sizes of infinite sets.

The size of sets is measured by cardinal numbers. 0, 1, 2, ... are cardinal numbers. For example, the empty set, $\emptyset$, has size 0, and the set that contains as its only element the set of natural numbers, $\{\mathbb{N}\}$, has size 1. The set of the natural numbers itself, $\mathbb{N}$, has size $\aleph_0$, which is the first infinite cardinal number (set theorists always start counting at 0). Of course, there are further cardinal numbers: $\aleph_1, \aleph_2, \ldots$ Now, set theorists have built models in which there are exactly $\aleph_1$ real numbers, and they have built models in which there are exactly $\aleph_2$ real numbers. Thus, the sentence "there are exactly $\aleph_1$ real numbers" (the famous Continuum Hypothesis (CH)) is an independent sentence. This can only be shown by building such models, and the most powerful technique to do this is forcing.

Forcing was introduced in 1963 by Paul Cohen,[16] who showed by this method that the Continuum Hypothesis is independent. The method was then adopted by the set theorists who have since then found (and continue to find) many independent

[15]Of course, any stronger assumption works as well, in particular any Large Cardinal Axiom.

[16]Cohen (1963, 1964).

statements. Different problems require different variations of the forcing method so that many kinds of forcing have been developed. This led to the formulation of Forcing Axioms. Such an axiom can be added to ZFC in order to facilitate the application of forcing. A Forcing Axiom for a certain kind of forcing states that any object that can be forced to exist by that kind of forcing already exists; it states that the forcing method already has been applied. These axioms are part of the new axioms in set theory.

It should be noted that the notion of a new axiom is not used by all set theorists. But in the philosophy of set theory, this notion includes all the axioms which are not part of the standard axiomatisation ZFC, but which are considered in set-theoretic practice.

In addition to the Forcing Axioms, there is another important class of new axioms—the Large Cardinal Axioms, which state that there exists a certain large cardinal. The smallest known large cardinal is an inaccessible cardinal. Other important large cardinals are measurable cardinals, Woodin cardinals, and supercompact cardinals (ordered by increasing strength). The existence of such large cardinals is not provable in ZFC (but for all we know, it might be that ZFC proves the non-existence of some of them!).

There are further statements, for instance determinacy statements, which are sometimes considered as new axioms, e.g. the Axiom of Determinacy (AD) which is consistent with ZF but contradicts the Axiom of Choice, and Projective Determinacy (PD) which is implied by the existence of infinitely many Woodin cardinals (Martin-Steel Theorem, 1985).

1.4 Some Important Forcing Results

This section briefly presents important steps in the development of the forcing technique. We first describe the first author's perspective on the moment of the introduction of forcing and the process of its adoption by set theorists. Second, we give an overview on subsequent inventions of different kinds of forcing, the conjectures they solved, and the formulation of Forcing Axioms.[17]

[17]Readers interested in the mathematical details of forcing are referred to Chow (2007) for an introduction, or Kunen (2011) for a classical presentation of forcing, or Shelah (2017) for a presentation of the forcing methods that are used today.

1.4.1 Cohen's Introduction of Forcing

Before the introduction of forcing by Paul Cohen in 1963,[18] there were no substantial independence results in set theory. It was already known that CH cannot be refuted since Gödel had constructed a model, L, of ZFC in which CH holds. L is an inner model of ZFC which is not obtained by the forcing method. M. Džamonja thinks that "many practising set theorists at that time were hoping, or were assuming, that CH or GCH[19] would be proven to be true." She points to a similar situation today: "Maybe just like now, we think that Large Cardinal Axioms are true in a sense, even though we cannot prove that they are." Today, Large Cardinal Axioms are an integral part of set-theoretic research. In comparison to other new axioms, they appear to be the most acceptable ones for set theorists. In general, most set theorists trust in the consistency of these axioms and do not believe that assuming them causes any harm. In other words, if they were forced to choose between the Large Cardinal Axioms and their negations, most set theorists would choose the former. Imagine now that this was the case with the Continuum Hypothesis before Cohen's result, which would mean that most set theorists did not expect that the negation of CH is a reasonable statement to consider.[20] This would explain why Cohen's proof was such a surprising result. In the following dialogue, we are speaking about this moment, the subsequent adoption of forcing by the set theorists, and, in particular, how forcing became a natural part of set theory.

D. Kant: *When Cohen's result was published, was it regarded as unnatural?*

M. Džamonja: *It was regarded as something very unnatural. There were many people who stopped working in set theory when they found out about this result. One of them was P. Erdös. He was a most prominent set theorist who had proved many interesting results, but he just didn't think that forcing was an interesting method or that it brings anything. Well, he has this famous statement that 'independence has raised its ugly head'. So, he didn't like it. He never learned the method. And, I think, in general, it was regarded of course as a big surprise. Cohen got a Fields medal for it. But it was very esoteric, and I think that even Cohen himself did not understand it the same way that we understand it now, after so many years, of course. Things become easier after many people had looked at them, and, yes, forcing was unnatural, totally unnatural. It is worth noting that one person, and one person only, had been entirely convinced that CH was going to be proven independent, and that person was Gödel. In spite of his*

[18] *See* Cohen (1963, 1964).

[19] General Continuum Hypothesis: For every ordinal α: $\aleph_{\alpha+1} = 2^{\aleph_\alpha}$.

[20] Amazingly, this was not the case of Gödel, who shortly after discovering L and proving the relative consistency of GCH stated that he believes in the independence of GCH, see the beginning of the interview for this. But then, Gödel was considered a logician and a philosopher, not a mathematician, by the peers of the time.

own proof of the relative consistency of CH he wrote as early as 1947 that CH is most likely to be independent.[21]

D. Kant: *And, then, people quickly understood this technique and applied it?*

M. Džamonja: *Some people did, yes. It started in California and, well, there was Solovay, and there were also Dana Scott and many other people around Stanford and Berkeley, for example Ken Kunen and Bill Mitchell among the younger ones. Paul Cohen was in Stanford and Ken was a student in Stanford. So, I think, forcing was localized to the United States for a while. But, then, just a year or two later, it spread around to Israel. Yes, people did understand, but, I think, it was rather slow. I mean, the specialists understood perhaps quickly, but it was slow and it was not published. Cohen's book*[22] *took time to be published and it is not easy to learn the method from this book.*

D. Kant: *What would you say, when or with which results did forcing become more natural?*

M. Džamonja: *I think a subject generally becomes natural when people start writing and reading books about it. In this case, it was quite late. For example, Kunen's 'Set Theory'*[23] *came out in 1980, and that is really where people learned this from, from a book. Jech's book*[24] *also came out at that time. Before that, well, if you were at the right place at the right time you learned something about it. But it wasn't a well spread method. For example, I came from a country [Yugoslavia] in which there was a considerable amount of set theory, combinatorial set theory. But nobody was really doing forcing. I finished my undergraduate degree in 1984 and I wanted to write a thesis (we needed to write a thesis at the end of our undergraduate studies) on forcing. But I couldn't find an advisor for this in Sarajevo. So, I think, this tells you that people knowing this subject were rare from the Yugoslav perspective. Maybe Kurepa knew it, in Belgrade. I don't even know if he learned this method. I don't think he published any papers on it. And he was probably one of the greatest set theorists of the previous generation. In Hungary, I think, it took quite a while. It was maybe only when Lajos Soukup and Peter Komjath worked on this that it was seriously understood, and that was in the 80s. So, it took some time. In other countries of Eastern Europe, there were people like Bukovsky and Balcar who worked on this already in the 1960s, but it was politically difficult for them. Their work was practically unknown to others because of the Cold War. And, finally, in Russia, this method just didn't came through. Moti Gitik came from Russia to Israel thinking that he had discovered a new method: the method of forcing. He discovered it on his own, because he didn't have access to the research already known in the West. Probably, for someone of your generation, it's very difficult to imagine that time.*

D. Kant: *Yes.*

[21] Gödel (1947).

[22] Cohen (1966).

[23] Kunen (1980).

[24] Jech (2003).

M. Džamonja: *But literature was a really big problem. Because of the Cold War, and the other thing was the cost of journals. It was incredibly expensive. Except for top universities you wouldn't find in your university library journals that publish this kind of thing. So, it was really very restricted.*

We see that there was not so much resistance to adopt the forcing method. There was rather a kind of disappointment and resignation on the one hand, and interest and enthusiasm on the other hand. In addition, the circumstances of that time made the adoption of forcing a slow process. Nowadays, most set theorists know of the forcing method, although not everyone works with it.

The forcing technique has been developed for application to many problems.

1.4.2 Important Forcing Results

This section is intended to illustrate the set-theoretic research on independence. It contains more mathematical details than the other parts. We present important conjectures, theorems, and Forcing Axioms such as Easton's theorem, Suslin's Hypothesis, Martin's Axiom, the Borel Conjecture and the Proper Forcing Axiom, and we define many notions involved.

With Cohen's original method, one can prove that $2^{\aleph_0}$—which is the size of the continuum—can be any regular uncountable cardinal (and, even more generally, any cardinal of uncountable cofinality). A regular cardinal κ is one which cannot be obtained as a supremum of a sequence of cardinals of length less than κ. Conversely, a singular cardinal κ is one that can be reached in less than κ many steps. For example, the cardinal $\aleph_\omega$, which is greater than ω, can be reached in ω many steps: $\aleph_0, \aleph_1, \aleph_2, \ldots, \aleph_\omega$.[25] So, given an uncountable regular cardinal κ, one can force the statement '$2^{\aleph_0} = \kappa$'.

After Cohen presented this method, he did not pursue different applications of it. Instead, other people learned the forcing method, and applied it to solve many open problems. Robert Solovay was one of the most important pioneers in forcing from the early 1960s on. One should, of course, also mention Jack Silver and several other important set theorists from that time.

1.4.2.1 Easton Forcing

Solovay had many excellent students, including Matthew Foreman, Hugh W. Woodin, and William Easton. Easton introduced a new kind of forcing in his PhD thesis. Where Cohen only used a set of forcing conditions, Easton used a

[25]The correct notion to distinguish between regular and singular cardinals is the notion of cofinality. The cofinality of a cardinal κ is the length of the shortest sequence of ordinals less than κ which converges to κ.

proper class of forcing conditions.[26] This allowed him to show that the GCH can be violated almost arbitrarily. The only restrictions for cardinalities of 2^κ for regular κ are given by the requirements that $\kappa < \lambda$ implies that $2^\kappa \leq 2^\lambda$, and by König's Theorem:

König's Theorem (for cofinalities), 1905: *For every cardinal* κ*:* $\mathrm{cf}(2^\kappa) > \kappa$.[27]
Given this, Easton was able to present what is now called *Easton Forcing* and proved the following theorem.

ZFC+¬GCH, Easton's Theorem, 1970: *Let* F *be a non-decreasing function on the regular cardinals, such that* $\mathrm{cf}(F(\kappa)) > \kappa$ *for every regular* κ*. Then, by a cofinality and cardinality preserving forcing, we can obtain a model in which* $F(\kappa) = 2^\kappa$ *for every regular* κ.[28]

1.4.2.2 Suslin's Hypothesis, Iterated Forcing and Martin's Axiom

An old hypothesis about properties of the real line has become very important in set theory. It is based on a question asked by Michaił J. Suslin.[29]

Suslin's Hypothesis (SH), 1920: Every dense linear order in which there are at most countably many disjoint open intervals is isomorphic to the real line.
Given the definition of a Suslin Tree, the Suslin Hypothesis states that there is no Suslin Tree. This version is often used in practice.[30]

Suslin's Hypothesis is independent of ZFC. It was shown, independently by Stanley Tennenbaum in 1968 and Thomas Jech in 1967 (see Jech (2003) for historical remarks and references) that SH cannot be proved in ZFC. Likewise, Ronald Jensen proved that SH is false in L. He proved that the axiom V=L implies the $\diamondsuit$-principle[31]:

[26] A proper class contains all sets that satisfy a given first order formula, but is itself not a set. (Thus, a proper class is defined by unrestricted comprehension over the universe of sets.)

[27] König (1905).

[28] Easton (1970, Theorem 1, pp.140-1). It is interesting to note that this is potentially class forcing. The Forcing Theorem of Cohen only applies to special cases of class forcing, so class forcing is less widely spread in applications. All of the forcing notions to follow will be set forcings.

[29] "(3) Un ensemble ordonné (linéairement) sans sauts ni lacunes et tel tout ensemble de ses intervalles (contenant plus qu'un élément) n'empiétant pas les uns sur les autres est au plus dénombrable, est-il nécessairement un continu linéaire (ordinaire)?" Sierpiński et al. (1920). Translation (by the authors): "Is a (linearly) ordered set without jumps nor gaps, such that every set of its non-overlapping intervals (containing more than one element) is at most countable, necessarily a linear continuum?"

[30] *See* Jech (2003, pp.114–116).

[31] 'Diamond-principle'.

$\diamondsuit$-principle, 1972: There exists a sequence of sets $\langle S_\alpha : \alpha < \omega_1 \rangle$ with $S_\alpha \subseteq \alpha$, such that for every $X \subseteq \omega_1$, the set $\{\alpha < \omega_1 : X \cap \alpha = S_\alpha\}$ is a stationary subset of ω_1.[32] The sequence $\langle S_\alpha : \alpha < \omega_1 \rangle$ is called a *$\diamondsuit$-sequence*.[33]
Jensen proved that $\diamondsuit$ implies that there is a Suslin Tree:

ZFC + ¬SH: In ZFC+V=L, one can prove the existence of a Suslin Tree.[34]

Using a new technique, *iterated forcing*, Solovay and Tennenbaum showed that there is a forcing extension of ZFC+SH. This is the second part of the independence proof for SH. Iterated forcing consists in the transfinitely iterated application of single forcing notions. To make this work, one needs a *preservation theorem* which guarantees that the iteration of single forcings all satisfying a certain important property has still that important property. Here, that important property is the countable chain condition:

> Roughly spaking, it [the preservation theorem] says that the transfinite iteration of a sequence of Cohen extensions satisfies the countable chain condition (c.c.c.) if every stage satisfies c.c.c..[35]

To define this condition, we need to know that a *forcing notion* $(P, <)$ is a partial order, that two elements $p, q \in P$ are called *incompatible* iff there is no $r \in P$ such that $p < r$ and $q < r$, and that an *antichain* is a set $A \subseteq P$ such that all of its elements are pairwise incompatible.

Countable chain condition (c.c.c.): A forcing notion $(P, <)$ satisfies the *countable chain condition* if every antichain is at most countable.

The countable chain condition is an important property because c.c.c. forcing notions preserve cardinals and cofinalities. In general, when using an arbitrary forcing notion, things may happen that are undesired in the given context. Cardinals can be collapsed, cofinalities can be changed etc. preventing the respective statement from being forced.

Now, iterating certain c.c.c. forcing notions permitted Solovay and Tennenbaum to construct a forcing extension in which both ZFC+SH and a Forcing Axiom, called Martin's Axiom (after Donald A. Martin), hold:

Martin's Axiom (MA): For every c.c.c. forcing notion $(P, <)$ and every family of dense sets $\mathscr{D}$ such that $|\mathscr{D}| < 2^{\aleph_0}$ there exists a $\mathscr{D}$-generic subset $G \subseteq P$.[36]

---

[32] A subset of ω_1 is called *stationary* if it intersects all closed and unbounded subsets $C \subseteq \omega_1$, where C is *closed* if for every sequence $(a_n)_{n<\omega} \subseteq C$ the limit $\bigcup\{a_n : n < \omega\}$ is also an element of C, and C is *unbounded* if for every $a \in C$ there is a $b \in C$ such that $b > a$.

[33] Jech (2003, p.191).

[34] Jensen actually showed a more general version of which this theorem is one instance Jensen (1972, Theorem 6.2 and Lemma 6.5, pp.292–5).

[35] Solovay and Tennenbaum (1971) refer to theorem 6.3 on p.228.

[36] G is $\mathscr{D}$-generic means that $G \cap D \neq \emptyset$ for every $D \in \mathscr{D}$.

Given CH, Martin's Axiom is provable, because in this case $\mathscr{D}$ can only be countable, and for countably many dense sets there is always a generic set. Therefore, when using Martin's Axiom, it is often additionally assumed that $2^{\aleph_0} > \aleph_1$. In a footnote of their article, Solovay and Tennenbaum explain that they first worked without Martin's Axiom, but that Martin noticed a possible formulation in axiomatic terms.[37] However, they prefer using Martin's Axiom because it gives rise to a general proof scheme that is easier to apply than the repeated iteration of c.c.c. forcings.

> Most of the applications of iteration to date may be presented in the following manner. One shows that M [MA] (or possibly M + "$2^{\aleph_0} > \aleph_1$") implies a theorem T. Then the consistency proof for ZF + AC + M yields a consistency proof for ZF + AC + T.

To follow this approach, a relative consistency proof of Martin's Axiom is needed. Solovay and Tennenbaum gave such a proof as well. Once they had shown that Martin's Axiom is relatively consistent, it was enough to show that Martin's Axiom implies some theorem T. For, both results combined show that T is relatively consistent to ZFC. Solovay and Tennenbaum apply this scheme to SH and prove

ZFC + SH: "*Suppose ZF is consistent. Then so is ZF + AC + SH.*"[38]

1.4.2.3 Laver Forcing

Laver Forcing was developed to show the independence of the Borel Conjecture (named after Émile Borel).

Borel Conjecture, 1919: Every strong measure zero set is countable.[39]
Wacław Sierpiński proved in 1928 that the Borel Conjecture can be false.[40] To show that the Borel Conjecture can also be true, Richard Laver developed a forcing to add specific reals (*Laver reals*), which he then iterated.

ZFC + BorelConjecture, Laver Forcing, 1976: "*If ZFC is consistent, then so is ZFC+Borel's Conjecture.*"[41]

[37] Solovay and Tennenbaum (1971, fn on p.232).

[38] Solovay and Tennenbaum (1971, Theorem 7.11 on p.242).

[39] Borel (1919). A *strong measure zero set* is a subset X of the reals such that for every sequence $\langle \varepsilon_n : n < \omega \rangle$ of positive real numbers there is a sequence $\langle I_n : n < \omega \rangle$ of intervals with $\text{length}(I_n) \leq \varepsilon_n$ such that $X \subseteq \bigcup \{I_n : n < \omega\}$.

[40] Sierpiński (1928).

[41] Laver (1976, Theorem on p. 152), *see* also Jech (2003, pp.564–8).

1.4.2.4 Proper Forcing and Proper Forcing Axiom

Proper Forcing was defined by Saharon Shelah.[42] He takes proper forcings to be well-behaved forcings, in particular because they behave nicely when they are iterated:

> When we iterate we are faced with the problem of obtaining for the iteration the good properties of the single steps of iterations. Usually, in our context, the worst possible vice of a forcing notion is that it collapses $\aleph_1$. The virtue of not collapsing $\aleph_1$ is not inherited by the iteration from its single components. As we saw, the virtue of the c.c.c. is inherited by the ... iteration from its components. However in many cases the c.c.c. is too strong a requirement. We shall look for a weaker requirement which is more naturally connected to the property of not collapsing $\aleph_1$, and which is inherited by suitable iterations.[43]

The weaker requirement Shelah is looking for is *properness*.

Proper Forcing: A forcing notion $(P, <)$ is *proper* if forcing with P preserves, for every uncountable cardinal λ, the stationary sets of $[\lambda]^\omega$.[44]
Many specific forcing notions can be shown to be proper. For instance, every c.c.c. forcing is proper.[45]

As for c.c.c. forcing, one also needs a preservation theorem to iterate proper forcing. This important result is due to Shelah.[46] Since iterating proper forcing works well, an axiom was formulated to simplify it.

Proper Forcing Axiom (PFA): For every proper forcing notion $(P, <)$ and every family of dense sets $\mathscr{D}$ such that $|\mathscr{D}| = \aleph_1$ there exists a $\mathscr{D}$-generic subset $G \subseteq P$.

Again, once the relative consistency of the Forcing Axiom is shown,[47] one can prove theorems as consequences of the Forcing Axiom without working through the details of the iterated forcing method. The consequences of a Forcing Axiom are then proven relatively consistent to ZFC. One important consequence of the Proper Forcing Axiom was proven by Boban Veličković and Stevo Todorčević:

In **ZFC + PFA**: *Assuming the Proper Forcing Axiom, one can prove* $2^{\aleph_0} = \aleph_2$.[48]
This means that in ZFC+PFA the Continuum Hypothesis is false.

Today, most forcings applied in practice are iterated forcings or the Forcing Axioms obtained by consistency proofs through iterated forcing. For example, a current research problem is to find properties of forcing notions which allow the existence of set-theoretic universes saturated for the generics for families of dense sets of size $\aleph_2$.

[42] Shelah (1982).

[43] Shelah (2017, p.90).

[44] $[\lambda]^\omega$ is the set of all countable subsets of λ.

[45] Jech (2003, Lemma 31.2 on p. 601).

[46] *See* Shelah (2017, III. §3.).

[47] Assuming that there is a supercompact cardinal, one can prove that there is a model of ZFC+PFA.

[48] Jech (2003, Theorem 31.23 on p. 609).

1.5 Philosophical Thoughts in Set Theory

For someone who is interested in philosophical questions set theory is an exciting subject matter. A question that arises in set theory is for example the truth question: Are the independent sentences in set theory neither true nor false? What is truth in set theory if it does not coincide with provability? The truth question is part of traditional philosophy of mathematics. So is the question whether sets exist. If we assume that sets exist then we can easily give an account of truth: A sentence is true if and only if it is true in the universe of sets.

A closely related definition is the following: A set-theoretic sentence is true if and only if it is true in the intended model of set theory. This definition does not presuppose the existence of a universe of sets because it refers to the technical term of an intended model. Such a model may be given formalistically, i.e., as part of a formal theory. This definition is usually used when considering the formal number theory PA.[49] The standard model $\mathfrak{N} = (\mathbb{N}, 0, S, +, \cdot, <)$ of number theory can be given in set-theoretic terms. So, it can be given in the formal theory ZFC. From a formalistic point of view, one could adopt the second definition (if one believes that there is an intended model) but not the first. And from the point of view of a platonist, the first and the second definition would correspond to each other—the intended model would be taken as the formal counterpart of the real universe of sets.

Such ideas about truth and existence are some of the philosophical thoughts that we find in set-theoretic practice. They are invoked, for example, as a justification of axioms. When the second author asked the first author about the meaning of the word "axiom", she had a clear answer: "For me, what it means is an obvious property of the intended universe," and she admitted that this "is a strong meaning because it implies the existence of an intended universe." This corresponds exactly to the above mentioned view of true set-theoretic sentences. But the word "axiom" is also used for the new axioms, which are either not generally accepted by set theorists or their acceptance is less immediate than the acceptance of the ZFC-axioms.

M. Džamonja: *When we think of axioms in the classical sense, we think of Euclid and his geometry, and the idea there is that the axioms are statements that are obvious. Obvious in the sense that we take some basic objects, which, I think, in Euclid's mind come from his intended application, which he takes as the only one, and then the axioms are certain statements that are obviously true about these basic objects. From these we build out further content. I think, the idea of axiomatic set theory was to do this but with mathematics in general. The basic objects are sets. So, certainly the Axiom of Pairing is obvious, even though, now with the Homotopy Type Theory, it's a complicated issue, but in the classical set theory, this type of axioms—Pairing, Union—are somehow clear. Well, some of the classical axioms are also less clear. Of course, an example is the notorious Axiom of Choice. It is not clear in what sense it is an axiom. And, in fact, maybe*

[49]PA stands for Peano Arithmetic which is the formal theory of the natural numbers.

that situation between ZF and ZF with Choice is in some sense similar to the situation between ZFC plus some Forcing Axiom or just ZFC because you might say that for some people the Axiom of Choice was not natural, they refused to work with it and they worked in ZF set theory. Or even, you can say that for those people working with the ZFC set theory it is also interesting to understand where one really needs the Axiom of Choice. So, to work somehow between ZF and ZFC. Certainly, then, one doesn't have to take an opinion of whether the Axiom of Choice is true philosophically or not, but can work in both ZF and ZFC and somehow take the neutral view that this is what I can do if I have the Axiom of Choice and this is what I can do otherwise.
So, if you take that view, then the Forcing Axioms are consistent extensions of ZFC. The first Forcing Axiom—the Martin Axiom—has the same consistency strength as ZFC. We do not need any extra Large Cardinal Axiom to prove its consistency. So, if we just concentrate on that one, if we look at ZFC vs. ZFC plus Martin Axiom, then we can say: 'Well, we don't have to take the view that Martin Axiom is true or not, but we could say that this is a possible axiom to add. Is this true or not, well, we don't know.' So, in that sense, it is reasonable to make this an axiom. Also, it was the first forcing-related statement that got in any way close to being, let us say, comprehensible to a large number of mathematicians and logicians. Once, we had the Martin Axiom, of course, the extensions of it started coming, like the Proper Forcing Axiom. They are extensions because, mathematically speaking, they are very similar to the Martin Axiom. Logically speaking they are not extensions because they require Large Cardinal Axioms, so they lose that property of equiconsistency with ZF that we had before. Considering ZFC or ZFC plus Proper Forcing Axiom for example, we have a much stronger consistency strength with PFA added. So, the two are not exactly at the same level. In the end, by extending the strength of these Forcing Axioms, we seem to get further and further from what an actual axiom might mean.

We have seen in these paragraphs that the word "axiom" can also be used as a rather technical term without any philosophical implications. This is also possible regarding the ZFC-axioms, but it is even more important to emphasise the possibility of such a *neutral view* regarding the new axioms. In this view, the use of the word "axiom" does not imply that this statement is in any way accepted as a statement itself. It is rather accepted as a reasonable, possible addition to ZFC. Thus, one can work in the corresponding theory to address mathematical questions without taking a stance on its acceptability.

But still, the answer that an axiom is either obviously true or just a possible statement to assume, does not seem to give the whole story. The role of Large Cardinal Axioms in set-theoretic practice could be a challenge to this view. Neither are they treated as on part with the axioms of ZFC, nor as mere opportune additions. For instance, every Large Cardinal Axiom (at least as strong as the existence of an inaccessible) implies the consistency of ZFC. This seems to support their acceptability.

D. Kant: *What do you say about the statement "ZFC is consistent"? Does it still play a role in set-theoretic practice?*

M. Džamonja: *Well, of course, in the beginning it was hoped that we will prove that ZFC is consistent. That was Hilbert's Program,*[50] *and then Gödel's results said: if we believe that ZFC is the basic theory, then we cannot, within that basic theory, prove that it is consistent. So, now, we have two choices: Either we accept just ZFC as our basic theory and then we have to take on faith that it is consistent, or we say 'Well, ok, I believe in large cardinals and then I get the consistency of ZFC for free', in the sense that, when we take the cumulative hierarchy and cut it at a large cardinal, we get a model of ZFC.*
This is actually the other side of your question of what is an axiom. If we have an axiom scheme that is supposed to be obvious, then it is supposed to be talking about the intended model. Now, there are people who do not believe in the intended model. I believe that there is a universe of sets, personally, this is my philosophical view. So, if there is such a thing, then the ZFC Axioms are the axioms of this universe. They are not the only axioms but they are the axioms that we accept. They describe this universe quite well, so, they have an intended model. They have other models as well. But, somehow, believing in the consistency comes back to thinking if there is this universe of sets or not. And, I think, this is now a philosophical question rather than a mathematical one.

D. Kant: *So, you would think that, among set theorists, the existence of a universe of sets is somehow subjective, and some believe in it and some do not?*

M. Džamonja: *Yes. For example, I think, Gödel was a very strong Platonist. Woodin confirms to be a very strong Platonist and he is searching for more complete axioms of set theory. I think Shelah also is a Platonist. But I know people, like Cummings for example, who told me some years ago that, for him, the question if there is an intended model or not is not at all interesting. What is interesting is that we get to do beautiful mathematics with these objects and if they exist or not is not that interesting. So, one can do the same mathematics independently of one's philosophical view. In fact, mathematicians in general, even set theorists who work in logic, do not always ask philosophical questions. Some do and some don't.*

In set theory, there are mathematicians who think about philosophical concepts such as platonism. However, it would be wrong to assume that every set theorist thinks about independence also in philosophical terms. Džamonja suggests that it is a matter of interest. There may be more people in set theory who are interested in philosophical questions than there are in other mathematical disciplines. Yet, not all set theorists are philosophically inclined.

In the above conversation, the possibility to believe in large cardinals is mentioned. This highlights an important difference between the Large Cardinal Axioms

[50] Hilbert wanted to prove the consistency of mathematics and focussed on axiomatisations of number theory. His program can be transferred to set theory, as set theory counts as a foundation of mathematics. Thus, of one would prove its consistency, Hilbert's aim would be resolved.

and the Forcing Axioms. We think that Forcing Axioms are often not seen as candidates for acceptance.[51] The function of a Forcing Axiom is not to capture a possible truth about the universe of sets, but rather to formalise a specific fruitful kind of forcing. The formulation of such an axiom makes the application of forcing easier because one does not have to build up the whole forcing machinery each time.

M. Džamonja also said explicitely that she believes in the existence of a universe of sets. She said that she believes the ZFC Axioms as well as the Large Cardinal Axioms. However, she made clear that these beliefs are relative to set theory. Since the universe of sets is an abstract reality, it is possible that it is not the unique reality for all of mathematics. This view is supported by research in Univalent Foundations and Homotopy Type Theory (HoTT). This mathematical field has its own concepts and methods, which differ significantly from other mathematical fields; it creates own content, and mathematics can be embedded in this theory. However, some set-theoretic principles do not generally hold there, e.g., the Axiom of Choice. With this in mind, M. Džamonja believes in the Axiom of Choice, but only restricted to set theory, not with respect to all of mathematics.[52] Both set theory and HoTT can serve as a foundation for much of mathematics. In a Platonist framework, both fields can have their own mathematical reality.

1.6 Set-Theoretic Intuition About Independence

Both the truth question and the existence question mentioned in the last section are part of classical philosophy of mathematics. However, these questions are not those which the second author seeks to answer by talking to set theorists. One can observe that, in set theory, one independence result is not similarly conceived of as another. Set theorists see differences in the value of the insights which they provide, or in the naturalness of independent statements. For example, some set theorists consider the axiom $V = L$ to be less natural than the existence of infinitely many Woodin cardinals, because $V = L$ is not compatible with the existence of many large cardinals. Woodin says: "[T]he axiom $V = L$ limits the large cardinal axioms which can hold and so the axiom is *false*."[53] The existence of infinitely many Woodin cardinals imply Projective Determinacy, which is an attractive statement for some set theorists, for example for Ralf Schindler: "The principle of projective determinacy, being independent from the standard axiom system of set theory, produces a fairly complete picture of the theory of 'definable' sets of reals."[54]

[51]Menachem Magidor certainly is an exception because he thinks that Forcing Axioms are natural axioms.

[52]For more on her view, see her article 'A New Foundational Crisis in Mathematics, is It Really Happening?' in this volume.

[53]His italics, Woodin (2010, p.504).

[54]www.math.uci.edu/node/20943 (06/05/2018)

In order to elaborate on such judgements, the authors talked about the question whether set theorists have an intuition about their subject matter which is based on their wide experience, but which is not necessarily backed up by proofs.

D. Kant: *I imagine that set theorists have gained a good intuition about what is provable in ZFC and what is independent. Would you say that you have a good intuition about this?*

M. Džamonja: *Yes, I think we do have a good intuition. Of course, not about everything, but about certain things, certain areas. I have a way that I see it. I think of the line of cardinal numbers. Certain areas of that line are well understood and we really have an intuition in that context, but other areas are murky.*

D. Kant: *Is there maybe something that you can say about these borders of ZFC? More concretely, is there something about these independent sentences that they have in common?*

M. Džamonja: *To start with, there are certain things that definitely cannot be independent because they are described by simple formulas and we have absoluteness theorems.*[55] *Things that are combinatorially close to them can likely be shown to be true or not true. So, descriptive set theory and things that go with it. We may find there some sort of mini-independence. They would be connected with certain classes of sets whose properties would be understood within a ZF bit which exhibits less absoluteness, such as analytic sets, or projective sets, etc. Sometimes, we can reflect independence to truth by restricting our attention to certain classes of sets. For example, suppose that we can use a Forcing Axiom to prove some statement about subsets of the reals in general. We can then hope to have the same statement hold about analytic sets without needing any additional axioms. Certain results that are obtained under PFA for general sets turn out to be true for analytic sets. For example, one can find this in the work of Todorcevic about gaps.*[56] *There are analytic gaps, there are general gaps, there is the p-ideal dichotomy, and then there is this dichotomy applied to analytic objects. Or in the work of Solecki. So, that is one border of independence.*

Another border is, as I mentioned, the line of cardinal numbers. We know that at successors of regular cardinals we can do a lot by forcing, especially at $\aleph_1$. We also know that at singular cardinals and their successors things are much more, let us say, resistant to forcing. This is so because we have pcf theory which shows that some things about singular cardinals are just true in ZFC, and, therefore, many statements that are implied by pcf theory are also just true.[57] *So, there*

---

[55] For every Δ_0-sentence φ (a sentence with only bounded quantifiers) and every transitive standard model M of ZFC, $\varphi \leftrightarrow \ulcorner M \models \varphi \urcorner$ (in ZFC). Under the assumption that there is a transitive standard model of ZFC, this means that Δ_0-sentences cannot be independent. They are either provable or refutable. There are other well-known absoluteness theorems, such as the Shoenfield's Absoluteness.

[56] See for example Avilés and Todorcevic (2015, 2016).

[57] Shelah (2000) and Burke and Magidor (1990).

are these two distinct regions on the cardinal numbers line. There are successors of regular cardinals, which have some behaviour, and then there are singular cardinals and their successors, which have another one, and then, of course, there are large cardinals.[58] *So, we do have a good intuition when we start from a certain kind of cardinal in which direction to try to start working. And then we also have a good intuition about the kind of sentences as explained above. Combinatorial set theory is almost always about unrestricted sets. So, there we can expect to have a lot of independence.*

D. Kant: *About independent sentences in history: have there been some surprising or unexpected results? So, that, at first, the sentence was thought to be true, and then it turned out to be independent, or something like that? That really set theorists ...*

M. Džamonja: *... were surprised?*

D. Kant: *Yes, were surprised about what came out?*

M. Džamonja: *Well, we have already said that the independence of CH came as a surprise to many mathematicians. But there is a recent example of just the opposite, when a statement was thought to be independent but at the end it turned out just to be true. I refer to a theorem by Malliaris and Shelah.*[59] *They proved a certain cardinal equality, that is, they proved that two cardinals,* $\mathfrak{p}$ *and* $\mathfrak{t}$, *which are cardinal invariants of the continuum, are actually just equal in ZFC. This was totally unexpected because there are many cardinal invariants known and they tend to be independent from each other. The independence of* $\mathfrak{p}$ *and* $\mathfrak{t}$ *was one of the last open questions and everybody expected they would behave like any other invariants, be independent—but they are not! The Malliaris-Shelah proof is also very ingenious, it mixes many different methods. That proof obtained an important prize in 2017, the Hausdorff medal that is given biannually by the European Set Theory Society for the most influential work published in the last five years.*

D. Kant: *But this does not happen very often?*

M. Džamonja: *No. That does not happen very often. Well, see, what I think is that, in mathematics, a huge percentage of results is proving something that is not so surprising. Everywhere in mathematics, including set theory, there are results that everybody suspects to hold, but if you want to be sure, you have to produce a proof. So, when somebody takes two new cardinal invariants and makes them independent of each other, that makes an ok PhD thesis but it does not make a huge surprise or a Hausdorff medal. Because we have seen such results very often. The opposite is surprising.*

This conversation should make clear that set theorists can say something more about the independence phenomenon than what they can prove. They can sometimes give probabilistic statements about the mathematical objects they work with. One

[58]For a philosophical discussion of these regions see Džamonja and Panza (2018).

[59]Malliaris and Shelah (2013).

example was mentioned: Two cardinal invariants are often independent. Of course, such general ideas are based on experience and could turn out to be wrong—just imagine that there will be found many cardinal invariants such that pairs of them can often shown to be equal. However, such probabilistic, experience-based statements seem to play a very important role in set-theoretic research. In addition, it is an interesting philosophical question whether and, if so, how such general ideas can be seen as a part of set-theoretic knowledge.

For a first exploration on this question, we would argue that experience-based statements are part of set-theoretic knowledge. Given potential future experiences, they would have to be relativised to a time-frame in which they correspond to beliefs of most set theorists. Describing this part of set-theoretic knowledge is valuable because it can explain the development of the discipline. Imagine that some day, a new axiom is accepted. Then this would be possibly seen as a surprising event when only looking at the theorems. However, it could possibly be explained by looking at the experience-based, probabilistic part of set-theoretic knowledge. Furthermore, normative judgements by set theorists also play an important role in this context. For instance, judgements concerning the naturalness or attractiveness of particular axioms. These might also be explainable via the informal part of set-theoretic knowledge.

1.7 Conclusion

We started with a philosophical perspective on the set-theoretic independence phenomenon. This mathematical phenomenon raises many questions and can appropriately be called independence *problem*. In mathematical terms, on the other hand, it is not clear whether the independence phenomenon is a problem or a mere mathematical fact.

Putting together our mathematical and philosophical perspectives, we gave an insight in contemporary set theory. We focussed on forcing in order to illustrate to what extent independence results determine today's research. After that, we presented Džamonja's views on various topics, such as the introduction of forcing in set theory, the use of new axioms in practice (distinguishing between ZFC-axioms, Forcing Axioms, and Large Cardinal Axioms), the notion of the universe of sets, and surprising events. With this, we attempted to grasp what Tao calls the "solid intuition"[60] of an expert mathematician on her/his field of expertise. This solid intuition is grounded in many years of set-theoretic research, which seems to make it unavailable to non-set theorists. For, it is often the set-theoretic formalism and rigour which make it hard for philosophers and other mathematicians to acquire an understanding of the topics of set-theoretic research. However,

[60] Tao (2018).

> it is only with a combination of both rigorous formalism and good intuition that one can tackle complex mathematical problems; one needs the former to correctly deal with the fine details, and the latter to correctly deal with the big picture.[61]

Given this, we can hardly hope to widely communicate the fine details of set-theoretic research. We can hope, however, to communicate a big picture in a comprehensible and correct way. It would be correct if it is consistent with set-theoretic practice. Such a big picture will include different perspectives of set theorists. They will differ on some aspects and agree on others.

To close this discussion, we want to leave the reader with a question and three hypotheses.

Question: What are the different aims/motivations for the uses of different axioms?

1. Hypothesis: Most set theorists were surprised by the introduction of the forcing method.
2. Hypothesis: Most set theorists think that forcing is a natural part of contemporary set theory.
3. Hypothesis: Most set theorists prefer an answer to a problem with the help of a new axiom of lowest possible consistency strength. And for most set theorists, a difference in consistency strength weighs much more than the difference between Forcing Axiom and Large Cardinal Axiom.

We distinguished a neutral view toward the use of Forcing Axioms on the one hand, and the use of accepted Large Cardinal Axioms on the other. This analysis can certainly be refined. The hypotheses are motivated because they correspond to Džamonja's view, which, we think, is representative for other set theorists as well. However, certain objections may be levelled. In contrast to the first hypothesis, one could also support a historical view according to which time had simply come for the forcing method to be introduced. The second hypothesis may be challenged by views of descriptive set theorists who rarely use forcing in their research. Finally, one might certainly find set theorists who would not agree with the third hypothesis (for example, for many years the school of the set theory of the reals did not accept the idea of large cardinals). Thus, we are not in the position to draw final conclusions. Rather, we encourage further research on set-theoretic practice which will bring further clarification.

Acknowledgements The second author wants to thank the first author for her generous time to share her knowledge and views. She is also grateful to Colin Rittberg for his encouragement and discussion, to the audiences in Brussels, Berlin, and Konstanz for their feedback and questions, which helped a lot to sharpen her approach, to the anonymous referee for his/her substantial and very helpful review, and to Stefan Steins for a rigourous improvement of the language.

[61] ibid.

References

Arrigoni, T., & Friedman, S.-D. (2013). The Hyperuniverse program. *The Bulletin of Symbolic Logic, 19*(1), 77–96.

Avilés, A., & Todorcevic, S. (2015). Finite basis for analytic multiple gaps. *Publications mathématiques de l'IHÉS, 121*(1), 57–79.

Avilés, A., & Todorcevic, S. (2016). Types in the n-adic tree and minimal analytic gaps. *Advances in Mathematics, 292*, 558–600.

Badiou, A. (2005). *Being and event*. London: Continuum.

Borel, E. (1919). Sur la classification des ensembles de mesure nulle. *Bulletin de la Société Mathématique de France, 47*, 97–125.

Burke, M. R., & Magidor, M. (1990). Shelah's PCF theory and its applications. *Annals of Pure and Applied Logic, 50*(3), 207–254.

Chow, T. Y. (2007). A beginner's guide to forcing. arXiv ID: 0712.1320.

Cohen, P. J. (1963). The independence of the continuum hypothesis. *Proceedings of the National Academy of Sciences of the United States of America, 50*(6), 1143–1148.

Cohen, P. J. (1964). The independence of the continuum hypothesis, II. *Proceedings of the National Academy of Sciences of the United States of America, 51*(1), 105–110.

Cohen, P. (1966). *Set theory and the continuum hypothesis*. New York: Benjamin.

Džamonja, M., & Panza, M. (2018). Asymptotic quasi-completeness and ZFC. To appear in the volume "Between Consistency and Inconsistency" of Trends in Logic (eds. W. Carnielli and J. Malinowski).

Easton, W. B. (1970). Powers of regular cardinals. *Annals of Mathematical Logic, 1*(2), 139 – 178.

Gödel, K. (1947). What is Cantor's Continuum problem? *The American Mathematical Monthly, 54*(9), 515–525.

Hamkins, J. D. (2012). The set-theoretic multiverse. *The Review of Symbolic Logic, 5*, 416–449.

Jech, T. (2003). *Set theory*. Berlin/Heidelberg: Springer. (The third millennium edition, revised and expanded edition.)

Jensen, R. (1972). The fine structure of the constructible hierarchy. *Annals of Mathematical Logic, 4*(3), 229–308.

König, J. (1905). Zum Kontinuum-Problem. *Mathematische Annalen, 60*(2), 177–180.

Kunen, K. (1980). Set theory: An introduction to independence proofs. In J. Barwise, H. J. Keisler, P. Suppes, & A. S. Troelstra (Eds.), *Studies in logic and the foundations of mathematics* (Vol. 102). Amsterdam: Elsevier Science Publishers.

Kunen, K. (2011). *Set theory*. Studies in logic. London: College Publications.

Laver, R. (1976). On the consistency of Borel's conjecture. *Acta Mathematica, 137*(1), 151–169.

Maddy, P. (2007). *Second philosophy: A naturalistic method*. Oxford: Oxford University Press.

Maddy, P. (2011). *Defending the axioms: On the philosophical foundations of set theory*. Oxford: Oxford University Press.

Malliaris, M., & Shelah, S. (2013). General topology meets model theory, on p and t. *Proceedings of the National Academy of Sciences of the United States of America, 110*(33), 13300–13305.

Rittberg, C. J. (2015). How woodin changed his mind: New thoughts on the continuum hypothesis. *Archive for History of Exact Sciences, 69*, 125–151.

Rouse, J. (1995). *Engaging science: How to understand its practices philosophically*. Ithaca: Cornell University Press.

Shelah, S. (1982). *Proper forcing*. Berlin: Springer.

Shelah, S. (2000). Applications of PCF theory. *Journal of Symbolic Logic, 65*(4), 1624–1674.

Shelah, S. (2017). *Proper and improper forcing* (Perspectives in Logic, 2nd ed.). Cambridge: Cambridge University Press.

Sierpiński, W. (1928). Sur un ensemble non dénombrable, dont toute image continue est de mesure nulle. *Fundamenta Mathematicae, 11*(1), 302–303.

Sierpiński, W., Knaster, B., Kuratowski, C., Souslin, M., Steinhaus, H., Mazurkiewicz, S., Lusin, N., & Felsztyn, T. (1920). Problèmes. *Fundamenta Mathematicae, 1*(1), 223–224.

Solovay, R. M., & Tennenbaum, S. (1971). Iterated Cohen extensions and Souslin's problem. *The Annals of Mathematics, 94*(2), 201.

Tao, T. (2018). There's more to mathematics than rigour and proofs. https://terrytao.wordpress.com/career-advice/theres-more-to-mathematics-than-rigour-and-proofs/, Requested on 28 May 18.

Woodin, W. H. (2010). Strong axioms of infinity and the search for V. In *Proceedings of the International Congress of Mathematicians* (Vol. 1, pp. 504–528).

Woodin, W. H. (2017). In search of Ultimate-L: The 19th Midrasha Mathematicae Lectures. *Bulletin of Symbolic Logic, 23*(1), 1–109.

Chapter 2
How to Choose New Axioms for Set Theory?

Laura Fontanella

Abstract We address the problem of the choice of new axioms for set theory. After discussing some classical views about the notion of axiom in mathematics, we present the most currently debated candidates for a new axiomatisation of set theory, including Large Cardinal axioms, Forcing Axioms and Projective Determinacy and we illustrate some of the main arguments presented in favour or against such principles.

2.1 Introduction

The development of axiomatic set theory originated from the need for a rigorous investigation of the basic principles at the foundations of mathematics. The classical theory of sets ZFC offers a rich framework, nevertheless many important mathematical problems (such as the famous continuum hypothesis) cannot be solved within this theory. Set theorists have been exploring new axioms that would allow one to answer such fundamental questions that are independent from ZFC. Research in this area has led to consider several candidates for a new axiomatization such as Large Cardinal Axioms, Forcing Axioms, Projective Determinacy and others. The legitimacy of these new axioms is, however, heavily debated and gave rise to extensive discussions around an intriguing philosophical problem: *what criteria should be satisfied by axioms?* What aspects would distinguish an axiom from a hypothesis, a conjecture and other mathematical statements? What is an axiom after all? The future of set theory very much depends on how we answer such questions. Self-evidence, intuitive appeal, fruitfulness are some of the many criteria that have been proposed. In the first part of this paper, we illustrate some classical views about the nature of axioms and the main difficulties associated with these positions. In the second part, we outline a survey of the most promising candidates for a new

L. Fontanella (✉)
Institut de Mathématiques de Marseille, Université Aix Marseille, Cachan, France
e-mail: laura.fontanella@univ-amu.fr

S. Centrone et al. (eds.), *Reflections on the Foundations of Mathematics*,
Synthese Library 407, https://doi.org/10.1007/978-3-030-15655-8_2

axiomatization for set theory and we discuss more specific arguments that were suggested in favor of these statements. In the end, we will not answer the question *How to chose new axioms for set theory?*, but we will discuss the main challenges associated with this problem. We assume basic knowledge of the theory ZFC.

2.2 Ordinary Mathematics

Before we start our analysis of the axioms of set theory and the discussion about what criteria can legitimate those axioms, we should address a quite radical view based on the belief that 'ordinary mathematics needs *much less than ZFC or ZF*'. This claim suggests that strong axioms such as the Axiom of Choice, or Infinity are not really needed for standard mathematical results, and certainly ordinary mathematics does not need new strong axioms such as Large cardinals axioms, Forcing Axioms etc. If so, then our goal of securing the axioms of ZFC and the new axioms would simply be irrelevant or a mere set theoretic concern (where set theory is not considered standard mathematics).

The issue with this view is to clarify what counts as 'ordinary mathematics'. In fact, the Axiom of Choice is heavily used in many fields such as algebra, general topology, measure theory and functional analysis. For instance, the Axiom of Choice is indispensable for the following claims and theorems (they are actually equivalent to the Axiom of Choice):

- For every equivalence relation there is a set of representatives.
- Every surjection has a right inverse.
- Every vector space has a basis.
- Krull's theorem: Every unital ring other than the trivial ring contains a maximal ideal.
- Tychonoff theorem.
- In the product topology, the closure of a product of subsets is equal to the product of the closures.
- Every connected graph has a spanning tree.

Other weaker consequences of the Axiom of Choice cannot be proven within ZF:

- Baire category theorem (which is equivalent to the Axiom of Dependent Choice).
- Hahn Banach theorem.
- Every Hilbert space has an orthonormal basis.
- The closed unit ball of the dual of a normed vector space over the reals has an extreme point.
- Every field has an algebraic closure.
- Stone's representation theorem for Boolean algebras.
- Nielse-Schreier theorem: every subgroup of a free group is free.
- Vitali theorem: there exists a set of reals which is not Lebesgues measurable.
- The existence of a set of reals which does not have the Baire property.
- The existence of a set of reals which does not have the perfect set property.
- Every set can be linearly ordered.

Thus, important applications of the Axiom of Choice can be found in many areas, hence rejecting the Axiom of Choice would come with a big price for the scope of 'ordinary mathematics'.

The analogous claim that ordinary mathematics does not need *more than ZFC* runs into a similar problem, as it is often the case that natural questions, that were raised in what one might consider a standard mathematical framework, turned out to be independent from ZFC, thus requiring stronger additional axioms to be settled. It is the case, for instance, for Whitehead problem in group theory: formulated in the 1950s, Whitehead problem was considered one of the most important open problems in algebra for many years, until S. Shelah showed in 1974 its undecidability in ZFC (see Shelah 1974); Whitehead conjecture is true if we accept the axiom of constructibility, namely that every set is constructible. A similar case is the Normal Moore Space Conjecture, a topological problem whose solution was eagerly sought for many years until strong large cardinal assumptions turned out to be indispensable for its solution. The reader is certainly familiar with the famous Fermat's conjecture recently demonstrated by Wiles who won the Abel prize for his outstanding result; what the reader might not be aware of, is that Wiles's proof relies on Grothendieck's universes whose existence requires large cardinals, namely strongly inaccessible cardinals (for more details, see McLarty 2010). It is generally believed that eventually we will be able to prove Fermat's conjecture in Peano Arithmetic (at the cost of a much more complicated organization of the proof), yet the only known proof today – more than 20 years after Wiles released it – uses Grothendieck universes in fact.

Independence results have always caused a certain embarrassment in the community of mathematicians. When a mathematical problem is proven to be independent from ZFC, suddenly it is labeled as 'just set theoretical' or 'vague' and no longer mathematical in the traditional sense. A precise definition of what 'ordinary mathematics' means should then take into account this attitude towards those problems which, at first, seem to emerge naturally as intrinsically relevant questions for mathematical research, then are dismissed after proven to require strong axioms. A simple move would be to claim that independent problems are legitimate mathematical questions that yet are 'unsolvable'. In this perspective, then, any attempt to answer such questions with stronger assumptions can only be seen as speculative. Surely, many mathematicians navigate these lines of thoughts. For instance, when Nykos (1980) proved in 1980 the consistency of the Normal Moore Space Conjecture from a strongly compact cardinal, he titled his paper 'A *provisional* solution to the Normal Moore Space Conjecture' (emphasis mine). However, if any result assuming large cardinals were just 'provisional', as Nykos' choice of words suggests, then Large Cardinals Axioms would be nothing more than mere *hypotheses*. Yet, despite the general skepticism towards the legitimacy of these principles, the mathematical community seems to acknowledge them a different status, a stronger role. In fact, we can point out that Wiles's proof of Fermat's conjecture was well accepted by the community of number theorists despite the fact that it relies on inaccessible cardinals. Imagine that his proof were assuming Riemann hypothesis instead, would his result even be published? The supporter of the view that ordinary mathematics can all be done in ZFC, or in a much weaker system than ZF, needs to clarify what should be the status of independent problems and of the additional assumptions needed for their solution.

2.3 Intrinsic Motivations

The word 'axiom' comes from the Greek $\alpha\xi\iota\omega\mu\alpha$ 'that which commends itself as evident'. To these days, most of the mathematicians would consider axioms to be a *self-evident* propositions requiring no formal demonstration to prove their truth, but received and assented as soon as mentioned. This is the ideal meaning of the word 'axiom'. The problem with this view is that what counts as obvious, self-evident, intuitive or inherently true is highly subjective.

> I can in no way agree to taking 'intuitively clear' as a criterion of truth in mathematics, for this criterion would mean the complete triumph of subjectivism and would lead to a break with the understanding of science as a form of social activity. (Markov 1962).

Moreover, the self-evidence criterion is quite restrictive. Not only the new axioms considered in contemporary set theory such as Large Cardinal axioms are far from self-evident (not even their strongest supporters claim they are self-evident), but even the axioms of ZFC are not strictly obvious. Certainly, the Axiom of Choice and the Axiom of Infinity were not immediately received and assented as soon as mentioned, on the contrary they were extensively debated and a mild skepticism still survives.

> The set theoretical axioms that sustain modern mathematics are self-evident in differing degrees. One of them – indeed, the most important of them, namely Cantor's axiom, the so-called axiom of infinity – has scarcely any claim to self-evidence at all. (Mayberry 2000, p. 10)

Problems occur even if we reformulate the self-evidence requirement and consider the following criterion that we may call '*intrinsic necessity*':

> An axioms must have some intuitive appeal, however the axiom may not be immediately obvious, but it should ultimately occur to us that what the axiom states is true and it could not be otherwise.

This reformulation may legitimate those controversial axioms, such as the Axiom of Choice, that were not immediately accepted but were eventually welcomed and employed. For instance, while the well-ordering principle mainly encountered reluctance, the equivalent statement 'the cartesian product of a collection of non-empty sets is non-empty' seems to be better accepted by the mathematical community as a fundamental truth. Unfortunately, the criterion of intrinsic necessity is still problematic. There is no strong reason for believing that what the Axiom of Choice states could not be otherwise. In fact, as proven by Banach and Tarski, the Axiom of Choice is actually paradoxical as it implies a quite counterintuitive statement (Banach-Tarski paradox): given a solid ball in a 3-dimensional space, there exists a decomposition of the ball into a finite number of disjoint subsets, which can then be put back together in a different way to yield two identical copies of the original ball. Thus the Axiom of Choice challenges basic geometric intuition, leaving a shadow on its alleged intrinsic necessity.

Maddy's analysis in Maddy (1988) shows that even the less controversial axioms of the theory ZF were not motivated by intrinsic reasons but rather practical ones. Consider for instance the Axiom of Foundation: first introduced in the form $A \notin A$ to block Russell's paradox, it is nowadays adopted in its stronger version "every set is well-founded". Reasons for reformulating the axiom in this way were not based on self-evidence, but originated from the belief that "no field of set theory or mathematics is in any general need of sets which are not well-founded" (Fraenkel et al. 1973, p. 88). Actually, non-well founded sets found applications in computer science to model non-terminating computational processes; there is a whole line of work in logic that deals with alternatives to the Axiom of Foundation, the work of Forti, Honsell (1983) and Aczel (1988) in this area was especially influential.

Today the Axiom of Foundation is better supported by the so-called '*Iterative conception*'. Roughly, this consists in the idea that sets must be obtained by an iterative process where at a first stage certain sets are secured 'immediately', then new sets can be obtained starting from the sets at the first level so to form a second level, and at each stage new sets can be defined from the ones introduced at the previous levels. Under the Axiom of Foundation, all sets can be obtained in this way, in fact the class of all sets V coincide with the *Von Neumann Universe* which is defined as follows. The level zero V_0 is the empty set, then the first level V_1 contains just the empty set, at each successor stage $\alpha + 1$, the level $V_{\alpha+1}$ is defined as the set of all subsets of V_α (in fact V_1 coincides with $\mathscr{P}(V_0)$), at limit stages λ, we let V_λ be the union of all V_α for $\alpha < \lambda$. The Von Neumann Universe is the class obtained from the union of all V_α's. The Axiom of Foundation is equivalent to V being equal to the Von Neumann Universe which is the main expression of the iterative conception just discussed. This is often considered to be an intrinsic justification for the Axiom of Foundation, yet it is not *obvious* that such an iterative process would exhaust all possible sets. On the other hand, the Von Neumann hierarchy certainly gives a very useful and elegant description of the class of all sets, thus the Axiom of Foundation has undoubtedly strong *practical* merits.

2.4 Extrinsic Motivations

We argued that intrinsic motivations such as self-evidence, intrinsic necessity etc. are subjective and restrictive. Those considerations led Maddy to claim that axioms are mainly supported by *extrinsic motivations*, namely by their success, or fruitfulness. The roots of this idea already appeared in Gödel (1947):

> Furthermore, however, even disregarding the intrinsic necessity of some new axiom, and even in case it had no intrinsic necessity at all, a decision about its truth is possible also in another way, namely, inductively by studying its "success", that is, its fruitfulness in consequences and in particular in 'verifiable' consequences, i.e., consequences demonstrable without the new axiom, whose proofs by means of the new axiom, however, are considerably simpler and easier to discover, and make it possible to condense into one proof many different proofs. (Gödel 1947, p. 521)

It is important to stress that Gödel and Maddy mean different things with the term 'fruitful'. Let us discuss first Gödel's view. For Gödel, the fruitfulness of an axiom is measured in terms of '*verifiable consequences*'. This is a delicate notion that deserves several comments. How can we verify a mathematical statement? Is this verification the result of an empirical process? Gödel believed in some sort of perception of mathematical entities analogous to our perception of physical objects. Thus, in Gödel's view, the truth of certain mathematical statements imposes on us to the extent that our intuition provides us with some sort of perception of the mathematical objects involved; other statements are not given to us immediately by mathematical intuition, but they are supported by the 'evidence' of their consequences.

In more recent work, Magidor considers this mathematical verification to be directly connected to our empirical knowledge of the physical world:

> As far as verifiable consequences, I consider the fact that these axioms [large cardinals] provide new Π^1_0 sentences which so far were not refuted. In some sense we can consider these Π^1_0 sentences as physical facts about the world that so far are confirmed by the experience. (Magidor 2012)

Whatever meaning we accord to the expressions 'mathematical verification' and 'mathematical evidence', we should note that, as for natural sciences, a plurality of verifiable consequences cannot secure the theory with certainty, since even inconsistent mathematical theories can, at first, appear to have many verifiable consequences. Thus, we can only say of a given axiom or theory that it was not refuted *so far*. In other words, paraphrasing Popper, mathematical theories are not strictly speaking verifiable, they are only *falsifiable*; but this is not different from physics, chemistry or other sciences.

A more challenging remark is that mutually incompatible theories can all be 'successful'. Consider, for instance, the theory ZFC plus the axiom of measurable cardinals versus ZF + V = L (more details about these axioms will be provided in the second part of this paper). These two theories are incompatible, yet none of their consequences was 'refuted' so far. In Gödel's view the axiom V = L should be rejected because it implies the continuum hypothesis which he believed was 'false' due to certain consequences that he considered 'highly implausible': for instance, CH implies the existence of a set of size $2^{\aleph_0}$ which has Lebesgue measure zero, but is not absolute zero. Yet, his feeling that such consequences would be implausible is not unanimously shared, thus we are left with an unclear notion of mathematical evidence which cannot guide us in the choice of one theory over another.

Maddy developed a wider conception of extrinsic justifications that she describes as based on *practical, inter-theoretic motivations*. In Maddy (2011), she explains the success of an axiom on the basis of the '*proper methods*' of the theory considered, namely the fruitfulness of an axioms is measured in terms of its effectiveness to achieve specific mathematical goals. So, for instance, Projective Determinacy came into considerations as the result of a broader research for new axioms that might settle certain problems in analysis and set theory that could not be solved within ZFC, accordingly certain Large Cardinals axioms are justified as they imply Projective Determinacy and settle other problems that are independent from ZFC.

Once again, distinct incompatible theories can be equally successful, even in this sense. For instance, if many strong Large Cardinals Axioms such as the axiom of measurable cardinals can be justified in this way, even the axiom of constructibility V = L can be viewed as an effective mean to achieve specific mathematical goals: V = L settles the continuum problem as it implies the Continuum Hypothesis and even GCH, it also implies the Axiom of Choice which can be used itself for proving classical fundamental theorems in mathematics, and it settles many other questions that are independent from ZF, for instance it implies the negation of the Suslin's hypothesis. Thus even this approach requires additional criteria for choosing a specific theory over the other. Maddy's suggestion is to appeal on the *maximality principle*. Roughly, this consists in the idea that we should prefer the theory that maximizes the concept of set. For instance, the concept of set underlying large cardinals seems to be wider than the one associated with the axioms of constructibility which is often ruled out as 'too restrictive'. Reference to this 'maximize rule' can be found for instance in Drake (1974), Moschovakis (1980) and Scott (1961). Nevertheless, the alleged restrictiveness of the axiom of constructibility was recently refuted by Hamkins (2014) who proved, roughly, that the axiom of constructibility is rich enough to allow one to talk about the concept of sets in the sense of large cardinals within a model of $V = L$. We will discuss this further in Sect. 2.5.

Finally, we can remark that the 'extrinsic approach' makes axioms depend on their consequences. This conflicts with the traditional view that considers axioms to be *the starting point for demonstration* from which, ideally, the truth of the other mathematical statements can be derived. Here, the situation is reversed: the consequences of an axiom legitimate the axiom, or they lead us to reject it when we have some 'counter-evidence' for such mathematical consequences. In this picture, then, any part of mathematics, including axioms, could be altered in the light of 'evidence'.

2.5 The Axiom of Constructibility

We now illustrate the main candidates for new axioms considered in contemporary set theory. We now disregard all the criticisms made so far of intrinsic and extrinsic motivations in general and we discuss more specific arguments that were suggested in favor or against these statements.

The oldest of these axioms is certainly the *Axiom of Constructibility* V = L that asserts that every set is constructible, namely every set belongs to Gödel's constructible universe L. L is inductively defined as follows:

- $L_0 = \emptyset$;
- $L_{\alpha+1}$ is the set of all subsets a of L_α that are definable with parameters in L_α (i.e. there is a formula $\varphi(x, a_1, \ldots, a_n)$ with parameters $a_i \in L_\alpha$ such that $a = \{x \in L_\alpha;\ L_\alpha \models \varphi(x, a_1, \ldots, a_n)\}$);

- when λ is a limit ordinal, $L_\lambda = \bigcup_{\alpha<\lambda} L_\alpha$.

Finally, $L := \bigcup_{\alpha \in Ord} L_\alpha$ (L is a class). The constructible universe was introduced by Gödel in 1938 to prove the consistency of the continuum hypothesis. In fact, the axiom of constructibility implies the generalized continuum hypothesis. Moreover, it implies the Axiom of Choice, and it settles many other questions that are independent from ZF, for instance it implies the negation of the Suslin's hypothesis.

Sentiment in favor of the Axiom of Constructibility can be found for example in Fraenkel et al. (1973). Nevertheless, the axiom of constructibility counts very few supporters among contemporary set theorists. We already mentioned that Gödel himself did not consider $V = L$ as a valid candidate axiom for set theory, as he believed that CH was actually false. On the other hand, there is no clear evidence for the Continuum Hypothesis or its negation that would count as a corroboration or a falsification for the Axiom of Constructibility. Maddy suggested that V = L should be rejected because it implies the existence of a an analytic (Δ^1_2) non-measurable set of reals,[1] but it is not clear on what ground this consequence should be considered implausible or unsuitable (we will discuss Lebesgue measurability in Sect. 2.8).

The strongest objections to the Axiom of Constructibility are related to its apparent restrictiveness. In fact, L is provably the smallest inner model of ZFC (i.e. the smallest class satisfying the axioms of ZFC and containing all the ordinals). It follows that, if we consider other axioms such as the Axiom of Measurable cardinals (a large cardinal axiom that establishes the existence of a measurable cardinal), despite the incompatibility of this axiom with V = L, it is always possible to talk about the concept of set in the sense of V = L within the theory ZFC+$\exists\kappa$ measurable:

> we can see why most set theorists reject V = L as restrictive: adopting it restricts the interpretative power of the language of set theory. The language of set theory as used by the believer in V = L can certainly be translated into the language of set theory as used by the believer in measurable cardinals, via the translation $\varphi \mapsto \varphi^L$. There is no translation in the other direction. (Steel (Feferman et al. 2000))

Actually, as Hamkins showed in (2014), there is a sense in which the converse is also possible.

> even if we have very strong large cardinal axioms in our current set-theoretic universe V, there is a much larger universe V^+ in which the former universe V is a countable transitive set and the axiom of constructibility holds.

This means that, even the Axiom of Constructibility is reach enough to allow us to talk about the concept of sets in the sense of large cardinals within a model of $V = L$.

---

[1] "There are also extrinsic reasons for rejecting V = L, most prominently that it implies the existence of a Δ^1_2 well-ordering of the reals, and hence that there is a Δ^1_2 set which is not Lebesgue measurable." (Maddy 1988)

2.6 Large Cardinals Axioms

Let us discuss now *Large Cardinals Axioms* – there is a whole hierarchy of large cardinals axioms, we will discuss just some notions–. A large cardinal is any uncountable cardinal κ which is at least *weakly inaccessible*, namely which satisfies the following two properties:

(1) for every cardinal $\gamma < \kappa$, we have $\gamma^+ < \kappa$;
(2) for every subset $X \subseteq \kappa$ of size $< \kappa$, we have $\sup X < \kappa$.

If we replace the condition 1 with the stronger 'for every cardinal $\gamma < \kappa$, we have $2^\gamma < \kappa$', then we have the notion of *strong inaccessible* cardinal; we will simply call *inaccessible* the strong inaccessible cardinals. We can observe that the properties above are all satisfied by $\aleph_0$, thus the axioms of weakly and strongly inaccessible cardinals establish that there are other cardinals than $\aleph_0$ with those properties. There is no precise definition of what a large cardinal axiom is, but we can say that all large cardinals axioms establish or imply the existence of weakly inaccessible cardinals.

The existence of a weakly inaccessible cardinal yields the consistency of ZFC, as if κ is a weakly inaccessible cardinal, then L_k is a model of ZFC. Since, by Gödel's incompleteness theorem, ZFC cannot prove its own consistency, it follows that the existence of large cardinals cannot be proven within ZFC; neither the *consistency* of large cardinals axioms can be proven from the consistency of ZFC. This marks an important difference between large cardinals axioms and other kind of axioms such as $V = L$ whose consistency can be proven relative to the consistency of ZFC.

The following motivations were suggested in favor of the existence of inaccessible cardinals:

- *Uniformity.* Roughly, this is the belief that the universe of sets should be uniform, in the sense that "it doesn't change its character substantially as one goes over from smaller to larger sets or cardinals, i.e., the same or analogous states of affairs reappear again and again (perhaps in more complicated versions)" (Wang 1974, pp. 189–90, see also Kanamori and Magidor 1978, Solovay et al. 1978, and Reinhardt 1974); $\aleph_0$ is the first cardinal with the properties (1) and (2) above, hence a cardinal with the same property must reappear at higher levels.
- *Inexhaustibility.* The universe of all sets is too rich to be exhausted by some basic operations such as power set or replacement, therefore there must be a cardinal which is not generated by those operations (see e.g. Gödel 1947, Wang 1974 or Drake 1974); such a cardinal can be proven to be inaccessible.
- *Reflection.* The universe of sets is too complex to be completely described by some property, hence anything that is true of the entire universe, must be true at some initial segment of it, it must 'reflect' at some V_κ. In particular there must be a V_κ which is also closed by the power set and replacement operations; then κ can be proven to be inaccessible.

All these arguments seem to rest on mathematical platonism in an essential manner, as they appeal on some specific conception of 'the universe of sets' as uniform, inexhaustible, indescribable and so on. But even assuming a platonic point of view, what reasons do we have to believe that the universe of sets has such features? Some issues arise, for instance, with the claim of uniformity. In fact, there are properties that do hold at $\aleph_0$ and do not occur at higher cardinals. For instance, Ramsey's theorem establishes that for every $n, m < \aleph_0$ and for every coloring of the n-tuples of $\aleph_0$ into m colors, we can find a set $H \subseteq \aleph_0$ of size $\aleph_0$ such that all the n-tuples of H have the same color, this is called a *homogeneous set*; on the other hand, it can be proven that no uncountable cardinal can satisfy the same property: if we replace $\aleph_0$ with an uncountable κ, we get a statement that is provably false in ZFC.

Typically, large cardinals generalize properties of $\aleph_0$. For instance, the notions of *Ramsey cardinal, Erdös cardinal, weakly compact cardinals* and others can be defined as special generalizations of the theorem of Ramsey that we just mentioned; some limitations are necessary because as we said the direct generalization of Ramsey Theorem to an uncountable cardinal is provably false in ZFC. We consider, for example, the axiom of weakly compact cardinals which establishes the existence of an uncountable cardinal κ such that for every coloring of the pairs of ordinals of κ into less than κ many colors there is a homogeneous set of size κ. Once again, we stress the fact that generalizations are dangerous as they may lead to inconsistencies as in the case above.

The axiom of weakly compact cardinals can also be defined as a generalization of Compactness theorem to the infinitary language $\mathscr{L}_{\kappa,\kappa}$. Given two infinite cardinals κ, λ, we denote by $\mathscr{L}_{\kappa,\lambda}$ the infinitary language that roughly allows conjunctions and disjunctions of less than κ many formulas, and quantifications over less than λ many variables. Thus, for instance $\mathscr{L}_{\omega,\omega}$ corresponds to first order logic. An uncountable cardinal κ is weakly compact if, and only if, whenever we have a theory T in $\mathscr{L}_{\kappa,\kappa}$ with at most κ non logical symbols, if T is $< \kappa$-satisfiable (i.e. every family of less than κ many sentences of T is satisfiable), then T is satisfiable. If we remove the restriction to 'theories that have at most κ non-logical symbols, we have the notion of *strongly compact cardinal*. Other large cardinals axioms can be defined as generalizations of compactness theorem. Such generalizations imply interesting 'compactness results', namely given some structure, we assume that all its smaller substructures satisfy a certain property and we deduce that the whole structure satisfy the same property. For instance assuming a strongly compact cardinal κ it is possible to prove that every abelian group of size at least κ is free abelian whenever all its smaller subgroups are free abelian. The axiom of constructibility on the other hand is the 'cemetery of compactness properties': for instance, compactness for the freeness of abelian groups is actually false in $V = L$. The analysis of such compactness or incompactness results gives us no strong motivation to support one theory over the other, as there is no cogent reason to deem compactness more suitable than incompactness or the converse. Not even Uniformity helps us in this case, as ZFC proves both compactness and incompactness results: for example,

König's lemma can be regarded as a compactness result,[2] but on the other hand its generalization to $\aleph_1$ is provably false in ZFC (there are Aronszajn trees).

We can see that the notion of weakly compact cardinal can be defined both as a combinatorial and a model-theoretic notion. The same occur for other large cardinals, in fact it is often the case that certain mathematical problems arising in completely different contexts and fields lead to the same large cardinal notions. This fact is sometimes considered to be an intrinsic motivation for large cardinals, but however remarkable this might seem, it is not clear how it can actually be considered as evidence for these axioms, rather than just a practical advantage.

The most powerful large cardinals axioms are the ones that can be defined as elementary embeddings of V into some inner model of ZFC. We discuss some of these notions in the next section.

2.7 Measurable Cardinals and Elementary Embeddings

In the history of large cardinals axioms the introduction of *measurable cardinals* was probably the most crucial step as it lead to the theory of elementary embeddings that are extremely useful in solving set theoretical problems and answering other mathematical questions. Let us discuss, then, these notions.

In 1902, Lebesgue formulated the measure problem: he asked whether there is a function that associates to every bounded set of reals a real number between 0 and 1 such that the function is not identically 0, it is translation invariant and countably additive. Motivated by this question, he introduced his famous Lebesgue measure (a function with these properties) and asked whether every bounded set of reals was Lebesgue measurable, namely whether his measure was defined over every bounded set of reals. Vitali soon found a counterexample under the Axiom of Choice, the problem was then reformulated by replacing the condition of translation invariance with 'every singleton must have measure 0', the minimal request for avoiding trivial solutions. The problem was still proven to be independent from ZF, in fact a counterexample can be built under CH. At this point Banach realized that the problem did not depend on the structure of $\mathbb{R}$, and it could be reformulated for a general set S : is there a function $\mu : \mathscr{P}(S) \to [0, 1]$ which is not identically 0, assigns to every singleton the value 0 and is countably additive? The solution of this problem comes down to the existence of certain large cardinal, the *real valued measurable cardinals*. A cardinal κ is real valued measurable if every set of size κ has a measure μ with the properties above which moreover is κ-additive, namely for every family $\{X_\alpha\}_\alpha$ of less than κ many sets, $\mu(\bigcup_\alpha X_\alpha) = \sum_\alpha \mu(X_\alpha)$. This is an example of how Maddy's approach to 'proper methods' works, namely real valued

[2]Given a tree of height ω whose levels are finite, if every finite subtree has a branch of the same length as the height subtree, then the whole tree also has a branch of the same length as the height of the tree.

measurable cardinals arose naturally as the solution to a specific mathematical problem.

Now, if we require that not only every set of size κ has a measure, but also the measure takes just two values 0 or 1, then we have *measurable* cardinals. In fact this notion has an extremely powerful characterization: κ is measurable if and only if one can define a non-trivial elementary embedding[3] $j : V \to M$ where M is a transitive class, such that κ is the least cardinal that is moved by j. By using this characterization, Scott was able to prove that if there is a measurable cardinal, then $V \neq L$. Thus, measurable cardinals, as well as any other stronger large cardinal, are incompatible with the axiom of constructibility.

Many powerful large cardinal notions can be defined in terms of elementary embeddings where we require the transitive class M to be 'closer' to V. The ultimate large cardinal notion expressible in terms of elementary embeddings is provably inconsistent with ZFC. This is the notion of *Reinhardt cardinal*, an uncountable cardinal κ for which there is a non trivial embedding j of V into itself where κ is the least cardinal which is moved by j.

Large cardinals axioms that establish the existence of elementary embeddings are more successfully justified by their fruitfulness, as they settle a number of questions that are independent from ZFC. The mort remarkable application of such cardinals is the theory of projective sets that under these cardinals gets a very elegant and exhaustive analysis. In fact, the existence of infinitely many Woodin cardinals implies that every projective set of reals is Lebesgue measurable, has the perfect set property and the Baire property, the so-called *regularity properties*. These considerations brings us to discuss Determinacy hypotheses, which is the object of the next section.

2.8 Determinacy Hypotheses

The study of regularity properties dates back to the earliest twentieth century from the work of the french analysts Borel, Baire and Lebesgue. Research in this area led to the development of an independent discipline, known as descriptive set theory. About 40 years later, it was shown that the open questions that descriptive set theorists were trying to solve (namely whether every set of real has the regularity properties above) could not be answered within ZFC (as we have seen for Lebesgue mesurability). In 1962 Mycielski and Steinhaus introduced the Axiom of Determinacy AD which was proven to solve such problems. AD is the assertion that every set of reals is *determined*, that means that for every set of reals A, one of the two players has a winning strategy in the following game of length ω. We regard A as a subset of ${}^{\omega}\omega$ (in set theory a real is an omega sequence of natural numbers),

[3]A function $j : V \to M$ is an elementary embedding if for every formula φ and parameters $a_1, \ldots, a_n$ one has $V \models \varphi(a_1, \ldots, a_n)$ if and only if $M \models \varphi(j(a_1), \ldots, j(a_n))$.

the two players I and II alternatively choose natural numbers $n_0, n_1; n_2, \ldots$ At the end of the game a sequence $\langle n_i; i \in \mathbb{N}\rangle$ is generated, player I wins if and only if the sequence belongs to A.

The Axiom of Determinacy implies that all sets of reals are Lebesgue measurable, have the perfect set property and the Baire property. Moreover, the statement that every set of reals has the perfect set property implies a weak form of the continuum hypothesis: every uncountable set of reals has the same cardinality as the full set of reals. On the other hand, AD implies the negation of the generalized continuum hypothesis. Despite its fruitfulness, AD was never seriously considered as a valid candidate new axiom for set theory as it contradicts the Axiom of Choice – here is a lucid example of the fact that often in mathematics the priority goes to the consequences rather than the axioms; but the consequences (here the Axiom of Choice) are themselves in need of justifications–. This led to investigate two distinct directions. The first approach was to assume AD in a quite natural subuniverse, namely $L(\mathbb{R})$, together with AC in the full universe V (L($\mathbb{R}$) is the smallest transitive inner model of ZF containing all the ordinals and the reals). The second approach was to consider a weakening of AD, called *Projective Determinacy*, PD. Projective Determinacy is the statement that every *projective* set of reals is determined. PD implies that every projective set of reals is Lebesgue measurable, has the perfect set property and the Baire property, and unlike AD, Projective Determinacy is not known to contradict the Axiom of Choice. Projective Determinacy follows from the existence of infinitely Woodin cardinals and this is the reason why this large cardinal assumption implies that every projective set of reals has the regularity properties above.

2.9 Ultimate L and Forcing Axioms

As we said, the Axiom of Constructibility and the Axiom of Determinacy both decide the continuum problem (the former implies GCH, the latter implies a weak form of CH, but it also implies the negation of the generalized continuum hypothesis). Large cardinals axioms, on the other hand, do not decide the size of the continuum. In this regard, a quite promising direction of research was considered which combine large cardinals with L and it may decide the size of the continuum, this approach is known as *V = Ultimate L.*

To understand this view, consider the intuition behind the Axiom of Constructibility: L is build up from a cumulative process where each stage is obtained from the previous one by a canonical operation, namely by taking the definable subsets of the previous stage; in this process only few ‘canonical’ sets are accepted at each stage. The idea behind V = Ultimate L is that, while we want large cardinals to exist in the universe of sets, we only want to include sets that are canonical or necessary after a fashion. Ultimate L, proposed by Woodin, is the alleged inner model for supercompact cardinals. Roughly this would be an L-like model where lives a supercompact cardinal. Such a model was not build yet and it is an open problem whether it can actually be found, but it can be proven that if the construction of the Ultimate L is successful, then it would contain also all the stronger large cardinals (i.e. stronger than supercompact cardinals). More importantly, V = Ultimate L would imply CH.

Magidor, however, expressed some doubts about this approach:

> It is very likely that the Ultimate L, like the old L, will satisfy many of the combinatorial principles like $\Diamond_{\omega_1}$. These principles are usually the reason that "L is the paradise of counter examples". They allow one to construct counter examples to many elegant conjectures. (The Suslin Hypothesis is a famous case). (Magidor 2012)

As for the Axiom of Constructibility, V = Ultimate L rests on the idea that sets are obtained through a cumulative process which is a way to allow only canonical sets of some sort, while other views rely on the opposite slogan that the concept of set should be as rich as possible. The most important example of such a liberal view is given by *Forcing axioms*. Forcing is the main tool for proving independence results in set theory. There are essentially two main approaches for building models of set theory and proving consistency results: one is through inner models which are obtained roughly by 'restricting' V into a subclass; the other is by using the Forcing technique where, conversely, V is expanded to a larger universe. Forcing axioms roughly establish that anything that can be forced by some 'nice' forcing notions (a forcing is simply a partially ordered set) is a set in the universe. For instance, the two most fruitful Forcing Axioms, PFA and MM, are the following statements.

The Proper Forcing Axiom PFA states that if P is a forcing notion that is proper and D is a collection of $\aleph_1$ many dense subsets of P, then there is a generic filter that meets all the dense sets in D.

Roughly, this says that anything that can be forced by a *proper forcing* is a set in the universe.

Martin's Maximum MM asserts that if P is a forcing notion that preserves stationary subsets of ω_1 and D is a collection of $\aleph_1$ many dense subsets of P, then there is a generic filter that meets all the dense sets in D.

Roughly, this says that anything that can be forced by a forcing that *preserves stationary subsets of* ω_1 is a set in the universe. We will not discuss these notions in the details, we should only point out that MM is the strongest possible version of a Forcing Axiom and it was proven to be consistent relative to the existence of a supercompact cardinal (this was suggested as another motivation for large cardinals axioms). Forcing Axioms settle many important questions that cannot be answered within ZFC, but more importantly they find remarkable applications in cardinal arithmetic. In fact, Foreman Magidor and Shelah proved in 1988 that Martin's Maximum settles the size of the continuum, it implies that the $2^{\aleph_0} = \aleph_2$. Later in 1992, Todorčević and Veličkovič showed that even the weaker axiom PFA implies that the size of the continuum is $\aleph_2$. Other remarkable applications of Forcing Axioms include the singular cardinals hypothesis (from PFA), the Axiom of Determinacy in $L(\mathbb{R})$, the statement that any two $\aleph_1$-dense subsets of $\mathbb{R}$ are isomorphic (from PFA), every automorphism of the Boolean algebra $\mathcal{P}(\omega)/fin$ is trivial (from PFA), the $\aleph_2$-saturation of the ideal of non stationary sets on ω_1 (from MM), and the reflection of stationary subsets of κ for any regular cardinal $\kappa \geq \omega_2$ (from MM).

2.10 Conclusion

We have adopted Maddy's distinction of intrinsic and extrinsic justifications and we have argued that both the self evidence and the 'fruitfulness' criterion are problematic. We have further analyzed more specific motivations proposed in favor of various new axioms candidates for set theory. This analysis should suggest that none of those arguments can unquestionably justify the choice of one theory of sets over the others.

Acknowledgements I would like to thank the anonymous reviewer for his careful reading and for his constructive comments. I am also indebted to Juliette Kennedy, Menachem Magidor, Neil Barton and Claudio Ternullo who gave me many useful suggestions that helped improve the quality of this paper.

References

Aczel, P. (1988). *Non-well-founded sets* (CSLI Lecture Notes: Number 14). Stanford: CSLI Publications.

Drake, F. (1974) *Set theory*. Amsterdam: North-Holland.

Feferman, S., Friedman, H., Maddy, P., & Steel, J. (2000) Does mathematics need new axioms? *Bulletin of Symbolic Logic, 6*(4), 401–446.

Fraenkel, A., Bar-Hillel, Y., & Levy, A. (1973). *Foundations of set theory* (2nd ed.). Amsterdam: North-Holland.

Forti, M., & Honsell, F. (1983). Set theory with free construction principles. *Annali Scuola Normale Superiore di Pisa, Classe di Scienze, 10*, 493–522.

Gödel, K. (1947). What is Cantor continuum problem. *The American Mathematical Montly, 54*(9), 515–525.

Hamkins, J. D. (2014). A multiverse perspective on the axiom of constructibility. In: C.-T. Chong (Ed.), *Infinity and truth* (vol. 25, pp. 25–45). Hackensack: World Science Publication.

Kanamori, A., & Magidor, M. (1978). *The evolution of large cardinal axioms in set theory*. In: G. H. Muller & S. D. Scott (Eds.), *Higher set theory* (Lecture notes in Mathematics, vol. 669, pp. 99–275). Berlin: Springer.

Maddy, P. (1988). Believing the axioms, I. *Journal of Symbolic Logic, 53*, 481–511; II, *ibid.*, 736–764.

Maddy, P. (2011). *Defending the axioms: On the philosophical foundations of set theory*. Oxford: Oxford University Press.

Magidor, M. (2012). *Some set theories are more equal*. Unpublished notes, available at http://logic.harvard.edu/EFIMagidor.pdf

Mayberry, J. P. (2000). *The foundations of mathematics in the theory of sets*. (Enciclopedia of mathematics and its applications, vol. 82). Cambridge: Cambridge University Press.

Markov, A. A. (1962). On constructive mathematics. *Trudy Matematicheskogo Instituta Imeni V. A. Steklova, 67*(8–14). Translated in *American Mathematical Society Translations: Series 2, 98*, 1–9.

McLarty, C. (2010). What does it take to prove Fermat's last theorem? Grothendieck and the logic of number theory. *The Bulletin of Symbolic Logic, 16*(3), 359–377.

Moschovakis, Y. N. (1980). *Descriptive set theory*. Amsterdam: North-Holland.

Nykos, P. J. (1980). A provisional solution to the normal moore space problem. *Proceedings of the American Mathematical Society, 78*(3), 429–435.

Reinhardt, W. N. (1974). *Remarks on reflection principle large cardinals and elementary embeddings. Axiomatic set theory* (Proceedings of symposia in pure mathematics, vol. XIII, pp. 189–205, Part II). Providence: American Mathematical Society.

Scott, D. S. (1961). Measurable cardinals and constructible sets, *Bulletin de l'Académie Polonaise des Sciences. Série des Sciences Mathématiques, Astronomiques et Physiques, 7*, 145–149.

Shelah, S. (1974). Infinite Abelian groups, whitehead problem and some constructions. *Israel Journal of Mathematics, 18*(3), 243–256.

Solovay, R. M., Reinhardt, W. N., Kanamori, A. (1978). Strong axioms of infinity and elementary embeddings. *Annals of Mathematical Logic, 13*, 73–116.

Wang, H. (1974) *The concept of set*, in Benacerraf and Putnam [1983], pp. 530–570.

Chapter 3
Maddy On The Multiverse

Claudio Ternullo

Abstract Penelope Maddy has recently addressed the set-theoretic multiverse, and expressed reservations on its status and merits (Maddy, Set-theoretic foundations. In: Caicedo et al (eds) Foundations of mathematics. Essays in honor of W. Hugh Woodin's 60th birthday. Contemporary mathematics. American Mathematical Society, Providence, pp. 289–322, 2017). The purpose of the paper is to examine her concerns, by using the interpretative framework of set-theoretic naturalism. I first distinguish three main forms of 'multiversism', and then I proceed to analyse Maddy's concerns. Among other things, I take into account salient aspects of multiverse-related mathematics, in particular, research programmes in set theory for which the use of the multiverse seems to be crucial, and show how one may provide responses to Maddy's concerns based on a careful analysis of 'multiverse practice'.

3.1 The Problem

The development of set theory has progressively brought to the fore the problem of whether set theory should be interpreted as the theory of a *single* universe of sets, V, or whether it should be viewed as a theory about *multiple* structures (universes), that is, about a set-theoretic 'multiverse'.

I wish to thank the anonymous referee for suggesting several improvements. I am also hugely indebted to Penelope Maddy for useful comments and further theoretical inputs, as well as for discussing and reading earlier drafts of this paper. Finally, I would like to thank the editors, in particular Deborah Kant and Deniz Sarikaya, for their support and help, both during the FOMUS conference at Bielefeld and the preparation of this work.

C. Ternullo (✉)
Department of Philosophy, University of Tartu, Tartu, Estonia
e-mail: claudio.ternullo@ut.ee

S. Centrone et al. (eds.), *Reflections on the Foundations of Mathematics*,
Synthese Library 407, https://doi.org/10.1007/978-3-030-15655-8_3

The universe/multiverse dichotomy is just an *ontological* (*semantic*) variant of the pluralism/non-pluralism dichotomy. The (proof-theoretic) pluralist believes that there is no unique, or preferred, theory T of sets, and that all theories express, to some extent, some properties of sets. For instance, the pluralist may believe that both ZFC+CH and ZFC+$\neg$CH are equally valid and interesting theories of sets, insofar as they express alternative, but, in principle, equally acceptable, properties of sets. *Multiversism* is pluralism in its ontological/semantic form. It should be noted that one may be both a pluralist *and* a multiversist: that is, one may both believe that there are many, equally valid theories of sets and also that there are many, equally valid set-theoretic structures. Non-pluralism and, at the ontological level, *universism*, hold, respectively, that there is a single correct theory T of sets, or that there is a *unique* structure embodying all set-theoretic truths.[1]

The issue of which, between universism and multiversism, suits best set theory and its goals is crucial both for the philosophy of set theory and, more generally, for the philosophy of mathematics, as arguments in favour of multiversism may be taken to count as arguments in favour of the *absolute undecidability* of some set-theoretic statements, and we have evidence that multiversism is precisely construed in this way by some authors.[2]

The terms of the conceptions sketched above, and their relevance for the philosophy of mathematics, are rather uncontroversial to anyone. What, however, does seem contentious is whether the universe/multiverse dichotomy, and the possible adoption of a specific conception of the set-theoretic multiverse, is relevant to set-theorists and to their mathematical work. We know that set-theorists work with a plurality of models, such as models obtained through forcing, ultrafilter constructions, elementary embeddings, inner models like L, and many others. Now, how does the use of one (or more) collection(s) of set-theoretic models, that is, of the 'multiverse(s)' of a given theory T, contribute to their work? Are there any practical and foundational advantages in taking some collections of models, such as, say, countable transitive models or set-generic extensions of ZFC, as being especially relevant to set-theoretic work, and also as acting as a foundational framework for set theory?

This paper aims to provide answers to these questions, by using a specific, and especially authoritative, point of view, that of Maddian naturalism, and, in particular, by responding to the concerns that, in Maddy (2017), Maddy has expressed about the value and usefulness of a 'multiversist' approach.[3]

---

[1]I am indebted to Koellner (2014) for this articulation of the pluralist/non-pluralist positions.

[2]This seems to be Väänänen's point of view in Väänänen (2014). Väänänen's main goal is to articulate a position which allows one to express the absolute undecidability of set-theoretic statements which are currently undecidable in several important theories (such as ZFC plus large cardinals). See also Sect. 3.2.1.

[3]It should be noted that the article mentioned is, in fact, an appraisal of *different* competing foundations of mathematics, also including set theory, and of the roles such competing foundations carry out. Only one specific section of the article is explicitly devoted to examining the prospects of the set-theoretic multiverse.

Over the years, Maddian naturalism has progressively, and coherently, come to the fore as one of the most influential positions in the philosophy of set theory, and in general, in the naturalist philosophy of mathematics. The position is generally associated to the following views:[4]

1. Metaphysical issues are irrelevant to the practical development of set theory, as well as to the justification of its internal techniques and results
2. The proper method of a naturalist philosophy of set theory consists in using rational methodologies attentive to intra-set-theoretic practice which altogether rule out extra-mathematical considerations
3. The justification and adoption of set-theoretic principles/axioms is also guided by methodologies of this sort

Especially points 1. and 2. in the summary above are crucial for set-theoretic naturalists: we ought not to evaluate mathematics, its goals and results, using extra-mathematical views or conceptions. The professed ideal of a 'second philosophy', that is of a philosophy of set theory which is especially attentive to *practice*, is the central tenet of the set-theoretic naturalist.[5]

Now, one further view distinctively associated to Maddy's naturalism is that our conception of the universe, V, is justified in light of set-theoretic practice, and in view of our set-theoretic purposes, insofar as set theory is pre-eminently guided by the strive to find a *unifying* account of mathematical phenomena. Therefore, if the 'unification' goal has priority over other goals, we ought to accept V as being the most suitable foundational framework for our set-theoretic investigations. This, in turn, sanctions the view that set theory essentially deals with proving facts about/establishing properties of V, a view which, as is clear, places Maddian naturalism in the universist camp.

In Maddy's works, the set-theoretic goal of unification has progressively taken the shape of the maxim 'unify'. Through fostering 'unify', set theory is thought to have been able to become the unique and far-reaching subject that it is today, and the maxim also serves as a spur to pursue further the search for solutions to the open set-theoretic problems. This point of view is articulated by Maddy as follows:

> If set-theorists were not motivated by a maxim of this sort, there would be no pressure to settle CH, to decide the questions of descriptive set theory, or to choose between alternative axiom candidates; it would be enough to consider a *multitude* [my italics] of alternative set theories. (Maddy 1997, p. 210)

[4]Throughout the paper, I shall use 'Maddian naturalism' and 'set-theoretic naturalism' (sometimes, just 'naturalism') interchangeably. Of course, there are many other ways to spell out naturalism in the philosophy of mathematics. An overview of all such positions is in Paseau (2016).

[5]Maddy herself has put forward and discussed the central aspects of set-theoretic naturalism as a form of 'second philosophy' in several works, starting with Maddy (1996). Second philosophy is further delineated in Maddy (2007), as well as in Maddy (2011).

As we know, the ‘independence phenomenon’ has introduced a rather different picture of the ontology of set theory, one which seems to be more compatible with a pluralist account of set-theoretic phenomena. This poses pressing questions for the Maddian naturalist described above: should she entirely disregard the issue of pluralism or actively engage with it? Moreover, since ‘unification’ concerns seem paramount in her account of set theory, on what alternative grounds may she adopt a pluralist picture? What will be of ‘unify’ within such a picture?

The paper also aims to explore these questions, with a view to providing arguments which may support the following two claims: the universe/multiverse dichotomy is relevant, in many ways, to the naturalist’s approach, and, secondly, the multiverse may be as acceptable as the universe, from a naturalist perspective, for the foundational purposes of set theory.

The structure of the paper is as follows. I first review several conceptions of the set-theoretic multiverse (Sect. 3.2), and provide a classification. I then proceed to summarise Maddy’s concerns (Sect. 3.3), which, overall, will take the shape of five main problems for the multiverse supporter. Finally, in the larger Sect. 3.4, I discuss aspects of the multiverse and of multiverse-related mathematics which seem to adequately respond to the issues raised in Sect. 3.3.

3.2 Multiverse Conceptions

First, we need to clarify what the set-theoretic multiverse consists in, and this may already be a daunting task. Even based on a minimal perusal of the existing literature, it is clear that there is no such thing as *one* mathematical conception of the set-theoretic multiverse, but rather *a bunch of* them, and, in addition, several, alternative research programmes which are variously connected to all such conceptions.

One main difficulty in addressing the set-theoretic multiverse, therefore, is precisely the absence of a shared framework wherein one may discuss results and methodologies. In what follows, I propose a classification of multiverse conceptions, which suits my specific goals. There is nothing compelling about the classification, nor is there any a priori need to classify multiverse conceptions, for that matter.[6]

[6] In Antos et al. (2015), the authors adopt a classification based on the realism/non-realism divide. Hamkins’ conception, for instance, counts as realist, whereas the Hyperuniverse Programme as non-realist. Väänänen proposes a different classification in Väänänen (2014), pp. 191–2: he divides conceptions into *countable* (Hyperuniverse Programme), *full* (Hamkins) and *set-generic* (Woodin, Steel).

3.2.1 *Naive Multiversism*

It is maybe noteworthy to mention that some form of multiverse thinking was already at work in the characterisation of the universe, since, historically, one could find the first description of a multiverse in Zermelo's characterisation of V. As is known, Zermelo proved the *(quasi-)categoricity* of his system of second-order set theory by showing that V_κ, where κ must be, at least, a strongly inaccessible cardinal, is, up to isomorphism, a model of the axioms ZFC_2 (that is, ZFC with second-order Separation and Replacement).[7] However, since there is an absolutely infinite collection of strongly inaccessible cardinals, one may generate V_κ's of increasing height, a 'tower' of universes, each the rank initial segment of the other. This could, in turn, be seen as a *height multiverse*, in which all universes have the same width (that is, no 'new' subsets may be added), but different heights ('new' ordinals, and ordinal-indexed stages, may be added).[8]

Leaving history aside, the most basic way to express a multiversist attitude is through taking note of the existence of many models of the axioms (of ZFC, for instance). As said at the beginning, we all know that set-theorists work with several kinds of models, through which they may, among other things, prove independence results. 'Naive multiversism' is just the idea that no single model $\mathcal{M}$ of a theory of sets T, should be viewed as 'special', as being *the* universe of sets, *the* collection of all sets.

Saharon Shelah, for instance, has compared models of the axioms to *individuals*, each, presumably, endowed with unique features, but all belonging to the *same* species.[9]

Naive multiversism could also be characterised in a different way, that is, as a conception which incorporates semantic pluralism in an ontologically *monist*

---

[7]See Zermelo (1930). An examination (and reprise) of Zermelo's proof is in Martin (2001).

[8]For Zermelo's height potentialism, see Linnebo (2013) and Ternullo and Friedman (2016). Of course, historically, Zermelo did not construe the ideas contained in Zermelo (1930) in the current multiversist terms. Väänänen suggests a different reason why Zermelo's characterisation of V ought not to be viewed as an instance of the multiverse. He notes that believing in the existence of an inaccessible κ in V means accepting the axiom: '$\exists\kappa$ inaccessible', which, although clearly independent from ZFC, is not indeterminate in the same sense as, say, CH is. According to Väänänen, the multiverse phenomenon takes place in the presence of statements which are indeterminate in the sense specified. In other terms, if V is V_κ, where κ is the least strongly inaccessible cardinal, then '$\exists\kappa$ inaccessible' is just false, and there is no 'parallel' V with inaccessibles. Cf. Väänänen (2014), p. 187. Finally, one could resist this interpretation of Zermelo's conception by asserting that the set-theoretic hierarchy is, in fact, fully *actual* in height and width. For a fuller examination of the actualism/potentialism dichotomy, see Antos et al. (2015) or Koellner (2009).

[9]Cf. Shelah (2003), p. 211. It is worth quoting the passage in full: 'My mental picture is that we have many possible set theories, all conforming to ZFC. I do not feel "a universe of ZFC" is like "the Sun", it is rather like "a human being" or "a human being of some fixed nationality"'.

framework. Recently, Väänänen seems to have articulated a conception of this sort. He says:

> ...we want two universes in order to account for absolute undecidability and at the same time we want to say that both universes are everything. We solve this problem by thinking of the domain of set theory as a multiverse of parallel universes, and letting variables of set theory range intuitively over each parallel universe simultaneously. The axioms of the multiverse are just the usual ZFC axioms and everything that we can say about the multiverse is in harmony with the possibility that there is just one universe. But at the same time the possibility of absolutely undecidable propositions keeps alive the possibility that, in fact, there are several universes. The intuition that this paper is trying to follow is that the parallel universes are more or less close to each other and differ only at the edges. (Väänänen 2014, p. 182)

The main reason for this choice is that:

> ..the intuition about the multiverse is not that everything that is logically possible should also happen in some universe (which would lead to the full multiverse), but that the multiverse is one universe the boundaries of which are verschwommen ("blurred"), as von Neumann wrote. (Väänänen 2014, p. 199)

Väänänen distinguishes set-theoretic statements which are *invariant* for all (potentially alternative, but, in fact, co-existing) V's, and those which are not. ZFC belong to the first class, whereas CH to the second one. 'Multiverse axioms' will capture this intuition: introducing a special logic for independence, and a special symbol for independent statements such as CH ($\neq$CH), one may yield the theory ZFC+ $\neq$CH, which, on the one hand, expresses the truth of ZFC over all of V's, and, on the other, the fact that there are V's which violate CH.[10]

3.2.2 *Instrumental Multiversism*

A second, more developed, approach to the multiverse may be called 'instrumental multiversism'. By this view, the multiverse is an important mathematical tool, whose philosophical (ontological as well as epistemological) status may not be so relevant.

We may identify the 'multiverse-as-a-tool' conception with two specific, and influential, research programmes in the foundations of set theory. The first is the set-generic multiverse, first addressed by Hugh Woodin, and then taken up and re-formulated by John Steel.[11] The second is the Hyperuniverse Programme (HP), initiated and investigated by Sy Friedman and his collaborators.[12] I start with the latter, which is the form of multiverse I know best.

---

[10]See Väänänen (2014), pp. 196–202.

[11]For Woodin's set-generic multiverse, see Woodin (2011), although earlier instances of multiverse thinking may also be found, as will be shown later in this paper, in Woodin (2001). Steel's conception of the set-generic multiverse is in Steel (2014).

[12]For this conception, see, essentially, Arrigoni and Friedman (2013), and Antos et al. (2015).

The hyperuniverse, as explained by the authors, is:

> ...the collection of all transitive countable models of ZFC. Within the programme, such models are viewed as a technical tool allowing set-theorists to use the standard model-theoretic and forcing techniques (the Omitting Types Theorem and the existence of generic extensions, respectively). The underlying idea is that the study of members of the hyperuniverse allows one to indirectly examine properties of the real universe V [...]. (Antos et al. 2015, p. 2484)

The hyperuniverse becomes indispensable once one realises that theories T expressing maximality principles which address 'extensions' of V have models if and only if V is countable.[13] The hyperuniverse, therefore, has no significance *per se*, but only as a way to develop the study of: (1) *alternative* maximality principles; (2) first-order consequences of maximality principles (or refinements thereof).

I now proceed to examine the set-generic multiverse. Woodin defines it as the smallest collection of universes $\mathbb{V}_M$ generated by a countable transitive model M of ZFC, such that, if $M \in \mathbb{V}_M$, then also all forcing extensions of M (and generic refinements thereof) are in $\mathbb{V}_M$. It should be noted, already, that Woodin, ultimately, rejects the 'multiverse position', as he shows that, modulo the Ω-conjecture and the assumption of the existence of class-many Woodin cardinals (for which see Sect. 3.4.2.1), reasonably formulated multiverse laws must be violated. However, later on (again, Sect. 3.4.2.1) I will show that multiverse thinking has proved fundamental for many of Woodin's mathematical goals.

Steel characterises the set-generic multiverse as a collection of 'worlds'. He uses a two-sorted language, one sort for sets and one for worlds. Steel's multiverse is modelled upon Woodin's $\mathbb{V}_M$, but there is a crucial difference between the two accounts: Steel's multiverse is expressed by a collection of axioms, which its author calls MV.[14] Now, what matters most to Steel is to 'maximise the interpretative power' of his theory, therefore the MV axioms prescribe that all worlds share, as basic multiverse truths, ZFC and large cardinals (LCs).[15] Where, obviously, worlds do not agree with each other is on the truth-value of independent statements such as CH. Such statements are provisionally deemed not to be 'expressible' in multiverse language, and one of the purposes of Steel's MV theory is precisely to explore the limits of 'expressibility' (and 'meaningfulness') within multiverse language (I shall return to this in Sect. 3.4.4).

---

[13]See Sects. 3.4.1 and 3.4.2 later in this paper. Further technical details concerning the use of V-logic and the 'reduction to the hyperuniverse' may be found in Antos et al. (2015) and in Friedman (2016).

[14]Moreover, some of these axioms (such as Axiom 2d [Amalgamation]) are not valid in Woodin's multiverse. Woodin has also tried to axiomatise the generic multiverse, by introducing the aforementioned 'multiverse laws'. However, these laws are so very much *ad hoc* that one would naturally refrain from viewing them as axioms.

[15]This is because, given any two theories T_1 and T_2 extending ZFC having the same consistency strength as ZFC+"exist infinitely many Woodin cardinals", then at least up to Π^1_ω 'statements' (second-order arithmetic), we have that either $T_1 \subseteq T_2$ or $T_2 \subseteq T_1$, and this result may be extended to slightly more complex sentences.

Overall, also the set-generic multiverse exemplifies a practice-oriented attitude to the multiverse: in the same way as HP is guided by the goal of exploring maximality principles in (countable) models satisfying ZFC, Steel's set-generic multiverse aims to capture further set-theoretic truths in all set-generic universes which think ZFC+LCs.

3.2.3 Ontological Multiversism

The last strand of multiversism I describe is very different from all those described so far, and may legitimately be called 'ontological multiversism', that is, the view that the multiverse is a determinate *reality*, consisting of particular *entities*, the models of set theory. Hamkins' broad (radical) multiverse conception is such a form of multiversism.

Hamkins characterises this view as follows:

> ..the fundamental objects of study in set theory have become the models of set theory, and set-theorists move with agility from one model to another. [...] This abundance of set-theoretic possibilities poses a serious difficulty for the universe view, for if one holds that there is a single absolute background concept of set, then one must explain or explain away as imaginary all of the alternative universes that set-theorists seem to have constructed. (Hamkins 2012, p. 418)

As to the issue of what universes there are in the multiverse, Hamkins says:

> The background idea of the multiverse, of course, is that there should be a large collection of universes, each a model of (some kind of) set theory. There seems to be no reason to restrict inclusion only to ZFC models, as we can include models of weaker theories ZF, ZF^-, KP, and so on, perhaps even down to second-order number theory, as this is set-theoretic in a sense. At the same time, there is no reason to consider all universes in the multiverse equally, and we may simply be more interested in the parts of the multiverse consisting of universes satisfying very strong theories, such as ZFC plus large cardinals. (Hamkins 2012, p. 436)

Thus, Hamkins' multiverse is so broad as to include even non-well-founded models, and models of any collection of set- and class-theoretic axioms. Therefore, Hamkins' view is a form both of *proof-theoretic* and *model-theoretic* (*ontological*) pluralism I hinted at in the opening of the paper (p. 1). One peculiar feature of this conception is worth stressing straight away: according to its author, universes in the broad multiverse should be viewed as independent *existents*, just like more 'standard' mathematical entities (numbers, shapes, transfinite ordinals, etc.). The following quote illustrates Hamkins' full avowal of ontological realism (Platonism):

> The multiverse view is one of higher-order realism–Platonism about universes–and I defend it as a realist position asserting actual existence of the alternative set-theoretic universes into which our mathematical tools have allowed us to glimpse. The multiverse view, therefore, does not reduce via proof to a brand of formalism. (Hamkins 2012, p. 417)

Let's take stock. There are at least three conceptions of the set-theoretic multiverse at hand, 'naive', 'instrumental' and 'ontological'. However, 'naive multiversism' does not seem to foster a sufficiently characterised notion of the set-theoretic multiverse. Therefore, in the rest of the paper, I will only be concerned with the other two forms of multiverse.

3.3 Maddy's Assessment of the Multiverse

I now proceed to summarise Maddy's concerns. For the sake of economy, I won't always quote Maddy's text in full, and, in some cases, I will just provide the references to the relevant pages in Maddy (2017).

Maddy notes that set-theoretic naturalists must acknowledge that set theory fulfils specific *foundational* roles, which are consistent with, and originate from, their picture of mathematics as being part of our *best* scientific theory of the world.[16] However, some of these roles are seen by her as spurious, others as appropriate.[17] Among the appropriate ones, Maddy lists 'Shared Standard', 'Generous Arena' and 'Metamathematical Corral'. 'Shared Standard' is the idea that set-theoretic proofs constitute the standard of 'proof in mathematics', whereas 'Metamathematical Corral' refers to the role played by set theory in allowing mathematicians to carry out metamathematical investigations, such as the search for consistency proofs.

'Generous Arena' is especially valuable to set-theoretic naturalists, insofar as it gives rise to and fully motivates the adoption of the meta-theoretic maxims 'unify' and 'maximize', which, in turn, justify the adoption of the axioms of set theory as our foundational theory, that is, a theory where all mathematical interactions among all mathematical objects take place.[18] Such a theory is ZFC (plus, possibly, LCs), interpreted as the theory of V.

Now, as a very general, overarching concern, one might legitimately doubt that the set-theoretic multiverse will fulfil the role of 'Generous Arena' equally well. The concern above may be summarised as follows:

Main Problem (Unification). *It is not clear whether and how the multiverse will fulfil the foundational role of a 'Generous Arena' (particularly, insofar as, at least*

[16]Some such foundational roles are also re-stated by Maddy in the paper published in the present volume.

[17]Rather unsurprisingly, among the spurious ones, Maddy mentions 'Metaphysical Insight', that is, the pretension that set theory provides us with an account of what mathematical objects *really* are, and 'Epistemic Source', *viz.* the idea that set theory provides us with an account of what mathematical knowledge is.

[18]The maxim 'maximize' was also introduced by Maddy in (1996). The application of the maxim to our conception of V implies that this has as many 'objects' as possible. It should be noted that, while, theoretically, maxims may be in tension with each other, they ought to be seen as having the same foundational (normative) content (cf. Maddy 1997, pp. 211–2). Among other things, this is shown by the fact that the iterative concept of set (which is generally taken to motivate V) is also naturally construed as being 'maximal' (for this, also see Wang 1974, Boolos 1971 or Gödel 1947). Cf. also Maddy (1996), in particular, pp. 507–12.

prima facie, *the multiverse provides us with a disconnected picture of set-theoretic phenomena).*

This concern can even become more general. Along with 'Generous Arena', there are further foundational roles expressed by set theory when 'standardly' construed as the theory of V, that the multiverse may not be able to fulfil. Maddy describes her further concerns as follows:

> The choice between a universe approach and a multiverse approach is justified to the extent that it facilitates our set-theoretic goals. The universe advocate finds good reasons for his view in the many jobs that it does so well, at which point the challenge is turned back to the multiverse advocate: given that we could work with inner models and forcing extensions from within the simple confines of V, as described by our best universe theory, what mathematical motivation is there to move to a more complex multiverse theory? (Maddy 2017, p. 316)

The troubles expressed in the quote above may be re-phrased as follows:

Problem 1 (Foundational Roles) *While we know what the universe can do for us, we do not know what jobs the multiverse can do for us, in particular whether it can successfully carry out all and the same (foundational) jobs that the universe does.*

As is clear, the Main Problem as well as Problem 1 strike both 'instrumental' and 'ontological multiversism'.

There is, however, one issue which relates specifically to 'ontological multiversism', which Maddy sees as particularly worrisome: metaphysics is so much involved in the characterisation of this position (one need only consider Hamkins' statements that his own view is one of 'higher-order realism, that is, Platonism about universes'), as to make it highly unsuitable to the set-theoretic naturalist. Therefore, we have:

Problem 2 (Metaphysics) *Multiversism heavily relies on metaphysics, in a way that the set-theoretic naturalist does not view as legitimate.*

One further foundational role of set theory is 'Conceptual Elucidation', a role that set theory has often held in replacing muddled and unclear mathematical concepts with sharper ones. As examples of 'Elucidation', Maddy mentions the formulation of the concept of 'continuity' in the nineteenth century, and 'the replacement of the imprecise notion of function with the set-theoretic version [...]'.[19] Now, can this foundational role also be carried out by the multiverse? The 'ontological multiverse' practitioner thinks that one of the roles associated to the multiverse is precisely that of exploring different 'concepts of set', by examining the *structures* which instantiate them.[20] But, for the Maddian naturalist, this is a sharply different

[19]Maddy (2017), p. 293.

[20]See Hamkins (2012), p. 417. Hamkins says: 'Often the clearest way to refer to a set concept is to describe the universe of sets in which it is instantiated, and in this article I shall simply identify a set concept with the model of set theory to which it gives rise. By adopting a particular concept of set, we in effect adopt that universe as our current mathematical universe; we jump inside and explore the nature of set theory offered by that universe.'

way of construing ‘Conceptual Elucidation’, a way which is strongly tied to the metaphysics evoked in Problem 2, as concepts, in this case, are also taken to be *objective* entities (p. 312, 315–6).

The naturalist multiversist, then, has to face up with one further problem, which, like Problem 2, fundamentally relates to ‘ontological multiversism’:

Problem 3 (Concepts) *‘Conceptual Elucidation’, construed as elucidation of* set concepts *instantiated by universes in the multiverse, should not be seen as a legitimate foundational role of set theory.*

There is one last, and fundamental, problem, with the multiverse, which strikes both ‘instrumental’ and ‘ontological’ multiversism. We have already seen how different multiverse conceptions are motivated by different research programmes in set theory. Now, the mathematics is, maybe, ok, but it is not clear, to the set-theoretic naturalist, that the multiverse construct is really essential to develop it, and, if yes, whether it can really act as a fully formal theory of sets. Maddy says:

> I’m in no position to evaluate the mathematics: my question is whether multiverse thinking is playing more than a heuristic role, whether there’s anything that couldn’t be carried out in our single official theory of sets. If not, then it’s not clear that these examples give us good reason to incur the added burden of devising and adopting an official multiverse theory as our preferred foundational framework. (Maddy 2017, p. 316)

We may re-phrase the concerns above as follows:

Problem 4 (Axioms) *It is presently not clear whether the multiverse is just a useful* heuristic *tool, or whether it is really* instrumental *for pursuing our set-theoretic investigations and, in particular, whether it will be able to replace the currently standard* axioms *of set theory.*

3.4 Addressing the Problems

3.4.1 Phenomenology of the Multiverse

I start with addressing Problems 2 (metaphysics) and 3 (Concepts). As we have seen, a recurring metaphysical claim made by ontological multiversists is that universes in the multiverse *really* exist. This claim implies, among other things, that ‘extensions’ of V exist. As we shall see, this is an especially problematic claim to make.

One further, equally problematic metaphysical claim is Hamkins’ assertion that there are different ‘concepts of set’ (as well as different concepts of ‘ordinal’, ‘cardinal’, ‘power-set’, and so on), construed, once more, as platonic objects instantiated by other platonic objects (set-theoretic structures).

3.4.1.1 Platonism and Existence

Hamkins' conception seems to re-state what has been known, for some time, as Full-Blooded Platonism (FBP), that is, that particular kind of Platonism which implies the existence of a *plenitude* of mathematical (set-theoretic) objects, as many as posited by all conceivable theories T of sets.[21] Let us now try to assess whether and how FBP really impacts on the mathematics of the multiverse. For instance, let us take into account what Hamkins calls the 'ontology of forcing'.

We know what the main problem with forcing is: set-theorists often use the notation $V[G]$ to refer to 'forcing extensions' of the universe, but, as is clear, V already has *all* sets, so it is not clear how it could possibly be extended. Clearly, FBP would have us view $V[G]$ as a fully meaningful (and existent) object, provided we can define it in a consistent way.

Surprisingly, though, Hamkins' account of forcing does not use any consciously FBP-inspired metaphysical principle, but rather what he calls a 'naturalist' account of forcing, based on purely mathematical facts. A full mathematical analysis of this is beyond the scope of this paper, but some details may be provided.

Hamkins proves that:

Theorem 1 (Hamkins) *Given the universe of sets V, it is possible to define an elementary embedding $j : V \to \bar{V}$, where $\bar{V}$ is a definable class* in V, *and a $\bar{V}$-generic filter G, such that $\bar{V} \subseteq \bar{V}[G]$, and $\bar{V}[G]$ is also a definable class* in V.

The crux of the theorem is that $\bar{V}$ and $\bar{V}[G]$ are definable classes *in* V and, thus, the naturalist account of forcing consists in showing that one can code extensions of V with subclasses of V itself. Already at this stage, the issue of the existence of such objects as $\bar{V}[G]$ becomes, in a sense, irrelevant. It is true that, given Theorem 1, one may use FBP to re-inforce the idea that such objects as $\bar{V}[G]$ *really* exist, but it is clear that the metaphysical content of FBP, is not instrumental, per se, for the proof of theorem.

Thus, the set-theoretic naturalist may simply want to take note of the methodology invoked by the theorem, but entirely disregard the metaphysical content attributed to it by FBP. In Hamkins' own words:

> This method of application, therefore, implements in effect the content of the multiverse view. That is, whether or not the forcing extensions of V actually exist, we are able to behave via the naturalist account of forcing entirely as if they do. In any set-theoretic context, whatever the current set-theoretic background universe V, one may at any time use forcing to jump to a universe $V[G]$ having a V-generic filter G, [...]. (Hamkins 2012, p. 425)

[21] FBP was introduced by Mark Balaguer in (1995). See also Balaguer (1998). Further details on Hamkins' use of FBP may also be found in Antos et al. (2015), pp. 2468–2470.

3.4.1.2 Concepts

There is a way to make sense of Hamkins'/the ontological multiversist's references to different concepts of 'set', 'ordinal', etc., as being instantiated by different set-theoretic structures, by bringing in a perspective wherein, ultimately, the metaphysical status of such concepts is irrelevant. In particular, I propose to construe the ontological multiversist's use of concepts within the framework of 'concept expansion'.

An especially promising and comprehensive account of concept expansion has recently been presented by Meir Buzaglo.[22] The main features of Buzaglo's account are: (1) concepts are flexible constructs; (2) the expansion of a concept is a law-like, forced process, that is, it is guided by the 'stretching' of some pre-established laws (axioms) which force the concept to evolve in a way which is *unavoidable* and, above all, (3) concept expansion gives rise to new objects.

Using such an account, Buzaglo shows that there are regulated ways to create new mathematical objects in an unavoidable way. At no point throughout this process, in Buzaglo's view, the mathematician has to assume that concepts are rigid, mind-independent *existents*: if they were such things, one could not make sense very easily of their 'extensibility'.[23]

Now, this account of concepts may be able to make sense of the ontological multiversist's invocation of different set concepts' being instantiated by alternative set-theoretic universes: the crux, here, is to see set concepts as mutually intertwined (that is, as arising from each other).

Again, the example of a forcing extension of the universe $V[G]$ will do. Using Buzaglo's account, the latter may just be seen as a 'new' set-theoretic object (and, correspondingly, set concept) arising from the regulated 'stretching' of the laws holding for the set concept of another object, V. After all, Buzaglo's account is not very far from Hamkins' own 'naturalist' account of forcing, whereby $V[G]$ is made fully *real* as a result of 'stretching' (laws holding for) V through forcing. Hamkins notes that this process may be compared to that relating to complex numbers, which are constructed by 'stretching' the square root function, in particular, by forcing it to have a value at $\sqrt{-1}$. Hamkins says:

> The case of forcing has some similarities [with that of complex numbers, *my note*]. Although there is no generic filter G inside V, there are various ways of simulating the forcing extension $V[G]$ inside V, using the forcing relation, or using the Boolean-valued structure $V^{\mathbb{B}}$, or by using the Naturalist account of forcing. None of these methods provides a full isomorphic copy of the forcing extension inside the ground model (as the complex numbers are simulated in the reals), and indeed they provably cannot–it is simply too much

---

[22]See Buzaglo (2002). Hints of a conception of concept expansion and evolution in mathematics may also be found in Lakatos (1976), which is also discussed by Buzaglo in the work mentioned.

[23]However, Buzaglo does not deny that realism about concepts may be compatible with concept change and evolution. See Buzaglo (2002), pp. 116–137, where the author examines Gödel's realist conception which, contrary to what has been stated above, takes concept evolution precisely as proof that concepts are objective constructs.

> to ask–but nevertheless some of the methods come maddeningly close to this. (Hamkins 2012, p. 420)

Now, this strategy may be extended to all other 'set concept-instantiating-universes' in the ontological multiverse, which may be construed as being new set-theoretic objects arising from stretching the laws (axioms) holding for concepts of other (previously established) set-theoretic objects.

3.4.1.3 Reality and Illusion

Leaving aside, for a moment, Hamkins' 'naturalist' interpretation of the multiverse, one could think that one of the purposes of the 'ontological multiverse' is that of making the illusion of 'living in separate, parallel *universes*' as fully real, which seems to involve some thorough-going metaphysical construal of set-theoretic practice.

However, in practice, what set-theorists working with the multiverse seem to be mostly interested in is something of a more definite mathematical character, that is, articulating a methodology which makes sense of 'accessing' other universes from a given universe. This, in particular, implies making sense of our experience of 'jumping', in Hamkins' own words, from one universe to another. This may be seen as a form of mathematical 'perspectivism', which may be described as follows:

Multiverse Perspectivism (MP) In working with one universe V, set-theorists are particularly interested in knowing: (1) what universes one may have access to from V and how, and (2) what V looks like from the point of view of those universes.

Now, as even multiversists may concede, one could certainly be a universist and implement MP within the single-universe conception, but MP is, trivially, facilitated by the adoption of the/a multiverse, as the multiverse precisely allows one to view (and refer to) universes as being 'separate', although somehow interrelated, constructs.

For instance, consider the following examples illustrating MP. The first two are multiverse axioms formulated by Hamkins:

Axiom 1 (Countability Principle) *Every universe V is countable* from the perspective of *a better universe W*.

Axiom 2 (Absorption into L) *Every universe V is a countable transitive model* in *another universe W satisfying $V = L$*.

One further example comes from the HP. As we have seen in Sect. 3.2.2, within the HP, it is essential that one can really refer to 'extensions of V'. In V-logic, one can prove the following fact:

Theorem 2 (S. Friedman) *If there is a small 'lengthening' of V, called $Hyp(V)$, then 'thickenings' of V may be viewed as outer models of V whose existence is* *implied by V-logic statements φ asserting: 'there is an outer model of V which satisfies T', where T is an extension of ZFC.*

So, it is only from the perspective of *Hyp(V)*, as defined in V-logic, that outer models of V can really be seen as existing.

In sum, the reference to 'reality' and 'illusion', thus, only serves to highlight more sharply the purposes and the extent of MP: the multiverse allows one to study inter-universe relationships by preserving our intuitive experience of this as a 'move' or a 'jump' from one universe to another.

3.4.2 Multiverse-Related Mathematics

I will now briefly present three case studies of 'multiverse-related' mathematics, which will help me illustrate that the multiverse may be able to fulfil many foundational jobs associated to set theory (which provides a response to Problem 1 (Foundational Jobs)), and also that it may not just be a useful heuristic, but rather a central construct in contemporary set theory, which partly responds to concerns expressed by Problem 4 (Axioms). I shall further address Problem 4 in Sect. 3.4.4.

3.4.2.1 Woodin's Set-Generic Multiverse and Ω-logic

I start with Woodin's results on CH in the set-generic multiverse. Woodin's guiding question was: is it possible to find an axiom which plays, for the structure $H(\omega_2)$, the same role as that played by PD for $H(\omega_1)$?[24] That is, is there any axiom which makes $H(\omega_2)$ 'well-behaved', as PD does with $H(\omega_1)$? Now, it turns out that it is relatively easy to force over properties of $H(\omega_2)$, which means that it is relatively easy to have different, mutually inconsistent pictures of $H(\omega_2)$. Therefore, Woodin identified the solution of the problem in identifying an axiom able to induce forcing-invariant properties of $H(\omega_2)$.

Now, it was known that many *forcing axioms* have this characteristic (that is, they are 'absoluteness axioms') and, moreover, that they imply the failure of CH. Therefore, Woodin's work was directed at identifying the appropriate forcing axiom which would make $H(\omega_2)$ well-behaved (and which, among other things, would also imply $\neg$CH), but the work carried out for this goal subsequently led to a parallel, equally fruitful, undertaking, that of defining a broader logical framework, wherein forcing invariance, in general, may be addressed. All this led Woodin to introduce a new logic, Ω-logic.

[24] $H(\kappa)$, for a cardinal κ, is the collection of all sets whose cardinality is hereditarily less than κ, that is, all sets whose elements and the elements of whose elements and so on, have cardinality less than κ.

Ω-logic is a logic in the full sense of the word, that is, a logical system which comes with its own definitions of semantic validity and logical consequence.[25] The semantics of Ω-logic is hinged on the use of Boolean-valued models $V^{\mathbb{B}}$ (where $\mathbb{B}$ is a complete Boolean algebra). The collection of all such models would, subsequently, become what Woodin defined the 'set-generic multiverse'.[26]

Now, it is important, for my purposes, to recall the definitions of validity and provability in Ω-logic. The definition of validity, with respect to a theory T, in Ω-logic (of $\models_\Omega$) is as follows:

Definition 1 (Ω-validity) $T \models_\Omega \phi$ if and only if, for all α ordinals, and all complete Boolean algebras $\mathbb{B}$, when $V_\alpha^{\mathbb{B}} \models T$, then $V_\alpha^{\mathbb{B}} \models \phi$.

The notion of provability is a lot more complex, as it uses *universally* Baire sets of reals, which cannot be addressed here.[27]

Definition 2 (Ω-provability) $T \vdash_\Omega \phi$ if and only if there exists an $A \subseteq \mathbb{R}$ universally Baire, such that $M \models \phi$, for every A-closed set M such that $M \models$ ZFC.

In turn, the Ω-conjecture is the conjecture that Ω-logic is complete (that is, that $\models_\Omega$ is equivalent to $\vdash_\Omega$).

As is known, in his (2001) Woodin, ultimately, focussed his attention on a specific forcing axiom, the $(\star)$ axiom, which allowed him to prove that, for all ϕ, ZFC+$(\star) \vdash_\Omega$ "$H(\omega_2) \models \phi$" or ZFC+$(\star) \vdash_\Omega$ "$H(\omega_2) \models \neg\phi$", precisely the kind of absoluteness result for $H(\omega_2)$ that Woodin was looking for.[28]

In the results mentioned, the foundational roles fulfilled by multiverse thinking are many. Woodin's 'Ω-logic solution' to CH looks very different from a 'standard', that is a solution consisting in showing that there is an axiom A which implies the truth or falsity of CH (something that Hamkins would subsequently define the 'dream solution' for CH).[29] The use of the Boolean-valued multiverse, or, more simply, of what would later become the set-generic multiverse, thus, stands out as an immensely successful way to elucidate statements of the complexity of CH, by

---

[25]Bagaria et al. (2006) is a comprehensive introduction to Ω-logic.

[26]Ω-logic makes its first appearance in Woodin (1999), and figures as a prominent tool in Woodin (2001). In those works, there is no direct reference to the set-generic multiverse, although the basic definitional ideas and concepts relating to it are already there.

[27]A rather accessible treatment of the provability relation in Ω-logic, and of universally Baire sets, is in Woodin (2011), p. 108.

[28]Moreover, Woodin was able to prove that:

Theorem 3 (Woodin) *ZFC*+$(\star) \vdash_\Omega$ "$H(\omega_2) \models \neg CH$".

It should be noted that the result requires the assumption of the existence of class-many Woodin cardinals, a particular strand of large cardinals having, as is known, far-reaching connections with Definable Determinacy Axioms, such as PD.

[29]See Hamkins (2012), p. 430.

showing, in particular, what such statements require in terms of 'solving resources', itself a way, in turn, to fulfil 'Conceptual Elucidation'.

Secondly, multiverse thinking leads to define a broader logical environment, that of Ω-logic, through which statements like CH (or $\neg$CH), in particular, proof-theoretic and semantic facts about them, can be represented. This, among other things, also implies a re-structuring of the notion of proof in set theory, something which should be viewed as being strongly connected with two further foundational roles, 'Shared Standard' and 'Metamathematical Corral'.

Finally, it should be noted that the kind of 'multiverse logic' inherent in the results mentioned is not only a specific 'tool' to be employed in representing facts about set-theoretic undecidability, but also a way to produce concrete mathematics, as shown by further work done on the Ω-conjecture.[30]

3.4.2.2 The Hyperuniverse Programme

As said in Sect. 3.2, the Hyperuniverse Programme (HP) has identified two main kinds of multiverse:

1. Zermelo's height multiverse[31]
2. The hyperuniverse $\mathbb{H}$, that is, the collection of all countable transitive models of ZFC

How did the programme get there? First came the proof that certain maximality principles have very important first-order consequences. For instance, take the IMH:

Definition 3 (IMH) For all ϕ, if ϕ is true in an inner model of an outer model of V, then ϕ is true in an inner model of V.

On the one hand, the IMH implies that there are no large cardinals in V (only in inner models of V), and that PD is false. On the other, refinements of IMH, like SIMH#, imply, among other things, that CH is false.[32]

Now, HP's maximality principles address extensions of V (in height and width). Therefore, one of the programme's main goals from the beginning has been to clarify what mathematical resources are needed in order to express principles which address extensions of V, like the IMH.

The answer consisted in adopting the multiversist position that we have mentioned above. In particular, in order to make maximality principles *mathematically* expressible, HP turned to taking into account:

[30]See, for instance, Viale (2016).

[31]See Footnote 8.

[32]Further details on all the different maximality principles explored by the HP may be found in Friedman (2016).

1. A partially *potentialist* view of V, whereby height extensions are admissible, but the width of the universe is fixed, which accounts for the introduction of the multiverse (1), or
2. A fully *potentialist* view of V, that is, a view whereby V is extendible in height and width, which accounts for the introduction of the multiverse (2)

By adopting (1), one may only state the IMH syntactically, as, by (1), outer models of V aren't really available, whereas, if one adopts (2), in particular, if one takes V to be countable, then one may have *real* 'thickenings' and, thus, *real* outer models satisfying the IMH. The latter choice leads to the introduction of $\mathbb{H}$.

As is clear, then, multiverse thinking is fully integrated in the mathematics of the HP, in the sense that it would be a lot more cumbersome to express maximality principles such as the IMH within a universist framework.

Moreover, we could say that, also within the HP, the multiverse helps one fulfil tasks which are associated to 'Conceptual Elucidation', such as elucidate what V is like, and also foster one's mathematical investigations on maximality principles and their consequences. Therefore, as in the case of Woodin's set-generic multiverse, the introduction of the hyperuniverse is not merely a way to represent facts about 'truth in V': it is a way to produce new mathematics, which subsequently leads to finding solutions to outstanding set-theoretic problems.

3.4.2.3 The Multiverse Case for $V = L$

One further, striking example of multiverse-related mathematics is Hamkins' multiversist construal of $V = L$. A very influential and widespread view concerning $V = L$ is that the axiom would be a sort of *minimality principle*, that is, a principle which implies that V is as small as possible.[33] Hamkins has attempted to challenge this point of view, by using mathematical facts which are deeply connected with the multiverse.

First, let us contrast the following two conceptions:

Conception 1 *There is an* absolute *background concept of set, and of other set-theoretic notions, such as set, ordinal, cardinal.*

Conception 2 *There is* no *absolute background concept of set and of other set-theoretic notions, such as set, ordinal, cardinal.*

[33]Of course, the fact that L is 'minimal' is simply a mathematical fact (insofar as L is the smallest inner model of V). As is widely known, the view that construes L as a 'minimality principle' has been expressed by Gödel in (1947), p. 478–9. In that work, Gödel explicitly contrasts $V = L$ to *maximum principles*. Maddy herself, as is known, has argued in favour of the claim that $V = L$ would be 'restrictive'. The full argument may be found in Maddy (1997), pp. 216–232.

The 'ontological multiverse' view, as has been said many times, is bound up with Conception 2, which, in turn, suggests the following facts: L may not have the same ordinals as V, insofar as the concept '$(\text{ordinal})^L$' may be different from the concept '$(\text{ordinal})^V$'. By adopting this approach, one may, then, proceed to establish further striking mathematical facts, all of which suggest that $V = L$ is not inherently restrictive. What follows is a summary of Hamkins' argument:

1. By Shoenfield absoluteness, statements such as 'T has a transitive model', which are Σ^1_2, are *absolute* between V and L. Therefore, even theories T which contradict $V = L$ (for instance, the theory ZFC+'$\exists$ a measurable cardinal') have transitive models in V if and only if they have transitive models in L.
2. All reals in V can be coded in a model M of ZFC+V=L.
3. Any transitive countable model M can be 'continued' such that it may, ultimately, become a model of ZFC+V=L. Thus, using *forcing*, one may first collapse any model to the countable, and then use such model to make it satisfy $V = L$.[34]

The upshot of this is remarkable: a case for the non-limitative character of V=L can be successfully made within multiverse-related mathematics in a way which is entirely in accordance with the set-theoretic naturalist's desiderata.

Again, while talk of concepts might look suspect to the naturalist, I have already construed their use in an essentially non-metaphysical way through 'concept expansion' (in Sect. 3.4.1.2 which addressed Problem 2), and, based on this perspective, we may elucidate such notions as that of 'constructibility' and also completely revolutionise our mathematical perspective on such axioms as $V = L$.

3.4.3 Opposing the Argument from Priority and a New Unification

The argument I will be reviewing in this subsection re-states concerns expressed by the Main Problem (Unification) and Problem 1 (Foundational Jobs), and, therefore, here I will mostly be concerned with these two problems.

In particular, one may see the Main Problem as being introduced by an argument, which I will call 'argument from priority', which states that the drive towards a *unifying* account of set theory was a primordial goal of set-theoretic research, already inherent in the process of axiomatisation carried out by such pioneers as Zermelo, Fraenkel, von Neumann and Skolem. By virtue of this fact, this goal should still be viewed as a major goal of set-theoretic research.

[34]This leads Hamkins to even surmise that: 'For all we know, our entire current universe, large cardinals and all, is a countable transitive model inside a much larger model of $V = L$.' (Hamkins 2012, p. 436).

The argument may more accurately be summarised as follows:

1. One main goal of the axiomatisation, since its emergence, was the drive towards producing a unifying account of set-theoretic phenomena[35]
2. The cumulative hierarchy V, which progressively emerged as the intended universe of discourse of the axioms, fully fulfilled this goal
3. Then came the 'multiverse phenomenon', triggered by the 'independence phenomenon'
4. The goal of unification, as expressed by 1., was viewed as *prioritary* over other goals in the early axiomatic set theory
5. Therefore, the universe view has *priority* over the multiverse view

There are, at least, two ways to resist the argument, both of which attack (4), and passing from (4) to (5). First of all, temporal priority is not adequate, as a criterion, to establish the absolute preferability of the single universe view over the multiverse view. In other terms, the fact that set-theorists *first* aimed at describing something like V and, *only later*, turned to considering a plethora of different structures does not imply that a single universe was (or is) something *inherently* more desirable.

A second way to dismantle the significance of temporal priority is provided by knowledge of further historical details. It is true that set theory was seen, already very early, as the 'theory of V', but the reasons which motivated this fact may be of a rather different kind than pre-supposed by the argument. One such reason was Zermelo's insistence on the determinacy of the concept of set, as expressed by the cumulative hierarchy and fully highlighted by his (quasi)-categoricity theorem. It should be noted that at the time when Zermelo presented the final version of the axioms there was no fully developed account or awareness of the potentialities of first-order logic, nor were axiomatic theories of other areas of mathematics (in particular, of arithmetic and analysis) naturally seen as being first-order theories.[36] The reason why the 'ontological problem' of set theory, so to speak, was seen by Zermelo as very pressing was the serious challenge to the determinacy of set-theoretic concepts posed by Skolem in the same years (something which had led to speculations about the 'relativity' of set theory). Therefore, (4), in the

[35]It is true, however, that this goal was not fully attained, arguably, until (Zermelo 1930). Of course, another prioritary goal of the axiomatisation was to prevent the formation of the paradoxes, something which Zermelo carried out through introducing the Axiom of Separation already in Zermelo (1908). For this and other aspects of the history of the axiomatisation of set theory, see Ferreirós (2010), in particular, pp. 297–324.

[36]At least, until first-order logic largely took over as a consequence, in particular, of Skolem's and Gödel's work. An exhaustive history of the triumph of first-order languages is provided by Shapiro in (1991), pp. 173–197. The relevance of 'categoricity arguments' for the 'triumph' of the single universe conception is also addressed in the same work, on pp. 250–259. A historically accurate reconstruction of the 'first-order proposal' for set theory is in Ferreirós (2010), pp. 357–64.

previous argument, should be assessed in the context of the fight between two early conceptions of set theory:

- Zermelian conception: there is one *determinate* concept of set, and one (associated) single structure instantiating it (this is the point of view of Zermelo in (1930))[37]
- Skolemian conception: there is no *unique* (and *fixed*) interpretation of set-theoretic concepts[38]

Now, while the dispute between these two viewpoints is, somehow, still alive, there is no need to see it as relevant to the Maddian naturalist's goals, and I'm rather inclined to think that adopting 'Generous Arena' based on Zermelo's concerns and conception would be, for such a naturalist, slightly embarrassing. Moreover, since, presumably, there is no need to follow Zermelo's strategy to successfully counter the argument for the relativity of set-theoretic concepts, one may also reject its main consequence, the support to the single-universe picture.

But then one could legitimately wonder what would be of 'unify' and unification concerns, if the latter ought not to be viewed as arising from the alleged determinacy of the concept of set and from the uniqueness of V. What follows aims to provide a tentative response to this.

There is a way to construe 'Generous Arena' along entirely different lines, which is suggested by Hamkins' foundational views. In a crucial quote from Hamkins (2012), Hamkins says:

> On the multiverse view, set theory remains a foundation for the classical mathematical enterprise. The difference is that when a mathematical issue is revealed to have a set-theoretic dependence, as in the independence results, then the multiverse response is a careful explanation that the mathematical fact of the matter depends on which concept of set is used, and this is almost always a very interesting situation, in which one may weigh the desirability of various set-theoretic hypotheses with their mathematical consequences. (p. 419)

Later on, he unpacks the methodology briefly sketched above, by addressing CH. He says that CH is a settled question, on the multiverse view. The reason would be that:

> The answer to CH consists of the expansive, detailed knowledge set theorists have gained about the extent to which it holds and fails in the multiverse, about how to achieve it or its negation in combination with other diverse set-theoretic properties. Of course, there are and will always remain questions about whether one can achieve CH or its negation with this or that hypothesis, but the point is that the most important and essential facts about CH are deeply understood, and these facts constitute the answer to the CH question. (p. 429)

[37]Of course this does not prevent one from interpreting Zermelo's conception of V in multiversist terms. See, again, Footnote 8.

[38]Although Skolemian relativism is a much stronger claim than this: it is the claim that set-theoretic concepts have no definite *theory-independent* meaning. Here, we cannot delve into the issue of whether Hamkins' FBP, ultimately, fosters such a form of anti-objectivism, as, for instance attributed by Field to Balaguer's FBP in Field (2001), pp. 334–5.

What Hamkins is setting out in the quote above is a fact already emerged in connection with Woodin's 'Ω-logic solution', that is, that finding a solution to CH implies taking into account the problem of what is needed, in terms of logical and mathematical resources, to solve it. However, this task cannot be executed, if one does not have a sufficiently broad collection of models (of a sufficiently strong theory) available, where the CH vs. $\neg$CH hypothesis may be tested, which is precisely what the multiverse provides us with.

Now, what I would like to highlight is the fact that the multiverse may, if viewed from such a foundational perspective, provide a different kind of unification, one which is needed to gain the sort of meta-theoretic 'knowledge' Hamkins is alluding to in the quotes above. In particular, one could suggest that the multiverse provides a different kind of 'Generous Arena' (a multiversist's 'Generous Arena'), one wherein all metamathematical interactions needed to provide us with knowledge about how to 'settle' problems of the same complexity, for instance, as CH, take place.

I argue that this may also be seen as a foundational role of set theory, which could be formulated as follows:

Multiversist's Generous Arena. *Set theory is also a systematic inquiry into the independence and unprovability phenomena, which provides us with knowledge about set-theoretic truth. In order to carry out such an inquiry, it is fundamental to have a unified metamathematical arena, where all relevant interactions take place.*

Of course, as already said, 'practically' one could still carry out such an inquiry within V. But should set-theorists agree that the Multiversist's Generous Arena is one further correct epistemological maxim for set theory to adopt, wouldn't the multiverse be the most natural candidate to fulfil it?

3.4.4 Relativism Reconsidered

Responses to the Main Problem (Unification), Problem 1 (Foundational Jobs) and, finally, Problem 4 (Axioms) are also provided by Steel's assessment of the goals of the multiverse in Steel (2014). We have already seen (in Sect. 3.2.2) that Steel's conception is an axiomatic version of the set-generic multiverse (MV).

Now, it is crucial, for Steel's purposes, to try to understand what MV really consists in. First, Steel shows that multiverse language is a sub-language of LST, that is, of the language of set theory. This is because a theorem proved by Woodin and Laver implies the following:

Theorem 4 (Woodin, Laver) *Given ϕ, $M^G \models \phi \leftrightarrow M \models t(\phi)$, where $t(\phi)$ is a formula saying: 'ϕ is true in some (all) multiverses obtained from M'.*[39]

[39]In Steel's notation, M^G is the set-generic multiverse containing all worlds satisfying MV.

The theorem says that there is a recursive translation from MV to LST given by $t(\phi)$, such that, given an MV-statement, there is always an LST-statement which also expresses it (whereas no inverse procedure to translate an LST-statement to an MV-statement is currently known). Now, there is a problem, however, with statements like CH. The reason is that CH, as is known, is not preserved by set-forcing in different worlds, therefore, CH cannot be expressed in MV.[40]

Although, on the one hand, this fact might be construed as showing that MV might lead to a loss of set-theoretic information, on the other hand, it might be seen as an indicator that the multiverse really is useful for us, insofar as it carries out an important task: that of drawing a line between ordinary set-theoretic statements which are 'meaningful', that is, are in the range of $t(\phi)$, and those which are not. But this is, again, a form of 'Conceptual Elucidation'!

But there's more. If Steel's MV really expresses the correct approach to set-theoretic truth, then one need abandon what Steel calls:

Strong Absolutism (SA) There is a reference universe, $\dot{V}$, which cannot be captured by MV.

As an alternative, one could take into account the following position:

Weak Relativism (WR) All propositions expressible in LST can also be expressed in multiverse language.

By WR, $\dot{V}$ makes sense if and only if it is expressible in multiverse language. However, WR still does not tell us whether one can really do that. The thesis that this is the case, a sort of combination of SA and WR, is called by Steel:

Weak Absolutism (WA) $\dot{V}$ makes sense, but as an *individual* definable world in the multiverse, which is *included* in all other worlds.

This latter standpoint looks particularly attractive to Steel, insofar as it introduces the view that the multiverse is reducible to one of its members, which he calls the *core*.

It should be noted that the acceptance of WA is, in turn, hinged on the acceptance of other mathematical conjectures. Steel has identified an axiom, the Axiom **H**, which singles out the core of the *multiverse* in a very detailed way.[41] With the addition of Axiom **H** to 'core truths', we might reach a situation wherein the

[40]However, it might still be the case that there are 'traces' of CH in MV, if one accepts additional hypotheses, such as the existence of a *core*, which is described later on, p. 22. See also Footnote 41.

[41]The Axiom says that V=HOD^M, where HOD is the class of all hereditarily definable sets, and M is a model of AD. Moreover, among other things, the axiom also implies CH. See Steel (2014), p. 171–177.

multiverse contains a preferred, 'reference' world. In practical terms, this is a way to revert to a partly universist account, which, in Steel's view, provides a more 'standard' unification of set-theoretic phenomena, than the one provided by Hamkins' multiverse addressed in Sect. 3.4.3.

Moreover, Problem 4 (Axioms) is also answered, in a sense, by Steel's theory, insofar as MV is an axiomatic theory of the multiverse. It is true, though, that the theory, in itself, has never been used as a tool to make concrete set-theoretic mathematics. Furthermore, the status of the MV axioms (in particular, of some LCs) may be controversial. But this should rather encourage us to continue our investigations on the multiverse than retreat to the single-universe picture, as it is reasonable to predict that more developed axiomatic theories of the multiverse will respond more fully to the concerns expressed by Problem 4.

3.5 Concluding Remarks

Let's take stock. My goal has been that of trying to explain away the Maddian naturalist's concerns about the set-theoretic multiverse by, essentially, adopting the following two points of view: (1) metaphysics, even when explicitly evoked by multiverse theorists, is not fundamental for a successful articulation of the multiverse and (2) multiverse practice has led to important mathematical discoveries, which have a robust foundational impact and, moreover, show that a multiverse theory may be able to fulfil the many foundational roles associated by the set-theoretic naturalist to set theory, including that of 'unification'.

This work has been concerned to a large extent with multiverse practice, without providing a sharp definition of it. It should be noted that 'multiverse practice' does not just consist in dealing with models of set theory, but rather implies manipulating mathematically a determinate multiverse framework from the beginning, with a view to pursuing a broad, but clearly set out, range of mathematical goals, such as the study of: (1) inter-universe relationships, (2) axioms for the multiverse itself, (3) principles which are formulated in a multiversist language, and (4) problems whose complexity requires a strictly multiversist construal.

In Sect. 3.2 we have seen that the multiverse may not be seen as a single strand of mathematical practices and methodologies. Furthermore, we do not currently know whether there will ever be an ultimate conception of the multiverse. In some respects, 'ontological multiversism' may be seen as such a conception, insofar as: (1) it is maximally broad, and (2) it is particularly flexible, as it allows for the highest amount of 'perspectivism'. However, 'ontological multiversism' is also the most controversial and problematic conception among those examined, for the reasons we have reviewed in Sect. 3.3.

From time to time, it has been made clear that there may be no specific task that the universist may not try to successfully emulate within their single V. However, it is, in my view, rather apparent that the multiverse enormously facilitates fundamental practical tasks, so why would the conscientious set-theoretic naturalist, ultimately, oppose this fact?

There is surely one main concern which has not been fully dispelled by the paper, that relating to the absence of an axiomatic theory of the multiverse. However, I have suggested that further mathematical research may eventually allow us to fully axiomatise the multiverse, along the lines of what has been attempted by Steel with MV. As is clear, this would help the Maddian naturalist to defeat more convincingly the concerns expressed by Problem 4.

However, even now, and notwithstanding the current absence of an axiomatic framework, I have suggested that the multiverse may already provide us with the same foundational benefits that the single-universe picture does, something which should help fully dissipate the set-theoretic naturalist's concerns about its usefulness.

Acknowledgement I would like to thank the University of Tartu for its support through ASTRA project PER ASPERA (financed by the European Regional Development Fund).

Reply to Ternullo on the Multiverse

Penelope Maddy

Doctor Ternullo raises a host of important issues, but I focus here on the central theme: his defense of the multiverse from the point of view of a naturalist, indeed, a naturalist of a particular variety that I call a 'second philosopher'. In a recent paper on the foundations of mathematics (a companion piece to my contribution to the present volume), I considered the possibility that some sort of multiverse theory could replace our current set theory in a range of foundational jobs now performed by ZFC + Large Cardinals (LCs). I concluded that for now, in the current state of knowledge, it isn't clear that this move is feasible or advisable. Ternullo apparently disputes this conclusion: 'the multiverse may be as acceptable as the universe for ... the foundational purposes of set theory' (Ternullo 2019, p. 46).

In an odd twist, though, Ternullo argues that one fundamental aspect of the foundational goal, the job I call Generous Arena,[1] is itself misguided, that the argument for it is, for my naturalist, 'slightly embarrassing' (Ternullo 2019, p. 63). As it happens, leading multiverse theorists don't see the situation this way; they embrace this foundational goal and argue that their theories meet it. For example, this from Hamkins:

> The multiverse view does not abandon the goal of using set theory as an epistemological and ontological foundation for mathematics, for we expect to find all our familiar mathematical objects ... inside any one of the universes of the multiverse. (Hamkins 2012, p. 419)

[1] Ternullo also uses the term 'Generous Arena' for a different idea (Ternullo 2019, pp. 63–64), but in the passage under discussion here, he's concerned with the sense delineated in my contribution to this volume.

P. Maddy (✉)
Department of Logic and Philosophy of Science, University of California Irvine, Irvine, CA, USA
e-mail: pjmaddy@uci.edu

And here's Steel:

> We want one framework theory [i.e., foundational theory], to be used by all, so that we can use each other's work. It is better for all our flowers to bloom in the same garden. If truly distinct frameworks emerged, the first order of business would be to unify them. ... The goal of our framework theory is to *maximize interpretive power*, to provide a language and theory in which all mathematics, of today and of the future so far as we can anticipate it, can be developed. (Steel 2014, pp. 164–165)

Ternullo seems to reject this approach, on the grounds that the argument for the foundational goal itself is flawed (Ternullo 2019, pp. 61–63).

That argument, as he sees it, rests on the claim that Generous Arena was present early on in the history of set theory – as it was – and that this temporal priority implies logical priority. I think we can all agree that this is a weak argument, but it's not the right argument, as I hope is clear from §I of my contribution here.[2] Regardless of when they first arose, the mathematical attractions of Generous Arena, along with Risk Assessment, Metamathematical Corral, and Shared Standard, remain as strong today as ever. In his willingness to forgo Generous Arena, Ternullo sounds a theme familiar, as we've seen, in category-theoretic and univalent foundations, but not one we hear from multiverse theorists like Hamkins and Steel. This creates some mismatch in the debate between Ternullo and me, since my analysis is explicitly addressed to the question of the multiverse's aptitude for filling this foundational role (among others) and he seems to think it needn't be filled. Still, there are some surrounding points well worth considering.

I begin, as Ternullo does, by attending to various versions of multiversism, though now with an eye not to the mathematical differences between them, but to the different philosophical or methodological stances one might take toward them. §II sketches my concerns about the multiverse in a foundational role, and §III considers other significant roles that multiverse thinking might play.

I. What Is a Multiverse View?

Ternullo begins with a straightforward characterization of the central distinction:

> Whether set theory should be interpreted as the theory of a *single* universe of sets ... or ... as a theory about *multiple* structures ... that is, about a set theoretic 'multiverse'. (Ternullo 2019, p. 43)

He alludes to Koellner's distinction between pluralism and non-pluralism:

> *pluralism* ... maintains that ... although there are *practical* reasons that one might give in favor of one set of axioms over another – say, that it is more useful for a given task –, there are no *theoretical* reasons that can be given ...
>
> *non-pluralism* ... maintains that the independence results merely indicate the paucity of our standard resources for justifying mathematical statements. ... theoretical reasons *can* be given for new axioms. (Koellner 2014, p. 1)

[2] See §I of [2017] for a bit more detail.

Multiversism, then, is a model-theoretic version of pluralism:

> There is not a single *universe* of set theory but rather a *multiverse* of legitimate candidates, some of which may be preferable to others for certain practical purposes, but none of which can be said to be the 'true' universe. (Koellner 2013, p. 3)

Ternullo agrees that multiversism is 'an *ontological* (*semantic*) variant' of pluralism (Ternullo 2019, p. 44). Presumably he also takes his multiversist to hold that, though there might be practical reasons for preferring one truth value for an indeterminate statement over another or one universe of the multiverse over another, there are no theoretical reasons for this – there is no determinate truth value, no 'true' universe.

Now Ternullo notes that unabashedly metaphysical views like these may be problematic for the naturalist, but he concludes that the metaphysics can be disregarded (Ternullo 2019, §4.1). He has his own reasons for saying this, but in any case, we agree on the underlying point, that metaphysics is largely irrelevant to the mathematical issues at hand. In the hope of clarifying some of these matters, let me sketch a rough taxonomy of philosophical stances on multiversism, beginning from the most ontologically or semantically loaded and moving on from there.

On the deeply metaphysical end of the spectrum, there's Hamkins's position:

> With forcing, we seem to have discovered the existence of other mathematical universes, outside our own universe, and the multiverse view asserts that yes, indeed, this is the case. (Hamkins 2012, p. 425) Each ... universe exists independently in the same Platonic sense that proponents of the universe view regard their universe to exist. (Ibid., pp. 416–417)

In Koellner's terms, presumably this Metaphysical Multiversism[3] takes 'theoretical' considerations to tell us something about the structure of its generous ontology, perhaps, for example, that every world thinks ZFC.[4] It might seem that the theoretical/practical distinction coincides with the intrinsic/extrinsic distinction familiar in the philosophical foundations of set theory – where intrinsic considerations are somehow intuitive, or self-evident, or contained in the concept of set, and extrinsic considerations involve attractive consequences or interrelations or something of that sort – but I think this can't be right. Koellner writes that ...

> given the current state of our knowledge a case can be made for being a non – pluralist about ZFC and large cardinal axioms (Koellner 2013, p. 4)

---

[3]I use this term in place of Ternullo's 'ontological multiversism' to leave room for a position that replaces objective entities with determinate truth values.

[4]Steel (2014) and Woodin (2011) both take ZFC to be true in every world of the multiverse, but Hamkins sometimes does not: 'There seems to be no reason to restrict inclusion only to ZFC models, as we can include models of weaker theories ZF, ZF^-, KP, and so on, perhaps even down to second-order number theory' (Hamkins 2012, p. 436). On the other hand, we've seen that he addresses the foundational goal like this: 'we expect to find all our familiar mathematical objects ... inside any one of the universes of the multiverse' (ibid., p. 419), which would seem to require at least ZFC (and if Risk Assessment is taken into account, large cardinals would be handy as well). In any case, Hamkins certainly embraces a number of objective truths ('multiverse axioms') about the multiverse in §9 of his [2012].

... in other words, in multiverse terms, for assuming these axioms across all worlds of the multiverse, and from his other writings, it's clear that some of this case is extrinsic. Similarly, Hamkins allows that

> the mathematician's measure of a philosophical position may be the value of the mathematics to which it leads (Hamkins 2012, p. 440).

So it appears that at least some extrinsic considerations must also yield information about the multiverse, must also be included under 'theoretical'.

Suppose, then, that a certain set-theoretic statement, perhaps a candidate for a new axiom to be true in all worlds, or perhaps another sort of general claim about the structure of the multiverse, has many mathematical advantages and no mathematical disadvantages. This wouldn't be enough for the Metaphysical Multiversist to endorse it, because we'd have to be confident that those mathematical merits produce theoretical support, not mere practical support. We'd have to be confident that our belief in that set-theoretic statement, however attractive it might be, isn't just wishful thinking, that the objective mathematical realm of the multiverse doesn't just happen to deny us something we'd very much like to have. Many philosophers, in the tradition of Benacerraf's famous challenge to Platonism, would ask how we could come to know that our beliefs are tracking the truth about an abstract realm. My naturalist asks a question that's logically prior to Benacerraf's: why should we demand more than mathematical merits? Why should those mathematical merits be held hostage to extra-mathematical metaphysics? Her answer is that this is wrong-headed, that the compelling mathematical reasons should be enough all by themselves.[5]

Though Metaphysical Multiversism is uncongenial to the naturalist, as Ternullo says, this isn't the end of the story; there are other varieties of multiversism. We could, for example, leave metaphysics aside and simply talk about theories. Set theory, then, isn't the project of describing an abstract mathematical realm; it's the project of forging a powerful mathematical theory to serve the foundation goal (among others).[6] The universist advocates ZFC and its extensions in this role; the multiversist proposes an alternative multiverse theory of sets and worlds to take its place. For these purposes, all extrinsic considerations would be on equal footing; there'd be no distinction between 'theoretical' or truth-tracking cases and 'merely practical' cases. This Theory Multiversism is an option entirely open to the naturalist, should the evidence point that way.

A final variant sees the multiverse as analogous, not to a universe ontology, not to a universe theory, but to the iterative conception. In universe thinking, the iterative conception serves as an intuitive picture that helps us see our way around in deriving consequences from the axioms or seeking new avenues for axiom choice. From

[5] See [2011], pp. 55–59. In that book, I propose an alternative metaphysical position, Thin Realism, that avoids this problem by essentially reading its ontology off the analysis of proper methods, including extrinsic methods, but I doubt this is what Hamkins or Koellner or Ternullo has in mind.

[6] We could think of this as the project of forming an optimally effective concept of set. Cf. the Arealism of [2011].

the naturalist's non-metaphysical perspective, intrinsic considerations based on this picture are potentially of great heuristic value; the history of the subject amply demonstrates what an immensely successful tool the iterative conception has been. But, for the naturalist, it's important to stress that the value of this intuitive picture rests on the great mathematical merits of the work it's inspired, in other words, on its extrinsic success.[7] If we were presented with an alternative intuitive picture that conceptualizes set-theory differently, if that alternative way of guiding the subject were more fruitful than the iterative conception, we should switch our allegiance without regret.[8] The Heuristic Multiversist[9] argues that the intuitive picture of a multiverse is just such an alternative; he might propose that ZFC and its extensions, guided by the iterative conception, should be replaced with a multiverse theory based on the new picture. This would be a version of Theory Multiversism, but other possibilities emerge in §III below. Either way, the basic suggestion is that the intuitive multiverse picture would guide the practice in new and different directions with important mathematical advantages.

There are no doubt other ways to frame a philosophical perspective for multiversism, and perhaps predictably, one prominent multiverse theory, the one due to Steel, doesn't fit squarely in any of the three bins just described. As Ternullo notes, Steel is out to explore whether CH is 'meaningful'; his multiverse theory is intended, not as an alternative subject matter (Metaphysical Multiversism), not exactly as an alternative theory (Theory Multiversism), but as a way of determining which, if any, sentences in the language of set theory (not the multiverse language of sets and worlds) are 'meaningless', pose 'pseudo-questions'. How he goes about this and what conclusions he draws are quite subtle matters that go well beyond the scope of this reply.[10] Still, I hope these three rough categories will help illuminate the debate between Ternullo and me. As this is an intramural debate between naturalists, we're focused primarily on multiversisms of the Theory and Heuristic varieties.

II. Naturalistic Concerns About Multiversism

In the paper Ternullo is discussing, I raise a number of questions about multiverse theories as potential alternatives to ZFC and its extensions as our basic foundational theory. The most fundamental of these is that a foundational theory, as we now understand it, has to be a theory, has to be an explicit set of axioms capable of

[7]See [2011], pp. 131–137.

[8]Something like this actually happened when the intuitive picture of sets as extensions of properties fell out of favor in light of its conflict with the extremely fruitful axiom of choice.

[9]Ternullo contrasts 'heuristic' with 'instrumental' (2019, p. 53). See §III below.

[10]Toby Meadows and I hope to clarify some of these matters in 'A philosophical reconstruction of Steel's multiverse', in preparation. I also neglect the hyperuniverse program, simply because I don't understand it well enough to comment.

doing the foundational jobs. Of the multiverse accounts on offer, only Steel's comes with a set of axioms, a fully explicit first-order theory of sets and worlds, but as noted, his goal is to evaluate the sentences of ordinary set theory, not to replace them with something different. So, flat-footed as it sounds, the general lack of an explicit multiverse theory strikes me as a serious obstacle to a new and different multiverse foundation.

Hamkins's stand on the foundational status of the multiverse was quoted above:

> We expect to find all our familiar mathematical objects ... inside any one of the universes of the multiverse. (Hamkins 2012, p. 419)

Roughly speaking, it seems any world of the multiverse can serve as our Generous Arena, and ZFC (satisfied by that world) as our Shared Standard. Presumably Risk Assessment is to be carried out in a world with large cardinals, that is, in ZFC+LCs. There's some question about Meta-mathematical Corral: if we only care about corralling a generous arena, we're once again thrown back on ZFC and its extensions; if we want to corral all of mathematics, it seems we'd need a theory of our multiverse, which we've seen Hamkins's doesn't provide. On Steel's view, ZFC+LCs turns up in the meaningful part of set-theoretic language and continues to carry out its usual foundational functions. For the most part, then, ZFC and its extensions retain their foundational roles – in that respect, no alternative is actually on offer. So it's hard to see a case for replacing a universe view with multiverse view for foundational purposes.

But this isn't the end of the story. Some version of the multiverse perspective may have such attractive mathematical features that we're moved to adopt it even if a familiar theory like ZFC remains our official foundation. Ternullo mounts a case along these lines.

III. Ternullo's Defense

One striking turn in Ternullo's discussion is his characterization of Zermelo's famous 'On boundary numbers and domains of sets' (Zermelo 1930) as 'the first description of a multiverse' (Ternullo 2019, p. 47).[11] If this were so, it would go a long way toward showing that multiversism has important and far-reaching mathematical consequences! Working in a strong implicit meta-theory, Zermelo presents an analysis of 'normal domains' characterized by second-order ZFC minus Infinity[12]: their 'boundary numbers' are inaccessible cardinals; they can be

[11] To be clear, Ternullo isn't claiming that the historical Zermelo understood his work in multiverse terms (see Ternullo 2019, p. 47, footnote 8). He holds, rather, that Zermelo is a 'height potentialist' and that this position can be seen as a kind of 'height multiversism' (see Footnote 13 below).

[12] He leaves out the axiom of infinity to allow for a 'finitary' normal domain acceptable to intuitionists (so for him ω is a boundary number). As he sees it, a generous store of normal domains makes set theory adaptable for a wide range of applications.

decomposed into ranks up to that number (this is touted as one of the extrinsic benefits of Foundation); any two with the same boundary number are isomorphic; for any two with different boundary numbers, one is an initial segment of the other. The question then arises: are there any normal domains, are there any boundary numbers? Zermelo mounts an argument in the meta-theory that for any ordinal α, there's a corresponding boundary number κ_α; in modern terms, he's argued for the Axiom of Inaccessibles:

> We must postulate the *existence of an unlimited sequence of boundary numbers* as a new axiom for the 'meta-theory of sets'. (Zermelo 1930, p. 429)

Though Zermelo does take second-order 'ZFC-Infinity' to characterize each of an unending series of normal domains, I see no evidence that he intends his second-order 'ZFC + a proper class of inaccessibles' in the meta-theory as anything other than a description of the single universe in which all these normal domains reside.[13] If including an axiom of inaccessibles is enough to qualify a list of axioms as a multiverse theory, then almost all set theorists these days are multiversists; this sets the bar far too low, renders the term useless. So it seems that Zermelo is best left out of this discussion.

Ternullo is on stronger ground when he extends the appeal to mathematical consequences into contemporary set theory. I'm happy to grant that, for example, Hamkins's multiverse thinking has led to a fruitful investigation of 'set-theoretic geology' or that Steel's approach has focused attention on important questions about the 'core'. Cases like these display a clear heuristic benefit to thinking in terms of an intuitive multiverse picture – on this Ternullo and I agree – but he goes on to insist that these benefits aren't purely heuristic, that they are actually 'instrumental'. He draws this distinction from a question raised in my paper: can all the welcome mathematics inspired by multiverse thinking be carried out in our familiar universe theory, that is, are these all theorems of ZFC and its extensions? If the answer to this question is no, then multiverse thinking would be more than merely heuristic – fully instrumental, in Ternullo's terms – but as far as I can tell, the answer is yes, which Ternullo seems to acknowledge:

> It has been made clear that there may be no specific task that the universist may not try to successfully emulate within their single V. However, it is, in my view, rather apparent that the multiverse enormously facilitates fundamental practical tasks. (Ternullo 2019, p. 66)

This is just to say that multiverse thinking is of great (purely) heuristic value.

Notice that we have here instances of Heuristic Multiversism different from what was envisioned in §I: the multiverse picture isn't being used to inspire new axioms toward a version of Theory Multiversism, but to inspire new mathematics, new

[13]Many observers see Zermelo as a potentialist, but I have my doubts. Though everyone, actualist and potentialist alike, uses a familiar range of metaphors – the universe is unending, etc. – it seems to me that the cash value of potentialism is the rejection of quantification over all sets. But this is exactly what Zermelo seems to do, for example, in arguing that there's an inaccessible for every ordinal.

concepts and methods, within our existing theory of ZFC and its extensions. And there's another potential contribution of Heuristic Multiversism, as well. Hamkins observes that

> There is no reason to consider all universes in the multiverse equally, and we may simply be more interested in parts of the multiverse consisting of universes satisfying very strong theories, such as ZFC plus large cardinals. (Hamkins 2012, p. 436)

Now the process of narrowing down to a restricted range of worlds may well be functionally equivalent to the process of adding new axioms to ZFC, so what's of interest here is the suggestion that thinking in multiverse terms could bring new and different considerations to bear on that process. In other words, multiverse thinking might help us to refine our official theory of sets. In fact, this may be the ultimate upshot of Steel's line of thought: a stretch of multiverse thinking leads him to propose a new axiom for ordinary set theory.

In sum, then, our naturalist has no straightforward form of Theory Multiversism, only Steel's set of axioms with a different motivation, but there seems to be ample room for Heuristic Multiversism to do significant mathematical work in a number of different ways. We can draw two morals. The first is that ZFC and its extensions aren't uniquely tied to the intuitive universe picture of the iterative conception. They could be thought of, instead, as the shared theory of a range of worlds in the multiverse, so that what the universist sees as adding new axioms about V, could instead be seen as a narrowing of the range of worlds we take to be of interest. The second moral is one that should appeal to Ternullo's naturalism: since these intuitive pictures, universist and multiversist, are playing a merely heuristic role, there's no reason at all not to exploit them both, no reason at all not to switch back and forth depending on which is more suggestive in a given context. In the end, set theorists should feel entirely free to think in *any* intuitive terms that can lead them to good mathematics!

References to Maddy On The Multiverse

Antos, C., Friedman, S.-D., Honzik, R., & Ternullo, C. (2015). Multiverse conceptions in set theory. *Synthese, 192*(8), 2463–2488.

Arrigoni, T., & Friedman, S. (2013). The hyperuniverse program. *Bulletin of Symbolic Logic, 19*(1), 77–96.

Bagaria, J., Castells, N., & Larson, P. (2006). An Omega-logic primer. In J. Bagaria & S. Todorcevic (Eds.), *Set theory* (Trends in mathematics, pp. 1–28). Basel: Birkhäuser.

Balaguer, M. (1995). A platonist epistemology. *Synthese, 103*, 303–25.

Balaguer, M. (1998). *Platonism and anti-platonism in mathematics*. Oxford: Oxford University Press.

Boolos, G. (1971). The iterative conception of set. *Journal of Philosophy, 68*(8), 215–231.

Buzaglo, M. (2002). *The logic of concept expansion*. Cambridge: Cambridge University Press.

Ferreirós, J. (2010). *Labyrinth of thought. A history of set theory and its role in modern mathematics*. Basel: Birkhäuser.

Field, H. (2001). *Truth and absence of fact*. Oxford: Oxford University Press.

Friedman, S. (2016). Evidence for set-theoretic truth and the hyperuniverse programme. *IfCoLog Journal of Logics and their Applications, 3*(4), 517–555.
Gödel, K. (1947). What is Cantor's continuum problem? *American Mathematical Monthly, 54*, 515–525.
Hamkins, J. D. (2012). The set-theoretic multiverse. *Review of Symbolic Logic, 5*(3), 416–449.
Koellner, P. (2009). On reflection principles. *Annals of Pure and Applied Logic, 157*(2–3), 206–219.
Koellner, P. (2014). Large cardinals and determinacy. In E. N. Zalta (Ed.), *The Stanford encyclopedia of philosophy* (Spring 2014 Edition). http://plato.stanford.edu/archives/spr2014/entries/large-cardinals-determinacy/
Lakatos, I. (1976). *Proofs and refutations*. Cambridge: Cambridge University Press.
Linnebo, Ø. (2013). The potential hierarchy of sets. *Review of Symbolic Logic, 6*(2), 205–228.
Maddy, P. (1996). Set-theoretic naturalism. *Bulletin of Symbolic Logic, 61*(2), 490–514.
Maddy, P. (1997). *Naturalism in mathematics*. Oxford: Oxford University Press.
Maddy, P. (2007). *Second philosophy: A naturalistic method*. Oxford: Oxford University Press.
Maddy, P. (2011). *Defending the axioms*. Oxford: Oxford University Press.
Maddy, P. (2017). Set-theoretic foundations. In A. Caicedo, J. Cummings, P. Koellner, & P. B. Larson (Eds.), *Foundations of mathematics. Essays in Honor of W. Hugh Woodin's 60th Birthday* (Contemporary mathematics, Vol. 690, pp. 289–322). Providence: American Mathematical Society.
Martin, D. (2001). Multiple universes of sets and indeterminate truth values. *Topoi, 20*, 5–16.
Paseau, A. (2016). Naturalism in the philosophy of mathematics. In M. Leng (Ed.), *Stanford encyclopedia of philosophy*. New York: Oxford University Press.
Shapiro, S. (1991). *Foundations without foundationalism. A case for second-order logic*. Oxford: Oxford University Press.
Shelah, S. (2003). Logical dreams. *Bulletin of the American Mathematical Society, 40*(2), 203–228.
Steel, J. (2014). Gödel's program. In J. Kennedy (Ed.), *Interpreting Gödel. Critical essays* (pp. 153–179). Cambridge: Cambridge University Press.
Ternullo, C., & Friedman, S.-D. (2016). The search for new axioms in the hyperuniverse programme. In F. Boccuni & A. Sereni (Eds.), *Philosophy of mathematics: Objectivity, realism and proof. Filmat studies in the philosophy of mathematics* (Boston studies in philosophy of science, pp. 165–188). Berlin: Springer.
Väänänen, J. (2014). Multiverse set theory and absolutely undecidable propositions. In J. Kennedy (Ed.), *Interpreting Gödel. Critical essays* (pp. 180–205). Cambridge: Cambridge University Press.
Viale, M. (2016). Category forcings, MM^{+++} and generic absoluteness for the theory of strong forcing axioms. *Journal of the American Mathematical Society, 29*(3), 675–728.
Wang, H. (1974). *From mathematics to philosophy*. London: Routledge & Kegan Paul.
Woodin, W. H. (1999). *The axiom of determinacy, forcing axioms and the non-stationary ideal*. Berlin: De Gruyter.
Woodin, W. H. (2001). The continuum hypothesis. *Notices of the American Mathematical Society, Part 1: 48*(6), 567–576; *Part 2: 48*(7), 681–690.
Woodin, W. H. (2011). The realm of the infinite. In W. H. Woodin & M. Heller (Eds.), *Infinity. New research frontiers* (pp. 89–118). Cambridge: Cambridge University Press.
Zermelo, E. (1908). Investigations in the foundations of set theory. In J. van Heijenoort (Ed.), *From frege to Gödel. A source book in mathematical logic, 1879–1931* (pp. 199–215). Harvard: Harvard University Press.
Zermelo, E. (1930). Über Grenzzahlen und Mengenbereiche: neue Untersuchungen über die Grundlagen der Mengenlehre. *Fundamenta Mathematicae, 16*, 29–47.

References to Reply to Ternullo on the Multiverse

Hamkins, J. (2012). The set-theoretic multiverse. *Review of Symbolic Logic, 5*(3), 416–449.

Koellner, P. (2013). Hamkins on the multiverse. EFI paper. http://logic.harvard.edu/efi.php#multimedia

Koellner, P. (2014). Large cardinals and determinacy. In E. N. Zalta (Ed.), *The Stanford encyclopedia of philosophy* (Spring 2014 Edition). https://plato.stanford.edu/archives/spr2014/entries/large-cardinals-determinacy/

Maddy, P. (2011). *Defending the axioms*. Oxford: Oxford University Press.

Maddy, P. (2017). Set-theoretic foundations. In: A. Caicedo, J. Cummings, P. Koellner, & P. Larson (Eds.), *Foundations of mathematics: Essays in honor of W. Hugh Woodin's 60th birthday* (pp. 289–322). Providence: AMS.

Steel, J. (2014). Gödel's program. In J. Kennedy (Ed.), *Interpreting gödel* (pp. 153–179). Cambridge: Cambridge University Press.

Ternullo, C. (2019) Maddy on the multiverse, this volume.

Zermelo, E. (1930). On boundary numbers and domains of sets. In: H.-D. Ebbinghaus, C. Fraser, & A. Kanamori (Eds.), *Ernst Zermelo: Collected Works* (Vol. I, 2010, pp. 400–431). Heidelberg: Springer.

Chapter 4
Proving Theorems from Reflection

Philip D. Welch

Abstract We review some fundamental questions concerning the real line of mathematical analysis, which, like the Continuum Hypothesis, are also independent of the axioms of set theory, but are of a less 'problematic' nature, as they can be solved by adopting the right axiomatic framework. We contend that any foundations for mathematics should be able to simply formulate such questions as well as to raise at least the theoretical hope for their resolution.

The usual procedure in set theory (as a foundation) is to add so-called strong axioms of infinity to the standard axioms of Zermelo-Fraenkel, but then the question of their justification becomes to some people vexing. We show how the adoption of a view of the universe of sets with classes, together with certain kinds of *Global Reflection Principles* resolves some of these issues.

4.1 Introduction

This essay falls into two distinct parts. We first look at some long-standing questions in mathematical analysis, from the Russian and French schools of the early twentieth century, and how they have, or have not, been answered since. Our purpose here is two-fold: to step away from the eternal recurrence of Cantor's Continuum Problem in debates of this kind, which is a question in third order number theory, to give examples in second order number theory, or what logicians would also call plain 'analysis'. Our second purpose is to here make the case that the questions considered are natural ones in the context of mathematical thought. Few mathematical analysts ever come across a problem where the continuum hypothesis, that $2^{\aleph_0} = \aleph_1$, is ever an important consideration, and they are aware of its independence from the other ZF axioms. Questions such as whether projections of co-analytic sets are Lebesgue measurable, for example, are much nearer their domains of interest. If a

P. D. Welch (✉)
School of Mathematics, University of Bristol, Bristol, England
e-mail: p.welch@bristol.ac.uk

S. Centrone et al. (eds.), *Reflections on the Foundations of Mathematics*,
Synthese Library 407, https://doi.org/10.1007/978-3-030-15655-8_4

mathematician wants to know whether such a set A, say is Lebesgue measurable, or has meagre symmetric difference from an open set, we cannot wish this question away by talking about a 'multiverse', or the dependence of its truth on some model of set theory obtained by forcing, or on some variant foundational theory or other: they want to know the answer.

Our not so hidden agenda then, is to make the point that any foundation of mathematics has to be able to both simply formulate these questions, since they are naturally occurring statements of mathematical significance, even to the extent of their being simply written, and moreover to give some succour at least to the possibility of their resolution.

The second part is rather different. One advance over the independence phenomena ushered in by Cohen, has been for set theorists to expand the axioms of Zermelo-Fraenkel set theory (ZF, or ZFC with the Axiom of Choice added) by so-called 'strong axioms of infinity' often phrased in terms of 'large cardinal' numbers (actually it is not their largeness, but the strong or exotic properties they bear, that yields their strength). The question of justification of the assumption of such axioms then looms large. (But perhaps it is only a larger worry for the foundationalist than for the mathematician: when Andrew Wiles was asked whether it would bother him if the unbounded class of Grothendieck universes (and hence a proper class of inaccessible cardinals), that *prima facie* had been invoked for his proof of Fermat's Last Theorem, turned out to be necessary for his argument, his reaction was a metaphorical shrug when not a literal one: not in the slightest. It was neither here nor there; in short he had a proof. The point of the story is that the mathematics was already convincing.)

We give a straightforward account of much of this that is familiar to set theorists, but perhaps not elsewhere, and in the second part (Sect. 4.4) we deal with a recent proposal that notions of 'reflection' on the universe of sets instituted by early researchers such as Ackermann and Bernays, and warmly endorsed by Gödel, can be expanded in ways to demonstrate the existence of such large cardinals that solve the problems we give in the first part.

We should like to emphasise that our contribution in the first part is limited only to exposition and is indebted, amongst others, to Woodin (2001a,b) and to any general history of descriptive set theory. The reader will find the descriptive set theory they need in Moschovakis (2009).

We should like to warmly thank the referee who saved us from more than one embarrassing infelicity.

4.2 The Task to Hand

We look at some problems in the *projective hierarchy* of Luzin. However first we give Borel's hierarchy. In the following a 'Polish space' is a separable, complete metrisable space. This includes the common examples of the reals $\mathbb{R}$ with the usual Euclidean metric, Baire space $\mathbb{N}^{\mathbb{N}}$, and Cantor space $2^{\mathbb{N}}$ with metrics derived from the standard product topologies.

Definition 4.2.1 (Lebesgue (1905) Borel Sets) Let T be a Polish space; let B_0 be the class of closed sets in T;

Let $B_\eta = \{\bigcup_{n\in\mathbb{N}} A_n \mid \neg A_n$ in some B_{η_n} for an $\eta_n < \eta\}$.
Let $\mathcal{B} = \bigcup_{\eta<\omega_1} B_\eta$.

Implicit in the definition above is that we perform complementation and union throughout all the countable ordinals, and the process finishes at stage ω_1 – the first uncountable cardinal: nothing further would result from continuing further. This analysis of a certain sequence of easily described sets into a hierarchy is a step in so-called 'descriptive set theory' that seeks to analyse the real line (or other nearby Polish space examples) in terms of a hierarchy of increasing complexity.

Definition 4.2.2 (Suslin 1917: Analytic Sets) Let T be a Polish space; let $\mathcal{B}$ be the class of Borel sets in $T \times T$; let

$$\mathcal{A} =_{df} \{A \mid \exists C \in \mathcal{B}(A = proj(C))\}$$

where $proj(C) = \{x \in T \mid \exists y \in T : C(x, y)\}$. Then $\mathcal{A}$ is called the class of *analytic* sets.

Theorem 4.2.1 (Suslin 1917) *Borel* $=$ $\mathcal{A}$ *& co-*$\mathcal{A}$*, that is the Borel sets in a space are precisely those analytic sets with analytic complement.*

So here we have written *co-*$\mathcal{A}$ for the class of sets whose complement is in $\mathcal{A}$ (and similarly will do so below "*co-S*" for other classes S). Descriptive set theorists would call the class of Borel sets the 'self-dual' class of $\mathcal{A}$. The study of the projective hierarchy was initiated by the discovery of Suslin (a student at the time) that Lebesgue had erred in assuming the projection of a Borel set was Borel. It was not. Indeed there was a hierarchy of sets to be investigated obtained by projection and complementation:

Definition 4.2.3 (Luzin 1925, Sierpiński 1925: The Projective Sets) Let T be a Polish space.

$$S_1 = \mathcal{A} \subseteq T^k \text{(in any dimension)};\ S_{n+1} = \{proj(D) \mid D \subseteq T^k \times T, D \in co\text{-}S_n\};\ PROJ = \bigcup_n S_n.$$

Lebesgue studied these and showed that they formed a proper hierarchy of increasing complexity as n increased. Sierpinsky later showed they were closed under countable unions and intersections. It is important to realise that these are the *definable* sets in analysis: the operations of projection and complementation in the above definition, correspond when written out even in informal notation to an existential quantification over the elements of T and to negation. With T equalling $\mathbb{R}$, this means that any definition of a set of reals the analyst writes down will fall inside the class $PROJ$.

The following intimates that the projective sets might be very *regular*: in this case that they can always be assigned a meaningful, length, area, volume...

Theorem 4.2.2 (Suslin 1917) *Any $D \in \mathcal{A}$ is Lebesgue measurable.*

However there the matter lay stuck. Attempts to ascend the projective hierarchy and establish, for example the Lebesgue property conspicuously failed.

Problem 5 (Lebesgue Measurability) *Are the sets in $PROJ$ Lebesgue measurable?*

It seemed intractable:

> (Luzin – 1925) One does not know and one will never know whether the projective sets are Lebesgue measurable.

4.2.1 The Baire and Perfect Subset Properties

A set U is said to have the *Baire Property (BP)* if it has meagre symmetric difference with some open set. (In turn a set is *meagre* if it is the countable union of nowhere dense sets. In some sense it is 'negligible'.) It was known (Lusin and Sierpiński 1923) that analytic sets (and so also their complements) had the Baire property.

Problem 6 (Property of Baire) *Do sets in $PROJ$ have the property of Baire (BP)?*

A perfect set is one which is closed but contains no isolated points. Since a perfect set has size the continuum, Cantor's continuum problem is settled for such sets.

Problem 7 (Perfect subset property (PSP)) *Does every uncountable set in $PROJ$ contain a perfect set?*

It was known (due to Suslin) that every uncountable analytic set contained a perfect subset. (This may fail for co-analytic sets, for example in the Gödel constructible hierarchy.)

4.2.2 Uniformisation

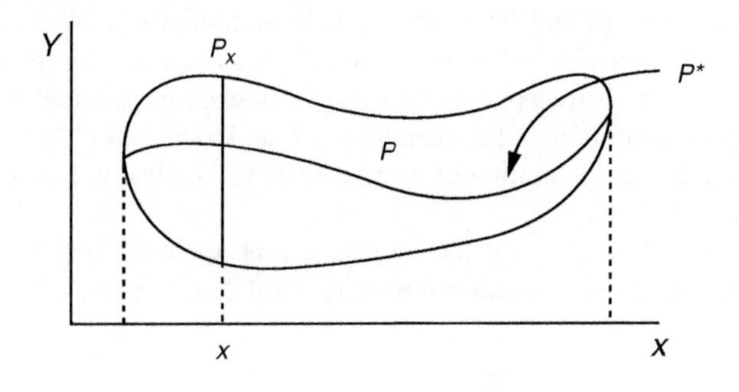

A function $P^* \subseteq P \subseteq \mathcal{X} \times \mathcal{Y}$ *uniformizes* P if

$$\forall x[\exists y(x, y) \in P \rightarrow \exists y' (P^*(x) = y' \wedge (x, y') \in P)].$$

A function P^* is *projective* if its graph is.

Problem 8 (Uniformization Property (*Unif*)) *Does every set P in $T \times T$ in $PROJ$ have a projective uniformizer? To abbreviate: $Unif(PROJ)$?*

For co-analytic sets a classical theorem yields that there is always a projective uniformizing function moreover one of the same complexity.

Theorem 4.2.3 (Novikov-Kondō 1937) *Every co-analytic subset of the plane has a co-analytic uniformizer.*

The above properties of the projective sets are called the *regularity properties*. For good measure we add one more.

4.2.3 The Banach-Tarski Property

Problem 9 (Banach-Tarski Problem) *Is there a paradoxical decomposition of the sphere in R^n into projective pieces?*

The original Banach-Tarski theorem states that it is possible to decompose a sphere into finitely many pieces and reassemble the pieces to form two spheres identical to the first. In fact 5 pieces are enough, but they cannot be Lebesgue measurable. Could there be then such a decomposition where the pieces are projective, that is definable in analysis? (See Wagon 1994.)

Discussion: to summarise, using some obvious abbreviations, we have a list of 5 Problems.

P1: $LM(PROJ)$
P2: $BP(PROJ)$
P3: $PSP(PROJ)$
P4: $Unif(PROJ)$
P5: *Banach Tarski with projective pieces.*

Each of these questions deals with subject matter that is familiar to mathematical analysts in the twenty-first century, and has been so since the early 20th. One of my points in introducing these is to make clear that such questions are themselves clear. The Continuum Problem is usually wheeled out to serve as a stalking horse for the difficulties of a realist view of set theory, or at least of the real continuum, that an author wishes to introduce. However in logical terms the Continuum Problem is a problem in *third order* number theory: one must use an existential quantifier ranging over subsets of the real line. The problems above are expressible without requiring such quantifications to take place, they are expressible in *second order number theory* or commonly called *analysis*: the quantifiers range over sets of numbers, or over functions from $\mathbb{N}$ to $\mathbb{N}$ and the complexity, that is the number of quantifier alternations, is the number of the rank of the set in the Lusin projective hierarchy being discussed (roughly speaking). The Real Continuum is often spoken of (cf. Feferman 2012) as having potentially an "indeterminate nature" since its third order statement relies upon the supposed mysteries of the power set operation when applied to an infinite set. "What is the cardinality of $\mathcal{P}(\omega)$?" The problems above are however of a more concrete nature. Analysts rarely come across questions that turn upon the cardinality of the continuum. They come across questions about the Lebesgue measurability of definable sets, that is sets within $PROJ$, on a daily basis. And they commonly recognise analytic and co-analytic sets as being tractable, as they enjoy these regularity properties. Thus the Problems listed are concrete problems within, and stated within, mathematical analysis.

4.3 Difficulties

We shall see that notwithstanding the 'simpler' logical second order definition of the concepts involved in these problems, they are subject to the same independence phenomena as the third order Continuum Problem, and, as we shall see, for roughly the same reasons. (That is: on the one hand there is an *inner model* of the universe of sets in which the continuum hypothesis was true, namely Gödel's L – such inner models are transitive subclasses of V which are models of the ZFC axioms; and on the other hand there are techniques derived from Cohen's *forcing method* where he showed the consistency of the negation of CH with the other axioms. The same dichotomy appears below: on one hand appeal to the model L to get one answer, and forcing techniques the consistency of the other.)

Thus: the Regularity Properties can consistently fail:

Theorem 4.3.1 (Gödel) *If ZF is consistent, then so is ZFC+*"There is a projective set that is not LM".

Indeed there is a *projection of a co-analytic* ("PCA") set that fails to be LM. This gives a negative "answer" to P1. The reason being that in Gödel's universe of constructible sets, L, with which he showed the consistency of the axioms of ZF together with CH, there are non-Lebesgue measurable sets at roughly the level of the complexity of the wellordering of the universe of L that he also demonstrated existed. Recall that the construction of the Vitali non-measurable set uses a wellordering of the continuum. Thus one expects a failure of Lebesgue measurability at roughly the same level of complexity as the wellordering of the continuum we have in L, which is used to construct a Vitali counter-example.

This use of the wellorder of L also leads to the following propositions relating to the problems above. In L:

P2: there is a PCA-set without the Baire property BP;

P3: there is a co-analytic set that is uncountable with no perfect subset;
P5: there is a paradoxical decomposition of the unit sphere in $\mathbb{R}^3$ using PCA-pieces.

For P4 the matter is slightly more nuanced: For co-analytic sets in the plane or higher dimensions, we have seen by the Novikov-Kondo theorem that they are uniformisable by co-analytic functions. For higher levels the wellorder of L ensures that sets in the projective classes $PCA, PCPCA\ldots$ etc. are all uniformisable by functions in the same class. (And if those hold for a suitable class Γ it is a straightforward result that it must fail for their complements in co-Γ.) A more delicate question for the Uniformisation Problem is to ask that the uniformising function come from the *very same* class or level in the projective hierarchy as the set being uniformised. Then by Novikov-Kondo this holds for co-analytic sets; in L this also holds for PCA sets (and for further classes on the repeated projected side: $PCPCA\ldots$ etc. .)

Whereas Gödel's construction of L gives a canonical *inner model* of V – the universe of all sets of mathematical discourse, there are various constructions based on extensions of Cohen's *method of forcing* which allow one to conclude that *consistent with* the axioms of ZF is the possibility that various levels of the projective hierarchy can be all Lebesgue measurable, or enjoy the other regularity properties.

Indeed a renowned theorem of Solovay allows that *all* sets are Lebesgue measurable and have the Baire property:

Theorem 4.3.2 (Solovay 1964,1965,1970) *If the theory* $ZF+$ "There is an inaccessible cardinal" *is consistent, then so is the theory* $ZF+DC+$ "Every set is LM and has the BP".

The above is quite remarkable. The DC is 'Dependent Choice' that allows for an infinite *sequence* of choices in any given relation $R(v_1, v_0)$ to be made. This is usually – but not always – all that an analyst requires. (The full Axiom of Choice is paradoxically usually only invoked to guarantee the existence of pathological sets, i.e. difficult sets that are not LM or do not have the regularity properties etc.) The extra assumption beyond ZFC of the inaccessible cardinal was queried for many years as to its necessity. Eventually Shelah (1984) showed that it was needed for Lebesgue measurability of all sets but not for the Baire property (thus breaking what had seemed a tight link, that what was true for sets of one kind was true of the other.)

However, these are only consistency results, and do not tell us about the facts of the matter in V. Notwithstanding this, mathematicians might simply have shifted to a view that all sets that they could write down and specify were LM and BP and used DC with comfort. But they seemingly have not.

Solovay also showed that, even retaining the full axiom of choice all definable sets of reals have strong regularity properties:

Theorem 4.3.3 (Solovay 1964,1965,1970) *If the theory* $ZF+$ "There is an inaccessible cardinal" *is consistent, then so is the theory* $ZFC+$ "Every set projective set is LM and has the BP".

4.4 Resolution and Reflection

4.4.1 Resolution

Are there principles that are somehow missing from the ZFC axioms, and that could resolve these problems? For many set theorists the assumption that all sets are in Gödel's L – which indeed resolves these problems in a somewhat negative direction – is unpalatable. The iterative conception of sets appearing ever upwards in increasing ranks in a hierarchy built along the ordinals using the power set operation:

$$V_0 = \varnothing; \quad V_{\alpha+1} = P(V_\alpha); \quad Lim(\lambda) \to V_\lambda = \bigcup_{\alpha<\lambda} V_\alpha$$

has an entirely mathematical feel to it. The construction of L replaces the successor step with only allowing sets definable in first order logic rather than the full power set. However why should this purely logical construction deliver all the sets that there are, sets which arise from a purely mathematical set theoretical conception?

It is a well known part of this story that Gödel himself allowed for the possibility that *strong axioms* might settle questions such as CH. However then a discussion then ensues about the justification of these strong axioms. At such a length of time since Zermelo's formulations of the axioms for sets, and with the additions of Skolem, and Fraenkel, it seems inconceivable that any basic fact of sets has been overlooked in the ZFC system. Any supplementing axioms may have to have a different set of justifications or grounds for acceptance. It is usual at this point to talk about *intrinsic* grounds that follow from the iterative conception of set and the V hierarchy as outlined as above, or more widely 'set-structure' concerning the whole of $(V, \in)$. These are to be contrasted with *extrinsic* grounds where the consequences of these hypotheses are so rich and so compelling that we feel we should to adopt them.

There is much to be said (and has been) at this point but we shall pass over this. Our targeted aim is that certain viewpoints of the universe $(V, \in)$ encourage a view that 'large cardinals' or 'strong axioms of infinity' can be invoked by 'reflecting' on the universe. Solovay delivered a striking clue in an early theorem relying on the assumption of a measurable cardinal:

Theorem 4.4.1 (Solovay 1969) *ZF proves that if there is a $<$-κ-additive 2-valued measure on some set of cardinality $\kappa > \aleph_0$ then $BP(PCA)$, $LM(PCA)$, $PSP(PCA)$.*

These conclusions are then quite in contradiction to the picture given in Gödel's L. There is also a mystery as to why the existence of measures, or equivalently ultrafilters on fields of sets quite remote from $V_{\omega+1}$ (which contains all the real numbers, or elements of Baire space or ...) should affect properties down at this very modest rank.

As oblique as the idea at first appears the determinacy of *two person perfect information games* implies much about the regularity properties of the real continuum.

Let $A \subseteq \mathbb{N}^{\mathbb{N}}$ (or some $X^{\mathbb{N}}$). The game G_A is defined as follows:

$$I \text{ plays } k_0 \quad k_2 \quad \ldots k_{2n} \ldots$$
$$II \text{ plays } \quad k_1 \quad k_3 \quad \ldots k_{2n+1} \ldots$$

- I *wins* if and only if $x = (k_0, k_1, \ldots) \in A$.
- G_A is *determined* if either Player has a winning strategy in this game.

Let us write, for a class of set Γ "$Det(\Gamma)$" for the statement that for every set $A \in \Gamma$ that the game G_A is determined. "$Det(PROJ)$" is then read as "Projective Determinacy" or sometimes "Definable Determinacy".

Theorem 4.4.2 (Mycielski 1964, 1966) *$Det(PROJ)$ implies Regularity for the projective sets.*

Thus the Solovay theorem from a measurable cardinal, and the results from assuming Definable Determinacy were leading in the same direction. The following indicated that these matters were no coincidence;

Theorem 4.4.3 (Martin 1970) *ZF proves that if there is a $<$-κ-additive 2-valued measure on some set of cardinality $\kappa > \aleph_0$ then $Det(Analytic)$.*

This was earlier than the landmark theorem of Martin:

Theorem 4.4.4 (Martin 1975) *ZF proves $Det(Borel)$.*

(This remains the most quotable theorem in mathematics that *requires* ZF – as H. Friedman had previously shown (Friedman 1970) that ω_1-many iterations of the power set operation together with appropriate instances of Replacement would be required.)

However ZFC is just not strong enough to prove $Det(Analytic)$ on its own: this is because $Det(Analytic)$ can prove the consistency of ZFC. (And we cannot contradict Gödel's Incompleteness Theorems.) After much effort the prize was won:

Theorem 4.4.5 (Martin-Steel 1989) *If there are infinitely many Woodin cardinals then $Det(PROJ)$ and hence Regularity for the projective sets.*

Theorem 4.4.6 (Woodin 1988) *If there are infinitely many Woodin cardinals and a measurable cardinal above them, then in $L(\mathbb{R})$, the Gödel closure of $\mathbb{R}$ through all the ordinals, G_A is determined for* every *$A \subseteq \mathbb{R}$. And hence Regularity for all sets in $L(\mathbb{R})$.*

Thus "AD", the axiom that games based on any sets are determined, and which thus implies the regularity properties for all sets, is consistent with DC (as it holds in $L(\mathbb{R})$) but not the full AC. We could note also, that as strategies for such games can themselves be construed as sets of integers, or reals, that AD holding in $L(\mathbb{R})$, is equivalent to the statement that all games that are definable in $L(\mathbb{R})$ are determined (in V). We thus may prove outright the regularity properties from sufficient large cardinals. But we may be thought to have replaced a collection of problems by problems yet more problematic: how to justify the existence of such cardinals in the universe $(V, \in)$?

4.4.2 Reflection

To say that the universe of all sets is an unfinished totality does not mean objective undeterminateness, but merely a subjective inability to finish it.
Gödel, in (Wang: "A Logical Journey: From Gödel to Philosophy").

We take the view that the ordinals for example, indeed form a determinate concept: they are the class of sets that are transitive and wellordered by set membership. They form a proper class as Cantor and Burali-Forti (the latter eventually) recognised. We denote by On the totality of all ordinals.

Historically *reflection principles* are associated with attempts to formulate the idea that no one notion, idea, statement can capture our whole view of $V = \bigcup_{\alpha \in On} V_\alpha$. Such reflection principles are usually formulated in some language (first or higher order) as positing that sentences φ (when interpreted in the appropriate way over V) that hold in $\langle V, \in, \ldots \rangle$, must also hold in some $\langle V_\beta, \in, \ldots \rangle$. Let us call this *sentential* reflection. This is again a broad subject, and the reader is directed to Koellner's article 'Reflection Principles' (Koellner 2009) for a more in-depth discussion of the possible scope and limitations of reflection principles. Koellner argues that principles that may be deemed of an 'intrinsic nature' are unable to deliver any strong axioms that are inconsistent with a view that $V = L$, and so are not strong enough by themselves to prove outright anything about the real continuum beyond what we can already *in* L.

We first review some of the traditional sentential reflection principles.

(1) Montague-Levy: First order Reflection.
(R_0) : *For any* $\varphi(v_0, \ldots, v_n) \in \mathcal{L}_{\dot{\in}}$

$$ZF \vdash \forall\alpha\exists\beta > \alpha\forall\vec{x} \in V_\beta[\varphi(\vec{x}) \leftrightarrow \varphi(\vec{x})^{V_\beta}].$$

First order Reflection is actually provable in ZF. Indeed if we drop Infinity and the Replacement Scheme from ZF, the resulting theory, when augmented by (R_0), gives back Infinity and Replacement. It is a theorem scheme and thus a metatheorem: it is a theorem only with one φ at a time. However by formalising a Σ_n-Satisfaction predicate we have:

For each n, ZF $\vdash \exists C_n[C_n \subseteq On$ *is a c.u.b. class so that for any* $\varphi \in \mathrm{Fml}_{\Sigma_n}$:

$$\forall\beta \in C_n\forall\vec{x} \in V_\beta[\varphi(\vec{x}) \leftrightarrow \varphi(\vec{x})^{V_\beta}]].$$

Informally we write this as $\forall\beta \in C_n : (V_\beta, \in) \prec_{\Sigma_n} (V, \in)$.

(2) Levy, Bernays Reflection.

Suppose we allow some *second order* methods and allow *proper classes* to enter the picture more actively. If we allow reflection on classes then we can deliver some modest large cardinals. Let $\Phi(D)$ be the assertion that:

$$D \textit{ is a function from On to On, but } \forall \alpha\, D``\alpha \textit{ is bounded in On}".$$

By the Axiom of Replacement for any class D, we have then:

$$(V, \in, D) \models \Phi(D).$$

If we allow the assumption that $\forall D \Phi(D)$ *reflects* to some V_κ we shall have:

$$\forall D \subseteq V_\kappa\ (V_\kappa, \in, D) \models \Phi(D).$$

This implies that κ is a *strongly inaccessible cardinal.* However strongly inaccessible cardinals are strongly inaccessible in L, and are thus consistent with "$V = L$". The strict ZF-ist will eschew such an argument as it quantifies over all classes and not all such are necessarily definable over $(V, \in)$.

Whilst Levy (1960) remained at the level of discussing Reflection to obtain inaccessible cardinals, and inductively defined hierarchies of such principles relating to the much earlier cardinals of Paul Mahlo, Bernays (1961) allowed Φ above to be any Π^1_n formula about some parameter D. The resulting strengthened reflection principle now goes by the name of an *indescribability* property: any Π^1_n-property may be reflected downwards. Indeed there are Π^1_{n+1} sentences, that if reflected over $(V, \in)$ to some $(V_\kappa, \in)$ ensure that $(V_\kappa, \in)$ itself is Π^1_n-indescribable in the same sense.[1] The point to note here is that we have firmly entered the realm of second order entities: we must use these to realise the second order variables of our language, and moreover we must have a *domain of quantification* for the string of quantifiers in such a sentence to vary over. It is quite possible to consider third, fourth, n'th order languages over $(V, \in)$ and the associated reflection principles. But then such a layering of ranks of classes above V leads to the inevitable question as to why we do not declare such layers to be inhabited by sets.

We shall see that it is part of our viewpoint to avoid even second order methods wherever possible. We swallow the logical necessity of the existence of classes, as Cantor, Russell, and Burali-Forti showed, and admit of two types of objects: the mathematical realm of *sets* which constitute the universe of mathematical discourse $(V, \in)$; but we consider classes as just the *parts* of V (in a mereological fashion), which themselves may or may not be sets.

---

[1] We have not exactly delineated modern indescribability properties here, which usually are defined with an extra free predicate symbol, but we only wish to give the flavour of this. See Kanamori (2003), I.6, for a fuller discussion.

Gödel again:

> *All the principles for setting up the axioms of set theory should be reducible to Ackermann's principle: The Absolute is unknowable. The strength of this principle increases as we get stronger and stronger systems of set theory.* The other principles are only heuristic principles. *Hence, the central principle is the reflection principle, which presumably will be understood better as our experience increases.* Meanwhile, it helps to separate out more specific principles which either give some additional information or are not yet seen clearly to be derivable from the reflection principle as we understand it now.
>
> (Section 8.7.9 of Wang 1996).

(Our italics.)

> The Universe of sets cannot be uniquely characterized (i.e. distinguished from all of its initial segments) by any internal structural property of the membership relation in it, which is expressible in any logic of finite or transfinite type, including infinitary logics of any cardinal number.
>
> (Wang - *ibid.*)

> Generally I believe that, in the last analysis, every axiom of infinity should be derivable from the (extremely plausible) principle that V is indefinable, where definability is to be taken in [a] more and more generalized and idealized sense. (Wang, *ibid.*, p. 285)

Gödel is presumed to be happy with considering logics of higher types, and thus Bernays may not be overstepping that mark. But the reflection we have proposed below is not one of a logical character, meaning in a logic of higher types, or in an infinitary language; it is a structural reflection that takes the above 'unknowability' of the first quote above, or we prefer: 'ineffability', of the whole universe of sets, together with its parts, and reflects on that structure to bring it down to a set sized substructure.

Strengthening Reflection Principles

As we alluded to above, Koellner (2009) has outlined a heuristic argument that intrinsic justifications of reflection will never produce a justification for a large cardinal that cannot reside in L. The cardinals or principles produced will all be consistent with $V = L$ (and he argues that the small large cardinals they could conceivably justify are technically weaker than an ω-Erdős cardinal).

The Challenge then: To justify a set-theoretic *reflection principle* that will ensure the existence of large cardinals (or strong axioms of infinity) that are sufficient to deliver the hypotheses needed for modern set theoretical principles.

We first mention some recent attempts at strengthening reflection. Notwithstanding Bernays' higher order reflection of sentences, Reinhardt pointed out that for formulae with third order parameters, the reflection scheme was inconsistent (Reinhardt 1974). Tait (2005) attempted to provide some relief from this by placing restrictions on the substitutions possible and defined syntactic classes of higher order formulae with parameters on which one could nevertheless reflect, and showed the consistency of some of these principles from a measurable cardinal, and left the consistency of others open. Koellner (2009) showed the latter inconsistent but proved the consistency of the former from a ω-Erdős cardinal, which we do not define here, but is a cardinal again that is consistent with $V = L$. He further gave a heuristic argument that any reflection principle that was based on the intrinsic iterative hierarchy of sets, and which include such sentential reflections, are all limited in their outcomes, would have to be intraconstructible as its consistency would be derivable from such an ω-Erdős cardinal. Such cardinals then would never be strong enough to prove outright anything about the real continuum beyond what we can *in* L.

To summarise we thus have:

- The Reflection Principles to date are all consistent with a view of the universe as being L the constructible universe: they are *intra-constructible.*
- However these are all motivated on a syntactic level.

The moral is thus: We need stronger Reflection Principles: those that generalise Montague-Levy are not up to the task of providing any justification for the large cardinals needed for modern set theory.

We have proposed (Welch and Horsten 2016, Welch 2012) a *Global Reflection Principle* to overcome this intraconstructibility limitation. This principle has as its inspiration the properties related to those of a subcompact cardinal, and is more of the nature of *structural reflection* rather than sentential or linguistic.

It is quite legitimate to ask first why do we need stronger reflection principles? Here is one very good reason.

Theorem 4.4.7 (Woodin) *Suppose there is a proper class of Woodin cardinals. Then $Th(L(\mathbb{R}))$ is immune to change by set forcing.*

The import of the theorem is that the perhaps baleful effects of Cohen's forcing method can have no effect on the fact of the matter as to which sentences are true in analysis, or indeed of any statement about the reals and ordinals: in $L(\mathbb{R})$ every object is definable from such, and as such it encompasses analysis, the projective hierarchy, and way beyond, using iterated definability through On. This is thus a strong *absoluteness* result. Whilst Woodin cardinals have a somewhat tricky definition (which is why we have not defined them here) it turns out that this notion is absolutely central to proving the consistency of many concepts in modern set theory. The supposition of their existence, or indeed that they are unbounded in the ordinals, is now ubiquitous in current theorems of set theory.

We shall therefore intend to define such a *Global Reflection Principle* (GRP) which will deliver an unbounded class of such large cardinals. We take an almost naive Cantorian stance, and consider the *absolute infinities* that he identified at that time: the absolute infinity of On the ordinals, $Card$ the class of cardinals, V itself etc. , and we collect these (without as yet being too precise as to what this means), into a family C, but our viewpoint will be that C is the collection of the mereological parts of V. Some of these will be set-sized and we simplify matters by simply identifying them with the corresponding sets. Those parts that are not sized are the interesting entities in C, and we think of these as the proper classes. We then consider a *structural reflection* of the whole universe $(V, \in, C)$ together with its parts to a small structure.

4.4.2.1 Global Reflection Principle: GRP

We take a small (meaning *set-sized*) substructure of $(V, \in, C)$, the universe with all of its parts, C, and ask that this is then isomorphic to a small part of V: namely some V_α together with all of its parts. The 'parts' of V_α are naturally those $D \subseteq V_\alpha$, that is elements of $V_{\alpha+1}$. The language $\mathcal{L}$ in which we wish to state the principle's reflection properties is the usual first order language for set theory, but augmented with predicate variables $X_0, X_1, X_2, \ldots$ that will vary over the collection C. It is important to note that there are no second order quantifers over these variables. We thus avoid having an explicitly demarked domain of quantification for the second order objects. We thus write Σ^0_ω for the class of formulae of $\mathcal{L}$. In the second line of the next definition the first structure is thus 'Σ^0_ω'-elementary in the second, meaning as usual that such formulae with substitutions for set variables from X and for predicate variables from C' have the same truth value in both structures.

Definition 4.4.1 (Global Reflection Principle – GRP) There is a set $X \subseteq V$ and a collection $C' \subseteq C$ with:

$$(X, \in, C') \prec_{\Sigma^0_\omega} (V, \in, C)$$

and:

$$(X, \in) = (V_\alpha, \in)$$

for some $\alpha \in On$, and so that

$$V_{\alpha+1} = \{D \cap V_\alpha \mid D \in C'\}.$$

This can be summarised as we have a transitivising isomorphism π so that

$$\pi : (X, \in, C') \longrightarrow (V_\alpha, \in, V_{\alpha+1})$$

with π the identity on X.

Hence we have

$$(V, \in, C) \text{ is reflected down to } (V_\alpha, \in, V_{\alpha+1})$$

Or to put it another way, we are thus requiring that there is set-sized simulacrum of $(V, \in, C)$ that is of the form $(V_\alpha, \in, V_{\alpha+1})$.

Indeed the inverse of π yields an *elementary embedding* in an equivalent formulation that is perhaps more congenial to set theorists:

There is an initial segment of the universe V_α, *and a nontrivial elementary embedding*

$$\pi^{-1} : (V_\alpha, \in, V_{\alpha+1}) \longrightarrow_{\Sigma^0_\omega} (V, \in, \mathcal{C})$$

with critical point α *(i.e.,* $\pi^{-1}(\alpha) > \alpha$ *whereas for* $z \in V_\alpha$, π^{-1} *is the identity:* $\pi^{-1}(z) = z$*).*

Thus all that π^{-1} does is move, or stretch, objects from $V_{\alpha+1}$ to objects in C. Equivalently, as there are no 'points' above V_α in X, π's collapsing action on any $X \in C$ satisfies $\pi(X) = X \cap V_\alpha$. Thus, for example the part of V which is the class of ordinals, On, is in C, and $\pi^{-1}(\alpha) = On$. (Where α here is considered a class over V_α and as an element of $V_{\alpha+1}$.) We thus have

$$\varphi(\vec{x}, \vec{X})^{(V_\alpha, \in, V_{\alpha+1})} \leftrightarrow \varphi(\pi^{-1}(\vec{x}), \pi^{-1}(\vec{X}))^{(V, \in, C)} \leftrightarrow \varphi(\vec{x}, \pi^{-1}(\vec{X}))^{(V, \in, C)}.$$

Why GRP? Define a field of classes U on $\mathcal{P}(\alpha)$ by

$$X \in U \leftrightarrow \alpha \in \pi^{-1}(X).$$

As $\mathcal{P}(\alpha) \subseteq V_{\alpha+1} \subseteq \mathrm{dom}(\pi^{-1})$ by Σ^0_1-elementarity (in π^{-1}), this is an ultrafilter. Standard arguments show that U is a normal measure on α, and thus α is a measurable cardinal. But then:

$\forall \beta < \alpha \, \langle V, \in \rangle \models$"$\exists \alpha > \beta(\alpha$ a measurable cardinal)" $\Rightarrow$
$\Rightarrow \langle V_\alpha, \in \rangle \models$" $\forall \beta \exists \lambda > \beta(\lambda$ a measurable cardinal)" $\Rightarrow$
$\Rightarrow \langle V, \in \rangle \models$"*There are unboundedly many measurable cardinals*".

It is an exercise in the appropriate definitions to show that the critical point α is also a Woodin (indeed a Shelah) cardinal. So we thus have:

Theorem 4.4.8 (GRP) $(V, \in) \models \forall \alpha \exists \lambda > \alpha(\lambda$ *a measurable Woodin cardinal).*

Corollary 4.4.9 *By the results of Martin-Steel and Woodin mentioned above,* GRP *then implies:*

- *(a) Projective Determinacy* $Det(PROJ)$ *and* $(AD)^{L(\mathbb{R})}$.
- *(b) (Woodin)* $\mathrm{Th}(L(\mathbb{R}))$ *is fixed: no set forcing notion can change* $\mathrm{Th}(L(\mathbb{R}))$, *and in particular the truth value of any sentence about reals in the language of analysis, thereby including* $Det(PROJ)$.

4.5 Discussion

There is a discussion to be had as to whether we have here a genuine reflection principle. As we have intimated, the nature of the reflection that is occurring is logical in as much as it relies on passing from the whole universe V (with its parts) to a substructure that preserves a certain amount of logical elementarity, namely Σ^0_ω-elementarity. But it is structural in that it requires the substructure to be isomorphic to an initial segment of the universe V_α together with *all* of V_α's classes, that is $V_{\alpha+1}$. It is quite possible to posit weaker reflection principles where the second order domain of π^{-1} is only a proper subset of $V_{\alpha+1}$. It might for example only include $P(\alpha)^L$ for example. Whereas this would be enough to deduce the existence of $0^\sharp$, that is the existence of a non-trivial embedding $\pi^{-1} : L \longrightarrow L$, we could not define the measure on $P(\alpha)$ as we did above.

We could view GRP as the natural limit of a series of principles where we demanded more and more classes of V_α to be in the image of π (whilst perhaps allowing α to vary to achieve this). Thus larger and larger inner models M would have M-measures defined on their $P(\alpha)^M$ as we just saw for L yielding $0^\sharp$ above.

We did not quantify over the collection C in any fashion. All we required of it was that it contained sufficient elements to allow the definition of the GRP. This allowed us to be somewhat vague as to what the collection C was. The status of C is discussed in Welch (2012) and Welch and Horsten (2016). There we discuss various approaches as to how to regard C, for example through the manœuvre of considering plurals and plural quantification. However we reject this in favour of a mereological approach. This fulfils the need to find a way to sufficiently distinguish sets from classes (see Maddy 1983 for a discussion on this). The argument often deployed against considering higher order types over the top of V is that in such a case 'there was no reason to stop building V at the level On' (or some such). By thinking of the ordinals as a determinate concept, we have a class of sets priorly given in a mathematical manner. In this viewpoint there are no 'ordinals beyond V': V is the universe of all mathematical objects, and ordinals are mathematical objects. Likewise there is a mathematical power set operation $P(x)$, but a 'power-class operation' acting on the parts of V would be something else altogether different, and would not be considered a mathematical operation.

We did not (yet) formalise GRP in any class theory. We think of the development of our intuitions concerning V and its parts as taking place in a pre-formalised state: these include our intuitions concerning the ineffability of $(V, \in, C)$, and are not yet formalised. We think simply as mathematicians do about the semantics, or structure of our concepts.

One could then proceed to formalise the statement of GRP in NBG – class theory. The assertion of the existence of such a π appears *prima facie* to be third order, but with usual coding tricks, in fact it can be regarded as an assertion that a certain kind of class exists and thus is a Σ^1_1 statement. It is not hard to come up with strengthenings: we could for example increase the amount of elementarity demanded to include that of a language with full quantification over the variables X_i. This 'mereological reflection' then has reflection involving quantified statements about the parts of V. This is natural, but a stronger principle than GRP. However then one must be specific about the domain of quantification, i.e. the class C. Our own inclination is to eschew second order methods whenever possible – being not entirely convinced of their coherence. One point that can be made is that with

$(V_\alpha, \in, V_{\alpha+1})$ being a natural model of Morse-Kelley class theory, if we have an enhanced version of GRP with this form of full second order reflection, we can carry this up to deduce that $(V, \in, C)$ must also form a model of Morse-Kelly.

A final technical word on the consistency of GRP. One can easily see that:

Theorem 4.5.1

$$Con(ZFC + \exists\kappa(\kappa \textit{ is 1-extendible})) \longrightarrow Con(NBG + GRP).$$

Indeed 1-extendibility is stronger than the enhanced versions of GRP mentioned above. Sam Roberts in (2017) has extended the above account of global reflection to yield a very flexible family of higher order reflection principles which, when suitably formulated, can yield supercompact cardinals and more. But these cross the philosophical threshold we have stopped short of: not to quantify over the parts of V. They also cross over the admittedly more technical set theoretical threshold into those cardinals, such as supercompacts, that imply the existence of embeddings j of the universe into an inner model that are discontinuous at the successor cardinal of the critical point (the first ordinal moved) of j. Such embeddings require representation by systems of ultrapowers known as 'long extenders', which we do not define here, but for a discussion of 'long' and 'short' extender types see Section 1 of Neeman (2007). The GRP here falls just short of justifying such. The GRP embedding is of a 'superstrong' type where On is the target of the critical point κ and is at the limit of those that can be expressed using short extenders. Here a cardinal κ is called *superstrong* if there is an embedding j, preserving elementarity in the usual first order language of set theory, with critical point κ, and an inner model M so that $j : V \longrightarrow M$ with $V^M_{j(\kappa)} = V_{j(\kappa)}$. Note now that $j \restriction V_{\kappa+1} : V_{\kappa+1} \longrightarrow V^M_{j(\kappa)+1}$ however without any assumption that the latter is $V_{j(\kappa)+1}$. (The latter would require 1-extendibility.) But taking $C = V^M_{j(\kappa)+1}$, it is easy to see $(V_{j(\kappa)}, \in, C)$ together with $j \restriction V_{\kappa+1}$ gives a set model of $NBG + GRP$.

Hence the last theorem can be modestly improved, in that the antecedent is a large cardinal that can be expressed by short extenders.

Theorem 4.5.2

$$Con(ZFC + \exists\kappa(\kappa \textit{ is superstrong})) \longrightarrow Con(NBG + GRP).$$

It is possible given such a j from GRP to define from it a directed system of short extenders from V, in such a way that this directed system expresses an ultrapower embedding which when restricted to $V_{\kappa+1}$ is just j. We thus get back j from this directed system (see the discussion in Kanamori 2003 Ch. 5 Sect. 26 for example.) Were long extender embeddings ever to be shown, as a class, inconsistent, then GRP is pretty much what we would be left with.

References

Bernays, P. (1961). Zur Frage der Unendlichkeitsschemata in der axiomatische Mengenlehre. In *Essays on the foundations of mathematics* (pp. 3–49). Hebrew University of Jerusalem: Magnus Press.

Feferman, S. (2012). Is the continuum hypothesis a definite mathematical problem? In: *EFI Workshop Papers*, Harvard.

Friedman, H. (1970). Higher set theory and mathematical practice. *Annals of Mathematical Logic, 2*(3), 325–327.

Kanamori, A. (2003). *The higher infinite* (Springer monographs in mathematics, 2nd ed.). New York: Springer.

Koellner, P. (2009). On reflection principles. *Annals of Pure and Applied Logic, 157*, 206–219.

Levy, A. (1960). Axiom schemata of strong infinity in axiomatic set theory. *Pacific Jouranl of Mathematics, 10*, 223–238.

Lusin, N., & Sierpiński, W. (1923). Sur un ensemble non measurable B. *Journal de Mathématiques, 9e serie, 2*, 53–72.

Maddy, P. (1983). Proper classes. *Journal for Symbolic Logic, 48*(1), 113–139.

Martin, D. A. (1970). Measurable cardinals and analytic games. *Fundamenta Mathematicae, 66*, 287–291.

Martin, D. A. (1975). Borel determinacy. *Annals of Mathematics, 102*, 363–371.

Martin, D. A., & Steel, J. R. (1989). A proof of projective determinacy. *Journal of the American Mathematical Society, 2*, 71–125.

Moschovakis, Y. N. (2009). *Descriptive set theory* (Studies in logic series). Amsterdam: North-Holland.

Mycielski, J. (1964). On the axiom of determinateness. *Fundamenta Mathematicae, 53*, 205–224.

Mycielski, J. (1966). On the axiom of determinateness II. *Fundamenta Mathematicae, 59*, 203–212.

Neeman, I. (2007). Determinacy in $L(\mathbb{R})$, chapter 22. In M. Magidor, M. Foreman, & A. Kanamori (Eds.), *Handbook of set theory* (vol. III). Berlin/New York: Springer.

Reinhardt, W. (1974). Remarks on reflection principles, large cardinals, and elementary embeddings. In T. Jech (Ed.), *Axiomatic set theory*, vol. 13 part 2 of *Proceedings of Symposia in Pure Mathematics* (pp. 189–205). Providence: American Mathematical Society.

Roberts, S. (2017). A strong reflection principle. *Review of Symbolic Logic, 10*(4), 651–662.

Shelah, S. (1984). Can you take Solovay's inaccessible away? *Israel Journal of Mathematics, 48*, 1–47.

Solovay, R. M. (1965). The measure problem (abstract). *Notices of the American Mathematical Society, 12*, 217.

Solovay, R. M. (1969). The cardinality of Σ^1_2 sets of reals. In J.J. Bulloff, T.C. Holyoke, & S.W. Hahn (Eds.), *Foundations of mathematics: Symposium papers commemorating the sixtieth birthday of Kurt Gödel* (pp. 58–73). Berlin: Springer.

Solovay, R. M. (1970). A model of set theory in which every set of reals is Lebesgue measurable. *Annals of Mathematics, 92*, 1–56.

Tait, W. W. (2005). Constructing cardinals from below. In W. W. Tait (Ed.), *The provenance of pure reason: Essays in the philosophy of mathematics and its history* (pp. 133–154). Oxford: Oxford University Press.

Wagon, S. (1994). *The banach-tarski paradox*. Cambridge: Cambridge University Press.

Wang, H. (1996). *A logical journey: From Godel to philosophy*. Boston: MIT Press.

Welch, P. D. (2012). Global reflection principles. *Isaac Newton Institute Pre-print Series.* Exploring the Frontiers of Incompleteness, EFI Workshop Papers, Harvard (INI12051-SAS).

Welch, P. D., & Horsten, L. (2016). Reflecting on absolute infinity. *Journal of Philosophy, 113*, 89–111.

Woodin, W. H. (1988). Supercompact cardinals, sets of reals, and weakly homogeneous trees. *Proceedings of the National Academy of Sciences of the United States of America, 85*(18), 6587–6591.

Woodin, W. H. (2001a). The continuum hypothesis, Part I. *Notices of the American Mathematical Society, 48*, 567–576.

Woodin, W. H. (2001b). The continuum hypothesis, Part II. *Notices of the American Mathematical Society, 48*, 681–690.

Part II
What Are Homotopy Type Theory and the Univalent Foundations?

Chapter 5
Naïve Type Theory

Thorsten Altenkirch

Abstract We introduce Type Theory, in its latest incarnation of Homotopy Type Theory, as an alternative to set theory as a foundation of Mathematics. We emphasize the naïve, intuitive understanding of Type Theory.

5.1 Introduction

The word *type theory* has at least two meanings:

- The theory of types in programming languages,
- Martin-Löf's Type Theory as a constructive foundation of Mathematics

We will be mainly concerned with the latter (which is emphasised by capitalising it), even though there are interactions with the design of programming languages as well.

Type Theory is the base of a number of computer systems used as the base of interactive proof systems and very advanced (functional) programming language. Here is an incomplete list:

NuPRL is maybe the oldest implementation of Type Theory, which was developed at Cornell. It is based on a different flavour of type theory which was called *Extensional Type Theory* but which is now referred to as *Computational Type Theory* (Constable et al. 1986).

Coq is maybe now the system most used in formal Mathematics and has been used for some impressive developments, including a formal proof of the Four Colour Theorem and the verification of an optimising C compiler (Bertot and Castéran 2013; Chlipala 2013).

T. Altenkirch (✉)
Functional Programming Laboratory, School for Computer Science, University of Nottingham, Nottingham, UK
e-mail: txa@cs.nott.ac.uk

S. Centrone et al. (eds.), *Reflections on the Foundations of Mathematics*, Synthese Library 407, https://doi.org/10.1007/978-3-030-15655-8_5

Agda is a sort of a twitter it can be used as a interactive proof assistant or as a dependently typed programming language (Norell 2008; Stump 2016).

Idris goes further on the programming language road by addressing more pragmatic concerns when using Type Theory for programming (Brady 2017).

Lean is developed at Microsoft Research with strong support for automatic reasoning (de Moura et al. 2015).

Cubical Cubical is a very new system and more a proof of concept but it is the only one (so far) that actually implements Homotopy Type Theory (Cohen et al. 2016).

One way to introduce Type Theory is to pick one system (I usually pick Agda) and then learn Type Theory by doing it. While this is a good way to approach this subject, and do I recommend to play with a system, I want to concentrate more on the conceptual issues and then I find that having to explain the intricacies of a particular system can be a bit of a distraction. Hence this course will be a paper based introduction to Type Theory.

This chapter can be viewed as a taster for my forthcoming book (Altenkirch 2019) or for the book on Homotopy Type Theory (HoTT) (The Univalent Foundations Program 2013) which was the output of a special year at the Institute for Advanced Study in Princeton.

The material presented in these notes was the base for a number of courses since my presentation at the FOMUS meeting:

- *Introduction to homotopy type theory*
  22nd Estonian Winter School in Computer Science (EWSCS), Palmse, Estonia, January 2017.
- *Naïve Type Theory*
  Midland Graduate School 2017 (MGS 2017), Leicester, UK, April 2017.
- *Introduction to homotopy type theory*
  EU Types summer school, Ohrid, Macedonia, July 2017

5.2 Type Theory vs Set Theory

I view Type Theory in the first place as a intuitive foundation of Mathematics. This is similar to how most Mathematicians use Set Theory: the have an intuitive idea what sets are but they don't usually refer back to the axioms of Set Theory. This is sometimes called *naïve Set Theory*[1] and similar what I am doing here can be called *naïve Type Theory*.

In Set Theory we write $3 \in \mathbb{N}$ to express that 3 is an element of the set of natural numbers. In Type Theory we write $3 : \mathbb{N}$ to express that 3 is an element of the type of natural numbers. While this looks superficially similar, there are important differences:

[1]This is also the title of a well known book by Halmos (1998).

- While $3 \in \mathbb{N}$ is a proposition, $3 : \mathbb{N}$ is a *judgement*, that is a piece of static information.
- In Type Theory every object and every expression has a (unique) type which is statically determined.[2]
- Hence it doesn't make any sense to use $a : A$ as a proposition.
- This is similar to the distinction between statically and dynamically typed programming languages. While in dynamically typed languages there are runtime functions to check the type of an object this doesn't make sense in statically typed languages.
- In Set Theory we define $P \subseteq Q$ as $\forall x.x \in P \to x \in Q$. We can't do this in Type Theory because $x \in P$ is not a proposition.
- Also set theoretic operations like $\cup$ or $\cap$ are not operations on types. However, they can be defined as operations on predicates, aka subsets, of a given type. $\subseteq$ can be defined as a predicate on such subsets.
- Type Theory is extensional in the sense that we can't talk about details of encodings.
- This is different in Set Theory where we can ask wether $\mathbb{N} \cap \text{Bool} = \emptyset$? Or wether $2 \in 3$? The answer to these questions depends on the choice of representation of these objects and sets.

Apart from the judgement $a : A$ there is also the judgement $a \equiv_A b$ which means that $a, b : A$ are *definitionally* equal. We write definitions using $:\equiv$, e.g. we can define $n : \mathbb{N}$ as 3 by writing $n :\equiv 3$. As for $a : A$ definitional equality is a static property which can be determined statically and hence which doesn't make sense as a proposition. We will later introduce $a =_A b$, *propositional equality* which can be used in propositions.

While Type Theory is in some sense more restrictive than Set Theory, this does pay off. Because we cannot talk about intensional aspects, i.e. implementation details, we can identify objects which have the same extensional behaviour. This is reflected in the *univalence axiom*, which identifies extensionally equivalent types (such as unary and binary natural numbers).

Another important difference between Set Theory and Type Theory is the way propositions are treated: Set Theory is formulated using predicate logic which relies on the notion of *truth*. Type Theory is self-contained and doesn't refer to truth but to evidence. Using the propositions-as-types translation (also called the Curry-Howard equivalence) we can assign to any proposition P the type of its evidence $[\![P]\!]$ using the following table:

$$[\![P \Rightarrow Q]\!] \equiv [\![P]\!] \to [\![Q]\!]$$

$$[\![P \wedge Q]\!] \equiv [\![P]\!] \times [\![Q]\!]$$

[2]We are not considering subtyping here, which can be understood as a notational device allowing the omission of implicit coercions.

$$
\begin{aligned}
[\![\text{True}]\!] &\equiv 1 \\
[\![P \vee Q]\!] &\equiv [\![P]\!] + [\![Q]\!] \\
[\![\text{False}]\!] &\equiv 0 \\
[\![\forall x : A.P]\!] &\equiv \Pi x : A.[\![P]\!] \\
[\![\exists x : A.P]\!] &\equiv \Sigma x : A.[\![P]\!]
\end{aligned}
$$

0 is the empty type, 1 is the type with exactly one element and + is the sum or disjoint union of types. $\rightarrow$ (function type) and $\times$ should be familiar but we will revisit all of them from a type theoretic perspective. Π and Σ are less familiar in Set Theory and we will have a look at them later.

We are using a typed predicate logic, in the lines for universal and existential quantification A refers to a type. Other connectives are defined: $\neg P$ is defined as $P \Rightarrow \text{False}$. Logical equivalence $P \Leftrightarrow Q$ is defined as $(P \Rightarrow Q) \wedge (Q \Rightarrow P)$. Careful, equivalence such as $\neg(P \wedge Q) \Leftrightarrow \neg P \vee \neg Q$ or $\neg(\forall x : A.P) \Leftrightarrow \exists x : A.\neg P$ do not hold in Type Theory, hence we cannot define $\vee$ or $\exists$ via $\neg$ and $\wedge$ and $\forall$.

Later we will see a refinement of the proposition as types translation, which changes the translation of $P \vee Q$ and $\exists x : A.P$.

5.3 Non-dependent Types

5.3.1 Universes

To get started we have to say what a type is. We could achieve this by introducing another judgement but instead I am going to use universes. A universe is a type of types. For example to say that $\mathbb{N}$ is a type, I write $\mathbb{N}$: **Type** where **Type** is a universe.

But what is the type of **Type**? Do we have **Type** : **Type**? It is well known that this doesn't work in Set Theory due to Russell's paradox (the set of all sets which does not contain itself). However, in Type Theory $a : A$ is not a proposition, hence it is not immediately clear wether the paradox still works.

However, it is possible to encode Russell's paradox in a Type Theory with **Type** : **Type** by using trees which can branch over any type. In this theory we can construct a tree of all trees which don't have themself as immediate subtree. This tree is a subtree of itself iff it isn't which enables us to derive a contradiction.

To avoid Russell's paradox we introduce a hierarchy of universes

$$
\mathbf{Type}_0 : \mathbf{Type}_1 : \mathbf{Type}_2 : \ldots
$$

and we decree that any type $A : \mathbf{Type}_i$ can be lifted to a type $A^+ : \mathbf{Type}_{i+1}$. Being explicit about universe levels can be quite annoying hence we are going to ignore them most of the time but try to make sure that we don't use universes in a cyclic way. That is we write **Type** as a metavariable for $\mathbf{Type}_i$ and assume that all the levels are the same unless stated explicitly.

5.3.2 Functions

While in Set Theory functions are a derived concept (a subset of the cartesian product with certain properties), in Type Theory functions are a primitive concept. The basic idea is the same as in functional programming: basically a function is a black box and you can feed it elements of its domain and out come elements of its codomain. Hence given A, B : **Type** we introduce the type of functions $A \to B$: **Type**. We can define a function explicitly, e.g. we define $f : \mathbb{N} \to \mathbb{N}$ as $f(x) :\equiv x + 3$. Having defined f we can apply it, e.g. $f(2) : \mathbb{N}$ and we can evaluate this application by replacing all occurrences of the parameter x in the body of the function $x + 3$ by the actual argument 2 hence $f(2) \equiv 2 + 3$ and if we are lucky to know how to calculate $2 + 3$ we can conclude $f(2) \equiv 5$.

One word about syntax: in functional programming and in Type Theory we try to save brackets and write $f\,2$ for the application and also in the definition we write $f\,x :\equiv x + 3$.

The explicit definition of a function requires a name but we should be able to define a function without having to give it a name – this is the justification for the λ-notation. We write $\lambda x.x + 3 : \mathbb{N} \to \mathbb{N}$ avoiding to have to name the function. We can apply this $(\lambda x.x + 3)\,2$ and the equality $(\lambda x.x + 3)\,2 \equiv 2 + 3$ is called β-reduction. The explicit definition $f\,x \equiv x+3$ can now be understood as a shorthand for $f \equiv \lambda x.x + 3$.

In Type Theory every function has exactly one argument. To represent functions with several arguments we use *currying*, that is we use a function that returns a function. So for example the addition function $g :\equiv \lambda x.\lambda y.x + y$ has type $\mathbb{N} \to (\mathbb{N} \to \mathbb{N})$, that is if we apply it to one argument $g\,3 : \mathbb{N} \to \mathbb{N}$ it returns the function that add 3 namely $\lambda y.3+y$. We can continue and supply a further argument $(g\,3)\,2 : \mathbb{N}$ which reduces

$$\begin{aligned}(g\,3)\,2 &\equiv (\lambda y.3 + y)\,2\\ &\equiv 3+2\end{aligned}$$

To avoid the proliferation of brackets we decree that application is left associative hence we can write $g\,3\,2$ and that $\to$ is left associative hence we can write $\mathbb{N} \to \mathbb{N} \to \mathbb{N}$ for the type of g.

When calculating with variables we have to be a bit careful. Assume we have a variable $y : \mathbb{N}$ hanging around, now what is $g\,y$? If we naively replace x by y we obtain $\lambda y.y+y$, that is the variable y got *captured*. This is not the intended behaviour and to avoid capture we have to rename the bound variable that is $\lambda x.\lambda y.x + y \equiv \lambda x.\lambda z.x + z$ – this equality is called α-congruence. After having done this we can

β-reduce. Here is the whole story

$$\begin{aligned} g\,y &\equiv (\lambda x.\lambda y.x + y)\,y \\ &\equiv (\lambda x.\lambda z.x + z)\,y \\ &\equiv \lambda z.y + z \end{aligned}$$

Clearly the choice of z here is arbitrary, but any other choice (apart from y) would have yielded the same result upto α-congruence.

5.3.3 Products and Sums

Given A, B : **Type** we can form their product $A \times B$: **Type** and their sum $A + B$: **Type**. The elements of a product are tuples, that is $(a, b) : A \times B$ if $a : A$ and $b : B$. The elements of a sum are injections that is $\mathrm{left}\,a : A + B$ if $a : A$ and $\mathrm{right}\,b : A + B$, if $b : B$.

To define a function from a product or a sum it is sufficient to say what the functions returns for the constructors, that is for tuples in the case of a product or the injections in the case of a sum.

As an example we derive the tautology

$$P \wedge (Q \vee R) \Leftrightarrow (P \wedge Q) \vee (P \wedge R)$$

using the propositions as types translation. We assume that P, Q, R : **Type**, we have to construct an element of the following type

$$\begin{aligned} &(P \times (Q + R) \to (P \times Q) + (P \times R)) \\ \times&((P \times Q) + (P \times R) \to P \times (Q + R)) \end{aligned}$$

We define:

$$\begin{aligned} &f : P \times (Q + R) \to (P \times Q) + (P \times R) \\ &f\,(p, \mathrm{left}\,q) :\equiv \mathrm{left}\,(p, q) \\ &f\,(p, \mathrm{right}\,r) :\equiv \mathrm{right}\,(p, r) \end{aligned}$$

$$\begin{aligned} &g : (P \times Q) + (P \times R) \to P \times (Q + R) \\ &g\,(\mathrm{left}\,(p, q)) :\equiv (p, \mathrm{left}\,q) \\ &g\,(\mathrm{right}\,(p, r))) :\equiv (p, \mathrm{right}\,r) \end{aligned}$$

Now the tuple (f, g) is an element of the type above.

In this case the two functions are actually inverses, but this is not necessary to prove the logical equivalence.

Exercise 1 *Using the propositions as types translation, try*[3] *to prove the following tautologies:*

1. $(P \wedge Q \Rightarrow R) \Leftrightarrow (P \Rightarrow Q \Rightarrow R)$
2. $((P \vee Q) \Rightarrow R) \Leftrightarrow (P \Rightarrow R) \wedge (Q \Rightarrow R)$
3. $\neg(P \vee Q) \Leftrightarrow \neg P \wedge \neg Q$
4. $\neg(P \wedge Q) \Leftrightarrow \neg P \vee \neg Q$
5. $\neg(P \Leftrightarrow \neg P)$

where $P, Q, R : \mathbf{Type}$ *are propositions represented as types.*

Exercise 2 *While the principle of excluded middle* $P \vee \neg P$ *(*tertium non datur*) is not provable, prove its double negation using the propositions as types translation:*

$$\neg\neg(P \vee \neg P)$$

If for a particular proposition P *we can establish* $P \vee \neg P$ *then we can also derive the principle of indirect proof (*reduction ad absurdo*) for the same proposition* $\neg\neg P \Rightarrow P$. *Hence show:*

$$(P \vee \neg P) \Rightarrow (\neg\neg P \Rightarrow P)$$

Note that the converse does not hold (what would be a counterexample?). However, use the two tautologies to show that the two principles are equivalent.

Functions out of products and sums can be reduced to using a fixed set of combinators called non-dependent eliminators or *recursors* (even though there is no recursion going on).

$$\mathrm{R}^{\times} : (A \to B \to C) \to A \times B \to C$$

$$\mathrm{R}^{\times}\ f\ (a, b) :\equiv f\, a\, b$$

$$\mathrm{R}^{+} : (A \to C) \to (B \to C) \to A + B \to C$$

$$\mathrm{R}^{+}\ f\ g\ (\mathrm{left}\, a) :\equiv f\ a$$

$$\mathrm{R}^{+}\ f\ g\ (\mathrm{right}\, b) :\equiv g\ b$$

[3] I didn't say they are all tautologies!

The recursor $\mathrm{R}^\times$ for products maps a curried function $f : A \to B \to C$ into its uncurried form, taking tuples as arguments. The recursor R^+ basically implements the *case* function performing case analysis over elements of $A + B$.

Exercise 3 *Show that using the recursor* $\mathrm{R}^\times$ *we can define the projections:*

$$\begin{aligned}
&\mathrm{fst} : A \times B \to A \\
&\mathrm{fst}\,(a, b) :\equiv a \\
&\mathrm{snd} : A \times B \to B \\
&\mathrm{snd}\,(a, b) :\equiv b
\end{aligned}$$

Vice versa: can the recursor be defined using only the projections?

We also have the case of an empty product 1, called the unit type and the empty sum 0, the empty type. There is only one element of the unit type: $() : 1$ and none in the empty type. We introduce the corresponding recursors:

$$\begin{aligned}
&\mathrm{R}^1 : C \to (1 \to C) \\
&\mathrm{R}^1\, c\, () :\equiv c
\end{aligned}$$

$$\mathrm{R}^0 : 0 \to C$$

The recursor for 1 is pretty useless, it just defines a constant function. The recursor for the empty type implements the logical principle *eq falso quod libet*, from false follows everything. There is no defining equation because it will never be applied to an actual element.

Exercise 4 *Construct solutions to Exercises* 1 *and* 2 *using only the eliminators.*

The use of arithmetical symbols for operators on types is justified because they act like the corresponding operations on finite types. Let us identify the number n with the type of elements $0_n, 1_n, \ldots (n-1)_n : \overline{n}$, then we observe that it is indeed the case that:

$$\begin{aligned}
\overline{0} &= 0 \\
\overline{m+n} &= \overline{m} + \overline{n} \\
\overline{1} &= 1 \\
\overline{m \times n} &= \overline{m} \times \overline{n}
\end{aligned}$$

Read $=$ here as *has the same number of elements*. This use of equality will be justified later when we introduce the univalence principle.

The arithmetic interpretation of types also extends to the function type, which corresponds to exponentiation. Indeed, in Mathematics the function type $A \to B$ is often written as B^A. And indeed we have:

$$\overline{m^n} = \overline{n} \to \overline{m}$$

5.4 Dependent Types

By a dependent type we mean a type indexed by elements of another type. For example the types of n-tuples A^n : **Type** their elements are $(a_0, a_1, \dots a_{n-1}) : A^n$ where $a_i : A$, or the finite type $\overline{n}$: **Type**. Indeed, tuples are also indexed by A : **Type**. We can use functions into **Type** to represent these dependent types:

$$\mathrm{Vec} : \mathbf{Type} \to \mathbb{N} \to \mathbf{Type}$$
$$\mathrm{Vec}\, A\, n :\equiv A^n$$

$$\mathrm{Fin} : \mathbb{N} \to \mathbf{Type}$$
$$\mathrm{Fin}\, n :\equiv \overline{n}$$

In the propositions as types view dependent types are used to represent predicates, e.g. $\mathrm{Prime} : \mathbb{N} \to$ **Type** assigns to any natural number $n : \mathbb{N}$ the type of evidence $\mathrm{Prime}\, n$: **Type** that n is a prime number. This does not need to be inhabited, e.g. $\mathrm{Prime}\, 4$ is empty. Using currying we can use this also to represent relations, e.g. $\leq : \mathbb{N} \to \mathbb{N} \to$ **Type**, $m \leq n$: **Type** is the type of evidence that m is less or equal to n.

5.4.1 *Π-Types and Σ-Types*

Π-types generalize function types to allow the codomain of a function to depend on the domain. For example consider the function zeroes that assigns to any natural number $n : \mathbb{N}$ a vector of n zeroes

$$\underbrace{(0, 0, \dots, 0)}_{n} : \mathbb{N}^n$$

We use Π to write such a type:

$$\mathrm{zeroes} : \Pi n : \mathbb{N}.\mathbb{N}^n$$
$$\mathrm{zeroes}\, n :\equiv \underbrace{(0, 0, \dots, 0)}_{n}$$

The non-dependent function type can now be understood as a special case of Π-types, $A \to B \equiv \Pi - : A.B$.

In the same vain, Σ-types generalize product types to the case when the 2nd component depends on the first. For example we can represent tuples of arbitrary size as a pair of a natural number $n : \mathbb{N}$ and a vector of this size A^n as an element of $\Sigma n : \mathbb{N}.A^n$. So for example $(3, (1, 2, 3)) : \Sigma n : \mathbb{N}.\mathbb{N}^n$ because $(1, 2, 3) : \mathbb{N}^3$. Indeed, this type is very useful so we give it a name:

$$\text{List} : \textbf{Type} \to \textbf{Type}$$

$$\text{List}\, A :\equiv \Sigma n : \mathbb{N}.A^n$$

In the propositions as types translation we use Π-types to represent evidence for universal quantification. For example a proof of $\forall x : \mathbb{N}.1+x = x+1$ is a dependent function of type $f : \Pi x : \mathbb{N}.1+x = x+1$, such that applying it as in $f\,3$ is evidence that $1 + 3 = 3 + 1$.

Similar, we use Σ-types to represent evidence for existential quantification where the first component is the instance for which the property is supposed to hold and the second component a proof that it holds for this particular instance. For example the statement $\exists n : \mathbb{N}.\text{Prime}\, n$ is translated to $\Sigma n : \mathbb{N}.\text{Prime}\, n$ and a proof of this is $(3, p)$ where $p : \text{Prime}\, 3$.

As for Π-types the non-dependent products arise as a special case of Σ-types: $A \times B :\equiv \Sigma - : A.B$.

To avoid clutter we sometimes want to omit arguments to a Π-type when it is derivable from later arguments or the first component of a Σ-type. In this case we write the argument in subscript as in $\Pi_{x:A} B\, x$ or $\Sigma_{x:A} B\, x$. For example if we define $\text{List}\, A :\equiv \Sigma_{n:\mathbb{N}}.A^n$ we can omit the length and just write $(1, 2, 3) : \text{List}\, \mathbb{N}$.

We can also define a dependent recursor or eliminator for Σ-types which allows us to define any dependent function out of a Σ-type. This eliminator is not just parametrised by a type but by a family $C : \Sigma x : A.B\, x \to \textbf{Type}$:

$$\text{E}^{\Sigma} : (\Pi x : A, \Pi y : B\, x.C\, (x, y)) \to \Pi p : \Sigma x : A.B\, x.C\, p$$

$$\text{E}^{\Sigma}\, f\, (a, b) :\equiv f\, a\, b$$

Exercise 5 *As in Exercise* 3 *we can define projections out of a Σ-type, let A :* **Type** *and $B : A \to$* **Type***:*

$$\text{fst} : \Sigma x : A.B\, x \to A$$

$$\text{fst}\, (a, b) :\equiv a$$

$$\text{snd} : \Pi p : \Sigma x : A.B\, x.B\, (\text{fst}\, p)$$

$$\text{snd}\, (a, b) :\equiv b$$

Note that the type of the 2nd projections is a dependent function type using the first projection.

Derive the projections using only the eliminator E^{Σ}*. Vice versa, can you derive the eliminator from the projections without making further assumptions?*

Exercise 6 *Using the propositions as types translation for predicate logic try to derive the following tautologies:*

1. $(\forall x : A.P\,x \wedge Q\,x) \Leftrightarrow (\forall x : A.P\,x) \wedge (\forall x : A.Q\,x)$
2. $(\exists x : A.P\,x \vee Q\,x) \Leftrightarrow (\exists x : A.P\,x) \vee (\exists x : A.Q\,x)$
3. $(\exists x : A.P\,x) \Rightarrow R \Leftrightarrow \forall x : A.P\,x \Rightarrow R$
4. $\neg\exists x : A.P\,x \Leftrightarrow \forall x : A.\neg P\,x$
5. $\neg\forall x : A.P\,x \Leftrightarrow \exists x : A.\neg P\,x$

where A, B : **Type** *and* $P, Q : A \to$ **Type**, R : **Type**, *represent predicates and a proposition.*

We have seen that Σ-types are related to products but they are also related to sums. Indeed we can derive $+$ from Σ using as the first component an element of $\mathrm{Bool} = \overline{2}$ and the second component is either the first or the 2nd component of the sum (assuming A, B : **Type**):

$$A + B : \mathbf{Type}$$
$$A + B :\equiv \Sigma x : \mathrm{Bool}.\mathrm{if}\, x \,\mathrm{then}\, A \,\mathrm{else}\, B$$

In the same way we can also derive $\times$ from Π by using dependent functions over the booleans which returns either one or the other component of the product.

$$A \times B : \mathbf{Type}$$
$$A \times B :\equiv \Pi x : \mathrm{Bool}.\mathrm{if}\, x \,\mathrm{then}\, A \,\mathrm{else}\, B$$

It is interesting to note that $\times$ can be viewed in two different ways: either as a non-dependent Σ-type or as a dependent function-type over the booleans.

Exercise 7 *Show that injections, pairing, non-dependent eliminators can be derived for these encodings of sums and products.*

Finally, we notice that the arithmetic interpretation of types extends to Σ and Π giving a good justification for the choice of their names, let $m : \mathbb{N}$ and $f : \overline{m} \to \mathbb{N}$:

$$\overline{\Sigma x : m.f\,x} = \Sigma_{x=0}^{x<m} \overline{f\,x}$$
$$\overline{\Pi x : m.f\,x} = \Pi_{x=0}^{x<m} \overline{f\,x}$$

5.4.2 Induction and Recursion

Following Peano the natural numbers are introduced by saying that 0 is a natural number ($0 : \mathbb{N}$), and if n is a natural number ($n : \mathbb{N}$) then its successor is a natural number ($\mathrm{suc}\, n : \mathbb{N}$), which is equivalent to saying $\mathrm{suc} : \mathbb{N} \to \mathbb{N}$. When defining a function out of the natural numbers, we allow ourselves to recursively use the function value on n to compute it for $\mathrm{suc}\, n$. An example is the doubling function:

$$\begin{aligned} &\mathrm{double} : \mathbb{N} \to \mathbb{N} \\ &\mathrm{double}\, 0 &&:\equiv 0 \\ &\mathrm{double}\, (\mathrm{suc}\, n) &&:\equiv \mathrm{suc}\, (\mathrm{suc}\, (\mathrm{double}\, n)) \end{aligned}$$

We can distill this idea into a non-dependent eliminator which is now rightfully called the recursor:

$$\begin{aligned} &\mathrm{R}^{\mathbb{N}} : C \to (C \to C) \to \mathbb{N} \to C \\ &\mathrm{R}^{\mathbb{N}}\, z\, s\, 0 &&:\equiv z \\ &\mathrm{R}^{\mathbb{N}}\, z\, s\, (\mathrm{suc}\, n) &&:\equiv s\, (\mathrm{R}\, z\, s\, n) \end{aligned}$$

We can now define double using only the recursor:

$$\mathrm{double}\, n :\equiv \mathrm{R}^{\mathbb{N}}\, 0\, (\lambda n.\mathrm{suc}\, (\mathrm{suc}\, n))$$

Exercise 8 *We define addition recursively:*

$$\begin{aligned} &+ : \mathbb{N} \to \mathbb{N} \to \mathbb{N} \\ &0 + n &&:\equiv n \\ &(\mathrm{suc}\, m) + n &&:\equiv \mathrm{suc}\, (m + n) \end{aligned}$$

Define addition using only the recursor $\mathrm{R}^{\mathbb{N}}$.

Exercise 9 *Not all recursive functions exactly fit into this scheme. For example consider the function that halves a number forgetting the remainder:*

$$\begin{aligned} &\mathrm{half} : \mathbb{N} \to \mathbb{N} \\ &\mathrm{half}\, 0 &&:\equiv 0 \\ &\mathrm{half}\, (\mathrm{suc}\, 0) &&:\equiv 0 \\ &\mathrm{half}\, (\mathrm{suc}\, (\mathrm{suc}\, n)) &&:\equiv \mathrm{suc}\, (\mathrm{half}\, n) \end{aligned}$$

Try to derive half *only using the recursor* $\mathrm{R}^{\mathbb{N}}$.

When we want to prove a statement about natural numbers we have to construct a dependent function. An example is the proof that half is the left inverse of double: $\forall n : \mathbb{N}.\text{half}\,(\text{double}\,n) = n$. I haven't introduced equality yet but we only need two ingredients to carry out this construction, given $A, B : \textbf{Type}$:

$$\text{refl} : \Pi x : A.x = x$$

$$\text{resp} : \Pi f : A \to B.\Pi_{m,n:\mathbb{N}} m = n \to f\,m = f\,n$$

Using those we can define a dependent function verifying the statement:

$$\text{halfDouble} : \Pi n : \mathbb{N}.\text{half}\,(\text{double}\,n) = n$$

$$\text{halfDouble}\,0 \quad :\equiv \text{refl}\,0$$

$$\text{halfDouble}\,(\text{suc}\,n) \quad :\equiv \text{resp}\,\text{suc}\,(h\,n)$$

As for Σ-types we can derive dependent functions out of the natural numbers using a dependent recursor or eliminator. Assume that we have a dependent type $C : \mathbb{N} \to$ **Type**:

$$\text{E}^{\mathbb{N}} : C\,0 \to (\Pi n : \mathbb{N}.C\,n \to C\,(\text{suc}\,n)) \to \Pi n : \mathbb{N}.C\,n$$

$$\text{E}^{\mathbb{N}}\,z\,s\,0 \quad :\equiv z$$

$$\text{E}^{\mathbb{N}}\,z\,s\,(\text{suc}\,n) \quad :\equiv s\,(\text{E}\,z\,s\,n)$$

Exercise 10 *Derive* halfDouble *using only the dependent eliminator* $\text{E}^{\mathbb{N}}$.

The type of $\text{E}^{\mathbb{N}}$ precisely corresponds to the principle of induction – indeed from the propositions as types point of view induction is just dependent recursion.

Exercise 11 *Show that the natural numbers with* $+$ *and* 0 *form a commutative monoid:*

1. $\forall x : \mathbb{N}.0 + x = x$
2. $\forall x : \mathbb{N}.x + 0 = x$
3. $\forall x, y, z : \mathbb{N}.x + (y + z) = (x + y) + z$
4. $\forall x, y : \mathbb{N}.x + y = y + x$

Not all dependent functions out of the natural numbers arise from the propositions as types translation. An example is the function zeroes $: \Pi n : \mathbb{N}.\mathbb{N}^n$ which we only introduced informally. We can make this precise by inductively defining tuples:

$$\text{nil} \quad : A^0$$

$$\text{cons} \quad : \Pi_{n:\mathbb{N}} A^n \to A^{\text{suc}\,n}$$

Using this we can define zeroes by recursion

$$\text{zeroes}\,0 :\equiv \text{nil}$$
$$\text{zeroes}\,(\text{suc}\,n) :\equiv \text{cons}\,0\,(\text{zeroes}\,n)$$

This can be easily translated into an application of the eliminator

$$\text{zeroes} :\equiv \mathrm{E}^{\mathbb{N}}\,\text{nil}\,(\text{cons}\,0)$$

Exercise 12 *We can also define the finite types in an inductive way, overloading* 0 *and* suc*:*

$$0 : \Pi_{n:\mathbb{N}} \overline{\text{suc}\,n}$$
$$\text{suc} : \Pi_{n:\mathbb{N}} \overline{n} \to \overline{\text{suc}\,n}$$

Using this and the inductive definition of A^n derive a general projection operator

$$\text{nth} : \Pi_{n:\mathbb{N}} A^n \to \overline{n} \to A$$

that extracts an arbitrary component of a tuple.

Exercise 13 *Suggest definitions of eliminators for tuples and finite types. Can you derive all the examples using them?*

5.4.3 The Equality Type

Given $a, b : A$ the equality type $a =_A b : \mathbf{Type}$ is generated from one constructor $\text{refl} : \Pi_{x:A} x =_A x$. That is we are saying that two things which are identical are equal and this is the only way to construct an equality. Using this idea we can establish some basic properties of equality, namely that it is an equivalence relation, that is a relation that is reflexive, symmetric and transitive. Moreover, it is also a congruence, it is preserved by all functions. Since we already have reflexivity from the definition, let's look at symmetry first. We can define symmetry by just saying how it acts on reflexivity:

$$\text{sym} : \Pi_{x,y:A} x = y \to y = x$$
$$\text{sym}\,\text{refl}_a :\equiv \text{refl}_a$$

The main idea here is that once we apply sym to refl we also know that two points $x, y : A$ must be identical and hence we can prove the result using refl again.

Exercise 14 *Provide proofs of*

$$\text{trans} : \Pi_{x,y,z:A} x = y \to y = z \to x = z$$
$$\text{resp} : \Pi f : A \to B.\Pi x, y : A.x = y \to f\,x = f\,y$$

using the same idea.

For equality there is a recursor and an eliminator, and for the examples above we only need the recursor because we have no dependency on the actual proofs of equality. However, there is some dependency because equality itself is a dependent type. We assume a family that depends on the indices of equality but not on the equality proofs themselves: $C : A \to A \to \mathbf{Type}$ then the recursor is:

$$\mathrm{R}^{=} : (\Pi x : A.C\,x\,x) \to \Pi_{x,y:A} x = y \to C\,x\,y$$
$$\mathrm{R}^{=}\,f\,(\mathrm{refl}_a) :\equiv f\,a$$

Exercise 15 *Derive* sym, trans, resp *using the recursor* $\mathrm{R}^{=}$.

What would be a statement that actually depends on equality proofs? It seems that equality is rather trivial since there is at most one proof of it and we should be able to prove this. This is called uniqueness of equality proofs and states that any two proofs of equality are equal and it has an easy direct definition exploiting exactly the fact that the only proof of equality is reflexivity:

$$\text{uep} : \Pi_{x,y:A} \Pi p, q : x = y$$
$$\text{uep}\,\mathrm{refl}_a\,\mathrm{refl}_a :\equiv \mathrm{refl}_{\mathrm{refl}_a}$$

We now define the dependent eliminator for equality which $\mathrm{E}^{=}$ which is also called J but we stick to our terminology. This time we use a family that does indeed depend on the equality proof $C : \Pi_{x,y:A} x = y \to \mathbf{Type}$

$$\mathrm{E}^{=} : (\Pi_{x:A} C\,(\mathrm{refl}_x)) \to \Pi_{x,y:A} \Pi p : x = y.C\,p$$
$$\mathrm{E}^{=}\,f\,\mathrm{refl}_x :\equiv f_x$$

Now we should be able to perform the usual exercise and reduce the direct definition of uep to one using only the eliminator. The first step is clear, by eliminating one argument we can reduce the problem to:

$$\Pi_{x:A} \Pi q : x = x.\mathrm{refl}_x = q$$

but now we are stuck. We cannot apply the eliminator because we need a family were both indices are arbitrary. Indeed, Hofmann and Streicher were able to show that uep is unprovable from the eliminator (Hofmann and Streicher 1998). In the next section we will discuss reasons why this is actually not a bad thing. However, at least based on our current understanding of equality it seems that this is an unwanted incompleteness. One which can actually fixed by introducing a special eliminator which works exactly in the case when we want to prove something about equality proofs where both indices are equal. That is we assume as given a family $C : \Pi_{x:A} x = x \to \mathbf{Type}$ and introduce

$$K : (\Pi_{x:A} C\, \mathrm{refl}_x) \to \Pi_{x:A} \Pi p : x = x.C\ p$$

$$K\ f\ \mathrm{refl}_x :\equiv f_x$$

This eliminator is called K because K is the next letter after J.

Exercise 16 *Derive uep using only* $\mathrm{E}^=$ *and* K*.*

Exercise 17 *Instead of viewing equality as a relation generated by* refl*, we can also fix one index* $a : A$ *and now define the predicate of being equal to a:* $a = - : A \to$ **Type**. *This predicate is generated by* $\mathrm{refl}_a : a = a$ *so no change here. However, the eliminator looks different. Let's fix* $C : \Pi x : A.a = x \to$ **Type**

$$J'_a : (C\, \mathrm{refl}_a) \to \Pi_{x:A} \Pi p : a = x.C\ p$$

$$J'_a\ f\ (\mathrm{refl}_a) :\equiv f$$

Show that $\mathrm{E}^=$ *and* J' *are interderivable, that is both views of equality are equivalent.*

5.5 Homotopy Type Theory

5.5.1 Proof Relevant Equality

If we are only use $\mathrm{E}^=$ aka J we cannot in general prove that there is only one proof of equality, but what can we prove? It turns out that we can indeed verify some equalities:

$$\begin{aligned}
\mathrm{trans}\ p\ \mathrm{refl} &= p \\
\mathrm{trans}\ \mathrm{refl}\ p &= p \\
\mathrm{trans}\ (\mathrm{trans}\ p\ q)\ r &= \mathrm{trans}\ p\ (\mathrm{trans}\ q\ r) \\
\mathrm{trans}\ p\ (\mathrm{sym}\ p) &= \mathrm{refl} \\
\mathrm{trans}\ (\mathrm{sym}\ p)\ p &= \mathrm{refl}
\end{aligned}$$

where $p : a =_A b$, $q : b =_A c$ and $r : c =_A d$.

It is easy to verify these equalities using only J because all the quantifications are over arbitrary $p : x = y$ and there is no repetition of variables.

Exercise 18 *Explicitly construct the proofs using J.*

A structure with these properties is called a groupoid. A groupoid is a category where every morphism is an isomorphism. Here the objects are the elements of the type A, given $a, b : A$ the homset (actually a type) is $a =_A b$, composition is trans, identity is refl and sym assigns to every morphism its inverse.

We can go further an observe that resp also satisfies some useful equalities:

$$\text{resp}\, f\, \text{refl} = \text{refl}$$

$$\text{resp}\, f\, (\text{trans}\, p\, q) = \text{trans}\, (\text{resp}\, f\, p)\, (\text{resp}\, f\, q)$$

where $f : A \to B, p : a =_A b, q : b =_A c$.

In categorical terms this means that f is a functor: its effect on objects $a : A$ is $f\, a : B$ and its effect on morphism $p : a =_A b$ is $\text{resp}\, f\, p : f\, a =_B f\, b$.

Exercise 19 *Why don't we prove that* resp f *also preserves symmetries?*

$$\text{resp}\, f\, (\text{sym} p) = \text{sym}\, (\text{resp}\, f\, p)$$

Indeed, the idea of Streicher's and Hofmann's proof is to turn this around and to show that we can generally interpret types as groupoids where the equality type corresponds to the homset. Interestingly we can also interpret J in this setting but clearly we cannot interpret K because it forces the groupoid to be trivial, i.e. to be an equivalence relation. However, in the moment there is no construction which generates non-trivial groupoids, but this will change once we have the univalence principle or if we introduce Higher Inductive Types (HITs).

However, groupoids are not the whole story. There is no reason to assume that the next level of equalities, i.e. the equality of the equality of equality proofs is trivial. We need to add some further laws, so called coherence laws, which are well known and we end up with a structure which is called a 2-Groupoid (in particular all the homsets are groupoids). But the story doesn't finish here we can go on forever. The structure we are looking for is called a weak ω-groupoid. Alas, it is not very easy to write down what this is precisely. Luckily, homotopy theoreticians have already looked at this problem and they have a definition: a weak ω-groupoid is a Kan complex, that is a simplicial set with all Kan fillers. This is what Voevodsky has used in his homotopical model of Homotopy Type Theory (Kapulkin and Lumsdaine 2012). However, it was noted that this construction uses classical principles, i.e. the axiom of choice. Thierry Coquand and his team have now formulated a constructive alternative which is based on cubical sets (Cohen et al. 2016).

5.5.2 What Is a Proposition?

Previously, we have identified propositions as types but it is fair enough to observe that types have more structure, they can carry more information by having different inhabitants. This becomes obvious in the next exercise:

Exercise 20 *Given A, B :* **Type** *and a relation $R : A \to B \to$* **Type** *the axiom of choice can be stated as follows:*

$$(\forall x : A.\exists y : B.R\,x\,y) \to \exists f : A \to B.\forall x : A.R\,x\,(f\,x)$$

Apply the propositions as types translation and prove the axiom of choice.

This is strange because usually the axiom of choice is an indicator of using some non-constructive principle in Mathematics. But now we can actually prove it in Type Theory? Indeed, the formulation above doesn't really convey the content of the axiom of choice because the existential quantification is translated as a Σ-type and hence makes the choice of the witness explicit. In conventional Mathematics propositions do not carry any information hence the axiom of choice has to build the choice function without any access to choices made when showing the premise.

To remedy this mismatch we are more specific about propositions: we say that a type P : **Type** is a proposition if it has at most one inhabitant, that is we define

$$\text{isProp} : \mathbf{Type} \to \mathbf{Type}$$

$$\text{isProp}\,A :\equiv \Pi x, y : A.x =_A y$$

I write P : **Prop** for a type that is propositional, i.e. we can prove isProp P. That is we are interpreting **Prop** as $\Sigma P : \mathbf{Type}.\text{isProp}\,P$ but I am abusing notation in that I omit the first projection (that is an instance of subtyping as an implicit coercion).

Exercise 21 *Show that equality for natural numbers is a proposition, that is establish:*

$$\forall x, y : \mathbb{N}.\text{isProp}\,(x =_{\mathbb{N}} y)$$

Looking back at the proposition as types translation we would like that for any proposition in predicate logic P we have that $[\![P]\!]$: **Prop**. That can be shown to be correct for the so called negative fragment, that is the subset of predicate logic without $\vee$ and $\exists$.

Exercise 22 *Show that if P, Q :* **Prop** *and $R : A \to$* **Prop** *where A :* **Type** *then*

1. $[\![P \Rightarrow Q]\!]$: **Prop**
2. $[\![P \wedge Q]\!]$: **Prop**
3. $[\![\text{True}]\!]$: **Prop**

4. $[\![\text{False}]\!] : \mathbf{Prop}$
5. $[\![\forall x : A.P]\!] : \mathbf{Prop}$

However this fails for disjunction and existential quantification: $[\![\text{True} \vee \text{True}]\!]$ is equivalent to Bool which is certainly not propositional since true $\neq$ false. Also $\exists x : \text{Bool}.\text{True}$ is equivalent to Bool and hence also not propositional.

To fix this we introduce a new operation which assigns to any type a proposition which expresses the fact that the type is inhabited. That is given $A : \mathbf{Type}$ we construct $\|A\| : \mathbf{Prop}$, this is called the *propositional truncation* of A. We can construct elements of $\|A\|$ from elements of A, that is we have a function $\eta : A \to \|A\|$. However, we hide the identity of a, that is we postulate that $\eta\, a = \eta\, b$ for all $a, b : A$. How can we construct a function out of $\|A\|$? It would be unsound if we would allow this function to recover the identity of an element. This can be avoided if the codomain of the function is itself propositional. That is given $P : \mathbf{Prop}$ and $f : A \to P$ we can lift this function to $\hat{f} : \|P\| \to A$ with $\hat{f}(\eta\, a) \equiv f\, a$.

Using $\|-\|$ we can redefine $[\![-]\!]$ such that $[\![P]\!] : \mathbf{Prop}$:

$$[\![P \vee Q]\!] :\equiv \|[\![P]\!] + [\![Q]\!]\|$$
$$[\![\exists x : A.P]\!] :\equiv \|\Sigma x : A.[\![P]\!]\|$$

Now the translation of the axiom of choice

$$(\forall x : A.\exists y : B.R\, x\, y) \to \exists f : A \to B.\forall x : A.R\, x\, (f\, x)$$

which is

$$\Pi x : A.\|\Sigma y : B.R\, x\, y\| \to \|\Sigma f : A \to B.\forall x : A.R\, x\, (f\, x)\|$$

is more suspicious. It basically say if for every $x : A$ there is $y : B$ with a certain property, but I don't tell you which one, then there is a function $f : A \to B$ which assigns to every $x : A$ a $f\, x : B$ with a certain property but I don't tell you which function. This sounds like a lie to me!

Indeed lies can make our system inconsistent or they can lead to classical principle, which can be viewed is a form of lying that is not known to be inconsistent. Indeed, assuming one additional principle, *propositional extensionality* we can derive the principle of excluded middle. This proof is due to Diaconescu.

By propositional extensionality we mean that two propositions which are logically equivalent then they are equal:

$$\text{propExt} : \Pi P, Q : \mathbf{Prop}.(P \Leftrightarrow Q) \to P = Q$$

This is reasonable because all what matters about a proposition is wether it is inhabited, and hence two proposition which are logically equivalent are actually indistinguishable and hence extensionally equal. Indeed, we will see that propExt is a consequence of the univalence principle.

Theorem 1 (Diaconescu) *Assuming the amended translation of the axiom of choice:*

$$\text{ac} : \Pi x : A.\|\Sigma y : B.R\,x\,y\| \to \|\Sigma f : A \to B.\forall x : A.R\,x\,(f\,x)\|$$

and propExt *we can derive the excluded middle for all propositions:*

$$\forall P : \mathbf{Prop}.P \vee \neg P$$

We are going to instantiate A with the type of inhabited predicates over Bool, that is

$$A :\equiv \Sigma Q : \text{Bool} \to \mathbf{Prop}.\exists b : \text{Bool}.Qb$$

$B \equiv \text{Bool}$ and the relation $R : A \to B \to \mathbf{Prop}$ is defined as follows:

$$R\,(Q, q)\,b :\equiv Q\,b$$

that is an inhabited predicate is related to the boolean it inhabits. Can we now prove the premise of the axiom?

$$\Pi x : A.\|\Sigma y : B.R\,x\,y\|$$

that is after plugging in the definition of A, B, R it becomes

$$\Pi(Q, q) : A.\|\Sigma b : B.Q\,b\|$$

and after some currying

$$\Pi Q : \text{Bool} \to \mathbf{Prop}.\|\Sigma b : \text{Bool}.Qb\| \to \|\Sigma b : B.Q\,b\|$$

it is quite obvious that we can.

Now let's look at the conclusion.

$$\|\Sigma f : A \to B.\Pi x : A.R\,x\,(f\,x)\|$$

Let's for the moment ignore the outermost $\|-\|$ and expand the types inside:

$$\begin{aligned}&\Sigma f : A \to \text{Bool}.\\&\quad\Pi(Q, q) : A.Q\,(f\,Q)\end{aligned}$$

We consider two special predicates $T, F : \text{Bool} \to \mathbf{Prop}$:

$$T\,b :\equiv b = \text{true} \vee P$$

$$F\,b :\equiv b = \text{false} \vee P$$

Where does the P come from? From the beginning of this section: it is the $P : \mathbf{Prop}$ for which we want to show $P \vee \neg P$.

Now applying the conclusion of the axiom to these predicates we obtain two booleans $f\,T, f\,F : \text{Bool}$ and from the second part we know

$$T\,(f\,T) \equiv (f\,T = \text{true}) \vee P$$

$$F\,(f\,F) \equiv (f\,F = \text{false}) \vee P$$

Now let's analyse all the possibilities: there are four combinations:

1. $f\,T = \text{true} \wedge f\,F = \text{false}$
2. $f\,T = \text{true} \wedge P$
3. $P \wedge f\,F = \text{false}$
4. $P \wedge P$

In 2–4 we know that P holds, the only other alternative in which P is not proven is 1. In this case we can show $\neg P$ that is $P \to 0$. For this purpose assume P, in this case both $T\,b$ and $F\,b$ are provable for any b because P is and this means $T\,b \Leftrightarrow F\,b$ which now using propositional extensionality implies $T\,b = F\,b$. But this means that $T = F$ using functional extensionality. This cannot be since we assumed that $f\,T = \text{true}$ and $f\,F = \text{false}$, now $T = F$ would imply $\text{true} = \text{false}$ which is false that is it implies 0. Hence we have shown $\neg P$ by deriving a contradiction from assuming P.

To summarise we have shown $P \vee \neg P$ because in the case 2–4 we have P and in 1 we have $\neg P$. But hang on what about the $||-||$ we have been ignoring? It doesn't matter since $P \vee \neg P$ is already a proposition and putting $||-||$ around it doesn't change anything.

What has really happened is that from the lie that we can recover information we have just hidden we can extract information as long as we hide the function doing the extraction. And this has nothing to do with Σ-types and existentials but with the behaviour of the hiding operation, or inhabitance $||-||$. Hence we can formulate a simpler version of the axiom of choice in Type Theory:

$$(\Pi x : A.\|B\,x\|) \to \|\Pi x : A.B\,x\|$$

This implies the revised translation of the axiom.

5.5.3 Dimensions of Types

We have classified types as propositions if they have at most one inhabitant. uep says that equality is a proposition. Even if we are not accepting uep, we can use this to classify types: we say that a type is a set if all its equalities are propositions.

$$\text{isSet} : \textbf{Type} \to \textbf{Type}$$
$$\text{isSet}\, A :\equiv \Pi_{x,y:A} \text{isProp}\,(x =_A y)$$

As for propositions we abuse notation and write A : **Set** if A : **Type** and we can show isSet A. uep basically says that all types are sets. The idea here is that sets are types which are quite ordinary, they reflect our intuition that equality is propositional.

Actually we can show that certain types are sets, e.g. Exercise 21 basically asks to prove that $\mathbb{N}$: **Set**. However, we can do much better, we can show in general that any type with a decidable equality is a set. This is a theorem due to Michael Hedberg.

Theorem 2 (Hedberg) *Given* A : **Type** *such that the equality is decidable*

$$d : \forall x, y : A.x = y \vee x \neq y$$

then we can show

$$\text{isSet}\, A$$

To show Hedberg's theorem we establish a lemma saying that if there is a constant function on equality types then the equality is propositional. That is we assume

$$f : \Pi_{x,y:A} x =_A y \to x =_A y$$
$$c : \Pi_{x,y:A} \Pi p, q : x =_A y \to f\ p = f\ q$$

Now for arbitrary $x, y : A$, $p : x = y$ we can show that $p = \text{trans}\,(f\ p)(\text{sym}\,(f\ \text{refl}))$ because using J this reduces to showing that $\text{refl} = \text{trans}\,(f\ \text{refl})\,(\text{sym}\,(f\ \text{refl}))$ which is one of the groupoid properties. Now given any $p, q : x = y$ we can show

$$\begin{aligned} p &= \text{trans}\,(f\ p)\,(\text{sym}\,(f\ \text{refl})) \\ &= \text{trans}\,(f\ q)\,(\text{sym}\,(f\ \text{refl})) \qquad \text{using } c \\ &= q \end{aligned}$$

And hence $=_A$ is propositional.

To construct the function f from decidability we assume $p : x = y$ and apply $d\,x\,y$. Either we have another proof $q : x = y$ and we return this one, or we have that $x \neq y$ but this contradicts that we already have a proof $p : x = y$ and we can use R^0. However, in either case the output doesn't depend on the actual value of the input and hence we can show that the function is constant.

Indeed, if we additionally assume the principle of functional extensionality

$$\mathrm{funExt} : \forall f, g : A \to B.(\forall x : A.f\,x = g\,x) \to f = g$$

we can strengthen Hedberg's theorem:

Theorem 3 *Given* A : **Type** *such that the equality is stable*

$$s : \forall x, y : A.\neg\neg(x = y) \to x = y$$

then we can show

$$\mathrm{isSet}\,A$$

This strengthening shows that function types are sets, e.g. $\mathbb{N} \to \mathbb{N}$: **Set** even though its equality is not decidable but it is stable.

To prove the stronger version we observe that for if equality is stable, we can construct the function $g : \Pi x, y : A.x = y \to x = y$ by composing s with the obvious embedding $x = y \to \neg\neg(x = y)$. Since $\neg\neg(x = y)$ is propositional, f must be constant.

Exercise 23 *Show that the equality of* $\mathbb{N} \to \mathbb{N}$ *is stable.*

We can extend the hierarchy we have started to construct with **Prop** and **Set**. For example a type whose equalities are sets is called a groupoid (indeed it corresponds to the notion of groupoid which we have introduced previously).

$$\mathrm{isGroupoid} : \mathbf{Type} \to \mathbf{Type}$$
$$\mathrm{isGroupoid}\,A :\equiv \forall x, y : A.\mathrm{isSet}\,(x =_A y)$$

We can also extend this hierarchy downwards, we can redefine a proposition as a type such that its equalities are types with exactly one element – these types are called *contractible*:

$$\mathrm{isContractible} : \mathbf{Type} \to \mathbf{Type}$$
$$\mathrm{isContractible}\,A :\equiv \Sigma a : A.\Pi x : A.x = a$$

$$\mathrm{isProp} : \mathbf{Type} \to \mathbf{Type}$$
$$\mathrm{isProp}\,A :\equiv \Pi x, y : A.\mathrm{isContractible}\,(x = y)$$

We can now define the hierarchy starting with contractible types. For historic reasons (i.e. to be compatible with notions from homotopy theory) we start counting with -2 and not with 0. I am calling the levels dimensions, they are also called truncation levels.

$$\begin{aligned}
&\text{hasDimension} : \mathbb{N}_{-2} \to \textbf{Type} \to \textbf{Type} \\
&\text{hasDimension}\,(-2)\,A \quad :\equiv \text{isContractible}\,A \\
&\text{hasDimension}\,(n+1)\,A \quad :\equiv \forall x, y : A.\text{hasDimension}\,n(x =_A y)
\end{aligned}$$

I am writing $\mathbb{N}_{-2}$ for a version of the natural numbers where I start counting with -2. To summarize the definitions so far:

Dimension	Name
-2	Contractible types
-1	Propositions
0	Sets
1	Groupoids

We also introduce the notation $n-$**Type** for A : **Type** such that we can show hasDimension n A.

To convince ourselves that this really is a hierarchy, that is that every $n-$**Type** is also a $n+1-$**Type** we need to show that the hierarchy actually stops at -2 that is that the equality of a contractible type is again contractible.

To show this assume is given a contractible type A : **Type** that is we have $a : A$ and $c : \Pi x : A.a = x$. Now we want to show that for all $x, y : A$ the equality $x =_A y$ is contractible, that is we have an element and all other elements are equal. We define $d : \Pi x, y : A.x =_A y$ as $d\,x\,y :\equiv \text{trans}\,(\text{sym}\,(c\,x))\,(c\,y)$. Now it remains to show that $e : \Pi x, y : A.\Pi p : x =_A y.d\,x\,y = p$. Using J we can reduce this to $d\,x\,x = \text{refl}\,x$ unfolding $d\,x\,x \equiv \text{trans}\,(\text{sym}\,(c\,x))\,(c\,x)$ we see that this is an instance of one of our groupoid laws.

As a corollary we obtain that hasDimension n A implies hasDimension $(n+1)$ A.

5.5.4 Extensionality and Univalence

As I have already mentioned in the introduction: extensionality means that we identify mathematical objects which behave the same even of they are defined differently. An example are the following two functions:

$$\begin{aligned}
&f, g : \mathbb{N} \to \mathbb{N} \\
&f\,x :\equiv x + 1 \\
&g\,x :\equiv 1 + x
\end{aligned}$$

We can show that $\forall x : \mathbb{N}. f\,x = g\,x$ but can we show that $f = g$? The answer is **no**, because if there were a proof without any assumption it would have to be refl and this would only be possible if $f \equiv g$ but it is clear that they are not definitionally equal. However, we cannot exhibit any property not involving this equality which would differentiate them. Another example are the following two propositions:

$$P, Q : \mathbf{Prop}$$

$$P :\equiv \text{True}$$

$$Q :\equiv \neg\text{False}$$

Again we cannot show that $P = Q$ even though there clearly is no way to differentiate between them.

We have already mentioned the two principles which are missing here':

$$\text{funExt} : \forall f, g : A \to B.(\forall x : A.f\,x = g\,x) \to f = g$$

$$\text{propExt} : \Pi P, Q : \mathbf{Prop}.(P \Leftrightarrow Q) \to P = Q$$

We note that the corresponding principles are true in set theory, which seems to contradict my statement that Type Theory is better for extensional reasoning than set theory. However, this shortcoming can be fixed since all constructions preserve these equalities. This can be made precise by interpreting the constructions in the setoid model, where every type is modelled by a set with an equivalence relation. In this case we can model function types by the set of functions and equality is extensional equality. The same works for **Prop**, a proposition is modelled by the set of propositions identified if they are logically equivalent.

However, the shortcoming of set theory becomes obvious if we ask the next question: when are two sets equal? For example

$$A, B : \mathbf{Set}$$

$$A :\equiv \overline{1} + \overline{2}$$

$$B :\equiv \overline{2} + \overline{1}$$

We have no way to distinguish two sets with 3 elements, hence following the same logic as above they should be equal. However, they are not equal in set theory under the usual encoding of finite sets and + and it would be hard to fix this in general.

When are two sets equal? Given $f : A \to B$ we say that f is an isomorphism if there it has an inverse, that is[4]

$$\text{isIso} : (A \to B) \to \mathbf{Type}$$

$$\text{isIso}\, f :\equiv \quad \begin{array}{ll} \Sigma & g : B \to A \\ & \eta : \Pi x : B.f\,(g\,x) = x \\ & \epsilon : \Pi x : A.g\,(f\,x) = x \end{array}$$

And we define $A \simeq B :\equiv \Sigma f : A \to B.\text{isIso}\, f$. Using isIso we can formulate a new extensionality principle: we want to say that there is an isomorphism between isomorphism of sets and equality of sets. Indeed, we can observe that the is a function from equality of sets to isomorphism because every set is isomorphic to itself using J.

Exercise 24 *Define* eq2iso : $\Pi_{A,B:\mathbf{Set}} A = B \to A \simeq B$

Using this we can state extensionality for sets

$$\text{extSet} : \text{isIso}\,\text{eq2iso}$$

As a corollary we get that $A \simeq B \to A = B$. Since $\overline{1} + \overline{2} \simeq \overline{1} + \overline{2}$ we can show that $\overline{1} + \overline{2} = \overline{1} + \overline{2}$

Exercise 25 *Show that*

1. $(A + B) \to C = (A \to C) \times (B \to C)$
2. $(A \times B) \to C = A \to B \to C$
3. $1 + \mathbb{N} = \mathbb{N}$
4. $\mathbb{N} \times \mathbb{N} = \mathbb{N}$
5. $\text{List}\,\mathbb{N} = \mathbb{N}$
6. $\mathbb{N} \to \mathbb{N} \neq \mathbb{N}$

assuming A, B, C : **Set**. *For which of the results do we not need* extSet*?*

Exercise 26 *An alternative to isomorphism is bijection, one way to say that a function is bijective is to say there is a unique element in the domain which is mapped to an element of the codomain. We define* $\exists!$ *(exists unique):*

$$\exists! x : A.P\,x :\equiv \exists x : A.P\,x \wedge \forall y : A, P\,y \Rightarrow x = y$$

and using this we define what is a bijection:

$$\text{isBij} : (A \to B) \to \mathbf{Prop}$$

$$\text{isBij}\, f :\equiv \forall y : B.\exists! x : A.f\,x = y$$

[4] I am using here a record-like syntax for iterated Σ-types.

Show that isomorphism and bijection are logically equivalent:

$$\forall f : A \to B.\text{isBij}\, f \Leftrightarrow \text{isIso}\, f$$

extSet is incompatible with uniqueness of equality proofs (uep) because there are two elements of $\overline{2} \simeq \overline{2}$, namely identity and negation. If we assume that there is only one proof of equality for $\overline{2} = \overline{2}$ then we also identify this two proofs and hence $0_2 = 1_2$ which is inconsistent.

So far we have only considered sets what about types in general? There is a twist: if equality is not propositional then the η and ϵ components of isIso are not propositional in general. Indeed, assuming extSet for all types is unsound. Instead we need to refine the notion of isomorphism by introducing an extra condition which relates η and ϵ. We call this *equivalence* of types.

$$\begin{aligned}
&\text{isEquiv} : (A \to B) \to \textbf{Type} \\
&\text{isEquiv}\, f :\equiv \quad \Sigma g : B \to A \\
&\qquad \eta : \Pi x : B.f\,(g\,x) = x \\
&\qquad \epsilon : \Pi x : A.g\,(f\,x) = x \\
&\qquad \delta : \Pi x : A.\eta\,(f\,x) = \text{resp}\, f\,(\epsilon\,x)
\end{aligned}$$

And we define $A \cong B :\equiv \Sigma f : A \to B.\text{isEquiv}\, f$.

Exercise 27 *Define* eq2equiv : $\Pi_{A,B:\textbf{Set}} A = B \to A \cong B$

We can now state extensionality for types, which is what is commonly called univalence.

$$\text{univalence} : \text{isEquiv eq2equiv}$$

As before for extSet a consequence is that equivalence of types implies equality $A \cong B \to A = B$ which is what most people remember about univalence. In the special case of sets equivalence and isomorphism agree because the type of δ is an equivalence, that is for sets we have $(A \simeq B) = (A \cong B)$. But more is true, even for types in general equivalence and isomorphism are logically equivalent, that is $(A \simeq B) \Leftrightarrow (A \cong B)$. While this may be surprising at the first glance, it just means that we can define functions in both directions but they are not inverse to each other. Indeed, in general there are more proofs of an isomorphism that of an equivalence, indeed isEquiv f is a proposition, while isIso f in general isn't (but it is if f is a function between sets). However, the logical equivalence of equivalence and isomorphism means that to establish an equivalence all we need to do is to construct an isomorphism. In particular all the equalities of Exercise 25 hold for types in general.

The definition of isEquiv looks strangely asymmetric. Indeed, we could have replaced δ with its symmetric twin:

$$\delta' : \Pi y : A.\text{resp}\, g\, \eta\, y = \epsilon\, (g\, y)$$

We can either use δ or δ' it doesn't make any difference. Why don't we use both? Indeed, this would exactly be the definition of an adjunction between groupoids. However, assuming both messes everything up, now isEquiv f is no longer a proposition and we need to add higher level coherence equations to fix this. Indeed, there is such an infinitary (coinductive) definition of equivalence. But we don't need to use this, the asymmetric definition (or its mirror image) does the job.

Exercise 28 *We can extend Exercise* 26 *to the case of equivalences by taking the equality proofs into account. That is we say that not only there is a unique inverse but that the pair of inverses and the proof that there are an inverse are unique that is contractible. We define*

$$\text{isEquiv}' : (A \to B) \to \mathbf{Type}$$
$$\text{isEquiv}'\, f :\equiv \Pi y : B.\text{isContr}\, (\Sigma x : A.f\, x = y)$$

We have observed that already extSet implies that there are non-propositional equalities, e.g. $\overline{2} = \overline{2}$. In other words the first universe $\mathbf{Type}_0$ is not a set. Nicolai Kraus and Christian Sattler have generalised this and shown that using univalence we can construct types which are non $n - \mathbf{Type}$ for any $n : \mathbb{N}$ (Kraus and Sattler 2015). I leave is an exercise to do the first step in this construction:

Exercise 29 *Show that the equalities of* $\Sigma X : \mathbf{Type}_0.X = X : \mathbf{Type}_1$ *are not always sets and hence* $\mathbf{Type}_1$ *is not a groupoid.*

5.5.5 *Higher Inductive Types*

In HoTT we view a type together with its equality types with the structure of an ω-groupoid. Hence when we introduce a new type we also have to understand what its equality types are. So for example in the case of $\to$ and $\times$ we have the following equations:

$$f =_{A \to B} g = \Pi x, y : A.x =_A y \to f\, x = g\, x$$
$$(a, b) =_{A \times B} (a', b') = (a =_A a') \times (b =_B b')$$

When defining types inductively, that is by given the constructors, we can also add constructors for equalities. The simplest example for this is propositional truncation

$\|A\|$, which can be inductively defined by the following constructors:

$$\eta : A \to \|A\|$$
$$\mathrm{irr} : \Pi x, y : \|A\|.x =_{\|A\|} y$$

To define a non-dependent function $f : \|A\| \to B$ we need an function $g : A \to B$ but also an equalities $i : \Pi x, y : B.x =_B y$ which correspond to saying that $B : \mathbf{Prop}$. f will satisfy the following equations:

$$f\ (\eta a) :\equiv g\, a$$
$$\mathrm{resp}\, f\ (\mathrm{irr}\, x\ y) :\equiv i\ (f\ x)\ (f\ x)$$

Hence the eliminator does not only determine the behaviour of f on elements but also on equalities. This makes perfect sense because in the view of types as groupoids functions are functors. We can also turn this into a non-dependent eliminator:

$$\mathrm{R}^{\|A\|} : (A \to B) \to (\Pi x, y : B.x = y) \to \|A\| \to B$$
$$\mathrm{R}^{\|A\|}\, g\, i\ (\eta a) :\equiv g\, a$$
$$\mathrm{resp}\, \mathrm{R}^{\|A\|}\, g\, i\ (\mathrm{irr}\, x\ y) :\equiv i\ (\mathrm{R}^{\|A\|}\, g\, i\, x)\ (\mathrm{R}^{\|A\|}\, g\, i\, y)$$

We can also derive a dependent eliminator, for which we need a dependent version of resp.

Another class for examples for higher inductive types are set quotients, that is we have a set $A : \mathbf{Set}$ and a relation $R : A \to A \to \mathbf{Prop}$, we define $A/R : \mathbf{Set}$ by the constructors

$$[-] : A \to A/R$$
$$[-]^{=} : \Pi x, y : A.Rxy \to [x] =_{A/R} [y]$$
$$\mathrm{set} : \Pi_{x,y:A/R} \Pi p, q : x =_{A/R} y.p =_{x=_{A/R} y} q$$

Note that the last constructor is necessary to make sure that the type we construct is actually a set. We do not require that R is an equivalence relation but obviously the equality on A/R always is.

Exercise 30 *Derive the non-dependent eliminator for set quotients.*

An example is the definition of the integer as a quotient of pairs of integers:

$$\mathrm{eq}_{\mathbb{Z}} : (\mathbb{N} \times \mathbb{N}) \to (\mathbb{N} \times \mathbb{N}) \to \mathbf{Prop}$$
$$\mathrm{eq}_{\mathbb{Z}}\, (a^+, a^-)\, (b^+, b^-) :\equiv a^+ + b^- = b^= + a^- \qquad \mathbb{Z} = (\mathbb{N} \times \mathbb{N})/\mathrm{eq}_{\mathbb{Z}}$$

The idea is that a pair of natural numbers (a^+, a^-) represents its difference $a^+ - a^-$ and we identify pairs which have the same difference.

Exercise 31 *Define addition and multiplication and additive inverses on* $\mathbb{Z}$. *How would you prove the group laws?*

Higher Inductive Types even for sets are more general than set quotients because we can define elements and equalities at the same time. An example for this are infinite branching trees with permutations. That is we define infinitely branching trees

$$\begin{aligned}
&\mathrm{Tree} : \mathbf{Type}\\
&\mathrm{leaf} : \mathrm{Tree}\\
&\mathrm{node} : (\mathbb{N} \to \mathrm{Tree}) \to \mathrm{Tree}
\end{aligned}$$

We inductively define a relation which identifies trees upto permutations of subtrees:

$$\begin{aligned}
&- \sim - : \mathrm{Tree} \to \mathrm{Tree} \to \mathbf{Prop}\\
&\mathrm{perm} : \Pi_{g:\mathbb{N}\to\mathrm{Tree}} \Pi f : \mathbb{N} \to \mathbb{N}.\mathrm{isIso}\, f \to \mathrm{node}\, f \sim \mathrm{node}\,(\lambda n.g\ f\ n)\\
&\mathrm{respNode} : \Pi_{g,h:\mathbb{N}\to\mathrm{Tree}} (\Pi i : \mathbb{N}.f\ i\ g\ i) \to \mathrm{node}\, f \sim \mathrm{node}\, g
\end{aligned}$$

Now trees with permutations can be defined as $\mathrm{PTree} :\equiv \mathrm{Tree}/\sim$. However, it turns out that this type is not as useful as we might hope. Let's say we want to lift the node construction to PTree that is we want to define

$$\mathrm{node}' : (\mathbb{N} \to \mathrm{PTree}) \to \mathrm{PTree}$$

such that $\mathrm{node}'(\lambda n.[f\ n]) = [\mathrm{node}\, f]$. However, we have a problem, to define node$'$ we need to commute the quotient definition and function type, that is we need to go from $\mathbb{N} \to \mathrm{Tree}/\mathrm{PermTree}$ to a quotient on $\mathrm{Nat} \to \mathrm{Tree}$. This corresponds to an instance of the axiom of choice, in this case this is called countable choice because the indexing type A is $\mathbb{N}$.

Exercise 32 *Show that we can define* node$'$ *using countable choice (i.e.* $(\Pi x : \mathbb{N}.\|B\,x\|) \to \|\Pi x : \mathbb{N}.B\,x\|$ *for* $B : \mathbb{N} \to \mathbf{Set}$*).*

However, using HITs we have an alternative: we can define permutable trees by introducing the elements and the equalities at the same time:

$$\begin{aligned}
&\mathrm{PTree} : \mathbf{Type}\\
&\mathrm{leaf} \quad : \mathrm{PTree}\\
&\mathrm{node} \quad : (\mathbb{N} \to \mathrm{PTree}) \to \mathrm{PTree}\\
&\mathrm{perm} \quad : \Pi_{g:\mathbb{N}\to\mathrm{Tree}} \Pi f : \mathbb{N} \to \mathbb{N}.\mathrm{isIso}\, f \to \mathrm{node}\, f = \mathrm{node}\,(\lambda n.g\ f\ n)
\end{aligned}$$

We don't need to add a constructor corresponding to respNode because we know that any function respects equality. We don't ned to derive node′ because this is already captured by node itself. In the HoTT book an instance of this idea is used to define the Cauchy Reals (The Univalent Foundations Program 2013), section 11.3. The naïve definition identifies converging sequences of rational numbers but we can't show that this definition is Cauchy complete that is that we can identify converging sequences of real numbers. Again this can be fixed by using countable choice or alternatively by a HIT that defines the reals and their equality at the same time, now Cauchy completeness is simply a constructor.

All the examples of HITs we have seen so far are not really *higher* because they work on the level of sets. The only good examples of true HITs I know come from *synthetic homotopy theory*, here we use HoTT to model constructions in Homotopy theory. In homotopy theory we look at the path spaces of geometric objects: we start with paths between points and then we can have paths between paths which are continuous transformation of paths. The main idea behing synthetic homotopy theory is to identify paths with equality types.

A simple example is the circle. The paths on a circle correspond to the integers, they measure how many times we move around the circle, we have to use integers because we can go forwards and backwards. In HoTT the circle can be defined as the following HIT:

$$S^1 : \mathbf{Type}$$

$$\text{base} : S^1$$

$$\text{loop} : \text{base} =_{S^1} \text{base}$$

Intuitively we imagine that the circle is generated by a point base and by a path loop from the point to itself. S^1 is not a set because there is no reason to assume that loop and refl are equal and also loop and sym loop are not equal. Indeed, we can define a function

$$\text{encode} : \mathbb{Z} \to \text{base} = \text{base}$$

$$\text{encode}\, n :\equiv \text{loop}^n$$

$$\text{encode}\, 0 :\equiv \text{refl}$$

$$\text{encode}\, (-n) :\equiv \text{loop}^{n-1}$$

To define a function in the other direction is a bit more tricky. We cannot define a function directly from base = base because none of the indices is variable. However, using the result from Exercise 17 it is enough if one of the indices is variable. That is we can use base $= x$. However, we also have to generalize the left hand side to depend on x. We define a family $X : S^1 \to \mathbf{Type}$ s.t. X base are the integers. We also need to interpret loop as an equality of types $\mathbb{Z} = \mathbb{Z}$. We don't want to use the trivial equality because the isomorphism should represent the number of times we have used the loop (and its direction). We observe that $\lambda x.x + 1, \lambda x.x - 1 : \mathbb{Z} \to \mathbb{Z}$ are inverses hence using univalence (actually set extensionality is enough) we can derive a non-trivial proof $p : \mathbb{Z} = \mathbb{Z}$. Hence to summarise we define

$$X : S^1 \to \mathbf{Type}$$
$$X\,\text{base} :\equiv \mathbb{Z}$$
$$\text{resp}\,X\,\text{loop} :\equiv p$$

We define our generalised decode function using equality elimination

$$\text{decode}' : \Pi_{x:S^1}\text{base} = x \to X\,x$$
$$\text{decode}'\,\text{refl} :\equiv 0$$

Indeed, $X\,\text{base} \equiv \mathbb{Z}$ and for refl we want to return 0. We obtain now decode by instantiating decode$'$:

$$\text{decode} : \text{base} = \text{base} \to \mathbb{Z}$$
$$\text{decode} :\equiv \text{decode}'_{\text{base}}$$

While it is well known in Homotopy Theory that the group of paths of the circle is the integers the particular proof is due to Dan Licata who formalised it in Agda, see Licata and Shulman (2013). Dan also uses dependent elimination to show that encode and decode are indeed inverse to each other. This shows that $(\text{base} =_{S^1} \text{base}) = \mathbb{Z}$. However, it is not always the case that the loop space of a HIT is a set it can be a higher dimensional type again. This is especially true once we move from the circle to the sphere S^2 which can be defined as follows:

$$S^2 : \mathbf{Type}$$
$$\text{base} : S^1$$
$$\text{loop} : \text{refl}_{\text{base}} =_{\text{base}=_{S^2}\text{base}} \text{refl}_{\text{base}}$$

The idea is that surface of the sphere can be generated by a higher path which maps the empty path on the pole onto itself. Now it may appear that the sphere is quite uninteresting because all the paths base $=_{S_2}$ base can be transformed into refl. However, we cannot prove that this type is contractible, that is has exactly on element. The reason is that it has non-trivial higher path spaces, and indeed lots of them. From Homotopy theory it is known that all higher path spaces of the sphere are non-trivial hence S^2 is not an $n-$**Type** for any n.

5.5.6 *Example: Definition of the Integers*

As a more mundane example we consider the definition of the integers ($\mathbb{Z}$: **Type**) as a HIT. Yes, we have already defined the integers as a set quotient but I think this definition is actually nicer.

We have the following element constructors

$$0 : \mathbb{Z}$$
$$\mathrm{suc} : \mathbb{Z} \to \mathbb{Z}$$
$$\mathrm{pred} : \mathbb{Z} \to \mathbb{Z}$$

To get the type we want we need to add some constructors for equalities, namely:

$$\mathrm{sucpred} : \Pi i : \mathbb{Z}.\mathrm{suc}\,(\mathrm{pred}\, i) =_{\mathbb{Z}} i$$
$$\mathrm{predsuc} : \Pi i : \mathbb{Z}.\mathrm{pred}\,(\mathrm{suc}\, i) =_{\mathbb{Z}} i$$

However we are not done yet! The type we have constructed so far is certainly not a set. For example there are two ways to prove

$$\mathrm{suc}\,(\mathrm{pred}\,(\mathrm{suc}\, 0)) = \mathrm{suc}\, 0$$

namely

$$\mathrm{sucpred}\,(\mathrm{suc}\, 0)$$
$$\mathrm{resp}\ \mathrm{suc}\,(\mathrm{predsuc}\, 0)$$

We certainly want the integers to be a set, in particular they should have a decidable equality. One way to achieve this would be to equate all proofs by adding another higher constructor:

$$\mathrm{isSet} : \Pi i, j : \mathbb{Z}.\Pi p, q : i =_{\mathbb{Z}} j \to p =_{i =_{\mathbb{Z}} j} q$$

We call such a first order HIT a Quotient Inductive Type (QIT).

As an example let us define addition for integers. I will do this in a pattern matching style. It is certainly a good idea to derive the eliminator and check that the definitions can be reduced to this but just writing down the eliminator is quite a mouthful.

To define $+ : \mathbb{Z} \to \mathbb{Z} \to \mathbb{Z}$ we start with the element constructors:

$$0 + i :\equiv i$$
$$\mathrm{suc}\, j + i :\equiv \mathrm{suc}\,(i + j)$$
$$\mathrm{pred}\, j + i :\equiv \mathrm{pred}\,(i + j)$$

To handle the equational constructor we overload functions so that we can also apply them to equalities. In this case we consider the function $+i$ for some $i : \mathbb{Z}$, now given an equality $p : j =_{\mathbb{Z}} k$ we write $p+i : j+i = k+i$ as a shorthand for resp $(+i)\ p$. Using this notation we can give definitions for the equational constructors as well:

$$\text{sucpred}\ j + i :\equiv \text{sucpred}\ (j + i)$$
$$\text{predsuc}\ j + i :\equiv \text{predsuc}\ (j + i)$$

We also have to deal with the isSet constructor and indeed we can extend the overload the notation also to deal with hight proofs. However, all what we need to notice that the constructor forces us to show that the codomain of the operation is a set, which it is since it is the integers themselves.

Exercise 33 *Define the additive inverse* $- : \mathbb{Z} \to \mathbb{Z}$ *in the same style.*

Now let's prove something about addition, for example that it is associative.

$$\text{assoc} : \Pi_{i,j,k:\int}.(i + j) + k =_{\mathbb{Z}} i + (j + k)$$

We are in for a pleasant surprise: since the codomain of associativity is a proposition we don't have to do anything for the equational constructors because the equations they induce hold trivially. Hence it is enough to only consider element constructors:

$$\text{assoc}_0 :\equiv \text{refl}$$
$$\text{assoc}_{\text{suc}\,i} :\equiv \text{suc}\ (\text{assoc}_i)$$
$$\text{assoc}_{\text{pred}\,i} :\equiv \text{pred}\ (\text{assoc}_i)$$

Note that in the definition above we again lift operations to equalities.

Exercise 34 *Show that the additive inverse behaves properly, that is*

$$\Pi_{i:\mathbb{Z}}(-\,i) + i = 0$$

To show that it is also the 2nd equation hold you need to prove the lemmas you need to prove commutativity. Hence as an extended exercise prove that the integers form a commutative group.

To show that the equality of integers is decidable we need to define a normalisation function. A reasonable notion of normal form is signed numbers, that is we define $\mathbb{Z}^{\text{nf}}$: **Type** by the following constructors:

$$0 : \mathbb{Z}^{\text{nf}}$$
$$+ : \mathbb{N} \to \mathbb{Z}^{\text{nf}}$$
$$- : \mathbb{N} \to \mathbb{Z}^{\text{nf}}$$

$\mathbb{Z}^{\mathrm{nf}}$ is a set because all inductive definitions which don't refer to higher types are automatically sets. We can easily embed the normal forms into the integers, by defining the functions pos, neg : $\mathbb{N} \to \mathbb{Z}^{\mathrm{nf}}$ which repeat the predecessor n-times

$$\begin{aligned} \mathrm{pos}\, 0 &:\equiv 0 \\ \mathrm{pos}\, (\mathrm{suc}\, n) &:\equiv \mathrm{suc}\, (\mathrm{pos}\, n) \\ \mathrm{neg}\, 0 &:\equiv 0 \\ \mathrm{neg}\, (\mathrm{suc}\, n) &:\equiv \mathrm{pred}\, (\mathrm{neg}\, n) \end{aligned}$$

Putting everything together we define emb : $\mathbb{Z}^{\mathrm{nf}} \to \mathbb{Z}$ by

$$\begin{aligned} \mathrm{emb}\, 0 &:\equiv 0 \\ \mathrm{emb}\, (+\, i) &:\equiv \mathrm{pos}\, (\mathrm{suc} i) \\ \mathrm{emb}\, (-\, i) &:\equiv \mathrm{neg}\, (\mathrm{suc} i) \end{aligned}$$

Exercise 35 *Define a function* nf : $\mathbb{Z} \to \mathbb{Z}^{\mathrm{nf}}$ *that is inverse to* emb, *i.e. establish the following propositions:*

$$\begin{aligned} &\Pi_{i:\mathbb{Z}} \mathrm{emb}\, (\mathrm{nf}\, i) = i \\ &\Pi_{i:\mathbb{Z}^{\mathrm{nf}}} \mathrm{nf}\, (\mathrm{emb}\, i) = i \end{aligned}$$

You may notice that using univalence we have established that $\mathbb{Z} = \mathbb{Z}^{\mathrm{nf}}$, hence what is the point of defining the integers as a HIT? It turns out that while the two objects are extensionally equivalent they are intensionally different and the HIT definition is much easier to work with.

Let's go back to the definition of the integers. When we added the constructor isSet we forced $\mathbb{Z}$ to be a set but by using a steam hammer. This is bad consequences when we want to define functions from the integers into a type which is not a set like S^1 earlier. The only way to do this would be going via the normal form because we can eliminate into any type from them. However, this spoils the advantages we just claimed from using the HIT. Luckily, there is another way which basically means we solve a coherence problem. Instead of isSet we add the following constructor:

$$\mathrm{coh} : \mathrm{sucpred}\, (\mathrm{suc}\, i) = \mathrm{resp}\, \mathrm{suc}\, (\mathrm{predsuc}\, i)$$

This is enough to turn $\mathbb{Z}$ into a set. Indeed, the constructors pred, predsuc, sucpred together just state that suc is an equivalence We can see this by adopting the normalisation proof to this setting. This can be substantially simplified by observing that the constructors that state that suc is an equivalence form a proposition.

We end with an open problem which suggested by Paolo Capriotti. When moving from natural numbers to lists we are defining the free monoid over an arbitrary set of generators. We can do the same with the integers and define the free group FG A over a type A by labelling suc and pred in the definition with elements of A. So in the way we have a cons and an anti-cons and equations that they cancel each other. As before we are saying that suc a is an equivalence. Now the question is given A : **Set** is it the case that FG A is a set? If A has a decidable equality we can still define a normalisation function and establish the property this way. However, we don't know how to do this in the general case.

Acknowledgements There are very many people to thank: my colleagues from whom I learned very much and the students who attended my talks and courses and provided important feedback and comments. I would like to single out two people, who sadly passed away recently well before their time: Vladimir Voevodsky and Martin Hofmann. Vladimir basically invented Homotopy Type Theory and I have profited from many conversations with him and from his talks. Martin is a very good friend of mine and made many important contributions to type theory, one of them the groupoid interpretation of type theory which is now viewed as an early predecessor of HoTT. Both will be sorely missed.

References

Altenkirch, T. (2019, to appear). *The tao of types*.

Bertot, Y., & Castéran, P. (2013). *Interactive theorem proving and program development: CoqArt – The calculus of inductive constructions*. Berlin: Springer Science & Business Media.

Brady, E. (2017). *Type-driven development with Idris*. Shelter Island: Manning Publications.

Chlipala, A. (2013). *Certified programming with dependent types: A pragmatic introduction to the Coq proof assistant*. Cambridge: MIT Press.

Cohen, C., Coquand, T., Huber, S., & Mörtberg, A. (2016). Cubical type theory: A constructive interpretation of the univalence axiom. arXiv preprint arXiv:1611.02108.

Constable, R., et al. (1986). Implementing mathematics with the Nuprl proof development system. Prentice Hall.

de Moura, L., Kong, S., Avigad, J., Van Doorn, F., & von Raumer, J. (2015). The lean theorem prover (system description). In *International Conference on Automated Deduction* (pp. 378–388). Springer.

Halmos, P. R. (1998). *Naive set theory*. Springer-Verlag New York.

Hofmann, M., & Streicher, T. (1998). The groupoid interpretation of type theory. *Twenty-Five Years of Constructive Type Theory (Venice, 1995), 36*, 83–111.

Kapulkin, C., & Lumsdaine, P. L. (2012). The simplicial model of univalent foundations (after voevodsky). arXiv preprint arXiv:1211.2851.

Kraus, N., & Sattler, C. (2015). Higher homotopies in a hierarchy of univalent universes. *ACM Transactions on Computational Logic (TOCL), 16*(2), 18:1–18:12.

Licata, D. R., & Shulman, M. (2013). Calculating the fundamental group of the circle in homotopy type theory. In *Proceedings of the 2013 28th Annual ACM/IEEE Symposium on Logic in Computer Science*, LICS'13 (pp. 223–232), Washington, DC. IEEE Computer Society.

Norell, U. (2008). Dependently typed programming in Agda. In P. Koopman, R. Plasmeijer, & D. Swierstra (Eds.), *International school on advanced functional programming* (pp. 230–266). Berlin/Heidelberg: Springer.

Stump, A. (2016). *Verified functional programming in Agda*. San Rafael: Morgan & Claypool.

The Univalent Foundations Program. (2013). *Homotopy type theory: Univalent foundations of mathematics* (1st ed.). Princeton: Univalent Foundations program.

Chapter 6
Univalent Foundations and the Equivalence Principle

Benedikt Ahrens and Paige Randall North

Abstract In this paper, we explore the 'equivalence principle' (EP): roughly, statements about mathematical objects should be invariant under an appropriate notion of equivalence for the kinds of objects under consideration. In set theoretic foundations, EP may not always hold: for instance, '$1 \in \mathbb{N}$' under isomorphism of sets. In univalent foundations, on the other hand, EP has been proven for many mathematical structures. We first give an overview of earlier attempts at designing foundations that satisfy EP. We then describe how univalent foundations validates EP.

6.1 The Equivalence Principle

What should it mean for two objects x and y to be equal? One proposal by Leibniz (1989), known as the "identity of indiscernibles", states that if x and y have the same properties, then they must be equal:

$$\forall \text{ properties } P, (P(x) \leftrightarrow P(y)) \quad \rightarrow \quad x = y.$$

For this proposal to be reasonable, then the converse, the "indiscernibility of identicals," should hold incontrovertibly. That is, if x and y are equal, then they must have the same properties:

$$x = y \quad \rightarrow \quad \forall \text{ properties } P, (P(x) \leftrightarrow P(y)) \,. \tag{6.1}$$

B. Ahrens (✉)
University of Birmingham, Birmingham, United Kingdom
e-mail: b.ahrens@cs.bham.ac.uk

P. R. North
The Ohio State University, Columbus, OH, United States

S. Centrone et al. (eds.), *Reflections on the Foundations of Mathematics*,
Synthese Library 407, https://doi.org/10.1007/978-3-030-15655-8_6

Indeed, one would be hard-pressed to find a mathematician who disagreed with this principle. However, in classical mathematics based on set theory, this principle is of limited usefulness: too few objects are equal. A group theorist, for example, would have little interest in a principle which required them to suppose that two groups are equal.

Instead, mathematicians are often interested in weaker notions of sameness and those properties that are invariant under such notions. A group theorist, for example, would have more interest in an analogous principle that described the properties of any pair of isomorphic groups G and H:

$$G \cong H \quad \rightarrow \quad \forall \text{ group theoretic properties } P, (P(G) \leftrightarrow P(H)) .$$

Similarly, category theorists would be more interested in a principle that described the properties of any pair of equivalent categories A and B:

$$A \simeq B \quad \rightarrow \quad \forall \text{ category theoretic properties } P, (P(A) \leftrightarrow P(B)) .$$

To generalize: mathematicians working in some domain $\mathscr{D}$ often utilize a stronger variant of the principle given in line (6.1) above, called the **equivalence principle**: for all objects x and y of domain $\mathscr{D}$:

$$x \sim_{\mathscr{D}} y \quad \rightarrow \quad \forall\, \mathscr{D}\text{-properties } P, (P(x) \leftrightarrow P(y)) \quad , \tag{6.2}$$

where $\sim_{\mathscr{D}}$ denotes a suitable notion of sameness for the domain $\mathscr{D}$.

We might consider a still stronger variant of the equivalence principle. A group theorist, for example, might not only want *properties of* groups to be invariant under isomorphism, but they might also want *structures on* groups to be invariant under isomorphism. For example, if the equivalence principle (6.2) holds in the domain of group theory and if two groups G and H are isomorphic, then the statements "G has a representation on V" and "H has a representation on V" are equivalent (for some fixed vector space V). However, it is actually the case that the isomorphism $G \cong H$ induces a bijection between the set of representations of G on V and the set of representations of H on V, which we regard as structures on G and H respectively. Such a variant of the equivalence principle has become known as the *Structure Identity Principle* (see Buss et al. 2011; Univalent Foundations Program 2013, Section 9.8; Awodey 2014).

Our goal in this paper is to describe how one can find the right notion $\sim_{\mathscr{D}}$ of sameness and the right class of '$\mathscr{D}$-properties and $\mathscr{D}$-structures' for some specific domains $\mathscr{D}$.

This right notion of sameness is not uniformly defined across different mathematical objects. However, we usually use the one already present in mathematical practice since we aim for the equivalence principle to capture mathematical practice. As a rule of thumb, it is usually considered to be

- *equality* when the objects naturally form a set—numbers, functions, etc.
- *isomorphism* when the objects naturally form a category—sets, groups, etc.
- *equivalence* when the objects naturally form a bicategory—e.g., categories.

The hard part will be in determining the right class of $\mathscr{D}$-properties and $\mathscr{D}$-structures for some specific domain $\mathscr{D}$. In usual mathematical practice, we can state properties which break the equivalence principle; that is, we can state properties of mathematical objects that are not invariant under sameness. We will seek to exclude such properties from our class of $\mathscr{D}$-properties and $\mathscr{D}$-structures.

Exercise 1 *Denote by* $2\mathbb{N}$ *the set of even natural numbers. Find a property of sets that is not invariant under the isomorphism* $\mathbb{N} \cong 2\mathbb{N}$ *given by multiplying and dividing by 2, respectively.*

Answer: *The statement '$1 \in \mathbb{N}$', is not invariant under this isomorphism.*

Exercise 2 *Find a property of categories that is true for one, but not for the other of these two, equivalent, categories.*

Answer: *Take, e.g., the predicate „The category $\mathscr{C}$ has exactly one object.".*

Thus, to assert an equivalence principle for sets or categories, we need to exclude these properties from our collection of 'set theoretic properties' and 'category theoretic properties'. M. Makkai (Makkai 1998) says

> The basic character of the Principle of Isomorphism is that of a constraint on the language of Abstract Mathematics; a welcome one, since it provides for the separation of sense from nonsense.

Put differently, establishing an equivalence principle means establishing a *syntactic criterion* for properties and structures that are invariant under sameness.

6.2 History

Look again at Exercise 2. There, we violated the equivalence principle for categories by referring to equality of objects. This might lead one to conjecture (correctly) that categorical properties which obey the equivalence principle cannot mention equality of objects.

However, the traditional definition of category mentions equality of objects. It usually includes the following axiom: for any two morphisms f and g such that the codomain of f equals the domain of g, there is a morphism gf such that domain of gf equals the domain of f and the codomain of gf equals the codomain of g.

To avoid mentioning equality of objects, one can express the composability of morphisms of that category via different means, specifically by having not one collection of morphisms but many *hom-sets*: one for each pair of objects. This idea, for instance explained in Mac Lane (1998, Section I.8) usually requires asking the hom-sets to be disjoint. This last requirement is automatic if we work instead in a typed language, where types are automatically disjoint.

A category is then given by

- a type O of objects,
- for each $x, y : O$, a type $A(x, y)$ of arrows from x to y,
- for each $x, y, z : O$ and $f : A(x, y)$, $g : A(y, z)$, a composite arrow $g \circ f : A(x, z)$, and
- for each $x : O$, an identity arrow $\mathrm{id}_x : A(x, x)$

such that

- for each $w, x, y, z : O$ and $f : A(w, x)$, $g : A(x, y)$, $h : A(y, z)$, there is an equality $h \circ (g \circ f) = (h \circ g) \circ f$ in $A(w, z)$,
- for each $x, y : O$ and $f : A(x, y)$, there is an equality $f \circ \mathrm{id}_x = f$ in $A(x, y)$, and
- for each $x, y : O$ and $f : A(x, y)$, there is an equality $\mathrm{id}_y \circ f = f$ in $A(x, y)$.

Note that when stating axioms, the only equality that is mentioned is the equality within a hom-set of the form $A(x, y)$, that is, between arrows of the same type.

By adding quantifiers, ranging over one type at a time, to this typed language, we obtain a language for stating properties of, and constructions on, categories. It turns out that the statements of that language are invariant under equivalence of categories:

Theorem 3 (Théorème de préservation par équivalence (Blanc 1978/1979)) *A property of categories (expressed in 2-typed first order logic) is invariant under equivalence if and only if it can be expressed in the typed language sketched above, and without referring to equality of objects.*

We do not give here the precise form of the typed language, but refer instead to Blanc's article for details. Note that Freyd (1976) states a similar result to Blanc's above, in terms of "diagrammatic properties".

Makkai (1998) develops notions of *signature* and *theory*, to specify mathematical structures. A theory is a pair (L, Σ) consisting of a signature L (specifying the shape of the structure) and a set Σ of "axioms" over L (specifying the axioms of the structure). A theory determines a notion of "model"—which is an L-structure satisfying the properties specified by Σ—and of "equivalence" of such models, called L-equivalence.

His *Invariance Theorem* gives a result similar to Theorem 3 for models of a theory: given an interpretation $T = (L, \Sigma) \to S$ of such a theory in a first-order logic theory, an S-sentence ϕ is invariant under L-equivalence if and only if it is expressible in First Order Logic with Dependent Sorts (FOLDS) over L.

In the following sections, we will see that similar results can be shown in univalent foundations. Specifically, not only properties but also constructions will be "invariant" under equivalence, and invariance of properties will be recovered as a special case via the propositions-as-some-types correspondence.

6.3 Univalent Foundations and Transport of Structures Along Equivalences

Starting in the 1970s, Per Martin-Löf designed several versions of dependent type theory, which are now called Martin-Löf Type Theories (Martin-Löf 1998). These were intended to be foundations of mathematics that, unlike set theory, have an inherent notion of computation built in. For decades, Martin-Löf type theories have formed the basis of computer proof assistants such as Coq and Agda.

One of the most mysterious features of this kind of type theory is its *equality* type $a =_X b$ of any two inhabitants a and b of a type X—see Altenkirch (2019, Section 5.4.3). Inhabitants of such an equality type behave, in many ways, like a proof of equality; in particular, they can be composed and inverted, corresponding to the transitivity and symmetry of equality. In one important respect, however, they behave differently: as explained in Altenkirch (2019, Sections 5.4.3 and 5.5.1), one can *not* show that any two inhabitants e, f of an equality type $a =_X b$ are equal—with their equality now being given by the iterated equality type $e =_{a=_X b} f$.

The lack of uniqueness of those terms has given rise to a new way of thinking about them and interpreting them into the world of mathematical objects. Instead of interpreting them as (set-theoretic) equalities between a and b in the set interpreting X, one can interpret them as *paths* from a to b in a space interpreting X.

This intuition is made formal in Voevodsky's *simplicial set model* Kapulkin and Lumsdaine (2012) which satisfies an additional interesting property: given two types X and Y, the interpretation of their equality type $X = Y$ is equivalent to the interpretation of their type of equivalences $X \simeq Y$ (for the definition of equivalence, see Altenkirch (2019, Section 5.5.4)). This observation motivated Voevodsky to add this property as an axiom to Martin-Löf type theory which he called the *Univalence Axiom*. The addition of the Univalence Axiom turns Martin-Löf type theory into univalent foundations.

Obtaining an equivalence principle was one of the main motivations for Voevodsky in designing his univalent foundations:

> [...] My homotopy lambda calculus is an attempt to create a system which is very good at dealing with equivalences. In particular it is supposed to have the property that given any type expression $F(T)$ depending on a term subexpression t of type T and an equivalence $t \to t'$ (a term of the type $Eq(T; t, t')$) there is a mechanical way to create a new expression F' now depending on t' and an equivalence between $F(T)$ and $F'(T')$ (note that to get F' one can not just substitute t' for t in F—the resulting expression will most likely be syntactically incorrect). [Email to Daniel R. Grayson, Sept 2006]

In the following sections, we describe how Voevodsky's goal is realized in univalent foundations.

6.3.1 Indiscernibility of Identicals in Type Theory

In Martin-Löf type theory (perhaps without the univalence axiom), identicals—that is, elements $x, y : T$ with an equality $e : x =_T y$ between them—are easily seen to be indiscernible. That is, for every type T and $x, y : T$, we can find a function

$$x =_T y \quad \to \quad \forall \text{ properties } P, (P(x) \leftrightarrow P(y)) \quad . \tag{6.3}$$

To better formulate this in the language of dependent type theory, (i) we will define this function for all x, y at once, (ii) we will understand 'properties P' to be functions $P : T \to \mathbf{Type}$, and (iii) we replace the logical equivalence $P(x) \leftrightarrow P(y)$ with the type-theoretic equivalence

$$(P(x) \simeq P(y)) :\equiv \left(\sum_{f:P(x) \to P(y)} \text{isEquiv}(f) \right)$$

(where isEquiv is defined in Section 5.5.4 of Altenkirch's article Altenkirch (2019)).
Our goal is hence to define a function

$$\text{transport} : \Pi_{(x,y:T)} \left(x =_T y \quad \to \quad \Pi_{P:T \to \mathbf{Type}} \left(P(x) \simeq P(y) \right) \right) \quad . \tag{6.4}$$

To this end, recall that in order to define a map out of an equality type, it suffices to define its image on $\text{refl}_x : (x =_T x)$ for each $x : T$. Therefore, it suffices to show that there is a term

$$\text{transport}(\text{refl}) : \Pi_{(x:T)} \left(\Pi_{P:T \to \mathbf{Type}} \left(P(x) \simeq P(x) \right) \right).$$

But then, for each $x : T$ and $P : T \to \mathbf{Type}$, we can set this to be the equivalence $1_{P(x)} : P(x) \simeq P(x)$ whose underlying function $P(x) \to P(x)$ is the identity function:

$$\text{transport}(\text{refl})(x)(P) :\equiv 1_{P(x)}.$$

The function transport shows that any 'property', or dependent type, $P : T \to$ **Type** is invariant under equalities in T. In particular, given an equality $x =_T y$, we obtain functions $P(x) \to P(y)$ and $P(y) \to P(x)$ which allow us to transport terms of $P(x)$ or $P(y)$ back and forth along this equality.

6.3.2 From Equality to Equivalence

We have just seen that in Martin-Löf type theory, identicals are indiscernible. Now we investigate how to expand this to get a full-blown equivalence principle from this fact. In short, we will see that in many circumstances, the equality type $x =_T y$ is itself equivalent to some structured equivalence appropriate for the type T. Then composing this equivalence with the transport function, we will obtain the equivalence principle (6.2).

To be precise, fix a type T. Given any notion $\sim_T$ of equivalence (or at least a reflexive relation) in a type T, we immediately obtain a function

$$\text{idtoequiv} : \Pi_{x,y:T}\ ((x = y) \to (x \sim_T y)) \tag{6.5}$$

by setting idtoequiv(x, x)(refl) to the reflexive term on x in $x \sim x$ (since to define a function out of an equality type, it is enough to define it just at every occurrence of refl).

Now we hope that for notions of equivalence $\sim_T$ already of interest to us, this function is actually an equivalence for all terms $x, y : T$, or more precisely, that the following type is inhabited:

$$\Pi_{x,y:T}\ \text{isEquiv}\ (\text{idtoequiv}(x, y))$$

If this type is inhabited, then for each $x, y : T$, we can take π_1(isEquiv (idtoequiv (x, y))), the backwards function $(x \sim_T y) \to (x = y)$, and compose it with transport to obtain a function

$$\Pi_{(x,y:T)}\left(x \sim_T y \quad \to \quad \Pi_{P:T\to\textbf{Type}}\,(P(x) \simeq P(y))\right)$$

which is our equivalence principle.

Thus, in the next sections, we just aim to show that for certain types T and notions of equivalence $\sim_T$, the function idtoequiv is indeed an equivalence.

6.3.3 The Univalence Principle

"Equality is equivalence for types" is the slogan made precise by Voevodsky's univalence principle. More precisely, the univalence principle asserts part of an equivalence principle for types: it states that the canonical map

$$\text{idtoequiv} : \Pi_{A,B:\textbf{Type}}\,((A = B) \to (A \simeq B)) \tag{6.6}$$

from equalities of types to equivalences of types is itself, for any types A and B, an equivalence. Then, composing idtoequiv with transport as in the last section, we obtain an equivalence principle for types.

$$\Pi_{(A,B:\textbf{Type})}\left(A \simeq B \quad \to \quad \Pi_{P:\textbf{Type}\to\textbf{Type}}\,(P(A) \simeq P(B))\right)$$

The univalence principle is not provable in pure Martin-Löf type theory (Martin-Löf 1998), but needs to be postulated as an axiom—hence it is sometimes also called the "univalence axiom". In extensions of Martin-Löf type theory, as in the recently developed cubical type theory (Cohen et al. 2018), the univalence principle can be derived.

Building upon the equivalence principle for types—whether it is given as an axiom or as a theorem—one can derive equivalence principles for other kinds of structures. Establishing that idtoequiv is an equivalence for other types and notions of equivalence is the subject of the next sections.

6.4 The Equivalence Principle for Set-Level Structures

Now we turn our attention away from the type of all types and towards types of more specific mathematical objects. It turns out that for types of simple objects like propositions, sets, and monoids, the univalence axiom is enough to show the equivalence principle for these types' usual notion of equivalence. More precisely, in the presence of the univalence axiom, the function idtoequiv discussed in the last section is itself an equivalence. For an exploration and formalization of these ideas, see Coquand and Danielsson (2013).

6.4.1 Propositions

We call *propositions* those types that have at most one inhabitant. We think of propositions as either being *true* (when they are inhabited) or *false* (when they are not inhabited). What should an equivalence of two propositions P, Q be? Experience might indicate that such an equivalence should just be two functions

$$f : P \leftrightarrows Q : g$$

so that P is inhabited if and only if Q is. In fact, this notion of equivalence is the right one in the sense that it will validate the equivalence principle.

To be precise, we define

$$\begin{aligned} &\text{isProp} : \mathbf{Type} \to \mathbf{Type} \\ &\text{isProp } A :\equiv \Pi_{(x,y:A)} x = y \end{aligned}$$

whose inhabitants can be thought of as proofs that a type A is a proposition. A proposition is then a pair (A, p) of a type A and a proof p : isProp A, that is,

$$\mathbf{Prop} :\equiv \Sigma_{A:\mathbf{Type}} \text{ isProp } A \ .$$

Now we have, for $P \equiv (A, p)$ and $Q \equiv (B, q)$,

$$
\begin{aligned}
P = Q &\simeq (A, p) = (B, q) \\
&\simeq \Sigma_{(e:A=B)}(\text{transport}^{\text{isProp}}(e, p)) = q && (6.7) \\
&\simeq \Sigma_{(e:A=B)} 1 && (6.8) \\
&\simeq A = B \\
&\simeq A \simeq B && (6.9) \\
&\simeq (A \to B) \times (B \to A) && (6.10)
\end{aligned}
$$

Equivalence (6.7) above uses the fact that an equality between pairs is the same as pairs of equalities, where the second equality is "heterogeneous", i. e., requires a transport along the first equality to make it well-typed. Equivalence (6.8) uses the fact that being a proposition is itself a proposition, so that equality types between proofs of a proposition are equivalent to the unit type. Equivalence (6.9) is given by the univalence principle, and equivalence (6.10) uses that A and B are propositions; a pair of maps back and forth between types that are propositions automatically forms an equivalence of types.

Altogether, this means that P and Q are equal exactly if their underlying types A and B are logically equivalent—the expected notion of equivalence for propositions.

6.4.2 Sets

We call *sets* those types whose equality types are propositions,

$$\mathbf{Set} :\equiv \Sigma_{A:\mathbf{Type}} \Pi_{x,y:A} \text{isProp}(x = y).$$

Then a set in the type theory is a collection of terms, the equality types among which are either empty or contractible.

Given two sets $X, Y : \mathbf{Set}$, where $X = (A, p)$ and $Y = (B, q)$, the equivalences of types

$$
\begin{aligned}
(X = Y) &\simeq (A \simeq B) && (6.11) \\
&\simeq \Sigma_{f:A \cong B} \text{isCoherent}(f) && (6.12) \\
&\simeq (A \cong B) && (6.13)
\end{aligned}
$$

can be constructed. Here, $A \cong B :\equiv \Sigma_{(f:A \to B)} \text{isIso} f$ is the type of *isomorphisms of types* between A and B, and $\text{isCoherent}(f)$ states an equality of equalities in A. When A is a set, the type $\text{isCoherent}(f)$ is contractible (see the discussion in Altenkirch (2019, Section 5.5.4)), and hence we obtain equivalence (6.13).

6.4.3 Monoids

The equivalence principle can be shown for many algebraic structures commonly encountered in mathematics, such as groups and rings. Before presenting a general result to that extent in Sect. 6.4.4, in this section, we study in detail the case of monoids (which was formalized in Coquand and Danielsson 2013). This particular case exemplifies many of the concepts and results used in general.

A *monoid* is a tuple $(M, \mu, e, \alpha, \lambda, \rho)$ where

1. M : **Set**
2. $\mu : M \times M \to M$ (multiplication)
3. $e : 1 \to M$ (neutral element)
4. $\alpha : \Pi_{(a,b,c:M)}\mu(\mu(a,b),c) = \mu(a,\mu(b,c))$ (associativity)
5. $\lambda : \Pi_{(a:M)}\mu(e,a) = a$ (left neutrality)
6. $\rho : \Pi_{(a:M)}\mu(a,e) = a$ (right neutrality)

Given two monoids $\mathbf{M} \equiv (M, \mu, e, \alpha, \lambda, \rho)$ and $\mathbf{M}' \equiv (M', \mu', e', \alpha', \lambda', \rho')$, a *monoid isomorphism* is a bijection $f : M \cong M'$ between the underlying sets that preserves multiplication and neutral element. We can derive an equivalence between the equality type and the isomorphism type between any two monoids as follows:

$$
\begin{aligned}
\mathbf{M} = \mathbf{M}' &\simeq (M, \mu, e) = (M', \mu', e') && (6.14)\\
&\simeq \sum_{p:M=M'} (\mathrm{transport}^{Y \mapsto (Y \times Y \to Y)}(p, \mu) = \mu') \\
&\qquad \times (\mathrm{transport}^{Y \mapsto (1 \to Y)}(p, e) = e') \\
&\simeq \sum_{f:M \cong M'} \left(f \circ m \circ (f^{-1} \times f^{-1}) = m'\right) && (6.15)\\
&\qquad \times (f \circ e = e') \\
&\simeq \mathbf{M} \cong \mathbf{M}'
\end{aligned}
$$

Here, the equivalence of types (6.14) uses the fact that the axioms of a monoid (the types of α, λ, and ρ) are propositions (compare also to (6.8) above). The equivalence (6.15) uses the univalence principle for types in the first component, replacing an equality of sets by a bijection. This translates to replacing "transport along the equality" by "conjugating by the bijection" in the second component.

The equivalence of types constructed above, from left to right, is pointwise equal to the canonical map

$$\Pi_{\mathbf{M},\mathbf{M}'}\mathbf{M} = \mathbf{M}' \to \mathbf{M} \cong \mathbf{M}' \tag{6.16}$$

defined by equality elimination, which shows that the latter is an equivalence of types. In other words, we have just proved the equivalence principle (6.5) for the equivalence $\mathbf{M} \cong \mathbf{M}'$.

6.4.4 Univalent Categories

We have seen in the preceding sections that the types of propositions, sets, and monoids all have a certain nice property—they validate the equivalence principle. However, it is natural to consider such objects as each belonging to a category. In this section, we discuss those categories whose objects validate the equivalence principle.

In Sect. 6.2, we saw that in order to avoid mentioning equality of objects, we can define a *category* A to consist of

1. a type A_0 of *objects*;
2. for each $a, b : A_0$, a set $A(a, b)$ of *arrows* or *morphisms*;
3. for each $a : A_0$, a morphism $1_a : A(a, a)$;
4. for each $a, b, c : A_0$, a function of type

$$A(a, b) \times A(b, c) \to A(a, c)$$

   denoted by $(f, g) \mapsto f \cdot g$;
5. for each $a, b : A_0$ and $f : A(a, b)$, we have $\ell_f : \mathrm{id}\, f\, 1_a \cdot f$ and $r_f : \mathrm{id}\, f\, f \cdot 1_b$;
6. for each $a, b, c, d : A_0$ and $f : A(a, b)$, $g : A(b, c)$, $h : A(c, d)$, we have $\alpha_{f,g,h} : f \cdot (g \cdot h) = (f \cdot g) \cdot h$.

The reason for asking the types of arrows to be *sets* rather than arbitrary types is so that these categories behave as classical categories (and not any kind of higher category) and, in particular, so that the axioms—which state equalities between arrows—are propositions, meaning that we do not need to state higher coherence axioms. There is prima facie no condition of that kind on the type of objects of the category. However, it will turn out that the objects of a *univalent category* form a groupoid (meaning that all of its equality types form sets).

A morphism $f : A(a, b)$ of the category A is an *isomorphism* if there is a morphism $g : A(b, a)$ that is left and right inverse to f, that is

$$\mathrm{isIso} f :\equiv \Sigma_{g:A(b,a)}(f \cdot g = 1_a) \times (g \cdot f = 1_b) \ .$$

We call $\mathrm{Iso}(a, b) :\equiv \Sigma_{f:A(a,b)}\mathrm{isIso} f$ the type of isomorphisms from a to b, and for any $a : A_0$ we have $1_a : \mathrm{Iso}(a, a)$. We can define a function

$$\mathrm{idtoiso} : \Pi_{x,y:A_0}(x = y) \to \mathrm{Iso}(x, y) \tag{6.17}$$

by setting $\mathrm{idtoiso}(x, x, \mathrm{refl}_x)$ to 1_x for every $x : A_0$ just as we did to define idtoequiv in Sect. 6.3.2.

Now we call the category A *univalent* if $\mathrm{idtoiso}_{x,y}$ is an equivalence of types for every $x, y : A_0$. To see why the adjective *univalent* is used, compare the function in Display (6.17) above to the one in Display (6.6) underlying the univalence principle. The univalence principle asserts that equality and equivalence of types are the same; here, we assert that equality and isomorphism of objects of a category are the same.

In asserting that a category A is univalent, we assert that the equality types $a = b$ among its objects are equivalent to the sets $\mathrm{Iso}(a, b)$ of isomorphisms among its objects. Since the property of "being a set" itself obeys the equivalence principle for types the equality types $a = b$ are themselves sets. When a type's equality types are sets, we call the type a *groupoid*.

A categorical equivalence between univalent categories A and B gives rise to an isomorphism between them—indeed, the type $A \simeq B$ of adjoint equivalences is equivalent to the type $A \cong B$ of isomorphisms of categories.

With a set-theoretic reading of the univalence condition in mind, one could think that only skeletal categories are univalent. However, one should keep in mind that in type theory, the equality type $x = y$ between two objects of a category can—and often does—have more than one element. Consequently, in type theory, a category being univalent usually signifies that its type of objects has many equalities. This difference is witnessed by the many examples of univalent categories given below, most of which are not skeletal.

With these definitions in place, the composite equivalence of types shown in Displays (6.11), (6.12), and (6.13) can be restated as "the category of sets is univalent". Similarly, the result of Sect. 6.4.3 can be restated as "the category of monoids is univalent".

Many categories that arise naturally are univalent, in particular,

- the category of sets;
- the categories of groups, rings, etc.;
- the functor category $[A, B]$ if the target category B is;
- a preorder, seen as a category, exactly if it is anti-symmetric.

To extend our list of univalent categories to other algebraic structures beyond monoids, we could simply redo constructions similar to those for monoids, for groups, rings, and other structures of interest. However, in doing so, we would observe that we are doing the same reasoning over and over again. For instance, looking back at monoids, we used that the category of sets is univalent to show that the category of monoids is univalent, in step (6.15). This is due to the fact that "monoids are sets with additional structure", and monoid isomorphisms are isomorphisms of sets preserving this structure. Similarly, "groups are monoids with additional structure", and we would expect to reuse the equivalence of Display (6.16) when building an equivalence between the equality types of groups on the one hand, and of group isomorphisms on the other hand. *Displayed categories* as presented in Ahrens and Lumsdaine (2017) are a convenient tool for such modular reasoning about categories built step-by-step from simpler ones. In particular, Proposition 43 and Theorem 44 of Ahrens and Lumsdaine (2017) allow one to show that a category built from a simpler one using the framework of displayed categories is univalent, provided the simpler one is univalent and the "extra data" making the difference between the two categories satisfies some condition. That result validates the *Structure Identity Principle* (Univalent Foundations Program 2013, Theorem 9.8.2).

6.5 The Equivalence Principle for (Higher) Categorical Structures

We saw in the previous sections that for types of simple structures like propositions, sets, and monoids, the equivalence principle comes along with the univalence axiom. Now we see that for more complication structures, like categories, the equivalence principle only holds for certain well-behaved categories.

The most common notion of equivalence between two categories A and B is unsurprisingly called an equivalence $A \simeq B$. It consists of two functors $F : A \leftrightarrows B : G$ and natural isomorphisms $1_B \cong FG$ and $1_A \cong GF$ (see Mac Lane 1998). An equivalence of categories "transports" categorical structures, such as limits, between categories, and is hence considered the right notion of sameness for categories in most contexts. Can we show that the equality type between two categories, $A = B$, is the same as the type of categorical equivalences $A \simeq B$? The answer is that while this is not the case for arbitrary categories, it is the case when A and B are univalent.

For any two categories A and B, the univalence axiom implies that the function from equalities to isomorphisms (a stricter notion of sameness of categories) given by equality elimination is an equivalence (Ahrens et al. 2015, Lemma 6.16):

$$(A = B) \xrightarrow{\simeq} (A \cong B) \tag{6.18}$$

Furthermore, if A and B are univalent categories, then the type of isomorphisms between them is equivalent to that of categorical equivalences (Ahrens et al. 2015, Lemma 6.15):

$$(A \cong B) \xrightarrow{\simeq} (A \simeq B) \tag{6.19}$$

Composing these two equivalences yields the desired equivalence of types between equalities and categorical equivalences.

The example of categories shows that, in order to obtain the equivalence principle for mathematical structures that naturally form bicategory, one needs to impose a "univalence" condition on those structures. Defining such a univalence condition for general structures is the subject of active research.

Acknowledgements We are very grateful to Deniz Sarikaya and Deborah Kant for their editorial work and their encouragement, and to an anonymous referee for providing valuable feedback. Furthermore, we would like to thank all the organizers of the FOMUS workshop—Balthasar Grabmayr, Deborah Kant, Lukas Kühne, Deniz Sarikaya, and Mira Viehstädt—for giving us the opportunity to discuss and compare different foundations of mathematics. This material is based upon work supported by the Air Force Office of Scientific Research under award numbers FA9550-16-1-0212 and FA9550-17-1-0363.

References

Ahrens, B., Kapulkin, K., & Shulman, M. (2015). Univalent categories and the Rezk completion. *Mathematical Structures in Computer Science, 25*, 1010–1039. https://doi.org/10.1017/S0960129514000486

Ahrens, B., & Lumsdaine, P. L. (2017). Displayed categories *(conference version)*. In D. Miller (Ed.), *2nd International Conference on Formal Structures for Computation and Deduction (FSCD 2017), Leibniz International Proceedings in Informatics (LIPIcs)* (Vol. 84, pp. 5:1–5:16). Leibniz-Zentrum für Informatik. https://doi.org/10.4230/LIPIcs.FSCD.2017.5

Altenkirch, T. (2019). Naïve type theory, In S. Centrone, D. Kant, & D. Sarikaya (Eds.), *Reflections on the foundations of mathematics*. Cham: Springer.

Awodey, S. (2014). Structuralism, Invariance, and Univalence. *Philosophia Mathematica (III), 22*, 1–11. https://doi.org/10.1093/phimat/nkt030

Blanc, G. (1978/1979). Équivalence naturelle et formules logiques en théorie des catégories. *Arch. Math. Logik Grundlag, 19*(3–4), 131–137. https://doi.org/10.1007/BF02011874

Cohen, C., Coquand, T., Huber, S., & Mörtberg, A. (2018). Cubical type theory: A constructive interpretation of the univalence axiom. In T. Uustalu (Ed.), *21st International Conference on Types for Proofs and Programs (TYPES 2015), Leibniz International Proceedings in Informatics (LIPIcs)* (Vol. 69, pp. 5:1–5:34). Schloss Dagstuhl–Leibniz-Zentrum fuer Informatik, Dagstuhl. https://doi.org/10.4230/LIPIcs.TYPES.2015.5, http://drops.dagstuhl.de/opus/volltexte/2018/8475

Coquand, T., Danielsson, N. A. (2013). Isomorphism is equality. *Indagationes Mathematicae, 24*(4), 1105–1120. https://doi.org/10.1016/j.indag.2013.09.002, http://www.sciencedirect.com/science/article/pii/S0019357713000694. In memory of N.G. (Dick) de Bruijn (1918–2012).

Freyd, P. (1976). Properties invariant within equivalence types of categories. In A. Heller, & M. Tierney (Eds.), Algebra, topology, and category theory (a collection of papers in honor of Samuel Eilenberg) (pp. 55–61). New York: Academic Press.

Kapulkin, C., & Lumsdaine, P. L. (2012). The simplicial model of univalent foundations (after Voevodsky). *Journal of the European Mathematical Society*. arXiv:1211.2851 (forthcoming)

Leibniz, G. W. (1989). Philosophical papers and letters. In *Synthese historical library* (Vol. 2). Springer. https://doi.org/10.1007/978-94-010-1426-7

Mac Lane, S. (1998). Categories for the working mathematician. In *Graduate texts in mathematics* (Vol. 5, 2nd ed.). New York: Springer.

Makkai, M. (1998). Towards a categorical foundation of mathematics. In Logic Colloquium'95 (Haifa), *Lecture notes logic* (Vol. 11, pp. 153–190). Berlin: Springer. https://doi.org/10.1007/978-3-662-22108-2_11

Martin-Löf, P. (1998). An intuitionistic theory of types. In G. Sambin, & J. M. Smith (Eds.), *Twenty-five years of constructive type theory (Venice, 1995)* (Oxford logic guides, Vol. 36, pp. 127–172). New York: Oxford University Press.

Samuel, B., Ulrich, K., & Michael, R. (2011). Mathematical logic: Proof theory, constructive mathematics. *Oberwolfach Reports, 8*, 2963–3002. https://doi.org/10.4171/OWR/2011/52

Univalent Foundations Program, T. (2013). Homotopy type theory: Univalent foundations of mathematics. http://homotopytypetheory.org/book, Institute for Advanced Study

Chapter 7
Higher Structures in Homotopy Type Theory

Ulrik Buchholtz

Abstract The intended model of the homotopy type theories used in Univalent Foundations is the ∞-category of homotopy types, also known as ∞-groupoids. The problem of *higher structures* is that of constructing the homotopy types needed for mathematics, especially those that aren't sets. The current repertoire of constructions, including the usual type formers and higher inductive types, suffice for many but not all of these. We discuss the problematic cases, typically those involving an infinite hierarchy of coherence data such as semi-simplicial types, as well as the problem of developing the meta-theory of homotopy type theories in Univalent Foundations. We also discuss some proposed solutions.

7.1 Introduction

Homotopy type theory is at the same time a foundational endeavor, in which the aim is to provide a new foundation for mathematics, and an area of mathematics and logic, in which the aim is to provide tools for the mathematical analysis of homotopical (higher dimensional) structures. Let us call the former Univalent Foundations (UF) and the latter Homotopy Type Theory (HoTT), understanding that HoTT encompasses many different particular type theories.

In the present chapter we use the issue of *higher structures* as a lens with which to study both of these aims and their relations to other foundational approaches.

To motivate the problem of higher structures, we need to recall that the intended universe of UF, and the principal model of HoTT, is the realm of ∞-groupoids, a homotopical kind of algebraic structures that have elements, identifications, identifications between identifications, etc. ad infinitum, and these identifications behave sensibly in that we can invert them, compose them, and whisker by them, but the expected laws only hold up to higher identifications. Grothendieck's *homotopy*

U. Buchholtz (✉)
Fachbereich Mathematik, TU Darmstadt, Darmstadt, Germany
e-mail: buchholtz@mathematik.tu-darmstadt.de

S. Centrone et al. (eds.), *Reflections on the Foundations of Mathematics*,
Synthese Library 407, https://doi.org/10.1007/978-3-030-15655-8_7

hypothesis tells us that ∞-groupoids are the same as *homotopy types*, so we shall use these terms interchangeably, with a slight preference for the latter, as then homotopy type theories are both theories of homotopy types as well as homotopical type theories.

A common misconception is that higher homotopy types only occur in, or only are relevant to, homotopy theory. That is very far from the case, as even the type of sets, as used in most of mathematical practice, is a 1-type. And higher structures now feature prominently in many areas ranging from geometry, algebra, and number theory, to the mathematics of quantum field theories in physics and concurrency in computer science. An introduction to homotopy types and the homotopy hypothesis is given in Sect. 7.2.

It is a key point of difference between UF and earlier approaches to foundations inspired by category theory that the former takes ∞-groupoids rather than various notions of higher categories to be the basic objects of mathematics, from which the rest are obtained by adding further structure. This insight was due to Voevodsky[1] who remarked that many natural constructions are not functorial in the sense of category theory. (Think for example of the center of a group.) However, every construction—if it is to have mathematical meaning—has to preserve the relevant notion of equivalence. (Isomorphic groups do have isomorphic centers, etc.)

Because UF aims to be a foundation for all of mathematics, it is necessary that its language, in the shape of the HoTT, provides the means of construction for all the homotopy types that are used in mathematics. For the construction of *sets*, this is not such a big problem, as most of the sets that occur in mathematics can be constructed from the type formers of Martin-Löf type theory. (But even here there are subtleties if we wish to remain in the constructive and predicative realm.)

The main problems appear when it comes to *higher dimensional* homotopy types. We discuss some positive results (structures that have already been constructed) as well as some open problems (structures that have not already been constructed) in Sect. 7.3.

We remark that although we expect some actual *negative* results (i.e., impossibility proofs) for some of the open problems with respect to some particular homotopy type theories, these have yet to appear. But anticipating that further means of construction will be necessary, we discuss potential solutions in Sect. 7.4.

For the remainder of this Introduction, we shall consider an analogy. Martin-Löf type theory can be considered as a formal system for making constructions. In fact, a variant with an impredicative universe was called the *Calculus of Constructions* (CoC), and a further extension, the *Calculus of Inductive Constructions* (CIC) is the basis for the proof-assistant Coq. And we shall be concerned with the question of the limits of the methods of construction available in constructive type

[1]In Voevodsky (2014) he wrote: "The greatest roadblock for me was the idea that categories are 'sets in the next dimension.' I clearly recall the feeling of a breakthrough that I experienced when I understood that this idea is wrong. Categories are not 'sets in the next dimension.' They are 'partially ordered sets in the next dimension' and 'sets in the next dimension' are groupoids."

theories. An obvious analogy presents itself, namely with euclidean geometry and the limits of the methods of geometric constructions using ruler and compasses. We shall (probably) find that, just as in the geometric case, certain objects are not constructible from the most basic constructions, and require further tools, such as the *neusis*, for their construction. However, we shall follow Pappus' prescription of parsimony and demand that everything that can be constructed with lesser means, should be so constructed. As a corollary, since a proof is a special case of a construction, we demand that if something can be proved in a weaker system, then it should be so proved.

Obviously we can include among the list of further means of construction such well-known principles as the law of excluded middle (LEM), Markov's principle (MP), the axiom of choice (AC), various kinds of transfinite induction (TI), as well as principles of impredicativity. Some of these, as well as weaker versions of these, are referred to as *constructive taboos* because admitting them is contrary to certain philosophical outlooks inspired by constructivism or intuitionism, and also because they cannot be mechanically executed at all, or only with greatly increased computational complexity.

A further aspect of the constructive taboos is that they reduce the number of *models* in which we can interpret the constructions. It is well-known that constructive systems admit many useful models, indeed, this is one reason why classical mathematicians may be interested in such systems. Non-homotopical constructive systems can often be modeled in *toposes*, more precisely, 1-toposes, which can be seen either as generalizations of Kripke models, as generalized spaces, or indeed as generalized worlds of sets.

It is suspected that HoTT can be modeled in higher toposes, more precisely, $(\infty, 1)$-*toposes*. These dramatically extend the usefulness of HoTT, for instance as explained in Schreiber (2016). Earlier extensions of Martin-Löf type theory often imposed axioms, such as the *uniqueness of identity proofs* (UIP), that rule out higher dimensional models. These contradict the univalence axiom and may be called *homotopical taboos*. More refined axioms may hold in $(\infty, 1)$-toposes corresponding to 1-toposes (the 1-localic $(\infty, 1)$-toposes (Lurie 2009, Sect. 6.4)), but not in more general $(\infty, 1)$-toposes. These are called *constructive-homotopical taboos*.

7.2 Infinity Groupoids and the Homotopy Hypothesis

The types in UF are supposed to be *homotopy types*, so let us dwell a bit on what they are, both from an intuitive point of view, and from the perspective of mathematics developed in set-theoretic foundations.

Intuitions are always hard to convey, and in the case of the notion of homotopy type, even more so. Intuition is, after all, best developed through practice and familiarity. One way to build an intuition for homotopy types is through working in a homotopy type theory, either on paper or with the help of a proof assistant.

Many young workers in HoTT/UF did this before learning about homotopy theory from a classical point of view.

As a first approximation we can say that types A are collections of objects together with for each pair of objects $a, b : A$, a type of *identifications* $p : a =_A b$, together with meaningful operations on these identifications, such as the ability to compose and invert them. And there should also be higher order operations that produce identifications between identifications, such as an identification $\alpha(p) : (p^{-1})^{-1} = p$ for any $p : a = b$. This description is meant to capture types in their incarnation as ∞-*groupoids*, and on this view, two types A, B can be identified if there is a (weak) functor $F : A \to B$ that is an equivalence of ∞-groupoids.

Another intuition comes from describing types as (nice) topological spaces up to homotopy equivalence. The objects are the points of the space, and the identifications are the paths between points.

The *homotopy hypothesis* is the idea that these separate intuitions capture the same underlying concept. It grew out of Grothendieck's homotopy hypothesis concerning a particular definition of ∞-groupoids (Grothendieck 1983). The modern terminology is due to Baez (2007).

In order to explain the subtlety of the situation, let us turn to the most common implementation of the idea of ∞-groupoids in the context of set-theoretic mathematics. Here, these are represented by *simplicial sets* satisfying a certain filling condition. These simplicial sets are called *Kan complexes* in honor of Kan (1956). A simplicial set is a functor $X : \Delta^{\mathrm{op}} \to \mathbf{Set}$, where Δ is the category of non-empty finite ordinals and order-preserving functions. This means concretely that a simplicial set consists of a set of n-simplices X_n for each $n = 0, 1, \ldots$ together with face and degeneracy maps satisfying laws called the *simplicial identities*. We think of the 0 simplices as points, the 1-simplices as lines between points, 2-simplices as triangles, etc.

The Kan filling condition says that if we are given n compatible $(n-1)$-simplices in X in the sense that they could be n of the $n+1$ faces of an n-simplex, then there exists some such n-simplex. This condition is illustrated in Fig. 7.1 in some low-dimensional cases. In each case, we can think of the given data as a map from a *horn*, a sub-simplicial set $\Lambda^n_k \subseteq \Delta^n$ of the standard n-simplex Δ^n consisting of the union of all the faces opposite the kth vertex, into X. A lift is some extension of this to a map from Δ^n to X, or equivalently, an n-simplex in X with the requisite faces.

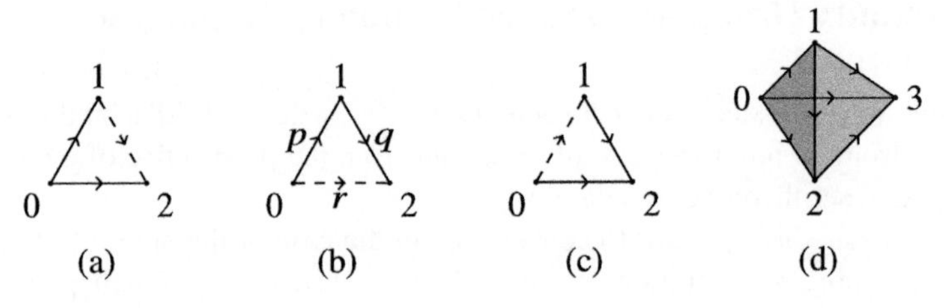

Fig. 7.1 The Kan filling condition in dimensions 2 and 3. **(a)** Λ^2_0 **(b)** Λ^2_1 **(c)** Λ^2_2 **(d)** Λ^3_0

For example, in Fig. 7.1b, if we are given two 1-simplices p and q in X with a common endpoint, then there exists some 2-simplex representing both a composite of p and q (the third face r) together with the interior representing the fact that r is the composite of p and q.

Note that Kan complexes give a *non-algebraic* notion of ∞-groupoid: there exist composites and higher simplicial identifications, but there are no operations singling out a particular composite.

Here we come to a potential pitfall: we cannot say that homotopy types *are* Kan complexes, for they have different criteria of identity: In usual mathematical practice we identify two simplicial sets if they are isomorphic (this is already a weaker notion of identity than that provided by set theory!), whereas an identification between Kan complexes X and Y *considered as homotopy types* should be a *homotopy equivalence*.

And this is perhaps an appropriate point at which to give a type-theoretic take on Quine's (1969) famous slogans[2]:

1. *To be is to be the value of a variable,* and
2. *No entity without identity.*

In 1 we require moreover that all variables be typed, so we say rather that to be an A is to be the value of a variable of type A and more importantly, *to be is to be an element of a type*, and in 2 we do not require any notion of identity between entities of different types, but we do require as an essential part of giving a type A that the identity type $x =_A y$, for $x, y : A$, is meaningful and correctly expresses the means of identifying elements of A: *no type without an identity type.*

The discrepancy between the notion of identity between the model objects (here Kan complexes) and the desired notion of identity (here homotopy equivalence) is usually addressed using *relative categories* as a tool. A relative category consists of a category equipped with a wide subcategory of *weak equivalences*. This is often refined by adding more properties (e.g., the weak equivalences satisfy the two-out-of-three or the two-out-of-six properties) or structure, such as fibrations and/or cofibrations interacting nicely with the weak equivalences. A particularly well-behaved notion is that of a *Quillen model category*, which does indeed contain both fibrations and cofibrations in addition to weak equivalences, and is assumed to be complete and cocomplete.

The category of simplicial sets **sSet** can be equipped with the structure of a Quillen model category in which the fibrant objects are the Kan complexes (these are also cofibrant as all objects are cofibrant) and the weak equivalences between Kan complexes are the homotopy equivalences. The category of topological spaces, or more precisely, for technical reasons, the category of compactly generated topological spaces, $\mathbf{Top}_{\mathrm{cg}}$, can likewise be equipped with a Quillen model structure in which the cofibrant objects are the nice spaces (technically, cell complexes; all

[2]The second has also been discussed from a univalent perspective in Rodin (2017) and Tsementzis (2017).

objects are fibrant) and the weak equivalences between the nice spaces are the homotopy equivalences.

Quillen (1967) proved that these two model categories give rise to equivalent *homotopy categories*. For this purpose he introduced the notion of (what is now called) a *Quillen equivalence* between model categories. Given a nice space X, the corresponding singular Kan complex $\Pi_\infty(X)$ has as n-simplices the continuous maps from the topological n-simplex into X, and given a Kan complex A, the corresponding space is the *geometric realization* $|A|$ given by gluing together topological simplices according to the face and degeneracy maps in A.

That the homotopy categories are equivalent is a first step towards getting what we actually want. We would actually like to show that the Quillen model categories of simplicial sets and topological spaces give rise to equivalent *homotopy types* (in both cases restricting to the objects that lie in a fixed Grothendieck universe). And which notion of homotopy type should we use here? It turns out not to matter, but it is easiest to make (large) Kan complexes out of either one.

The way this is achieved is by enhancing both **sSet** and $\mathbf{Top}_{\text{cg}}$ to simplicially enriched categories (the latter via the singular Kan complex construction on the level of mapping spaces) such that they become *simplicial model categories*, and then taking the homotopy coherent nerves of the subcategories of homotopy equivalences between bifibrant objects, i.e., between the model objects on both sides.

Notice that to get a good theory of homotopy types in the classical set-up we seem to need also a good theory of $(\infty, 1)$-categories, that is, categories (weakly) enriched in homotopy types, in order to also get a good hold on the *universe* of homotopy types, which is another name of course anticipating the type-theoretic notion of a universe, and which consists of the homotopy types that are small relative to some Grothendieck universe.

There is another model structure on simplicial sets whose bifibrant objects are the *quasi-categories*, those that satisfy a weakening of the Kan filling conditions that make them suitable as models of $(\infty, 1)$-categories. This notion was introduced by Boardman and Vogt (1973) and the resulting theory of $(\infty, 1)$-categories has been studied extensively by Joyal (2002) and Lurie (2009) (see also the appendix of Lurie (2009) for details on simplicial model categories as discussed above).

My point in bringing out these technicalities is not only to explain how homotopy types are defined and handled in set-theoretic mathematics, but also to give a sense of the subtleties involved. It has taken many years to give a good account of how to treat higher structure in set-theoretic mathematics (often by working in a 1-category-theoretic layer), and there are still many open questions about which constructions and properties are invariant under weak equivalences inside a model category and under Quillen equivalences between model categories. For instance, it was just recently established that a Quillen adjunction always induces an adjunction between underlying quasi-categories, and hence an adjunction of the presented $(\infty, 1)$-categories (Mazel-Gee 2016). Another line of open questions concerns the possibility of *algebraic* models for ∞-groupoids, where composition, inverses, etc., are given by operations rather than merely assumed to exist. It is quite possible that type theory will be influential in this area, see for instance the suggestion of Brunerie (Brunerie 2016, Appendix).

Thus it should come as no surprise that there are still open questions about how to treat higher structures in HoTT/UF, which is a much younger endeavor. These are the matters we shall now turn to.

7.3 Higher Structures in HoTT/UF

When Voevodsky proposed using type theory as a foundation for mathematics, he based this on the insight that higher structures in mathematics are *not* always naturally objects of a higher *category*, but they *are* always naturally objects of a higher *groupoid*.

Among the ∞-groupoids we find *truncated* higher groupoids, those whose structure is concentrated in a finite range of dimensions. At the lowest level (truncation level -2) we find the contractible types, those that only have one element up to identifications. Secondly, we have propositions. These are types all of whose identity types are contractible.

Moving up in the dimensions, we find next the sets, all of whose identity types are propositions, and the 1-groupoids, all of whose identity types are sets, and so on. We recall from Altenkirch's Chapter that these truncation levels have a natural formalization in HoTT in terms of predicates

$$\text{hasDimension} : \mathbb{N}_{-2} \to \mathbf{Type} \to \mathbf{Prop},$$

and that we have corresponding types of n-truncated types, $n-\mathbf{Type}$.

Not all types are truncated. The 2-sphere, for example, has structure in all dimensions, so it's not an n-type, for any n.[3]

The n-types are related to the universe of all types, **Type**, via the truncation construction that maps a type X to its closest n-type $\|X\|_n$. There is a construction $|-|_n : X \to \|X\|_n$ giving rise to an equivalence

$$- \circ |-|_n : (\|X\|_n \to Y) \to (X \to Y)$$

for any n-type Y.

When we go to discuss higher structure, it is often the untruncated types that are the hardest to construct. The principal reason is that we can often construct truncated types in a top-to-bottom fashion, dimensionwise. To construct a proposition, we can just specify the type of evidence P that the proposition is true and then if necessary take the propositional truncation $\|P\|_{-1}$.

[3]This celebrated result of Serre (1953) has not yet been formalized in HoTT, but it will soon be within reach using the HoTT version of the Serre spectral sequence (van Doorn 2018).

I want to emphasize at this point that the sets we discussed above (and in the Chapters of Altenkirch and Ahrens-North) are *not* the sets of set theory! Following Quine's dictum, these are different notions because they have different notions of identity. Let us temporarily use subscripts to differentiate, and write set_1 for a set theorist's set and set_2 for a structuralist/homotopy theorist's set (this is also the model theorist's notion). There is even a third notion of set, set_0, which is arguably more fundamental than either set_1 or set_2, and which is the one taught in elementary education.

From a type-theoretic point of view, a set_0 is simply a subset of a fixed universal set_2, X. That is, we have the type $\mathbf{Set}_0(X) :\equiv \mathrm{P}(X) :\equiv (X \to \mathbf{Prop})$ representing the powerset of X. We have an elementary membership relation, $\in_0 : X \times \mathbf{Set}_0(X) \to \mathbf{Prop}$, and two sets_0 are equal if they have the same elements in this sense.

This is of course not the set-theorist's notion of set, according which sets_1 are elements (rather than subsets) of a universe of discourse U (itself a set_2) that is equipped with a membership relation $\in_1 : U \times U \to \mathbf{Prop}$ satisfying the axiom of extensionality (and preferably many other set-theoretic axioms).

The naive set-theoretical hope would be to solve the equation $U = \mathrm{P}(U)$ (as an identification of sets_2, i.e., an isomorphism). This is impossible because of Cantor's diagonal argument, but it can be approximated by the cumulative hierarchy V, a construction that can be performed in HoTT via a higher inductive type (Univalent Foundations Program 2013, Sect. 10.5). Here V is a large set_2 that is the least solution of the equation $V = \mathrm{P}_{\text{small}}(V)$, where $\mathrm{P}_{\text{small}}(V) :\equiv \Sigma A : \mathbf{Set}.\Sigma f : A \to V.\text{isInjective}(f)$ is the type of small subsets of V. Such sets_1 can be thought of as certain well-founded trees, and their study has a quite combinatorial flavor.

The default notion of set in HoTT/UF is set_2 given by the 1-type **Set**, and this seems to be the one most often used in mathematical practice outside of set theory. For instance, in almost all mathematical contexts, each set can be replaced by an isomorphic copy without changing the meaning of anything. Of course, sets_0 as elements of powerset 0-types/sets_2 also occur throughout mathematical practice, but for these, the set theorist's and the structuralist's notion coincide.

Likewise the notion of category splits into several distinct notions: I will denote by *precategory* the notion defined in Sect. 6.4.4 of the Chapter by Ahrens-North, and leave the unadorned term *category* for a univalent precategory. Indeed, in most category-theoretic contexts, each category can be replaced by an equivalent while preserving the meaning. It is also useful to have the term *strict category* (Univalent Foundations Program 2013, Sect. 9.6) for a precategory whose type of objects is a set. From the perspective of set-theoretic mathematics, the 2-type of categories arise from a Quillen model category structure on the 1-category of strict categories.

Most of category theory can be formalized in HoTT/UF using the univalent definition of category. A precategory can be thought of as a category with extra structure, namely equipped with a functor from an ∞-groupoid. For a strict category, this functor has as domain a 0-dimensional homotopy type. In set-theoretic foundations, it will automatically be the case that every category can be equipped with such a strict structure, but in UF this is an extra assumption, indeed a constructive-homotopical taboo.

7.3.1 Analytic and Synthetic Aspects of HoTT/UF

A foundational theory must be synthetic, in that it describes how to construct and reasons with its fundamental objects in terms of postulated rules. It couldn't be otherwise, for if it described the "fundamental" objects in terms of other, more fundamental, objects, and derived its rules from the properties of those, it would hardly be foundational.

Homotopy type theories are synthetic theories of ∞-groupoids. The approach is deeply *logical*, where we think of *logic as invariant theory* as pioneered by Mautner (1946) and later developed by Tarski in a 1966 lecture (Tarski 1986). Both Mautner and Tarski were inspired by the approach to geometry given in Klein's *Erlangen Program* (Klein 1872). The idea is that the logical notions are those that are invariant under that maximal notion of symmetries of the universes of discourse. If the universe of discourse is a set, then the corresponding symmetry group is the *symmetric group* consisting of bijections of the set with itself, but if it is a higher homotopy type, then it is the (higher) *automorphism* group consisting of all self-homotopy equivalences.

In analogy with the synthetic theories of various notions of geometry (euclidean, affine, projective, etc.), homotopy type theories are synthetic theories of homotopy types (and set theories are synthetic theories of sets_1), cf. also Awodey (2018).

The analytic aspect is that all the rest of mathematics, all mathematical objects, their types, and their structure, needs to be developed in terms of homotopy types. And a key criterion for success of a formalized notion is that it satisfies what Ahrens-North call the *principle of equivalence*, and which I linked to Quine's dictum above: that the identity type captures the intended notion of identifications between the mathematical objects that we are modeling.

One novel aspect of doing this analytical work in HoTT is when defining a structured object, it can be a challenge already to get the correct *carrier type*. In set-theoretic foundations, any carrier set of the correct cardinality will do, but in HoTT we are more discerning.

We do reap some benefits of this extra care. For instance, any construction (which, remember, could be proving a proposition, inhabiting a set, etc.) we perform on a generic category is guaranteed to be invariant under equivalence of categories, and we can use the rules of identity types to transport the construction along any equivalence.

Compare this to the situation in set-theoretic foundations: there we have to prove invariance under equivalence for any construction on types of dimension greater than 0. For sets, this is not necessary, because if we are given a set for which the notion of identification between the elements is given by an equivalence relation, we can take the quotient. In this way, the situation in set theory is marginally better than that of type theory pre-HoTT, where the set-quotient construction was not generally available, leading to what some practitioners have called "setoid hell". But in set theory, the same problem arises for any mathematical type of dimension greater than zero, so we may surmise that formalizations based on set theory will run into "higher groupoid hell".

7.3.2 Some Constructions That Are Possible

Let us finally take a look at some constructions that are possible in HoTT. Many of these are already discussed in Univalent Foundations Program (2013); references are provided in other cases. I will structure this discussion according to the means of construction used. Firstly, there are those that only use the basic constructions in Martin-Löf type theory, namely Σ- and Π-types, identity types, universes, as well as (finitary) inductive types such as the natural numbers, disjoint unions, the empty type, and the unit type.

Next, there are those that use in addition the univalence axiom. Following that, there are those that can be reduced to one particular higher inductive type, the (homotopy) pushout.

Finally, we find those constructions that seem to require more advanced higher inductive types, and in the next subsection I shall discuss those for which there is no known construction at the time of writing.

In *Basic Martin-Löf type theory* (MLTT) we can already define many important notions such as homotopy fibers and other pullbacks, the predicate hasDimension : $\mathbb{N}_{-2} \to \mathbf{Type} \to \mathbf{Type}$ and the types $n-\mathbf{Type}$. We have the types of categories and †-categories (cf. Univalent Foundations Program 2013, Sect. 9.7), as well as many other types of mathematical objects occurring outside of homotopy theory. But we are severely limited in our ability to construct inhabitants of these types, or prove properties about them. For instance, we cannot prove that hasDimension is valued in propositions (this requires function extensionality), nor can we prove that anything is not a set (such as the type **Set** itself), since there are models of MLTT in which every type is a set. We cannot construct set quotients, and, perhaps most embarrassingly, we cannot even define the logical operations of disjunction and existential quantification, as these require propositional truncation!

With *univalence* we get function extensionality (as shown by Voevodsky), and we can now prove many structural properties. Besides hasDimension landing in **Prop**, we can prove that $n-\mathbf{Type}$ is an $(n + 1)$-type, and we get the equivalence principle for the types of algebraic structures and for categories as mentioned in the Chapter by Ahrens-North. (See also Awodey 2014.) We can also prove that the nth universe is not an n-type for any external natural number n (Kraus and Sattler 2015).

At this point we can explore an intermediate route: instead of adding higher inductive types, we can assume the *propositional resizing axiom* (Univalent Foundations Program 2013, Axiom 3.5.5), stating that the inclusion map $\mathbf{Prop}_i \to \mathbf{Prop}_{i+1}$ of propositions in the ith universe into the propositions in the $(i + 1)$st universe is an equivalence. This makes the theory impredicative, and it allows us to mimic many impredicative tricks known from (constructive) set theory. For instance, the propositional truncation can be defined as $\|A\| :\equiv \Pi P : \mathbf{Prop}.(X \to P) \to P$. (If we have the law of excluded middle, then we can just define $\|A\| :\equiv \neg\neg A$.)

The (homotopy) *pushout* type is a simple, but versatile example of a higher inductive type. It generalizes the disjoint union. Its inputs are three types A, B, and C, together with functions $f : C \to A$ and $g : C \to B$. (Such a configuration is called a *span*.) The pushout is a new type $D :\equiv A \sqcup^C_{f,g} B$ (often written $A \sqcup^C B$ if f and g can be deduced from the context) together with *injections* left $: A \to D$ and right $: B \to D$ fitting together in a square

$$\begin{array}{ccc} C & \xrightarrow{g} & B \\ {\scriptstyle f}\downarrow & & \downarrow{\scriptstyle \text{right}} \\ A & \xrightarrow[\text{left}]{} & D \end{array}$$

whose commutativity is given by a constructor glue $: \Pi x : C.\, \text{left}(f\, x) = \text{right}(g\, x)$. (See (Univalent Foundations Program 2013, Sect. 6.8) for the elimination and computation rules.)

Now a quite remarkable phenomenon appears. Most of the higher inductive types that are commonly used can be constructed just from pushouts and the other constructions in MLTT with univalent universes. These include joins and suspensions (and therefore, spheres), cofibers (and thus smash products), sequential colimits, the propositional truncation (van Doorn 2016; Kraus 2016) and all the higher truncations (Rijke 2017), set quotients, and in fact, all non-recursive HITs specified using point-, 1-, and 2-constructors by a construction due to van Doorn et al. (2017). We also get cell complexes (Buchholtz and Favonia 2018), Eilenberg-MacLane spaces (Licata and Finster 2014), and projective spaces (Buchholtz and Rijke 2017), and so a lot of algebraic topology can be developed on this basis, and even a theory of ∞-groups (Buchholtz et al. 2018) and spectra (and thus homology and cohomology theory), culminating in a proof that $\pi_4(S^3) = \mathbb{Z}/2\mathbb{Z}$ (Brunerie 2016), and a formalized proof of the Serre spectral sequence for cohomology (van Doorn 2018).

Another important construction enabled by pushouts is the Rezk completion, which turns a precategory into the category it represents. This can in fact be done using univalence alone (Ahrens et al. 2015), at the cost of going to a larger universe. This can be avoided either using pushouts or using the propositional resizing axiom.

Because so many things can be developed on the basis of univalence, pushouts, and propositional resizing, Shulman suggested that we define an *elementary* $(\infty, 1)$-*topos* to be a finitely complete and cocomplete, locally cartesian closed $(\infty, 1)$-category with a subobject classifier and object classifiers (Shulman 2017). Let me correspondingly introduce the term *elementary HoTT* for MLTT with univalence, pushouts, and propositional resizing.

Finally, let me mention some of the known constructions that seem to require more than the above means, but that can nonetheless be effected via more general higher inductive types. First, there is the cumulative hierarchy as mentioned above (Univalent Foundations Program 2013, Sect. 10.5), the Cauchy-complete real numbers (Univalent Foundations Program 2013, Sect. 11.3), as well as the partiality monad (Altenkirch et al. 2017). I won't say more about these, since this Chapter is supposed to be about *higher* structure. For these it is more relevant to mention *localizations* at a family of maps (Rijke et al. 2017). For example, if we localize at a family of maps of the form $P(a) \to 1$, for $a : A$, where each type $P(a)$ is a proposition, then we obtain an inner model of type theory in itself, in this case a *topological localization*.

Of course, the number of things that have been constructed and proved in HoTT grows every day, so undoubtedly I'll have left some out. Many of these constructions have already been formalized in proof assistants for HoTT. In the next section we move to constructions that we don't yet know how to perform (and which may perhaps require new means of construction).

7.3.3 Some Constructions That Seem Impossible

Because proving propositions is in HoTT/UF a special case of making constructions, any currently open problems count as constructions we don't yet know how to perform.[4] But for some of these, it is expected that the difficulty is not just that the construction is tricky to perform with the currently available means of construction, but rather that we conjecture that entirely new means of construction will be necessary.

The prime example that I will focus on is that of $(\infty, 1)$-categories, and the related notions of (semi-)simplicial types. Intuitively, an $(\infty, 1)$-category C consists of a type of objects C_0, for every pair of objects $a, b : C_0$ a type of morphisms $C_1(a, b)$, operations for identities and composition, operations that witness the unit- and associative laws, operations that witness higher laws that these must satisfy, and so on *ad infinitum*. The problem is to come up with a way of specifying all these higher coherence operations in a single type.

The basic example of an $(\infty, 1)$-category from the point of view of type theory is the category of types S that has as type of objects the universe **Type**, and as morphisms from A to B the type of functions from A to B, $S_1(A, B) :\equiv A \to B$. Here, the evident identity and composition operations satisfy all the laws and higher laws *definitionally*, so it ought to be particularly easy to show that S is an $(\infty, 1)$-category, that is, if we knew how to define the type of $(\infty, 1)$-categories, $(\infty, 1)-$**Cat**.

It is crucial for the success of UF that we have a working definition and theory of $(\infty, 1)$-categories. Even if they are not *fundamental* in the sense that everything is built out of them, they are still *fundamental* in the sense that they are a key tool in the development of modern higher algebra, geometry, and topology.

The problem of defining $(\infty, 1)$-categories is equivalent to the problem of defining the type of simplicial types, **sType**. A *simplicial type* is just a functor $X : \Delta^{\mathrm{op}} \to S$, so if we know how to define $(\infty, 1)-$**Cat** and the type of functors between any two $C, D : (\infty, 1)-$**Cat**, then we can define **sType**. On the other hand,

[4] See https://ncatlab.org/homotopytypetheory/show/open+problems for an up-to-date list of open problems.

$(\infty, 1)-$**Cat** itself can be defined as the subtype of **sType** consisting of *complete Segal types* (also called *Rezk types*) (Riehl and Shulman 2017).

The problem of defining $(\infty, 1)-$**Cat** can also be reduced to another, apparently simpler problem, namely that of defining the type of semi-simplicial types, **ssType**. A *semi-simplicial type* is a functor $X : \Delta_+^{\mathrm{op}} \to \mathrm{S}$, where Δ_+ is the subcategory of Δ with the same objects but only injective functions. However, Δ_+ is a *direct category*, viz. a 0-truncated category where the relation "x has a nonidentity arrow to y" is a well-founded relation on the *set* of objects, so we can give a more direct description as follows: A semi-simplicial type X consists of:

- a type of 0-simplices X_0, and
- for every pair of 0-simplices $a_0, a_1 : X_0$, a type of 1-simplices from a_0 to a_1, $X_1(a_0, a_1)$, and
- for every triple of 0-simplices $a_0, a_1, a_2 : X_0$ and 1-simplices $a_{01} : X_1(a_0, a_1)$, $a_{02} : X_1(a_0, a_2)$, and $a_{12} : X_1(a_1, a_2)$, a type of 2-simplices with boundary a_{01}, a_{02}, a_{12},
- *and so on . . .*

Again we have the problem that it seems that infinitely much data is needed, but here it seems more plausible that an inductive (or coinductive) approach could work. Only, no-one has figured out how to do it, and at an informal poll of HoTT-researchers in Warsaw in 2015 a majority believed that it is impossible.

We can define $(\infty, 1)$-categories in terms of semi-simplicial types, as the complete semi-Segal types (Capriotti and Kraus 2017). This work was inspired by analogous work in the classical setting (Harpaz 2015). Thus, the problems of defining $(\infty, 1)$-categories, simplicial types, and semi-simplicial types are equivalent, but currently just beyond reach.

It is interesting to contrast the case of $(\infty, 1)$-categories with that of ∞-groups: An $(\infty, 1)$-category structure on a pointed, connected type of objects is the same as an ∞-monoid (the type of which we also don't know how to construct). But the type of ∞-groups is simply that of pointed, connected types, with the type of group elements being the identity type $\Omega A :\equiv (\mathrm{pt} =_A \mathrm{pt})$, with $\mathrm{pt} : A$ the designated point.

While the above problems concern "large" types, there are also problems of higher structure concerning "small" types. For example, the three-sphere as a type S^3 should carry the structure of an ∞-group, because it is the homotopy type of the Lie group $SU(2)$. Thus, we should be able to construct the homotopy type of the classifying space $BSU(2)$ with $\Omega BSU(2) = S^3$, but so far we've not been able to do so. (We have the H-space structure, which is a first approximation (Buchholtz and Rijke 2018).) In this case, however, we expect that no new means of construction are needed.

An obvious approach would be to construct in the usual way a simplicial set whose homotopy type is $BSU(2)$. If we then had the realization operation $|\cdot| :$ **sSet** $\to$ **Type** that turns a simplicial set into the homotopy type it represents, then we'd be done. But such a realization operation itself seems impossible to construct in elementary HoTT!

As a final, important, but more open-ended problem, let me mention the problem of developing the meta-theory of HoTT inside HoTT/UF. This has two sides, one relatively easy, and one quite hard. The relatively easy side is the syntactic one, but even here there are difficulties. We can represent *extrinsic* untyped syntax and corresponding transformations familiar from compiler theory: picking a surface syntax, lexing and parsing this syntax, and then type-checking it. The result should be *intrinsic* syntax containing only well-typed elements. The intrinsic syntax can be modeled by *quotient inductive types* (QIT) (Altenkirch and Kaposi 2016), already mentioned in Altenkirch's Chapter. The main difficulty here is one of software-engineering: how do we structure both the intrinsic and extrinsic syntax, and the transformations between them, in sufficient generality to cover all the kinds of type theory we are interested in.

The more difficult side is the semantic one. We want to define interpretations of the intrinsic syntax in inner models, first of all the canonical model where syntactic types $\vdash A$ are mapped to types $[\![A]\!]$, syntactic terms $\vdash a : A$ are mapped to terms $[\![a]\!] : [\![A]\!]$, and so on. (For proof-theoretic reasons, we expect to only be able to represent the interpretation locally, for instance type theory with n universes inside the $(n + 1)$st universe, or using stronger principles in the target type theory than are in the source type theory.) Shulman has called this problem "making HoTT eat itself" (Shulman 2014), for which the term *autophagy* suggests itself.

The problem is that if we use the QIT intrinsic syntax, then everything syntactic is a (homotopy) set and the elimination rule will only allow us to eliminate into sets, whereas for the canonical model we're eliminating into **Type**. And if we try to formalize the intrinsic syntax using a non-truncated HIT, then it seems we need infinitely many layers of coherence (reminding us of our problems above with semi-simplicial types and the homotopical realization of simplicial sets).

7.4 Possible Further Means of Construction

Now that we have seen concretely both the range of constructions that are currently possible in (elementary) HoTT, and some prominent problems that seem out of reach, let us take stock.

The first conclusion is that we'd very much like to *prove* that the problem of semi-simplicial types cannot be solved in elementary HoTT. But assuming that, the next conclusion is that elementary HoTT is by itself too incomplete to serve its foundational role as the basis for UF. Further means of constructions need to be added. But which ones, and how do we decide which to add?

In some sense the situation is analogous to the question of new axioms for set theory, but there are two main differences: First, we want to use type theory as a *programming language* and that means that for any proposed extension, we should say how the new constructions *compute* when combined with the other constructions of type theory. The univalence axiom is a sore point in this regard, as it has been a long-standing open problem to give it a computational meaning.

This is now close to being solved via various *cubical type theories* (Angiuli et al. 2017; Bezem et al. 2014; Cohen et al. 2018), but there remains a question of whether the corresponding model structures on various categories of cubical sets model ∞-groupoids (we know that the *test model structures* do Buchholtz and Morehouse 2017). It is still completely open whether the propositional resizing axiom can be given computational meaning.

Secondly, we want to use HoTT also in other models than ∞-groupoids. It is conjectured that elementary HoTT can be interpreted in any $(\infty, 1)$-topos: a left-exact localization of the functor category $C^{op} \to S$ for a small $(\infty, 1)$-category C. (It would take us too far afield to give the exact formulation and up-to-date status of this conjecture; see Schreiber and Shulman 2019.)

Some of the most interesting targets are given by *cohesive* $(\infty, 1)$*-toposes*, whose objects can be thought of as geometrically structured ∞-groupoids. (See also Schreiber (2016)) For example, in the cohesive $(\infty, 1)$-topos of smooth ∞-groupoids, **Smooth∞Gpd**, we find all smooth manifolds among the 0-truncated objects. And we certainly want to be able to reason about smooth $(\infty, 1)$-categories using a HoTT interpreted in **Smooth∞Gpd**. So it will not do to propose a construction of the type $(\infty, 1)-\mathbf{Cat}$ that can't be performed meaningfully in any $(\infty, 1)$-topos.

In contrast, for the problem of interpreting type theory in internal models, including these cohesive $(\infty, 1)$-toposes, we need not require that the means of doing so themselves are available in arbitrary models. It seems sufficient to be able to do this "at the top level". But however we solve this problem, it should probably be with computationally meaningful (constructive) means, so that we're able to do proofs by reflection inside these models. (Of course not all models of interest will be definable constructively.)

Summing up, we expect there to be a stratification of homotopy type theories,

- $\mathrm{HoTT_{ua}}$: MLTT plus the univalence axiom and propositional resizing,
- $\mathrm{HoTT_{el}}$: $\mathrm{HoTT_{ua}}$ plus pushouts,
- $\mathrm{HoTT_{el++}}$: $\mathrm{HoTT_{el}}$ plus constructions needed for $(\infty, 1)-\mathbf{Cat}$,
- $\mathrm{HoTT_{UF}}$: $\mathrm{HoTT_{el++}}$ plus reflective constructions.

Here, $\mathrm{HoTT_{ua}}$ is the basis for Voevodsky's UniMath formalization effort (Voevodsky 2015; Voevodsky et al. 2014). We know that $\mathrm{HoTT_{el}}$ is strictly stronger (in the sense of having fewer models; not in the sense of proof-theoretic strength), because it is consistent with $\mathrm{HoTT_{ua}}$ that the nth universe is an n-type, while if the universes are closed under pushouts, then they are not truncated. Since we don't have impossibility proofs regarding the constructions of $(\infty, 1)-\mathbf{Cat}$ nor for autophagy, it is still conceivable that we can take $\mathrm{HoTT_{el++}}$ and $\mathrm{HoTT_{UF}}$ to be $\mathrm{HoTT_{el}}$.

For each of these we can consider adding classical axioms such as the law of the excluded middle (LEM) or the axiom of choice (AC). These can be seen as constructions that *we* don't know how to perform in general, but that an omniscient being would be able to perform. I will not here take any sides in the debate about whether mathematics is better done with or without these principles—I'll only say that it seems to me that a foundational theory should give its users the choice. (And in any case, we want theories without these constructive taboos for reasoning about other toposes of interest.)

We can also remove the resizing axiom to get (generalized) predicative systems, and for both the predicative and impredicative systems we may add various generalized inductive types to increase the proof-theoretic strength if needed, while keeping the systems constructive and without changing their class of $(\infty, 1)$-topos models. Thus, the stratification above is meant primarily to distinguish the (intended) models of the theories, and not their proof-theoretic strength, which is an orthogonal concern.

With all that in mind, let us discuss some possible further means of construction that we might add either for $\mathrm{HoTT_{el}}$ or $\mathrm{HoTT_{UF}}$.

7.4.1 Simplicial Type Theory

For any $(\infty, 1)$-topos C, we can consider the simplicial objects in C, i.e., functors $\Delta^{\mathrm{op}} \rightarrow$ C, and this is again an $(\infty, 1)$-topos. As mentioned above, we find therein a full subcategory of $(\infty, 1)$-categories relative to C. This is the basis for the suggestion by Riehl and Shulman (2017) for a synthetic type theory for $(\infty, 1)$-categories. In their type theory, let's call it sHoTT, types are interpreted as *simplicial types*, and they give definitions for *Segal* and *Rezk types* with the latter representing $(\infty, 1)$-categories. They can also define a type of *discrete* simplicial types, representing ordinary types/∞-groupoids, but this type is not a Rezk type and so not the representation of the $(\infty, 1)$-category of ordinary types, S. Indeed, it is not clear at this moment, whether this is at all representable in their system.

Much work remains before we can judge how useful this type theory is for reasoning about $(\infty, 1)$-categories. But from a philosophical point of view it cannot be satisfactory to view sHoTT as a foundational theory, for instance playing the role of $\mathrm{HoTT_{UF}}$. (And it is of course not intended as such!) Because simplicial types are begging to be analyzed as such: simplicial objects in a category of types, and not to be taken as unanalyzed in themselves. We really want to be able to *define* simplicial types inside a theory where types are the fundamental objects.

7.4.2 Two-Level Type Theories

Another approach to solving the problem of defining simplicial types is to have another layer above the univalent type theory in which to reason about infinitary strict constructions, including (semi-)simplicial types. One proposal is Voevodsky's Homotopy Type System (HTS) (Voevodsky 2013). This system has a distinction between fibrant and non-fibrant types, and it has two identity types: the usual homotopical identity type that only eliminates into fibrant types as well as a new non-fibrant strict equality type that satisfies the reflection rule: if $e : a \overset{\mathrm{s}}{=}_A a'$ is an inhabitant of the strict equality type, then a and a' are definitionally equal. This rule of course makes type-checking undecidable, necessitating a further language of evidence for typing derivations.

Another proposal is the two-level type theory (2LTT) (Altenkirch et al. 2016; Annenkov et al. 2017). This is similar to HTS in that it distinguishes between fibrant and non-fibrant types (the latter are called *pretypes*), but instead of the reflection rule for the strict equality type it adds the rule for uniqueness of equality proofs (Streicher's K) and function extensionality as an axiom. Thus, type checking is decidable, but the function extensionality axiom breaks computation.

In order to define simplicial types in 2LTT an extra principle beyond the basic set-up is needed. This can be the assumption that the fibrant and non-fibrant natural numbers coincide, or the more technical assumption that Reedy fibrant diagrams of fibrant types indexed by a strict inverse category have fibrant limits. The former limits the class of available models severely, while the latter does not.

However useful the two-level type theories may turn out to be, they also seem unsatisfactory from a foundational perspective. Because what is a pretype? Pretypes can only be motivated via the models of HoTT as described in set-theoretic mathematics, where they arise as the objects of a model category presenting an $(\infty, 1)$-topos. But they are not preserved by an equivalence of $(\infty, 1)$-toposes, so they don't have a presentation-independent meaning. They seem to be merely a tool of convenience.

7.4.3 *Computational Type Theories*

If we limit ourselves to one constructive model, then there is a principled way of making sense of new constructions. This is via the paradigm of computational type theories in the Nuprl tradition (Constable et al. 1986). Here we consider a particular model to give *meaning explanations* for the judgments of type theory. For a certain notion of cubical sets, this has been done by Harper's group (Angiuli et al. 2017, 2018). The benefit of the approach is that it guarantees that all constructions are computationally meaningful and make sense in the model. The downside is that it is tied to a particular model (though by being judicious with which primitives are added to the computational language this downside can be minimized) and that it also leads to a type theory without decidable type checking, so a separate proof theory or language of evidence is needed.

7.4.4 *Presentation Axioms*

It may have perhaps occurred to some readers that the problems discussed in Sect. 7.3 should be solved in the same way that they are solved in homotopical mathematics based in set-theory, namely by working with set-based presentations.

We already mentioned geometric realization, an operation that produces the underlying homotopy type of a given simplicial set or topological space. We could consider adding $|\cdot| : \mathbf{sSet} \to \mathbf{Type}$ as a basic construction and the axiom stating that $|\cdot|$ is surjective, meaning that every type is merely equivalent to the geometric realization of some simplicial set. And perhaps we should add a further axiom stating that every function $A \to B$ between types arises (merely) as the geometric realization of a function between representing simplicial sets.

Something like this may indeed be appropriate at the level of $\mathrm{HoTT}_{\mathrm{UF}}$ if it could be given a computational meaning. But certainly not at the level of $\mathrm{HoTT}_{\mathrm{el++}}$, because the axioms would severely restrict the range of models (they are constructive-homotopical taboos).

Even the much weaker axiom *sets cover* (SC), stating that every type admits a surjection from a set admits a simple counter-model $(\infty, 1)$-topos (Schreiber and Shulman 2018).

On the other hand, SC (or something like it) is necessary in order to describe the semantics of HoTT (with universes) in presheaf toposes. Indeed, the universe in a presheaf topos is built from certain *sets* covering the 1-types of presheaves of small sets.

7.5 Conclusion

Higher structures are at once the raison d'être and so far, the Achilles' heel, of HoTT/UF from a foundational perspective. HoTT can handle with ease many important higher structures, such as the 1-type of sets and the 2-type of categories, that can only imperfectly be represented in other foundational systems. But so far it cannot define the (untruncated) type of $(\infty, 1)$-categories, and this is a major impediment to the foundational aspirations of HoTT/UF. Of course, HoTT can be (and has been) used successfully to reason about (structured) homotopy types. In this way, the various homotopy type theories function as *domain specific languages* (DSLs).

To be foundational, however, we need to find a compelling construction of, and theory of, $(\infty, 1)$-categories, and of the semantics of HoTT-DSLs, inside homotopy type theory itself. It appears that new methods of construction are needed, but it is at this time not clear what they should be.

A dramatic possibility, not mentioned in Sect. 7.4, is that we should take $(\infty, 1)$-categories to be fundamental after all, and build a synthetic type theory where the types are $(\infty, 1)$-categories rather than ∞-groupoids. This would be a *directed* type theory. Such a thing would undoubtedly be quite complicated due to the need to keep track of variances (see Licata and Harper (2011) and Nuyts (2015) for some preliminary attempts), and it would represent a return to the old ways of thinking about categorical foundations, albeit updated to account for a homotopical perspective. We would carve out the ∞-groupoids as those $(\infty, 1)$-categories all of whose morphisms are invertible rather than trying to build $(\infty, 1)$-categories out of

∞-groupoids. In any case, directed type theories should be useful also as DSLs for reasoning about $(\infty, 2)$-toposes.

Personally, I think we'll find some solution that allows us to stay at the level of ∞-groupoids for the foundational theory. Perhaps there is a kind of two-level type theory that allows us to capture the strict nature of the $(\infty, 1)$-category of types without postulating a bunch of meaningless pretypes.

An analogy can perhaps be made with the foundations of stable homotopy theory. The $(\infty, 1)$-category of spectra is a symmetric monoidal stable $(\infty, 1)$-category, and from a foundational point of view, this is the correct viewpoint, since spectra should be identified when they are weakly equivalent. However, it was discovered that this $(\infty, 1)$-category can be presented by symmetric monoidal Quillen model categories (i.e., very strict structures), and this has been very important in facilitating computations in stable homotopy theory (Elmendorf et al. 1995; Mandell et al. 2001). (I should mention that there is work-in-progress by Finster-Licata-Morehouse-Riley on developing a HoTT-DSL for stable homotopy theory targetting the cohesive $(\infty, 1)$-topos of parametrized spectra: this captures the strict monoidal structure of spectra in a type theory.)

And so it may be, that in order to realize the foundational potential of HoTT/UF, we shall need to capture the strict structure of type theory itself, perhaps by reflecting more of judgmental structure at the level of types.

I'm confident that a good solution will be eventually found. The field is still young, and it will be exciting to see what the future brings.

References

Ahrens, B., Kapulkin, K., & Shulman, M. (2015). Univalent categories and the Rezk completion. *Mathematical Structures in Computer Science, 25*(5), 1010–1039. https://doi.org/10.1017/S0960129514000486

Altenkirch, T., & Kaposi, A. (2016). Type theory in type theory using quotient inductive types. In *Proceedings of the 43rd Annual ACM SIGPLAN-SIGACT Symposium on Principles of Programming Languages, POPL'16* (pp. 18–29). New York: ACM. https://doi.org/10.1145/2837614.2837638

Altenkirch, T., Capriotti, P., & Kraus, N. (2016). Extending homotopy type theory with strict equality. In Computer science logic 2016, *LIPIcs. Leibniz International Proceedings in Informatics* (Vol. 62, pp. Art. No. 21, 17). Schloss Dagstuhl. Leibniz-Zent. Inform., Wadern.

Altenkirch, T., Danielsson, N. A., & Kraus, N. (2017). Partiality, revisited. In J. Esparza & A. S. Murawski (Eds.), *Foundations of Software Science and Computation Structures: 20th International Conference, FOSSACS 2017, Held as Part of the European Joint Conferences on Theory and Practice of Software, ETAPS 2017, Uppsala, 22–29 Apr 2017, Proceedings* (pp. 534–549). Berlin/Heidelberg: Springer. https://doi.org/10.1007/978-3-662-54458-7_31

Angiuli, C., Harper, R., & Wilson, T. (2017). Computational higher-dimensional type theory. In *POPL'17: Proceedings of the 44th Annual ACM SIGPLAN-SIGACT Symposium on Principles of Programming Languages* (pp. 680–693). ACM. https://doi.org/10.1145/3009837.3009861

Angiuli, C., & Harper, R. (2018). Meaning explanations at higher dimension. *Indagationes Mathematicae, 29*(1), 135–149. https://doi.org/10.1016/j.indag.2017.07.010. L.E.J. Brouwer, fifty years later.

Annenkov, D., Capriotti, P., & Kraus, N. (2017). Two-level type theory and applications. https://arxiv.org/abs/1705.03307. Preprint.

Awodey, S. (2014). Structuralism, invariance, and univalence. *Philosophia Mathematica (3), 22*(1), 1–11. https://doi.org/10.1093/philmat/nkt030

Awodey, S. (2018). Univalence as a principle of logic. Indagationes Mathematicae. https://doi.org/10.1016/j.indag.2018.01.011

Baez, J. (2007). The homotopy hypothesis. http://math.ucr.edu/home/baez/homotopy/. Lecture at Higher Categories and Their Applications, Thematic Program on Geometric Applications of Homotopy Theory, Fields Institute, Toronto.

Bezem, M., Coquand, T., & Huber, S. (2014). A model of type theory in cubical sets. In *19th International Conference on Types for Proofs and Programs (TYPES 2013), LIPIcs. Leibniz International Proceedings in Informatics* (Vol. 26, pp. 107–128). Schloss Dagstuhl. Leibniz-Zent. Inform., Wadern. https://doi.org/10.4230/LIPIcs.TYPES.2013.107

Boardman, J., & Vogt, R. (1973). *Homotopy invariant algebraic structures on topological spaces* (Vol. 347). Cham: Springer.

Brunerie, G. (2016). On the homotopy groups of spheres in homotopy type theory. Ph.D. thesis, Laboratoire J.A. Dieudonné. https://arxiv.org/abs/1606.05916

Buchholtz, U., & Morehouse, E. (2017). Varieties of cubical sets. In P. Höfner, D. Pous, & G. Struth (Eds.), *RAMICS 2017: Relational and algebraic methods in computer science*, *Lecture notes in computer science* (Vol. 10226). Cham: Springer. https://doi.org/10.1007/978-3-319-57418-9_5

Buchholtz, U., & Rijke, E. (2017). The real projective spaces in homotopy type theory. In *32nd Annual ACM/IEEE Symposium on Logic in Computer Science (LICS 2017)* (pp. 1–8). New York: IEEE. https://doi.org/10.1109/LICS.2017.8005146

Buchholtz, U., & Favonia (Hou), K. B. (2018). Cellular cohomology in homotopy type theory. https://arxiv.org/abs/1802.02191. Accepted for Proceedings of Logic in Computer Science (LICS 2018).

Buchholtz, U., & Rijke, E. (2018). The Cayley-Dickson construction in homotopy type theory. To appear in Higher Structures.

Buchholtz, U., van Doorn, F., & Rijke, E. (2018). Higher groups in homotopy type theory. https://arxiv.org/abs/1802.04315. Accepted for Proceedings of Logic in Computer Science (LICS 2018).

Capriotti, P., & Kraus, N. (2017). Univalent higher categories via complete semi-Segal types. https://arxiv.org/abs/1707.03693. Preprint.

Cohen, C., Coquand, T., Huber, S., & Mörtberg, A. (2018). Cubical type theory: A constructive interpretation of the univalence axiom. In *21st International Conference on Types for Proofs and Programs (TYPES 2015), LIPIcs. Leibniz International Proceedings in Informatics*, Schloss Dagstuhl. Leibniz-Zent. Inform., Wadern. https://doi.org/10.4230/LIPIcs.TYPES.2015.5

Constable, R. L., Allen, S. F., Bromley, H. M., Cleaveland, W. R., Cremer, J. F., Harper, R. W., Howe, D. J., Knoblock, T. B., Mendler, N. P., Panangaden, P., Sasaki, J. T., Smith, S. F. (1986). Implementing mathematics with the Nuprl proof development system. Prentice-Hall. http://www.nuprl.org/

Elmendorf, A., Kříž, I., Mandell, M. A., & May, J. (1995). Modern foundations for stable homotopy theory. In *Handbook of algebraic topology* (pp. 213–253). Amsterdam: North-Holland.

Grothendieck, A. (1983). Pursuing stacks. http://thescrivener.github.io/PursuingStacks/. Manuscript.

Harpaz, Y. (2015). Quasi-unital ∞-categories. *Algebraic & Geometric Topology, 15*(4), 2303–2381. https://doi.org/10.2140/agt.2015.15.2303

Joyal, A. (2002). Quasi-categories and Kan complexes. *Journal of Pure and Applied Algebra, 175*(1–3), 207–222. https://doi.org/10.1016/S0022-4049(02)00135-4

Kan, D. M. (1956). Abstract homotopy. III. *Proceedings of the National Academy of Sciences of the United States of America, 42*, 419–421. https://doi.org/10.1073/pnas.42.7.419

Klein, F. (1872). Vergleichende Betrachtungen über neuere geometrische Forschungen. Verlag von Andreas Deichert, Erlangen.

Kraus, N. (2016). Constructions with non-recursive higher inductive types. In *Proceedings of the 31st Annual ACM/IEEE Symposium on Logic in Computer Science (LiCS'16)* (pp. 595–604). New York: ACM. https://doi.org/10.1145/2933575.2933586

Kraus, N., & Sattler, C. (2015). Higher homotopies in a hierarchy of univalent universes. *ACM Transactions on Computational Logic, 16*(2), 18:1–18:12. https://doi.org/10.1145/2729979

Licata, D. R., & Harper, R. (2011). 2-dimensional directed type theory. *Electronic Notes in Theoretical Computer Science, 276*, 263–289. https://doi.org/10.1016/j.entcs.2011.09.026. *Twenty-seventh Conference on the Mathematical Foundations of Programming Semantics (MFPS XXVII).*

Licata, D. R., & Finster, E. (2014). Eilenberg-MacLane spaces in homotopy type theory. In *Proceedings of the Joint Meeting of the Twenty-Third EACSL Annual Conference on Computer Science Logic (CSL) and the Twenty-Ninth Annual ACM/IEEE Symposium on Logic in Computer Science (LICS), CSL-LICS'14* (pp. 66:1–66:9). New York: ACM. https://doi.org/10.1145/2603088.2603153

Lurie, J. (2009). Higher topos theory, *Annals of mathematics studies* (Vol. 170). Princeton: Princeton University Press. https://doi.org/10.1515/9781400830558

Mandell, M., May, J., Schwede, S., & Shipley, B. (2001). Model categories of diagram spectra. *Proceedings of the London Mathematical Society (3), 82*(2), 441–512. https://doi.org/10.1112/S0024611501012692

Mautner, F. (1946). An extension of Klein's Erlanger Program: Logic as invariant-theory. *American Journal of Mathematics 68*(3), 345–384. https://doi.org/10.2307/2371821

Mazel-Gee, A. (2016). Quillen adjunctions induce adjunctions of quasicategories. *New York Journal of Mathematics, 22*, 57–93. http://nyjm.albany.edu/j/2016/22-4.html

Nuyts, A. (2015). Towards a directed homotopy type theory based on 4 kinds of variance. Ph.D. thesis, KU Leuven. https://people.cs.kuleuven.be/~dominique.devriese/ThesisAndreasNuyts.pdf

Quillen, D. (1967). *Homotopical algebra* (Vol. 43). Cham: Springer. https://doi.org/10.1007/BFb0097438

Quine, W. V. O. (1969). *Ontological relativity and other essays*. New York/London: Columbia University Press.

Riehl, E., & Shulman, M. (2017). A type theory for synthetic ∞-categories. *Higher Structures, 1*(1), 147–224. https://journals.mq.edu.au/index.php/higher_structures/article/view/36

Rijke, E. (2017). The join construction. https://arxiv.org/abs/1701.07538. Preprint.

Rijke, E., Shulman, M., & Spitters, B. (2017). Modalities in homotopy type theory. https://arxiv.org/abs/1706.07526. Preprint.

Rodin, A. (2017). Venus homotopically. *IfCoLog Journal of Logics and Their Applications, 4*(4), 1427–1445. http://collegepublications.co.uk/ifcolog/?00013. Special Issue Dedicated to the Memory of Grigori Mints. Dov Gabbay and Oleg Prosorov (Guest Editors).

Schreiber, U. (2016). *Modern Physics formalized in modal homotopy type theory*. https://ncatlab.org/schreiber/show/Modern+Physics+formalized+in+Modal+Homotopy+Type+Theory, nLab

Schreiber, U., & Shulman, M. (2019). Model of type theory in an $(\infty, 1)$-topos. https://ncatlab.org/homotopytypetheory/show/model+of+type+theory+in+an+(infinity,1)-topos Revision 19, Homotopy Type Theory wiki.

Schreiber, U., & Shulman, M. (2018). *n*-types cover. https://ncatlab.org/nlab/show/n-types+cover. Revision 6, nLab.

Serre, J. P. (1953). Cohomologie modulo 2 des complexes d'Eilenberg-MacLane. *Commentarii Mathematici Helvetici, 27*, 198–232. https://doi.org/10.1007/BF02564562

Shulman, M. (2014). Homotopy type theory should eat itself (but so far, it's too big to swallow). https://homotopytypetheory.org/2014/03/03/hott-should-eat-itself/. Blog post.

Shulman, M. (2017). Elementary $(\infty, 1)$-topoi. https://golem.ph.utexas.edu/category/2017/04/elementary_1topoi.html. Blog post.

Tarski, A. (1986). What are logical notions? *History and Philosophy of Logic, 7*(2), 143–154. https://doi.org/10.1080/01445348608837096. Ed. by J. Corcoran.

Tsementzis, D. (2017). Univalent foundations as structuralist foundations. *Synthese, 194*(9), 3583–3617. https://doi.org/10.1007/s11229-016-1109-x

Univalent Foundations Program (2013). Homotopy Type Theory: Univalent Foundations of Mathematics. http://homotopytypetheory.org/book/, Institute for Advanced Study.

van Doorn, F. (2016). Constructing the propositional truncation using non-recursive hits. In *Proceedings of the 5th ACM SIGPLAN Conference on Certified Programs and Proofs* (pp. 122–129). https://doi.org/10.1145/2854065.2854076

van Doorn, F. (2018). On the formalization of higher inductive types and synthetic homotopy theory. Ph.D. thesis, Carnegie Mellon University.

van Doorn, F., von Raumer, J., & Buchholtz, U. (2017). Homotopy type theory in Lean. In M. Ayala-Rincón & C. Muñoz (Eds.), *ITP 2017: Interactive theorem proving, Lecture notes in computer science* (Vol. 10499). Cham: Springer. https://doi.org/10.1007/978-3-319-66107-0_30

Voevodsky, V. (2013). A simple type system with two identity types. https://ncatlab.org/homotopytypetheory/files/HTS.pdf. Started 23 Feb 2013. Work in progress.

Voevodsky, V. (2014). The origins and motivations of univalent foundations. The IAS Institute Letter. https://www.ias.edu/ideas/2014/voevodsky-origins

Voevodsky, V. (2015). An experimental library of formalized mathematics based on the univalent foundations. *Mathematical Structures in Computer Science, 25*(5), 1278–1294. https://doi.org/10.1017/S0960129514000577

Voevodsky, V., Ahrens, B., Grayson, D., et al. (2014). UniMath – a computer-checked library of univalent mathematics. Available at https://github.com/UniMath/UniMath

Chapter 8
Univalent Foundations and the UniMath Library

The Architecture of Mathematics

Anthony Bordg

Abstract We give a concise presentation of the Univalent Foundations of mathematics outlining the main ideas, followed by a discussion of the UniMath library of formalized mathematics implementing the ideas of the Univalent Foundations (Sect. 8.1), and the challenges one faces in attempting to design a large-scale library of formalized mathematics (Sect. 8.2). This leads us to a general discussion about the links between architecture and mathematics where a meeting of minds is revealed between architects and mathematicians (Sect. 8.3). On the way our odyssey from the foundations to the "horizon" of mathematics will lead us to meet the mathematicians David Hilbert and Nicolas Bourbaki as well as the architect Christopher Alexander.

8.1 The Univalent Foundations and the UniMath Library

8.1.1 The Univalent Foundations of Mathematics

The *Univalent Foundations* (Voevodsky 2014) of mathematics designed by Vladimir Voevodsky builds upon Martin-Löf type theory (Martin-Löf 1982), a logical system for constructive mathematics with nice computational properties that makes mathematics amenable to proof-checking by computers (i.e. by a piece of software called a proof assistant). Certified or type-checked proofs should not be mistaken for

Work on this paper was supported by grant GA CR P201/12/G028, by the European Research Council Advanced Grant ALEXANDRIA (Project 742178) and by the Centre for Advanced Study (CAS) in Oslo, Norway.

A. Bordg (✉)
Department of Computer Science and Technology, University of Cambridge, Cambridge, UK
https://sites.google.com/site/anthonybordg/home
e-mail: apdb3@cam.ac.uk

S. Centrone et al. (eds.), *Reflections on the Foundations of Mathematics*,
Synthese Library 407, https://doi.org/10.1007/978-3-030-15655-8_8

automated proofs. Even if proof assistants come with various levels of automation, either built-in for elementary steps or user-defined via the so-called tactics for less basic steps, the proof assistant only checks that man-made proofs written with it are correct.

8.1.1.1 The Univalence Axiom

The main characters in Martin-Löf type theory (MLTT for short) are types and elements of these types. If T is a type, then the expression $t\colon T$ denotes that t is an element of T. In particular, if T is a type and t, t' are elements of T there is a new type called the identity type of t and t' denoted $t =_T t'$. Sometimes for convenience we will omit the type information and we will simply write $t = t'$. When one considers only a single element t, i.e. t' is definitionaly equal to t, the identity type $t =_T t$ has always at least one element denoted $\mathrm{idpath}\, t$, i.e. the expression $\mathrm{idpath}\, t\colon t =_T t$ is well-formed. This term idpath is called a constructor and the identity types belong to a particular class of types called *inductive types*. Indeed, besides its constructors (an inductive type can have either a single constructor or many constructors), a family of types defined inductively (like the identity types are when introduced formally) obeys an induction principle. In the case of identity types, this induction principle states that given a type T, an element $t\colon T$, a family F of types indexed by an element $t_0\colon T$ and an element $p_0\colon t =_T t_0$, if there is an element $f\colon F\, t\, (\mathrm{idpath}\, t)$ (the family F instantiated with the terms t and $\mathrm{idpath}\, t$), then for any elements $t'\colon T$, $p\colon t =_T t'$ there is an element of the type $F\, t'\, p$, and moreover this element is f itself when t' and p are definitionaly equal to t and $\mathrm{idpath}\, t$, respectively. Of course, one can iterate the process of building identity types, namely given p and q two elements of the identity type $t =_T t'$, one can form the identity type $p =_{t =_T t'} q$ and so on. As it happens, these identity types lead to a very rich mathematical structure and there is a surprising connection between homotopy theory and MLTT (the latter being also coined Martin-Löf *dependent type theory* in reference to these dependent types, i.e. dependent on previous types for their definition which may be inductive, like in the case of identity types, or not). Roughly, one can think of T as a space, two elements t and t' of T as points of this space, two elements p and q of the type $t =_T t'$ as paths from t to t' in the space T, and the elements of $p =_{t =_T t'} q$ as homotopies between the paths p and q and so on (the elements of the successive iterated identity types being higher homotopies). Under this correspondence $\mathrm{idpath}\, t$ is the identity path between a point t and itself in the given space. Each type bearing the structure of a weak ∞-groupoid obtained from the tower of iterated identity types over that type. Moreover, given two types A and B there is also a new type denoted $A \to B$ for the type of functions between A and B. Among these functions some of them have a distinctive property, namely their homotopy fibers[1] are contractible,[2] and they are called weak equivalences.

[1]The definition of the homotopy fibers of a map is given later in Sect. 8.1.2.1.

[2]The fundamental concept of contractibility is defined later in Sect. 8.1.1.2.

Again, one forms a new type for the weak equivalences between two types A and B, denoted $A \simeq B$. Voevodsky found an interpretation of the rules of MLTT using Kan simplicial sets where an additional axiom, the so-called *Univalence Axiom*, is satisfied. The Univalence Axiom (UA for short) states a property of a universe type U (interpreted as the base of a universal Kan fibration), a type whose elements are themselves types called "small" types. More specifically, first note that given two small types A and B, by applying the induction principle of identity types (take $T := U$, $t := A$, and the family F such that $F\,B\,p_0$ is $A \simeq B$ in the statement of the induction principle above) one defines a function *eqweqmap* $A\,B$, from $A =_U B$ to $A \simeq B$, that maps the identity path to the identity equivalence when B is definitionally equal to A. The Univalence Axiom states that for any two small types A and B the function (eqweqmap A B) is a weak equivalence, giving the correct notion of equality (or path under the connection alluded to above) in the universe.

8.1.1.2 The Homotopy Levels

Note that in the function type $A \to B$ introduced above the type B does not depend on the type A. Now, we can replace the type B by a family of (small) types indexed by the type A, namely an element F of type $A \to U$ (where U is a universe), in this case we get a new type, the cartesian product of the family of types F, denoted $\prod_{x:A} F\,x$. Given two elements $f, g\colon \prod_{x:A} F\,x$, we could also ask if there is an equivalence between the identity type $f = g$ and the dependent product $\prod_{x:A}(f(x) = g(x))$. This equivalence (or rather the non-obvious implication) is known as *function extensionality* and it does not hold in MLTT. Fortunately, UA does imply function extensionality, i.e. given $A\colon U$, $F\colon A \to U$ and $f, g\colon \prod_{x:A} F\,x$, using UA one produces a term of the type $(\prod_{x:A} f(x) = g(x)) \simeq (f = g)$. Thus, the Univalence Axiom can be seen as a strong form of extensionality and the Univalent Foundations are a powerful and elegant way to achieve extensional concepts in Martin-Löf dependent type theory.

Without surprise another very important type is the type of natural numbers denoted nat. This is a second example of an inductive type. The type nat has two constructors, 0 of type nat and s of type nat $\to$ nat that corresponds to the successor function. The induction principle of nat is what one expects, namely an element of the type

$$\prod_{P:\mathrm{nat}\to U} P\,0 \to (\prod_{n:\mathrm{nat}} P\,n \to P\,(s\,n)) \to (\prod_{n:\mathrm{nat}} P\,n)\,.$$

Finally, we would like to introduce an additional dependent type called the dependent sum type. Given a type A and an element $B\colon A \to U$, we form the type of dependent pairs (x, y) with $x : A$ and $y : B\,x$, denoted $\sum_{x:A} B\,x$. Given a small type A, the type A might have the property that it has an element cntr: A together with for every element $x : A$ a path from x to cntr, i.e. an element of $\prod_{x:A} x =_A \text{cntr}$. The dependent sum allows us to form the type of such elements, namely

$$\sum_{\text{cntr}:A} \prod_{x:A} (x =_A \text{cntr}),$$

shortened to *iscontr A*, that corresponds to the type of proofs that A seen as a space is contractible. If this last type is inhabited, i.e. if it has an element, the type A is said to be *contractible* and cntr is called *a center of contraction*. We are now equipped with all the tools we need to introduce the very important concept of *homotopy levels*, the so-called *h-levels*, that intuitively capture the fact that at some point in the tower connected with a type the iterated identity types might be contractible. First, we need to know that one is allowed to define functions over inductive types, in particular over the type of natural numbers nat. Hence, we will define an element denoted *isofhlevel* of type nat $\rightarrow U \rightarrow U$. To achieve this, it is enough to define isofhlevel $0\,X$ to be iscontr X and isofhlevel $(s\,n)\,X$ to be $\prod_{x:X} \prod_{y:X}$ isofhlevel $n\,(x =_X y)$, where X is a small type. Given a small type X and a natural number n, if the type isofhlevel $n\,X$ is inhabited, then one says that X is of h-level n. The type of all types of h-level n is $\sum_{X:U}$ isofhlevel $n\,X$.[3] The types of h-level 1 are called *propositions*, they are the types in which any two elements are equal. The types of h-level 2 are called *sets*. For $n \geq 3$ the types of h-level n are higher analogs of sets.[4] It is possible to prove for instance that given a type X and an element n: nat the type isofhlevel $n\,X$ is a proposition, that the type nat is a set, or that the type $\sum_{X:U}$ isofhlevel $n\,X$ is of h-level $n + 1$. Moreover, the Univalence Axiom is consistent with respect to the Law of Excluded Middle for propositions and the Axiom of Choice for sets, hence not diminishing our ability to reason about propositions or sets but increasing our ability to work with higher analogs of sets.

Informed by homotopy theory, the main merits of the Univalent Foundations are the realization that types in MLTT are interpreted by homotopy types (topological spaces up to weak homotopy equivalences), their corresponding stratification according to the h-levels, and the ability that types give us to build weak higher groupoids through the tower of their iterated identity types. Moreover, the Univalence Axiom gives us the ability to reason formally about structures on these higher groupoids by enforcing an equivalence principle that makes two equivalent types

[3] This type is small with respect to a higher universe. This technical detail is unimportant for people unfamiliar with type theory.

[4] This analogy is explained in the quote from Voevodsky that ends the current Sect. 8.1.1.2.

indistinguishable in the Univalent Foundations. Indeed, let U_0, U_1 be two universes with $U_0 : U_1$ and U_0 being univalent. Given any family $P : X \to U_1$, there exists two terms $\mathrm{transportf}^{\mathrm{P}} : (x =_X y) \to Px \to Py$ and $\mathrm{transportb}^{\mathrm{P}} : (x =_X y) \to Py \to Px$. In particular, if one takes U_0 for X, then using the univalence axiom for U_0 one derives two terms of types $(A \simeq B) \to PA \to PB$ and $(A \simeq B) \to PB \to PA$, respectively.

The Univalent Foundations realizes the following vision of Voevodsky:

> First note that we can stratify mathematical constructions by their "level". There is element-level mathematics - the study of element-level objects such as numbers, polynomials or various series. Then one has set level mathematics - the study of sets with structures such as groups, rings etc. which are invariant under isomorphisms. The next level is traditionally called category-level, but this is misleading. A collection of set-level objects naturally forms a groupoid since only isomorphisms are intrinsic to the objects one considers, while more general morphisms can often be defined in a variety of ways. Thus the next level after the set-level is the groupoid-level - the study of properties of groupoids with structures which are invariant under the equivalences of groupoids. From this perspective a category is an example of a groupoid with structure which is rather similar to a partial ordering on a set. Extending this stratification we may further consider 2-groupoids with structures, n-groupoids with structures and ∞-groupoids with structures. Thus a proper language for formalization of mathematics should allow one to directly build and study groupoids of various levels and structures on them. A major advantage of this point of view is that unlike ∞-categories, which can be defined in many substantially different ways the world of ∞-groupoids is determined by Grothendieck correspondence, which asserts that ∞-groupoids are "the same" as homotopy types. Combining this correspondence with the previous considerations we come to the view that not only homotopy theory but the whole of mathematics is the study of structures on homotopy types. (Voevodsky 2015).

8.1.2 *The UniMath Library*

Nowadays a fraction of the community working on the Univalent Foundations is involved in the design of a library of mathematics based on the Univalent Foundations, the UniMath (Voevodsky et al. 2014) project, using the Coq proof assistant.

The UniMath library was born in 2014 by merging three previous repositories, the repository *Foundations* written by Vladimir Voevodsky, the repository *rezk_completion* of Ahrens, Kapulkin and Shulman, and the repository *Ktheory* by Daniel Grayson. As of 6 September 2018, the library has the following packages (arranged in alphabetical order): Algebra, CategoryTheory, Combinatorics, Folds, Foundations, HomologicalAlgebra, Induction, Ktheory, MoreFoundations, NumberSystems, PAdics, RealNumbers, SubstitutionSystems, Tactics, Topology.

I will briefly focus on the three packages that implement the main ideas of the Univalent Foundations as outlined in the first Sect. 8.1.1, namely the Foundations package, the MoreFoundations package, and the CategoryTheory package. The latter grew out of the rezk_completion repository, hence our choice of packages covers two of the three original repositories. However, we underline that we do not do justice to the contents of the library.

8.1.2.1 The Foundations Package

The Foundations package is the very core of the UniMath library. Hence, it is intended to become a very stable part of the library. For that reason this package is more or less locked. The code in that package is highly organized, and Voevodsky thought of writing it as a mathematical "textbook" about the Univalent Foundations.

The file *PartA* therein contains, among other things, the correct formulation of the type of proofs that a given function $f: X \to Y$ is a weak equivalence (see Sect. 8.1.1.1), denoted *isweqf*, which is the type $\prod_{y:Y} \text{iscontr}(\text{hfiber}\ f\ y)$,[5] where *hfiberfy* is the type $\sum_{x:X} fx = y$ of pairs consisting of an element $x: X$ together with a path in Y from $f(x)$ to y. The point being that the type isweq f is a proposition,[6] i.e. is of h-level 1. So, that type is proof-irrelevant, hence it becomes contractible, i.e. of h-level 0, when it is inhabited. Here, proof-irrelevance means that given a proposition P, there exists an equality $p: p_1 =_P p_2$ for all elements p_1, p_2 of P, i.e. for all proofs p_1, p_2 that the proposition P holds. As we have seen, the property of being a weak equivalence is instrumental in formulating the Univalence Axiom.

This last axiom is stated in the file *UnivalenceAxiom*, and it consists in postulating that the type isweq (eqweqmap A B) is inhabited for all types A and B of a given universe type U, it means that we assume that the map (eqweqmap A B) is a weak equivalence for all elements A and B of the universe type U. We recall (see Sect. 8.1.1.1) that (eqweqmap A B) is the canonical map from $A =_U B$, the type of equalities between A and B, to $A \simeq B$, the type of weak equivalences between A and B. In this last file, one also finds various consequences of the Univalence Axiom. In that file, these consequences are themselves formulated as axioms after the corresponding implications have been proved. It is an interesting choice of design. Indeed, it allows the relevant use of the command "Print Assumptions" of the Coq proof assistant in order to track the weakest axioms needed to prove a given statement. In particular, among the consequences of the Univalence Axiom proved in that file, one should mention function extensionality (see Sect. 8.1.1.2) which reassures any mathematician that pointwise equal functions are equal, a fact that any mathematician takes for granted.

The file *UnivalenceAxiom2* is devoted to the proofs that the types corresponding to the statements of these various axioms are propositions. Hence, those types are proof-irrelevant as it should be. Last, the file *PartB* of the Foundations package contains the all-important formulation of the h-levels that we have covered in Sect. 8.1.1.2. The stratification of types according to their h-levels discriminates well behaved propositions and sets, hence it allows elegant formalizations of mathematics at the levels of sets and categories with a proof assistant.

---

[5]Given a small type A, see Sect. 8.1.1.2 for the definition of the type iscontr A.

[6]In the UniMath library any type whose name starts by the prefix "is" is a proposition.

8.1.2.2 The MoreFoundations Package

The MoreFoundations package is less fundamental than the Foundations one, but it is still very basic. There are constructions in the Foundations package that admit many variants, those variants are not necessarily interesting in themselves, at least on a purely mathematical level, but these slight variants might come in handy in a particular context. The developers of UniMath handle this discrepancy between a core mathematical idea and its various incarnations that are used as a convenient scaffolding in the process of formalization by using the MoreFoundations package as a home for those variants. It is a nice way to avoid losing the sharp structure of the Foundations package.

We have already underlined in Sect. 8.1.1.2 that the Univalent Foundations (MLTT + UA) are consistent with the Law of Excluded Middle (LEM) and the Axiom of Choice (AC) if properly stated, namely if stated at the levels of propositions and sets, respectively. The type of propositions, namely $\sum_{X:U} \text{isofhlevel}\, 1\, X$, is shortened by *hProp* in the UniMath library. Then, the type corresponding to LEM in UniMath[7] is roughly $\prod_{P:\text{hProp}} P \amalg \neg P$, where $P \amalg \neg P$ denotes the coproduct of P and $\neg P$. To be more specific, it is the type $\prod_{P:\text{hProp}} \text{decidable\_prop}\, P$, where $\text{decidable\_prop}\, P$ is a pair consisting of the type $P \amalg \neg P$ together with a proof that it is a proposition, allowing to prove that the type-theoretic LEM is itself a proposition.

The type of sets, namely $\sum_{X:U} \text{isofhlevel}\, 2\, X$, is shortened by *hSet* in UniMath. Then, the type corresponding to AC in UniMath[8] is roughly

$$\prod_{X:\text{hSet}} \prod_{P:X\to U} ((\prod_{x:X} \text{ishinh}(P\, x)) \to \text{ishinh}(\prod_{x:X} P\, x)) \,,$$

where given a small type Y, $\text{ishinh}\, Y$ is the type $\prod_{P:hProp} ((Y \to P) \to P)$. Again, the exact type is $\prod_{X:\text{hSet}} \text{ischoicebase}\, X$, where $\text{ischoicebase}\, X$ is a pair consisting of the type $\prod_{P:X\to U} ((\prod_{x:X} \text{ishinh}(P\, x)) \to \text{ishinh}(\prod_{x:X} P\, x))$ together with a proof that this type is a proposition, allowing to prove that our type-theoretic AC is itself a proposition.

---

[7] See the file *DecidablePropositions* of the package MoreFoundations.

[8] See the file *AxiomOfChoice* in the package MoreFoundations.

In the UniMath library the type-theoretic versions of LEM and AC introduced above are not used as axioms but only as types in order to force them to be mentioned explicitly as additional assumptions whenever they are used to prove a statement.

Finally, in the file entitled *AxiomOfChoice* there is a proof that the type-theoretic AC implies the type-theoretic LEM, a result originally due to Radu Diaconescu.

8.1.2.3 The CategoryTheory Package

The file *Categories* of the package *CategoryTheory* gives an analog of the univalence property for categories that mimics the pattern described for the Univalence Axiom. Indeed, given a category $\mathcal{C}$ and two elements a and b of $\mathrm{ob}\,\mathcal{C}$, the type of objects of $\mathcal{C}$, one defines by induction a function *idtoiso*, from $a =_{\mathcal{C}} b$ to the type $\mathrm{iso}\,a\,b$ of isomorphisms from a to b, that maps the identity path to the identity morphism when b is definitionaly equal to a. The category $\mathcal{C}$ is univalent if for any two elements $a, b\colon \mathrm{ob}\,\mathcal{C}$ the above function is a weak equivalence. As a consequence, in the Univalent Foundations all category-theoretic constructions and proofs are invariant under isomorphism of objects of a univalent category and under equivalence of univalent categories.

In a subdirectory entitled *categories* (with a lower-case c, it should not be confused with the file of the same name), one proves that the category of sets as well as many categories of structured sets (monoids, groups, rings, modules, discrete fields, all standard algebraic structures formalized in the Algebra package) are univalent.[9]

Moreover, any category is equivalent to a univalent category called its Rezk completion as established in the eponymous file *rezk_completion*. Note that the univalence property itself for categories is not invariant under equivalence of categories, only under isomorphism of categories.

8.2 Some Challenges to Achieve Large-Scale Libraries of Formalized Mathematics

8.2.1 Toward Massive Collaborations in Mathematics

Formalized mathematics leads to a high level of certification and collaboration in mathematics, in the case of the UniMath library using the Coq proof assistant for the former and the open source distributed revision control system *Git*, the

[9]A general theorem that "isomorphism is equality" for a large class of algebraic structures (assuming the Univalence Axiom) was proven by Thierry Coquand and Nils Anders Danielsson in 2013 (Coquand and Danielsson 2013). Closely related, is the formulation of the more abstract *Structure Identity Principle* due to Peter Aczel, see Chapter 9 of The Univalent Foundations Program (2013).

development platform *GitHub* and a Google Group *Univalent Mathematics* (Univalent Mathematics) for the latter. Note that certification is a prerequisite for true massive collaboration in mathematics that is otherwise hardly possible, since one would need to check by hand developments by others in order to rely on them for his own proofs and developments. The advantage of certification allows the UniMath developers to focus on the quality of the code in the library. Usually, when someone submits a contribution then volunteers make remarks and suggestions to improve the formalization, sometimes several rounds of rewriting are undertaken. However, so far the formalization of mathematics has not changed by a few orders of magnitude the traditional way of collaborating between mathematicians, which typically involves at most 2 or 3 mathematicians, as illustrated by the celebrated collaboration between Hardy and Littlewood.

Moreover, mathematicians have blamed the formalization of mathematics for its tediousness. Regarding an influential old formal language called Automath, N.G. de Bruijn wrote in "A survey of the project Automath" (de Bruijn 1994):

> A very important thing that can be concluded from all writing experiments is the constancy of the loss factor. The loss factor expresses what we loose in shortness when translating very meticulous "ordinary" mathematics into Automath. This factor may be quite big, something like 10 or 20, but it is constant: it does not increase if we go further in the book. It would not be too hard to push the constant factor down by efficient abbreviations.

So, de Bruijn notes two things. First, formal proofs are longer, sometimes to an inadmissible point, as measured by the loss factor. Second, the loss factor is constant and it does not increase beyond some threshold. Regarding the first point, Freek Wiedijk in Wiedijk (2000)[10] gives interesting data for the AutoMath, Mizar, and HOL Light systems. In several cases, what Wiedijk calls the de Bruijn Factor, roughly the ratio of a formalized text to a TeX encoding of its informal counterpart, is around 4. It would be interesting to have similar data for Isabelle, a system with more automation, and for a more expressive system based on a dependent type theory like Coq, these systems might compare favourably. A de Bruijn factor equal or less than 2 might be more acceptable to a mathematician, and it is certainly a goal one should strive for. In order to succeed, in addition to more efficient support for notations as pointed out by de Bruijn, we certainly need more automation to handle the most obvious and boring parts of formal proofs. In the meantime, given enough hands and eyeballs can any substantial formalization effort be made shallow enough? We hope that the formalization of well-known mathematics, at the undergraduate or even graduate level, is parallelizable and could benefit from a divide-and-conquer approach to build comprehensive libraries that the working mathematician could use to do research-level mathematics. In this perspective, the formalization of mathematics might be more suited to massive collaborations than a project that focuses exclusively on research-level open problems like the Polymath Project.[11]

[10] See also http://www.cs.ru.nl/~freek/factor/

[11] https://en.wikipedia.org/wiki/Polymath_Project

8.2.2 The Challenge of Scalability

With formalized mathematics one faces the challenge of *scalability*. As suggested by the second point of de Bruijn, the constancy of the loss factor, the problem is not so much about the increase of de Bruijn factor with the length of a text, but about other aspects whose scalability might be problematic.

We will give a simple example. With the growth of the library, its index for search becomes huge. If there is no homogeneity when possible for the names of the definitions, lemmas and theorems, then it becomes very difficult for the user to check whether some item useful for his goal has already been formalized and if so to find it in the library, for instance by guessing easily its name.

One could hope that the tools of machine learning could offer in the near future for instance more intelligent search support for definitions, lemmas and theorems in libraries as well as some other useful automated tools. But this perspective should not prevent developers from being very careful with the design of their library of formalized mathematics to achieve something whole.

8.2.3 The Foundations: A Never-Ending Work or an Horizon

In the twenty-first century something new of the same magnitude as the Bourbaki project could wait for mathematicians. However, Bourbaki told us the following:

> If formalized mathematics were as simple as the game of chess, then once our chosen formalized language had been described there would remain only the task of writing out our proofs in this language, just as the author of a chess manual writes down in his notation the games he proposes to teach, accompanied by commentaries as necessary. But the matter is far from being as simple as that, and no great experience is necessary to perceive that such a project is absolutely unrealizable: the tiniest proof at the beginning of the Theory of Sets would already require several hundreds of signs for its complete formalization. Hence, from Book I of this series onwards, it is imperative to condense the formalized text by the introduction of a fairly large number of new words (called abbreviating symbols) and additional rules of syntax (called deductive criteria). By doing this we obtain languages which are much more manageable than the formalized language in its strict sense. Any mathematician will agree that these condensed languages can be considered as merely shorthand transcriptions of the original formalized language. But we no longer have the certainty that the passage from one of these languages to another can be made in a purely mechanical fashion: for to achieve this certainty it would be necessary to complicate the rules of syntax which govern the use of the new rules to such a point that their usefulness became illusory; just as in algebraic calculation and in almost all forms of notation commonly used by mathematicians, a workable instrument is preferable to one which is theoretically more perfect but in practice far more cumbersome. (Bourbaki 1968, Introduction, p. 10).

As it happens, the end of the twentieth century gave us such unforeseen powerful theories and proof assistants, actually not so "complicated" as anticipated by Bourbaki, for instance under the form of the so-called Calculus of Inductive Constructions (a dependent type theory extended with various features) as embodied in Coq, equipped with notational support to handle notations even including LaTeX and unicode characters, incorporating automatic tools like tactics, and being able to automatically generate typeset documents. Contrary to Bourbaki's expectations, packaged this way these theories have rendered the formalization of mathematics more feasible. It opens new possibilities for learning and teaching mathematics,[12] doing mathematical research, or using mathematics in industry. Thus, time may be ripe for Bourbaki's abandoned dream. However, one should keep in mind the distinction between the formalization of mathematics using proof assistants and some ultimate foundations of mathematics. The foundations of mathematics may be a never-ending work, what Bourbaki called the "horizon" (Guedj 1985). Hence, the importance of a third technical challenge (in addition to massive collaborations using proof assistants, and the scalability of libraries), the *migration* of libraries, for instance from a system to a more evolved system and this is why the UniMath library uses for its development only a small subset of the Coq language. Given the numerous proof assistants and libraries of formalized mathematics on the market, migration is an important issue and old code for new proof assistants should be reused as easily as possible to become the scaffolding for new achievements. The UniMath library is intended to be a whole scalable migration-friendly library of formalized mathematics with certified proofs. With respect to large-scale formalization, one very interesting aspect of Bourbaki's project consists in noticing that its members faced large-scale architectural problems well before us, since they aimed in their series of books at a rigorous, general and self-contained, reformulation of the whole of mathematics known at the time. Hence, in this regard one can learn from Bourbaki's epoch-making project, and Armand Borel's article "Twenty-Five Years with Nicolas Bourbaki 1949–1973" (Borel 1998) and Pierre Cartier's article "The Continuing Silence of Bourbaki" (Cartier 1998) are informative.

8.3 Architecture and Mathematics

Alexander Grothendieck was a third-generation member of Bourbaki and when reading Grothendieck's "Récoltes et Semailles" (Grothendieck) one can wonder

[12]The key step towards the widespread use of formalized mathematics could be to start teaching mathematics with the help of proof assistants, not to try very hard to gain the support of the working mathematicians. Given that present-day students are the mathematicians of tomorrow, the latter could be a consequence of the former. Moreover, a mathematical education using, at least occasionally, proof assistants could be solidly grounded in rigorous proofs and the students could benefit from it. Unfortunately, most proof assistants might be still too difficult to use except for graduate students. Note that the Isabelle proof assistant, which offers more automation, was used recently by a group of undergraduate students in Germany, under the supervision of three advisers, to formalize partly the DPRM theorem motivated by Hilbert's Tenth Problem. See their joint paper Aryal et al. presented during the FLOC 2018 conference in Oxford.

why the architectural metaphor is recurrent.[13] Actually, there is a meeting of minds between great architects and great mathematicians linked by an abstract approach of space with surprisingly at the same time a feeling of its organic life. This abstract approach of space is remarkable in the great architectural theoreticians, like for instance Frank Lloyd Wright, Le Corbusier, or Christopher Alexander, in the sense that one can naively believe their prime business is the 3-dimensional space embodied in a house, a building or a city, it is certainly true but it goes beyond. We notice that architects have been facing large-scale problems for long, they have been challenging them and they have offered their thoughts. Christopher Alexander in "The Nature of Order" (Alexander 2002) develops what he calls *wholeness* to answer these challenges. Wholeness is precisely what is lacking in most libraries of formalized mathematics despite the fact it is an important feature of mathematics. Note that *wholeness* in Alexander's work is a specific concept defined by 15 properties.[14] We will not discuss here each of those properties, but only a few that seem more relevant with respect to mathematics, since it will be probably hopeless to search for a precise dictionary between Alexander's properties and some corresponding features of mathematics. These properties are an interesting attempt to capture what "organic" and "life" could mean for man-made artefacts like architectural works which are Alexander's main concern. In his 1900 address to

[13]I will give a few examples : "Je me sens faire partie, quant à moi, de la lignée des mathématiciens dont la vocation spontanée et la joie est de construire sans cesse des maisons nouvelles. Chemin faisant, ils ne peuvent s'empêcher d'inventer aussi et de façonner au fur et à mesure tous les outils, ustensiles, meubles et instruments requis, tant pour construire la maison depuis les fondations jusqu'au faîte, que pour pourvoir en abondance les futures cuisines et les futurs ateliers, et installer la maison pour y vivre et y être à l'aise. Pourtant, une fois tout posé jusqu'au dernier chêneau et au dernier tabouret, c'est rare que l'ouvrier s'attarde longuement dans ces lieux, où chaque pierre et chaque chevron porte la trace de la main qui l'a travaillé et posé. Sa place n'est pas dans la quiétude des univers tout faits, si accueillants et si harmonieux soient-ils - qu'ils aient été agencés par ses propres mains, ou par ceux de ses devanciers. D'autres tâches déjà l'appelant sur de nouveaux chantiers, sous la poussée impérieuse de besoins qu'il est peut-être le seul à sentir clairement, ou (plus souvent encore) en devançant des besoins qu'il est le seul a pressentir." (Grothendieck, 2.5 Les héritiers et le bâtisseur); and "Comme le lecteur l'aura sans doute deviné, ces "théories", "construites de toutes pièces", ne sont autres aussi que ces "belles maisons" dont il a été question précédemment : celles dont nous héritons de nos devanciers et celles que nous sommes amenés à bâtir de nos propres mains, à l'appel et à l'écoute des choses. Et si j'ai parlé tantôt de l' "inventivité" (ou de l'imagination) du bâtisseur ou du forgeron, il me faudrait ajouter que ce qui en fait l'âme et le nerf secret, ce n'est nullement la superbe de celui qui dit: "je veux ceci, et pas cela !" et qui se complaît à décider à sa guise; tel un piètre architecte qui aurait ses plans tout prêts en tête, avant d'avoir vu et senti un terrain, et d'en avoir sondé les possibilités et les exigences." (Grothendieck, 2.9); and again "C'était peut-être là la principale raison pour laquelle les maisons que je prenais plaisir à construire sont restées inhabitées pendant le longues années, sauf par l'ouvrier maçon lui-même (qui était en même temps aussi l'architecte, le charpentier etc.)." (Grothendieck, 18.2.8.3 Note 135).

[14](1) Levels of scale (2) Strong centers (3) Boundaries (4) Alternating repetition (5) Positive space (6) Good shape (7) Local symmetries (8) Deep interlock and ambiguity (9) Contrast (10) Gradients (11) Roughness (12) Echoes (13) The void (14) Simplicity and inner calm (15) Not-separateness.

mathematicians *Mathematical Problems* Hilbert mentioned this organic feature of mathematics in the following perceptive insights:

> The problems mentioned are merely samples of problems, yet they will suffice to show how rich, how manifold and how extensive the mathematical science of today is, and the question is urged upon us whether mathematics is doomed to the fate of those other sciences that have split up into separate branches, whose representatives scarcely understand one another and whose connection becomes ever more loose. I do not believe this nor wish it. Mathematical science is in my opinion an indivisible whole, an organism whose vitality is conditioned upon the connection of its parts. [...] We also notice that, the farther a mathematical theory is developed, the more harmoniously and uniformly does its construction proceed, and unsuspected relations are disclosed between hitherto separate branches of the science. So it happens that, with the extension of mathematics, its organic character is not lost but only manifests itself the more clearly. (Hilbert 1902).

If wholeness is a feature of mathematical science according to Hilbert, Alexander regrets its absence in most of modern, dead and dull, architectural works while this property is shining in some of the great artistic works of the past. Alexander's concerns may not be widely shared by present-day architects, but they are not without resonances among other great architects as testified by the organic architecture of Frank Lloyd Wright or Tadao Ando's obsession with making light vibrant, alive, through the use of concrete. If Alexander does not mention mathematics, a mathematician cannot help but think that mathematical entities and mathematics as a whole ("*la* mathématique" of Bourbaki, using a singular on purpose) display to a great extent this pervasive organic character, a life of their own, a wholeness.

Mathematical entities are like the *centers* of Alexander, the elementary components of any system that make it alive, but with the subtlety that a center cannot be isolated from other centers but needs to be understood in a mutual recursive relation with other centers (Alexander 2002, p.116). We can think about prime numbers in terms of what Alexander calls *strong centers* (Alexander 2002, p.151), centers that focus our attention and engage us, the set of primes numbers being described by Alain Connes as the heart of mathematics (Stéphane Dugowson et Anatole Khelif 2014) (compare Connes's vivid organic metaphor in the interview with the dull mechanic metaphor of the interviewers using a coffee machine).

In mathematics strong centers are not only displayed as specific mathematical entities but also in proofs, proving being at the core of the activity of the working mathematician. Indeed, any good mathematical proof has its own architecture. This architecture revolves around the main ideas that provide the flesh of the proof. In a given proof there are as many strong centers as there are main ideas, fitting together thanks to *boundaries* which are the glue of the inner workings of the mathematical mind, the hypotheses and the conclusion being the initial boundary and the last boundary, respectively. Simple proofs have usually only one center, more elaborate proofs may have many centers, but it does not matter, centers are always what make things click. One can define the strong centers in a proof as the main ideas such that handed to any mathematician with the appropriate training he will not fail to reconstruct the proof on his own. In most proof assistants, formal proofs have no structure.[15] It is an important issue. This is the case for instance in the Coq proof assistant, and as a consequence in the UniMath library, where a formal proof is basically a sequence of tactics lacking the structure of its informal counterpart. Even if one would not intend to read formal proofs, for instance to get pedagogical insights, but only to get certificates of correctness from them, then one still needs to maintain on a regular basis the code in a library to take into account revisions that might have been proposed. Since changes pushed in a library can break some proofs, the task of repairing broken code (and in particular broken proofs) is made harder by the lack of structure in formal proofs that makes them barely legible.

Also, there might be some *roughness* in the sense of Alexander in the distribution of prime numbers mentioned above, roughness being an elusive property:

> Things which have real life always have a certain ease, a morphological roughness. This is not an accidental property. It is not a residue of technically inferior culture, or the result of hand-craft or inaccuracy. [...] It is an essential feature of living things, and has deep structural causes. [...] Roughness does not seek to superimpose an arbitrary order over a design, but instead lets the larger order be relaxed, modified according to the demands and constraints which happen locally in different parts of a design. (Alexander 2002, p.210).

Moreover, it suffices to quote Hilbert again to find traces in mathematics of other properties of Alexander like *local symmetries*, *deep interlocking* and *echoes*:

> For with all the variety of mathematical knowledge, we are still clearly conscious of the similarity of the logical devices, the relationship of the ideas in mathematics as a whole and the numerous analogies in its different departments. (Hilbert 1902).

The libraries of formalized mathematics are annoyingly lacking echoes. Often proof assistants miss some nice built-in features. For instance, these libraries have no counterpart as convenient as the index of a book, the easy search function of a PDF file or the table of contents of a book,[16] let alone clickable keywords for pop-up windows to remind the reader about definitions.[17]

So far, formalized mathematics has focused only on making impossible to write faulty proofs, in doing so it has done nothing for making proofs easier to read. Quite the contrary, formalized mathematics is much harder to read than everyday mathematics and this can explain why it has encountered considerable resistance from mathematicians. One should not forget that mathematicians spend a lot of time reading mathematics, not only doing or writing it. Formalized mathematics has forgotten the communication function of written mathematics, and this is a problem not only with respect to mathematicians but also for students, especially if

---

[15]Some proof assistants like Isabelle have structured proofs (in the case of Isabelle thanks to an additional layer called the Isar language), but there is still room for improvement.

[16]The good practices of writing a short table of contents at the top of a file starting a new formalization and a bibliography at the end are surprisingly not even included in the style guide (https://github.com/UniMath/UniMath/blob/master/UniMath/README.md) of UniMath as of 6 September 2018.

[17]Again, the Isabelle theorem prover and its bundled editor jEdit, even if not perfect, have built a competitive advantage with search support and clickable keywords.

one believes that teaching mathematics with proof assistants may have pedagogical value and is a necessary milestone towards the widespread use of formalized mathematics.[18] Formalized mathematics is not responsive to the reader, this strong center of the subjective experience of mathematics, that changes across the mathematical community. In this sense, it lacks the property that Alexander coined *gradients*, the adaptive result in design when conditions vary. A simple solution should be to have expensible/collapsible parts in proofs, so that every reader, while reading a proof, can set for himself the level of details according to his background and ability. Hopefully, this feature would allow to hide very low-level details that make reading formal proofs cumbersome. Few libraries of formalized mathematics are really designed with the reader in mind.

Finally, could it be that the last fifteenth property of Alexander, *not-separateness*, the experience of "a living whole as being at one with the world" (Alexander 2002, p.230), is the "unreasonable effectiveness of mathematics in the natural sciences" emphasized by Eugene Wigner (Wigner 1960)?[19]

In the same way an architect try to realize the unfolding in space of a form through *levels of scale* (Alexander 2002, p.145), i.e. the property that consists in the presence of centers at a wide range of scales, mathematicians unfold their axioms through mathematical entities and theorems. The levels of scale are apparent for instance in the definition/lemma/theorem/corollary structure of a mathematical book or article. Both architects and mathematicians are happy when this unfolding looks like the unfolding of an organism from the seed within. The axioms of mathematics are the seeds, the labour of mathematicians are the ground that nurtures the seeds, and the mathematical entities and theorems are the resulting landscape with its wide open horizon. Some parts of this landscape are *jardins à la française*, some others are English gardens, both with their respective supporters. Most parts of this landscape secretly aspire to the inner peace of Japanese gardens, natural but neither artificial nor wild. This living unfolding, from the axioms to the theorems, is the stuff mathematical objects are made from. In the case of the Univalent Foundations of mathematics the unfolding of shapes, namely types of various h-levels, a concrete example of *levels of scale* relevant in our context, what appears less directly and less smoothly in sets-based mathematics as homotopy types, is remarkable. The enlarged notion of life, of living structures, coined *wholeness* by Alexander, can help to understand in particular where the platonistic attitude of mathematicians comes from.

As pointed out earlier, Alexander underlines the interplay of centers with the use of *boundaries* (Alexander 2002, p.150) which separate a center from others and at the same time unite them. For a second example, think in mathematics about the locus where two topics or two theories meet, share some methods and that could possibly merge in the future as a result. But Alexander seems to miss the point that

[18]See also the Footnote 6 on that point.

[19]I have discovered a truly remarkable answer to this question which this footer is too small to contain.

sometimes the life of some parts may be at the expense of others. This full dynamics was noted by Hilbert:

> [...] let me point out how thoroughly it is ingrained in mathematical science that every real advance goes hand in hand with the invention of sharper tools and simpler methods which at the same time assist in understanding earlier theories and cast aside older more complicated developments. It is therefore possible for the individual investigator, when he makes these sharper tools and simpler methods his own, to find his way more easily in the various branches of mathematics than is possible in any other science. (Hilbert 1902).

The result of these sharper tools and simpler methods, won after the struggle, that ease the orientation of mathematicians in the whole of their science is the *simplicity and inner calm* put forward by Alexander, and described by him as the

> quality [...] which is essential to the completion of the whole. [...] The quality comes about when everything unnecessary is removed. (Alexander 2002, p.226).

The regular clean-up and reorganizations in mathematics mentioned by Hilbert above might be the analog of evolution in the biological world and a condition for a renewal of creativity, biological systems being a paradigm of wholeness. Of course, biological evolution as understood by modern biology is a blind process, while reorganizations are made on purpose by mathematicians and some mathematicians have the platonistic feeling to be guided by an independent architectural principle of some kind, to discover rather than to invent mathematical objects. One could think this prompted Darwin to say

> I have deeply regretted that I did not proceed far enough at least to understand something of the great leading principles of mathematics, for men thus endowed seem to have an extra sense. (Darwin 1958).

But one could also see in this quote a reference to the complementary ability of the mathematician, like the artist, to tap into the subconscious mind as pointed out by Jacques Hadamard (1996).

8.4 Conclusion

In this article, following Alexander's approach, we have tried to underline a few strong centers in the foundations of mathematics, namely the Univalent Foundations, the UniMath library, the Bourbaki's cathedral of mathematics, twisting some architectural threads in a mathematical landscape. However, we have only sketched the boundaries between these centers to allow for at least some wholeness in the odyssey promised in the abstract.

Some mathematicians are afraid that formalization could disrupt their flow of work, their inner music, and this may be indeed a real danger if one is not able to cleverly design organic libraries to allow smooth reorganizations on a regular basis. But if we are sensitive to the wholeness of our library this danger could be avoided. By facing new large-scale challenges in design formalized mathematics could offer new opportunities. The Alhambra (close to Alexander's heart) located in Granada

(Spain), started in 889, still stands shadowing our mortality, in the same way can the libraries of formalized mathematics stand the test of time?

Acknowledgements The author would like to thank Benedikt Ahrens, Thierry Coquand, and an anonymous referee for their useful comments and suggestions.

References

Alexander, C. (2002). *The nature of order: An essay on the art of building and the nature of the universe. Book I The Phenomenon of Life*. The Center for Environmental Structure, Berkeley.
Aryal, D., Bayer, J., Ciurezu, B., David, M., Deng, Y., Devkota, P., Dubischar, S., Hassler, M. S., Liu, Y., Oprea, M. A., Pal, A., & Stock, B. (2018). Hilbert Meets Isabelle. Formalization of the DPRM Theorem in Isabelle/HOL. https://doi.org/10.29007/3q4s
Borel, A. (1998). Twenty-five years with Nicolas Bourbaki, 1949–1973. *Notices of AMS, 45*, 373–380.
Bourbaki, N. (1968). *Elements of mathematics. Theory of sets*. London: Addison-Wesley.
Cartier, P. (1998). The continuing silence of Bourbaki. *The Mathematical Intelligencer, 1*, 22–28.
Coquand, T., & Danielsson, N. A. (2013). Isomorphism is equality. *Indagationes Mathematicae, 24*(4), 1105–1120.
Darwin, C. (1958). Cambridge life, chp.2. In *The autobiography of Charles Darwin* (p. 58). Darwin Online http://darwin-online.org.uk/
de Bruijn, N. G. (1994). A survey of the project Automath *Studies in Logic and the Foundations of Mathematics, 133*, 141–161.
Grothendieck, A. Récoltes et Semailles, Réflexions et témoignages sur un passé de mathématicien.
Guedj, D. (1985). Nicolas Bourbaki, collective mathematician, an interview with Claude Chevalley. *The Mathematical Intelligencer, 7*(2), 18–22.
Hadamard, J. (1996). The Mathematician's mind: The psychology of invention in the mathematical field. Princeton University Press, Princeton.
Hilbert, D. (1902). Mathematical problems. Lecture delivered before the International Congress of Mathematicians at Paris in 1900 (trans: Dr. Mary Winston Newson). *Bulletin of the American Mathematical Society, 8*(10), 437–479.
Martin-Löf, P. (1982). Constructive mathematics and computer programming. In *Logic, methodology and philosophy of science, VI*, Hannover, 1979 (Studies in logic and the foundations of mathematics, Vol. 104, pp. 153–175). Amsterdam: North-Holland.
Stéphane Dugowson et Anatole Khelif Interview d'Alain Connes, IHES, 5 février 2014. https://www.youtube.com/watch?v=rHkhez4OxPU.
The Univalent Foundations Program. (2013). Homotopy type theory: Univalent foundations of mathematics. Institute for Advanced Study. Available online https://homotopytypetheory.org/book.
Univalent Mathematics. https://groups.google.com/forum/#!forum/univalent-mathematics.
Voevodsky, V. (2014). *Paul Bernays Lectures*, at ETH Zurich, 9–10 Sept 2014. Available at https://www.math.ias.edu/vladimir/lectures.
Voevodsky, V. (2015). An experimental library of formalized mathematics based on the univalent foundations. *Mathematical Structures in Computer Science, 25*(5), 1278–1294.
Voevodsky, V., Ahrens, B., Grayson, D., & others (2014). UniMath: Univalent Mathematics. Available at https://github.com/UniMath.
Wiedijk, F. The De Bruijn factor (2000). http://www.cs.ru.nl/F.Wiedijk/factor/factor.pdf.
Wigner, E. (1960). The unreasonable effectiveness of mathematics in the natural sciences. *Communications on Pure and Applied Mathematics, XIII*, 1–14.

Chapter 9
Models of HoTT and the Constructive View of Theories

Andrei Rodin

Abstract Homotopy Type theory and its Model theory provide a novel formal semantic framework for representing scientific theories. This framework supports a constructive view of theories according to which a theory is essentially characterised by its methods. The constructive view of theories was earlier defended by Ernest Nagel and a number of other philosophers of the past but available logical means did not allow these people to build formal representational frameworks that implement this view.

9.1 Introduction

During the second half of the twentieth century Evert Beth, Patrick Suppes, Bas van Fraassen followed by a large group of other contributors developed an approach to formal representation and formal logical analysis of scientific theories, which became known under the name of *semantic view of theories*.[1] This approach was proposed as a replacement for a different approach to the same subject developed several decades earlier by the so-called *logical positivists* including Rudolf Carnap, Carl Gustav "Peter" Hempel and Ernest Nagel. The proponents of the new *semantic view* dubbed the older approach *syntactic*. They used resources of the recently emerged Model theory and argued that a formal representation of scientific theories should first of all account for the intended semantics of these theories and leave aloof many syntactic details studied by their predecessors. Model theory helped

The work is supported by Russian Foundation for Basic Research, research grant number 19-011-00799.

[1]For an overview of the early history of the semantic view see Suppe (1989), Part 1, Prologue.

A. Rodin (✉)
Institute of Philosophy, Russian Academy of Sciences and Saint-Petersburg State University, Moscow, Russia
e-mail: andrei@philomatica.org

S. Centrone et al. (eds.), *Reflections on the Foundations of Mathematics*,
Synthese Library 407, https://doi.org/10.1007/978-3-030-15655-8_9

these people to develop and defend a "new picture of theories" (van Fraassen 1980, p. 64) where the semantic aspect played the leading role.

As a number of authors remarked more recently, from the viewpoint of today's theories of formal semantics it appears that the proponents of the "semantic" approach at that time did not properly understand the involved relationships between syntactic and semantic aspects of formal theories, and for this reason often conceived of these relationships as a form of concurrence. As it became clear more recently the model-theoretic concepts used by the proponents of the semantic view could not be rigorously defined in an open air without taking related syntactic issues into consideration, see Halvorson (2016), Krause and Arenhart (2016), and Muller (2011).

Nevertheless the "fight against the syntax" led by the proponents of the semantic view did not reduce to a mere rhetoric. Suppes had a strong insight according to which a formal representation of given scientific theory should not involve arbitrary syntactic choices. Suppes' epistemological emphasis on the invariant character of (non-formalized) theories in Physics and Geometry makes it evident where this insight comes from; in particular, Suppes is motivated here by the geometrical Erlangen Program (Suppes 2002, p. 99). It is clear that the idea of invariant formal representational framework for scientific theories cannot be realised via disregarding syntactic issues; on the contrary, this project calls for new syntactic inventions.[2]

Let us conclude this Introduction with a short statement that expresses our take on the debate between the syntactic and the semantic views of theories. Following Halvorson we take this issue to be mostly historical. In order to evaluate the contribution of the semantic view properly this approach should be compared with earlier attempts to use formal logical methods in science and in the philosophy of science but not with an artificially construed "syntactic view" which has been hardly ever defended by anyone. These earlier attempts did not involve and could not involve anything like formal semantics, which emerged later. When Nagel and his contemporaries considered possible applications of mathematical logic in the representation of scientific theories, the formal part of their work was necessarily limited to syntactic issues. The intended semantics of formal theories these authors described informally by associating with appropriate symbols and symbolic expressions certain empirical and mathematical contents in the same traditional manner in which logicians did this since Aristotle. Suppes and other pioneers of the new "semantic" approach used a mathematical theory of semantics, viz. Tarski's set-theoretic semantics, as an additional intermediate layer of formal representation between the symbolic syntax of formal theory F and the contents

[2]Halvorson and Glymour rightly observe that the realisation of such a project requires, in particular, to develop an appropriate notion of equivalence of theories, which is still missing (Halvorson 2012; Glymour 2013). Interestingly, the pioneering works of Bill Lawvere in categorical logic published in the 1960s were equally driven by the desire to avoid the syntactic arbitrariness in logic; in addition to examples of invariant theories borrowed from Geometry and Physics Lawvere was motivated by Hegel's notion of *objective logic* in this research (Rodin 2014, 5.8).

of scientific theory T which F formally represents. In order to recognise the importance of this formal semantic layer one doesn't need to downplay the role of syntax. In *that* conservative sense the semantic view is clearly more advanced than its "syntactic" predecessor. However we argue in what follows that the "semantic turn" also left aside some important insights of Nagel's "syntactic" approach, which we are trying to save from the oblivion in this paper by using a new formal technique.

9.2 Constructive View of Theories

The *once-received* (Halvorson 2016) aka syntactic view of theories and the more recent *semantic view* developed by Suppes and his followers were equally motivated by their contemporary developments in logic and foundations of mathematics: while logical positivists of the older generation were motivated by the rise of new symbolic logic on the edge between the 19-th and the twentieth centuries, the younger generation of researchers in the 1950s was motivated by the rise of Model theory and formal semantics. In this paper we stick to the same pattern and base our proposal on the recently emerged Homotopy Type theory (HoTT) and the closely related research on the Univalent Foundations of mathematics. A view of theories that we arrive at is a version of the *non-statement* view according to which a theory, generally, does not reduce either to a bare class of sentences or to a class of statements ordered by the relation of logical consequence. But this proposed view, which hereafter we call *constructive*, differs from the usual semantic view in how it answers the question "What are fundamental constituents of a theory (except its sentences)?" A proponent of the standard semantic view's answers "models". A proponent of the constructive view answers "methods".

As a general philosophical view on scientific theories the constructive view is not new: the fundamental role of methods in science has been stressed in the past by many philosophers from René Descartes to Ernest Nagel (see Sect. 9.3.1 below). The fact that various sorts of methods—including mathematical, experimental and observational methods—are abound in today's science hardly needs a special justification. The specification of relevant methods called by scientists a "methodology" (which should not be confused with the methodology of science in the sense used by philosophers) makes an important part of any piece of established scientific knowledge but not only a part of an open-ended scientific research. However we still lack convenient formal logical frameworks for representing methods as proper elements of scientific theories. In Sect. 9.4.1 we explain why standard logical and semantic means are not appropriate for the task. In Sect. 9.5 we show how HoTT helps to solve this problem. In the next Sect. 9.3 we take a closer look at Nagel's and Suppes work, which gives us some additional motivations for our proposed notion of constructive theory.

9.3 Ernest Nagel, Morris Cohen and Patrick Suppes on Logic and the Scientific Method

9.3.1 Ernest Nagel and Morris Cohen

According to Nagel and Cohen the most distinctive and valuable feature of science is its method:

> [T]he method of science is more stable and more important to men of science than any particular result achieved by its means (Nagel and Cohen 1934, p. 395);

which they, in fact, identify with logic:

> [T]he constant and universal feature of science is its general method, which consists in the persisting search for truth, constantly asking: Is it so? To what extent is it so? Why is it so? [...] And this can be seen on reflection to be the demand for the best available evidence, the determination of which we call logic. Scientific method is thus the persistent application of logic as the common feature of all reasoned knowledge (Nagel and Cohen 1934, p. 192).

What kind of logic Nagel and Cohen have here in their minds? Informally Nagel and Cohen describe their core conception of logic as the "study of what constitutes a proof, that is, complete or conclusive evidence" (Nagel and Cohen 1934, p. 5). However there is a wide gap, as we shall see, between this general conception of logic and the construal of logic in form of symbolic calculus, which the authors introduce in the same book.

By an axiomatic theory Nagel and Cohen understand a Hilbert-style formal theory[3]; they are aware of and stress the fact that such a theory has no bearing on the material truth of its axioms and theorems. The authors describe an axiomatic theory as a hypothetico-deductive structure, which represents objective relations of logical consequence between its sentences and only eventually may "conform" to the "world of existence". Whether or not a given axiomatic theory conforms to the real world is an empirical question that cannot be answered within this given theory:

> The deduction makes no appeal whatsoever to experiment or observation, to any sensory elements. ... Whether anything in the world of existence conforms to this system requires empirical knowledge. ... That the world does exemplify such a structure can be verified only within the limits of the errors of our experimental procedure. (Nagel and Cohen 1934, p. 137)

Thus Nagel and Cohen make it clear that logical means, which make part of an axiomatic theory, don't help one to determine a conclusive evidence for and thus establish the truth of any scientific statement. But this, recall, is exactly what logic is supposed to do in science according to the authors' general conception. Facing this problem Nagel and Cohen go as far as to suggest that Hilbert's conception

[3]This part of Nagel and Cohen's exposition is very outdated and, if judged by today's standard, also mostly informal. Since this historical detail has no bearing on the following arguments the reader is invited to think here about a Hilbert-style axiomatic theory in its modern form.

of *proof* as a formal deduction from axioms is unwarranted and that a more reasonable terminological choice would be to talk (in context of axiomatic theories) about *deductions* without calling such deductions *proofs*. "But so habitual—say Nagel and Cohen—is the usage which speaks of 'proving' theorems in pure mathematics that it would be vain to try to abolish it" (Nagel and Cohen 1934, p. 7). Needless to say that this linguistic compromise is very far from being innocent. Moreover the last quoted remark that the authors use as an excuse for Hilbert's identification of proofs with deductions is hardly sound. Mathematicians indeed colloquially speak of "proving" theorems. However they typically do *not* present their proofs in form of formal logical deductions. The identification of proofs with deductions is a strong assumption that cannot be downplayed by referring to the colloquial mathematical parlance. Apparently Nagel and Cohen want here to use Hilbert's axiomatic approach without being agree with Hilbert on the nature and scope of logic and, more specifically, on what constitutes a proof. Thus Nagel and Cohen clearly see the gap between their general conception of logic and the available formal logical techniques. They describe a possible epistemic role of these techniques in science as follows:

> The evidence for propositions which are elements in a [hypothetico-deductive] system accumulates more rapidly than that for isolated propositions. The evidence for a proposition may come from its own verifying instances, or from the verifying instances of other propositions which are connected with the first [i.e., the given proposition] in a system. It is this schematic character of scientific theories which gives such high probabilities to the various individual propositions of a science. (Nagel and Cohen 1934, p. 395)

Thus the logical deduction supported by formal techniques available to Nagel and Cohen can play at best an auxiliary role in science helping one to accumulate evidences more rapidly. By no means it can perform the fundamental methodological role, which Nagel and Cohen reserve for logic in their informal treatment of this question.

9.3.2 Patrick Suppes

Unlike Nagel and Cohen Suppes doesn't expose an informal conception of logic independent of symbolic and mathematical techniques. He regards the first-order logic and Tarskian Model theory as representational tools for scientific theories, which help to clarify the logical structure of these theories and their mutual relationships. Suppes builds a scene where various scientific methods can be used but he doesn't attempt to represent such methods at the formal level. See, for example, his chapter on *Representations of Probablity* (Suppes 2002, Ch. 5), which provides a foundation for statistical methods in science.

In the Introduction to the same book, which summarizes Suppes' work on formal analysis of science during four decades, Suppes briefly considers, without providing references, a number of alternative approaches to formalization of scientific theories, which he qualifies as "instrumentalist" (Suppes 2002, pp. 8–10) even if, as far as we can see, these approaches don't really commit one to the instrumentalist view on science in a strong epistemological sense.

According to one such proposal, "[t]he most important function of a theory [. . .] is not to organize or assert statements that are true or false but to furnish material principles of inference that may be used in inferring one set of facts from another." According to another proposal, "[t]heories are [. . .] methods of organizing evidence to decide which one of several actions to take." Suppes does not object these proposals on philosophical and epistemological grounds but rejects them for technical reasons stressing the fact that none of these proposal is supported with a developed formal technique. In what follows we argue that today we are in a better position because some techniques appropriate for the task since recently are available.

Thus we can conclude that neither the old-fashioned axiomatic representations of theories, which associates empirical contents to the (appropriate part of) symbolic syntax directly, nor the more advanced semantic approach, which uses the Tarski-style set-theoretic semantics and Model theory, have resources for representing scientific methods in a formal setting. This concerns methods of justification of the ready-made scientific knowledge as well as heuristic methods. Arguably a formal representation of ready-made scientific theories does not need to take into account heuristic methods and may safely leave them to the *context of discovery*. But the same argument does not apply to methods of justification, which make part of any scientific theory that deserves the name.

9.4 How to Represent Methods with Symbolic Calculi

Knowing a method amounts to knowing *how* to perform certain operations, which bring certain wanted outcome. One can distinguish between different senses in which a particular method can be known by an epistemic agent. It is possible that an agent is aware about certain instruction and can repeat it by heart but is unable to implement it in the appropriate circumstances. It is also possible that an agent is capable to perform an operation and get the wanted outcome without being aware of a formal linguistic description of this operation. For a contingent historical reason the current epistemological discussion on knowledge-how is mainly focused on examples of this latter sort such as one's knowledge how to ride a bicycle (Fantl 2017). Here, on the contrary, we focus on *explicit* knowledge-how, that is, the sort of knowledge-how, which is representable in the form of explicit rules and algorithms. In this paper we leave aside the analysis of epistemic attitudes and relations to syntactically represented methods and focus on the form of their representation.

9.4.1 Why Hilbert-Style Axiomatic Architecture and Tarski's Model-Theoretic Semantic of Logical Inference Are Not Appropriate for the Task

Syntactic derivations in suitable symbolic calculi are natural candidates for the role of symbolic representatives of various mathematical and material procedures regulated by the corresponding methods. However the familiar logical calculi (such as Classical first-order logic) and Tarskian set-theoretic semantics of these calculi don't support such an interpretation of syntactic derivations as we shall now see. Let us first fix some basic concepts for the further discussion.

A syntactic *rule* is a partial function of form $\mathcal{F}^n \rightarrow \mathcal{F}$ where $\mathcal{F}$ is a distinguished set of expressions of the given formal language L called *well-formed formulae*. In what follows we use word "formula" as a shorthand for "well-formed formula". The usual logical notation for rules is this:

$$\frac{A_1, \ldots, A_n}{B} \tag{9.1}$$

where the inputed formulae $A_1, \ldots, A_n$ are called *premises* and the outputed formula B is called *conclusion*.[4]

A syntactic *derivation* d (aka *formal proof*) is a finite sequence of formulae $F_1, \ldots, F_k$ where each formula F_i is either an *axiom* or a *hypothesis* or is obtained from some of preceding formulae $F_1, \ldots, F_{i-1}$ according to some of the available syntactic rules.

Axioms are fixed distinguished formulae, which serve as generators for syntactic derivations. It is common to identify axioms with rules with the empty set of premises (which requires a more general concept of rule than we use here). However since the conceptual difference between rules and axioms plays an important rule in our argument, we shall not use such a formal identification.

Hypotheses are also distinguished formulae, which in derivations function like axioms. The difference between axioms and hypotheses concerns their mutual roles in the formal framework (that typically has a form of formal *theory* in which derivations are exhibited): while the set of axioms is fixed the sets of hypotheses vary from one derivation to another.

The last formula F_k in d is conventionally called a *theorem*; it is said that F_k is derived from the axioms and hypotheses, which belong to (or *are used in*) d, according to syntactic rules applied in d. Notice that the sequence of formulae $F_1, \ldots, F_k$ does not provide the full information about the syntactic derivation as a fully determined syntactic procedure. In order to make the full information explicit

[4]This definition of syntactic rule leaves aside non-functional rules with multiple conclusions, which are also considered in some logical theories. However this limitation is not essential for the purpose of this paper.

each formula F_i in $F_1, \ldots, F_k$ needs to be supplied with a commentary, which specifies whether it is an axiom or a hypothesis or, by default, which rule has been used for obtaining it and with which premises.

In standard logical calculi including Classical propositional and first-order calculi well-formed formulae are supposed to express certain *statements*; this is the intended preformal semantics of these formulae. Accordingly, the syntactic rules used in these calculi such as *modus ponens* are interpreted as logical rules, which allow one, given some true statements (axioms), to deduce some other true statements (theorems). The same rules allow one to make conclusions on the basis of certain hypotheses. This general remark alone points to the fact that the syntactic rules and syntactic procedures of standard logical calculi are not apt for representing directly *extra-logical* scientific methods such as methods of conducting specific physical experiments or even mathematical extra-logical methods such as methods of calculating integrals.

The usual idea behind the project of axiomatizing science and mathematics is to reduce (in the sense, which is explained shortly) such extra-logical methods to logical methods via an appropriate choice of axioms. Consider a typical physical theory T that includes descriptions of observations and experiments, which verify empirically its theoretical statements. Think of Newton's *Principia* for a concrete historical example. An ideal formal version of this theory T_F axiomatized in Hilbert's vein allows one to deduce results of all the observations and experiments from its axioms (properly interpreted). Instead of using extra-logical observational and experimental methods specified in T one applies now logical methods specified in T_F for making the corresponding deductions. This does not mean, of course—and has been never meant by Hilbert or any other serious enthusiast of axiomatizing Physics—that logical and empirical methods perform in this situation the same epistemic role and that empirical methods are redundant in science. Logical methods in such a formal framework serve for clarifying logical relations (the precise character of which will be discussed shortly) between different theoretical statements but they cannot by themselves help one to verify these sentences. The task of verification aka proof[5] of theoretical statements is still reserved to empirical methods associated with T, which do not belong to T_F but may play a role in a meta-theory of T_F, which shows that T_F is empirically adequate or inadequate.

[5]The word "proof" used in such a context is ambiguous because it can also refer to formal deductions of theorems from axioms in T_F. It might seem to be a reasonable terminological decision, which would be more in line with the current logical parlance, to reserve term "proof" for formal deductions and talk of verifications or justifications in more general contexts including empirical ones. However following Nagel Sect. 9.3.1 and Prawitz (1986) we believe that such a terminological habit hides the problem rather than helps to solve it. Under certain semantic conditions a syntactic derivation aka formal proof $F_1, \ldots, F_k$ of formula F_k indeed may represent a proof of statement expressed by this formula, i.e., may represent a conclusive evidence in favour of this statement. In order to study and specify such semantic conditions one needs to evaluate whether or not a given interpretation of one's syntax makes syntactic derivations into proofs. The bold identification of proofs with syntactic derivations makes such a critical analysis impossible.

Thus T_F can be called a physical theory only in a peculiar sense of the word, which leaves central issues concerning the verification of a given theory outside this very theory. This certainly does not correspond to how the word "theory" is colloquially used by working scientists. It might be argued that even if T_F fails to represent adequately all epistemologically relevant aspects of physical theory T it still does satisfactorily represent its logical aspects. However this claim is also objectionable because it takes for granted a view of logic, which is not uncontroversial to say the least. One who shares Nagel and Cohen's understanding of the scope of logic (Nagel and Cohen 1934) can argue that since T_F leaves aside issues concerning the verification/proof/evidencing of T-statements it fails to capture the very logical core of T.

Does the same argument apply to theories of pure mathematics? The claim that computing an integral reduces to (or even "ultimately is") a logical deduction from axioms (say, the axioms of ZFC) and some additional hypotheses appears more plausible than a parallel claim about physical measurements; in spite of the fact that this way of computing integrals is not used in the current mathematical practice its theoretical possibility can be reasonably explained and justified by referring to standard set-theoretic foundations of the Integral Calculus. On such grounds one may argue that in the case of pure mathematics the above argument does not apply.

However in fact the standard Hilbert-style axiomatic framework combined with (some version of) axioms of Set theory doesn't support the evidencing of mathematical statements just as it doesn't provide the evidencing of empirical statements. *Firstly*, it leaves to a general philosophical discussion the question of truth of set-theoretic axioms (under their appropriate interpretation). *Secondly*, such foundations of mathematics in their usual form don't provide for a satisfactory epistemological account of valid logical *inference* but reduce that crucial concept to a meta-theoretical concept of logical *consequence* (this point will be explained shortly). *Thirdly*, there is the problem of gap between the technical notion of (formal) proof as logical deduction from axioms of Set theory and the colloquial concept of mathematical proof used in today's mathematics.

Some people see this latter problem as merely pragmatic or "practical" and having no logical and epistemological significance. We agree that logic and epistemology are normative disciplines and for that reason references to past and present mathematical practices cannot constitute by themselves conclusive arguments in these fields. However such references may point to real logical and epistemological problems, which is, in our view, indeed happens in the given case. As far as deductions from ZFC are treated as ideal mathematical (or meta-mathematical) objects their epistemological role as evidences for certain mathematical statements remains unclear. In this form such deductions cannot be used as an effective tool for proof-checking. A theoretical possibility of set-theoretic formalization (justified with some meta-mathematical arguments) in practical contexts is used rather counter-positively. If some mathematical argument can be shown to be not formalizable in ZFC this is a reason to regard it as problematic (which, however, doesn't generally imply its dismissal since the given argument can be shown to be formalizable in some appropriate axiomatic extension of ZFC). If, on the contrary, a mathematical argument is shown to be formalizable in ZFC this doesn't imply that the corresponding informal proof is correct because its formal counterpart may possibly not qualify as a valid deduction. Unless such a formal proof is effectively exhibited it is impossible to check this. The fact that the standard foundations of mathematics do not provide for an effective proof-checking was the main motivation of Voevodsky behind his project of developing alternative foundations that he called Univalent Foundations (see Sect. 9.6 below).

Let us now see what Tarski style Model theory brings into this picture. *Firstly*, it provides for a precise semantics of logical deduction as follows. Formula B is called a semantic consequence of formulas $A_1, \ldots, A_k$ just in case every interpretation of non-logical terms in $A_1, \ldots, A_k$ that makes all A_i into true statements (i.e., every *model* of $A_1, \ldots, A_k$) also makes formula B into a true statement (equiv. is also a model of B), in symbols $A_1, \ldots, A_k \models B$. Then it is proved that if there exists a syntactic deduction from $A_1, \ldots, A_k$ to B, in symbols $A_1, \ldots, A_k \vdash B$, then $A_1, \ldots, A_k \models B$ (soundness). This clarifies our point concerning the lack of proper semantics for logical inference in this setting: it interprets the *existence* of syntactic deduction in terms of a meta-theoretical statement that expresses a logical relation between theoretical statements but it provides no precise semantics for individual deductions. We shall see in Sect. 9.4.3 how this problem is treated with various versions of the *proof-theoretic* logical semantics.

Secondly, Tarski's Model theory allows one to associate with a given first-order Hilbert-style axiomatic theory a canonical class of its set-theoretic models (Hodges 2013). For an easy example think of usual axioms for (definition of) algebraic group and their model in form of set G with a binary operation $\otimes$ on it, which satisfies the appropriate axioms. The canonical character of this construction allows one to define a group from the outset as a set with an additional structure, namely, with a binary operation on this set that has certain required formal properties (associativity, existence of unit and of inverse elements). This is the "semantic" semi-formal presentation used systematically by Bourbaki (1974, p.30), which has been adopted by Suppes and his followers for the representation of scientific theories via the identification of appropriate sets with physical systems such as sets of physical particles or sets of points of the physical space-time. Importantly, the same structure type (such as the group structure type) can be determined by, i.e., be a canonical model of different sets of axioms. This is in accord with Suppes' "semantic" view according to which a scientific theory must be identified with a particular class of models rather than a particular syntactic structure.

The semantic set-theoretic presentation of mathematical theories developed by Bourbaki and a similar representation of scientific theories developed by Suppes and other enthusiasts of the *semantic view* provides a theory with a sense of objecthood: it represents the content of a given theory in terms of its objects such as algebraic groups or systems of physical particles. In this way the semantic representation supports a quasi-constructive reasoning about sets and set-theoretic structures. For example, one may think of forming the powerset $P(A)$ of given set A as a construction after the pattern of traditional geometrical constructions like the construction of a circle by its given radius. This helps one to develop within the set-theoretic mathematics a form of helpful intuition but the problem is that such set-theoretic "constructions" lack formal *rules*, which could be used for representing extra-logical mathematical and possibly scientific *methods*. When one needs to justify the "construction" of powerset in this framework one refers to the powerset *axiom* of ZFC, which is an existential statement that guarantees the existence of powerset for any given set. One may argue that in such contexts the difference between constructive rules and existential axioms is only a matter of taste or fashion. In the next Section we argue that this is not the case.

9.4.2 Two Axiomatic "Styles"

Hilbert and Tarski after him conceive of a theory T as an ordered set of formal sentences satisfied by a class of intended models of this theory and, ideally, not satisfied by any non-intended interpretation. (The latter is a desideratum rather than a definite requirement.) An interpretation of a given sentence s in this context is an assignment of certain semantic values to all non-logical symbols that belong to s. Thus this approach assumes that one distinguishes in advance logical and non-logical symbols of the given alphabet. This requirement reflects the epistemological assumption according to which logic is epistemologically prior to all theories, which are "based" on this logic. In other words the axiomatic method in its Hilbert-Tarski version requires that one first fixes logical calculus L and then applies it in an axiomatic presentation of some particular non-logical theory T.[6] Suppes assumes this basic scheme and says that a theory admits a "standard formalization" when L is the Classical first-order logic with identity (Suppes 2002, p. 24).

All existing approaches to formal representation of scientific theories use this familiar axiomatic architecture. However it is not unique. In 1935 Hilbert's associate Gerhard Gentzen argued that

> The formalization of logical deduction, especially as it has been developed by Frege, Russell, and Hilbert, is rather far removed from the forms of deduction used in practice in mathematical proofs. (Gentzen 1969, p. 68)

and proposed an alternative approach to syntactic presentation of deductive systems, which involved relatively complex systems of rules and didn't use logical tautologies. In Gentzen (1969) Gentzen builds in this way two formal calculi known today as Natural Deduction and Sequent Calculus.

---

[6]We refer here to a version of axiomatic method described by Tarski in (1941). Tarski's view of this method builds on Hilbert's but does it in a particular and original way. The exact relationships between Tarski's semantic conception of the axiomatic method and Hilbert's original ideas about this method is a matter of historical study, which is out of the scope of this paper.

Gentzen's remark quoted above constitutes a pragmatic argument but hardly points to an original epistemological view on logic and axiomatic method. However his further remark that

> The introductions [i.e. introduction rules] represent, as it were, the 'definitions' of the symbol concerned. (Gentzen 1969, p. 80)

is seen today by some authors as an origin of an alternative non-Tarskian conception of logical consequence and alternative logical semantics more generally, which has been developed in a mature form only in late 1990s or early 2000s and is known today under the name of *proof-theoretic* semantics (PTS) (Schroeder-Heister 2016; Piecha and Schroeder-Heister 2016).

Unlike Hilbert Gentzen never tried apply his approach beyond the pure mathematics and today the formal representation of scientific theories mostly follows in Hilbert's and Tarski's steps. However as we argue below the Gentzen style rule-based architecture of formal systems can be also applied in empirical contexts for representing various scientific methods, which include but are not limited to logical and mathematical methods.

9.4.3 Constructive Logic, General Proof Theory, Meaning Explanation and Proof-Theoretic Semantics

The conception of logic that hinges on the concept of proof rather than on that of truth has a long history that can be traced back to Aristotle; here we review only relatively recent developments. In a series of publications that begins in early 1970s Dag Prawitz formulates and defends the idea of *general proof theory* (GPT) as an alternative or a complement to the proof theory understood after Hilbert and Bernays as a meta-mathematical study of formal derivations (Hilbert and Bernays 1939). Prawitz defends here the traditional conception of proof as evidence and argues that Tarski's semantic conception of logical consequence (Tarski 1956) does not provide, by itself, a theoretical account of proofs in mathematics and elsewhere:

> In model theory, one concentrates on questions like what sentences are logically valid and what sentences follow logically from other sentences. But one disregards questions concerning how we know that a sentence is logically valid or follows logically from another sentence. General proof theory would thus be an attempt to supplement model theory by studying also the evidence or the process – i.e., in other words, the proofs – by which we come to know logical validities and logical consequences. (Prawitz 1974, p. 66)

Prawitz calls this conception of logic *intuitionistic* and, as the name suggests, sees Brouwer and Heyting as its founding fathers. Following Kolmogorov and some other important contributors to the same circle of ideas we prefer to call this approach to logic *constructive*. As we have seen in Sect. 9.3.1 Nagel and Cohen's informal conception of logic also falls under this category. Since in this paper we discuss applications of logic in empirical sciences we would like also to point to the interpretation of proofs in constructive logic as *truth-makers* (Sundholm 1994), which may be material objects and events either artificially produced or occurred naturally.

In 1972 Per Martin-Löf accomplished a draft (published only in 1998 as Martin-Löf 1998) of a constructive (or intuitionistic as the author calls it) theory of types, which implements the general constructive approach outlined above. In a more mature form this system known as MLTT is presented in Martin-Löf (1984). MLTT is a Gentzen-style typed calculus that comprises no axiom. The meaning of its syntactic rules and other syntactic elements is provided via a special semantic procedure that Martin-Löf calls the *meaning explanation*. In Martin-Löf (1985) the author compares meaning explanation with a program compiler, which translates a computer program written in a higher-level programming language into a lower-level command language. According to Martin-Löf a similar translation of MLTT syntactic rules into elementary logical steps gives these rules their meaning and simultaneously validates them (Martin-Löf 1996). Martin-Löf's meaning explanation is an informal version of the proof-theoretic semantics mentioned in Sect. 9.4.2 above (Schroeder-Heister 2016).

In the late 2000s MLTT started a new life as the syntactic core of the emerging Homotopy Type theory (HoTT) (Univalent Foundations Program 2013). In order to describe a feature of HoTT, which is relevant to the present discussion, we need first to expose few basic concepts of MLTT.

9.5 From MLTT to HoTT

MLTT comprises four types of *judgements* (which should not be confused with propositions, see Sect. 9.7.3 below).

(i) $A : TYPE$;
(ii) $A \equiv_{TYPE} B$;
(iii) $a : A$;
(iv) $a \equiv_A a'$

In words (i) says that A is a type, (ii) that types A and B are the same, (iii) that a is a term of type A and (iv) that a and a' are the same term of type A. Let me leave (i) and (ii) aside and provide more details on (iii) and (iv).

Martin-Löf offers four different informal readings of (iii) (Martin-Löf 1984, p. 5):

(1) a is an element of set A
(2) a is a proof (construction) of proposition A ("propositions-as-types")
(3) a is a method of fulfilling (realizing) the intention (expectation) A
(4) a is a method of solving the problem (doing the task) A (BHK semantics)

The author argues that these interpretations of judgement form (iii) not only share a logical form but also are closely conceptually related. The correspondence between (2) and (4) is based on the conceptual duality between problems and theorems, which dates back to Euclid (Rodin 2018); among more recent sources (4) refers to Kolmogorov's *Calculus of Problems* (Kolmogorov 1991) and the so-called *BHK-semantics* (after Brouwer, Heyting and Kolmogorov) of the Intuitionistic propositional calculus. (3) is an explanation of judgement in terms of Husserl's Phenomenology. The correspondence between (1) and (2) at the formal level is the Curry-Howard correspondence. In Martin-Löf's view this latter correspondence is not *only* formal but also contentful: the author argues that in the last analysis the concepts of set and proposition are the same:

> If we take seriously the idea that a proposition is defined by lying down how its canonical proofs are formed [...] and accept that a set is defined by prescribing how its canonical elements are formed, then it is clear that it would only lead to an unnecessary duplication to keep the notions of proposition and set [...] apart. Instead we simply identify them, that is, treat them as one and the same notion. (Martin-Löf 1984, p. 13)

Let us now turn to judgement type (iv). It says that terms a, a', both of the same type A, are equal. This equality is called *judgemental* or *definitional* and does *not* qualify as a proposition; the corresponding *propositional* equality writes as $a =_A a'$ and counts as a type on its own ($a =_A a' : TYPE$) called an *identity type*. In accordance to reading (2) of judgement form (iii) a term of identity type is understood as a proof (also called a witness or evidence) of the corresponding proposition. MLTT validates the rule according to which a judgemental equality entails the corresponding propositional equality:

$$\frac{a \equiv_A a'}{refl_a : a =_A a'}$$

where $refl_x$ is the canonical proof of proposition $a =_A a'$.

The *extensional* version of MLTT also validates the converse rule called *equality reflection rule*:

$$\frac{p : a =_A a'}{a \equiv_A a'}$$

but HoTT draws on an *intensional* version of MLTT that does not use such a principle and, in addition, allows for multiple proofs of the same propositional equality (the proof-relevant conception of equality) (Univalent Foundations Program 2013, p. 52).

Let p, q be two judgmentally different proofs of proposition saying that two terms of a given type are equal:

$$p, q : P =_T Q$$

it may be the case that p, q, in their turn, are propositionally equal, and that there are two judgmentally different proofs p', q' of this fact:

$$p', q' : p =_{P=_T Q} q$$

This and similar multi-layer syntactic constructions in MLTT can be continued unlimitedly. Before the rise of HoTT it was not clear that this syntactic feature of the intensional MLTT can be significant from a semantic point of view. However it became the key point of the homotopical interpretation of this syntax. Under this interpretation

- types and their terms are interpreted, correspondingly, as spaces and their points;
- identity proofs of form $p, q : P =_T Q$ are interpreted as paths between points P, Q of space T;
- identity proofs of the second level of form $p', q' : p =_{P=_T Q} q$ are interpreted as homotopies between paths p, q;
- all higher identity proofs are interpreted as higher homotopies;

(which is coherent since a path $p : P =_T Q$ counts as a point of the corresponding path space $P =_T Q$, homotopies of all levels are treated similarly).

In order to get an intuitive picture the non-mathematical reader is advised to think of homotopy h between paths $p, q : P =_T Q$ as a curve surface bounded by the two paths, which share their endpoints P, Q. Given two such surfaces h, i two-homotopy f between h and i is a solid bounded by these surfaces. Picturing higher homotopies of level > 2 is more difficult because it requires a geometrical intuition that extends beyond the three spatial dimensions.

Thus the homotopical interpretation makes meaningful the complex structure of identity types in the intensional MLTT. It makes this structure surveyable, as we shall now see. Consider the following definition:

Definition 1 Space S is called *contractible* or space of h-level (-2) when there is point $p : S$ connected by a path with each point $x : A$ in such a way that all these paths are homotopic (i.e., there exists a homotopy between any two such paths).

In what follows we refer to contractible spaces "as if they were effectively contracted" and identify such spaces with points. A more precise mathematical formulation involves the notion of *homotopy equivalence*, which provides a suitable identity criterion for spaces in Homotopy theory. In view of the homotopy interpretation of MLTT outlined above such spaces (defined up to the homotopy equivalence) are also called "homotopy types".

Definition 2 We say that S is a space of h-level $n + 1$ if for all its points x, y path spaces $x =_S y$ are of h-level n.

The two above definitions gives rise to the following stratification of types/spaces in HoTT by their h-levels:

h-level -2: single point pt;

h-level -1: the empty space $\emptyset$ and the point pt: truth-values aka (mere) propositions
h-level 0: sets (discrete point spaces)
h-level 1: flat path groupoids: no non-contractibe surfaces
h-level 2: 2-groupoids: paths and surfaces but no non-contractible volumes
•
•
h-level n: n-groupoids
• ...
h-level ω: ω-groupoids

A general space is of h-level ω: this is just another way of saying that the ladder of paths, homotopies and higher homotopies of all levels is, generally, infinite. The above stratification classifies special cases when this ladder is finite in the sense that after certain level n spaces of all higher-order homotopies (or paths if $n < 1$) are trivial, i.e., such homotopies (paths) are no longer distinguished one from another at the same level. For example a *set* aka *discrete space* (h-level 0) is a space that either does or does not have a path between any pair of its points; different paths between two given points are not distinguished. If two points are connected by a path they are contracted into one point as explained above. At the next h-level 1 live flat path groupoids that comprise sets of points connected by paths, which are, generally, multiple and well-distinguished *up to homotopy*: homotopical paths are identified and non-homotopical paths count as different. Thus in a flat groupoid the space of paths between any two given points is a set. This inductive construction proceeds to all higher levels.

Notice that h-levels are not equivalence classes of spaces. The homotopical hierarchy is cumulative in the sense that all types of h-level n also qualify as types of level m for all $m > n$. For example pt qualifies as truth-value, as singleton set, as one-object groupoid, etc. Hereafter we call a n-type a type, which is of h-level n but not of level $n - 1$.

A space of h-level l (now including the infinite case $l = \omega$) can be transformed into a space of h-level $k < l$ via its *k-truncation*, which can be informally described as a forced identification of all homotopies (paths) of levels higher than k. In particular the (-1)-truncation of any given space S brings point pt when S is not empty and brings the empty space $\emptyset$ otherwise.

A more mathematically precise explanation of HoTT basics can be found in Univalent Foundations Program (2013) and other special literature.

The h-stratification of types in MLTT suggests an important modification of the original preformal semantics of this theory introduced in the beginning of this section. This modification is not a mere conservative extension. Recall that in Martin-Löf (1984, p. 5) Matin-Löf proposes multiple informal meaning explanations for judgements and suggests that they conceptually converge. In particular, Martin-Löf proposes here to think of types as sets or as propositions and suggests that in the last analysis the two notions are the same. HoTT in its turn provides a geometrically motivated conceptions of set as 0-type and proposition (called in

(Univalent Foundations Program 2013, p. 103) *mere* proposition) as (-1)-type. The latter is motivated as follows. Recall that there are just two (-1)-types (up to the homotopical equivalence): the empty type $\emptyset$ and the one-point type pt. HoTT retains the idea that propositions are types and the idea that terms of such types are truth-makers (proofs, evidences, witnesses). The adjective "mere" used in (Univalent Foundations Program 2013, p. 103) expresses the fact that (-1)-type interpreted as proposition does not distinguish between its different proofs but only reflects the fact that such a proof exists (in case of pt) or the fact that the given proposition has no proof (in case of $\emptyset$). In other words, the adjective "mere" expresses the fact that mere propositions are proof-irrelevant. For the same reason (-1)-types can be also described as truth-value types, where truth is understood constructively as the existence of proof/evidence/truth-maker.

Whether or not the proof-irrelevant conception of proposition is fully satisfactory as a working conception of proposition *tout court* has been recently a matter of controversy. One may argue that in certain context people tend to think of and speak about propositions in a proof-relevant sense.[7] In our opinion this fact does not constitute a strong objection to the identification of propositions with (-1)-types because all proof-relevant aspects of propositions are taken into account at higher levels of the homotopical hierarchy as follows. General n-type S^n where $n > -1$ comprises:

- the "propositional layer" S^{-1}, which is a proposition extracted from S^n via its propositional truncation, and
- the higher-order structure H present on all h-levels > -1, which is a (possibly multi-layer) structure of *proofs* of proposition S^{-1}.

This observation suggests the following possible application of HoTT in the formal representation of theory T:

- extra-logical rules of T (including rules for conducting observations and experiments) are MLTT rules applied at h-levels > -1
- logical rules are MLTT rules applied to (-1)types.

This gives us a preliminary answer to the question posed in Sect. 9.4 How to represent scientific methods (including extra-logical methods) with symbolic calculi? (A more elaborated version of this answer will be given in Sect. 9.7.3 below). Although HoTT does not comprise different sets of logical and extra-logical rules it allows one to distinguish between logical and extra-logical applications of its rules as follows. *Logical* application of rules can be isolated from the extra-logical via (-1)-truncation. The application of HoTT rules to S^n comprises a logical application of these rules at h-level (-1) *and* an extra-logical application of the same rules at all higher h-levels. Such a joint application of MLTT rules makes immediately evident the logical function of extra-logical (i.e., higher order)

[7] As some time ago argued Michael Shulman in several blogs and in a private correspondence.

structures and constructions in S^n, which is truth-making (or *verification* if one prefers this term borrowed from the philosophy of science) of proposition S^{-1}.

The criterion of logicality that we use here is based on the h-hierarchy and the identification of (-1)-types with propositions. It is an interesting project to evaluate it against other criteria of logicality discussed in the philosophical literature. This, however, is out of the scope of this paper. For our present purpose it is sufficient to explain how set-based and higher order construction in HoTT work as truth-makers of propositions. The crucial aspect of the shift from Martin-Löf's original preformal semantics for MLTT to the new preformal semantics based on the h-hierarchy (which hereafter we shall call h-semantics for short) can be described in the jargon of Computer Science as the shift from the propositions-as-types paradigm to the propositions-as-*some*-types paradigm. Further implications of this shift are discussed in Sect. 9.7.3 below.

9.6 Univalent Foundations

Univalent Foundations (UF) use a version of HoTT-based framework as a new purported foundation of pure mathematics. The *univalence property* of type-theoretic universes plays a central role in this project. We shall not define and discuss this property in this paper but only point to its relevance to the choice between the two axiomatic styles mentioned above. The 2013 standard version of UF uses the *Univalence Axiom* (Axiom 2.10.3 in Univalent Foundations Program 2013), which is in fact the only axiom added on the top of *rules* of MLTT. This added axiom makes the axiomatic architecture of UF mixed in the sense that it comprises both rules and axioms. However a more recent work by Th. Coquand and his collaborators on Cubical Type theory (Cohen et al. 2016) provides a "constructive explanation for UA", i.e., a formal deduction of the univalence property in a Gentzen style rule-based formal system other than MLTT. At the present stage of development[8] such a constructive explanation of univalence is not yet achieved in all its desired generality but we take a liberty to bypass this technical issue in the present discussion and consider the Gentzen style constructive axiomatic architecture as a proper feature of UF.

Our proposed *constructive view* of theories motivated and informally described in Sect. 9.2 above applies the same axiomatic architecture for representing theories beyond the pure mathematics. A straightforward approach to representing scientific theories with HoTT consists in an identification of certain HoTT structures with structures coming from theories of Physics and other scientific disciplines. A dummy example of this sort is presented and discussed in Rodin (2017) where a fragment of HoTT is used for proving the identity of Morning Star and the Evening Star; the observed continuous trajectory of this celestial body is used here as an evidence (proof) of the identity statement. Urs Schreiber formalized in UF

[8]As presented by Thierry Coquand in his talk given at the EUTypes meeting, Nijmegen, Netherlands, 22–24 January 2018.

a significant body of up-to-date mathematical physics and made a new original contributions in this area using UF-based mathematics (Schreiber 2013). Even if none of this qualifies as a full-blood (representation of) physical theory such attempts show that the Gentzen-style axiomatic architecture and, more specifically, UF as a paradigm mathematical theory that uses this architecture can be used in science and support a constructive computer-based representation of scientific theories.

9.7 Modelling Rules

The main lesson of the debate between the "syntactic view" and the "semantic view", as one can see this matter after the critical analysis briefly reviewed in the above Introduction, is that the formal semantic aspect of one's representational framework is at least as much important as the syntactic one. The *formal* semantic of HoTT—which should not be confused with the informal notion of meaning explanation introduced in Sect. 9.4.3 above or with the suggestive description of types as spaces used in Univalent Foundations Program (2013) and other introductions to HoTT—by the present date is largely a work in progress. The following discussion concerns only the Model theory of HoTT rather than its formal semantics in a more general sense and it does not intend to cover all relevant developments.

Since one leaves the familiar territory of first-order theories and their set-theoretic semantics the very notions of theory and model become problematic and require certain conceptual revisions, which are not necessarily innocent from a philosophical viewpoint as we shall now see. By a *theory* we shall understand here a set of formal rules as described in Sect. 9.1 from Sect. 9.4.1 together with the set of all derivations supported by these rules. What is a *model* of a given theory presented as just described?

Alfred Tarski designed his Model theory back in early 1950-ies having in mind Hilbert-style axiomatic theories. Let us repeat its basic definitions. A model of (uninterpreted) axiom A is an interpretation m of non-logical terms in A that makes it into a true statement A_m; if such m exists A is called *satisfiable* and said to be satisfied by m. Model M of uninterpreted axiomatic theory T is an interpretation that makes all its axioms and theorems true. When rules of inference used in T preserve truth (soundness) it is sufficient to check that M satisfies all axioms of T for claiming that it also satisfies all its theorems.

Since we deal with modelling a theory presented in Gentzen style rather than in Hilbert style we need a notion of modelling a rule rather than modelling an axiom. Although such a notion is not immediately found in standard textbooks on Model theory (such as Hodges 1997) it can be straightforwardly construed on this standard basis as follows. We shall say that interpretation m is a model of rule R (of general form Sect. 9.1), in symbols

$$\frac{A_1^m, \ldots, A_n^m}{B^m} \tag{9.2}$$

when the following holds: whenever $A_1^m, \ldots, A_n^m$ are true statements B^m is also true statement. Arguably this notion of modelling a rule is implicit in Tarski's Model theory, so it can be used for modelling HoTT without revisions (albeit with some add-ons, see below). There are however several problems with this straightforward approach, which we list below and suggest some tentative solutions.

9.7.1 Logical Inference Versus Semantic Consequence Relation

In the standard setting considered by Tarski (Hilbert-style axiomatic theories and their set-theoretic semantics) the notion of modelling a rule is redundant because the syntactic rules used in this setting such as *modus ponens* preserve truth under *all* interpretations of their premises and conclusions. Moreover, one can argue in Prawitz's line (see Sect. 9.4.3) that the Tarski-style logical semantics provides a formal semantics for the relation of logical consequence between the premises and the conclusion of rule Sect. 9.2 but no semantics for this rule itself. In the passage quoted in Sect. 9.4.3 Prawitz makes an emphasis on the epistemological aspect of the above argument. However a form of this argument remains valid even if one doesn't accept the idea that the semantics for rules should be necessarily epistemic. To make this argument work it is sufficient to assume that a proper semantics for rules should necessarily involve the concept of "how", which also makes part of many non-epistemic semantic contexts such as computational models of some natural processes. Thus one can argue that the above notion of modelling a rule is unsatisfactory without referring to the epistemic role of rules.

As it has been mentioned in Sect. 9.4.3 Martin-Löf's notion of meaning explanation provides an informal semantics for syntactic rules, which is independent of Model theory. In our view, keeping the formal semantics for rules apart from Model theory can hardly be a tenable strategy today. From a mathematical point of view meaning explanations give rise to *realizability models* of Type theory (Reus 1999; Hofstra and Warren 2013) and the conceptual distinction between the two is not so easy to formulate and in any event it is not stably kept in the current literature.[9] This is why it makes more sense, in our view, to revise some conceptual foundations of Model theory in the light of new developments rather than leave the conceptual core of this theory untouched and develop a formal semantics for HoTT independently.

[9]For example Tsementzis (2017) prefers to call preformal meaning explanation what otherwise can be called a semi-formal model of HoTT, moreover that the author interprets rules of MLTT in Tarskian style according to our scheme (9.2) (above in the main text).

9.7.2 Logical and Non-logical Symbols

As we have already mentioned above in Sect. 9.4.2 in the standard setting by interpretation of formula one understands an interpretation of non-logical symbols in this formula; the distinction between logical and non-logical symbols is supposed to be set beforehand. Interpreting HoTT one deals with a very different situation, where such a distinction between logical and extra-logical elements does not apply at the syntactic level. At the semantical level the question of logicality (Which fragment of HoTT belongs to logic proper?) remains sound but does not have an immediate answer. Certain concepts of HoTT such as the basic concepts of type and term may qualify as logical in a broad sense but also receive an informal interpretation in terms Homotopy theory, which is normally not thought of as a theory of pure logic.

This has the following effect on the interpretation of syntactic rules. In the standard setting these rules are supposed to serve as logical rules and work equally for all legitimate interpretations of formulas. So in this setting the syntactic rules admit a default interpretation in terms of logical consequence relation. The situation is very different in HoTT and akin theories where no fixed distinction between logical and non-logical symbols is specified at the syntactic level. So finding an appropriate semantics for MLTT/HoTT syntactic rules constitutes a special problem of Model theory *of HoTT* but not only a problem of general Model theory. Below we describe such a semantics informally.

9.7.3 Extra-Logical Rules

In scheme Sect. 9.2 assumptions $A_1^m, \ldots, A_n^m$ and conclusion B^m are statements. Recall that in MLTT these formulas are interpreted as *judgements* (Sect. 9.5). This interpretation combines with scheme Sect. 9.2 only at the price of ignoring certain conceptual subtleties. Frege famously describes an act judgement as "the acknowledgement of the truth of a thought" (Frege 1984, p. 355–356) or an "advance from a thought to a truth-value" (Frege 1892, p. 159) understanding by a thought (roughly) what we call today a proposition. Frege warns us, continuing the last quoted passage, that the above "cannot be a definition. Judgement is something quite peculiar and incomparable" (ib.). Having this Frege's warning in mind one can nevertheless think of judgements as meta-theoretical propositions and, in particular, qualify formulas $A_1^m, \ldots, A_n^m, B^m$ in Sect. 9.2 both as judgements (internally) and meta-propositions belonging to Model theory (i.e., externally). This provides a bridge between Tarski's and Martin-Löf's semantic frameworks and allows one to interpret rules of MLTT via scheme Sect. 9.2. It should be stressed that this bridge has a formal rather than conceptual character.

A more difficult problem arises when one intends to use in the Model theory the internal preformal semantics of HoTT based on the homotopical hierarchy of types (h-semantics). Recall that h-semantics qualifies as propositions only types of certain h-level, namely, only (-1)-types—while Martin-Löf's early intended semantics for MLTT admits thinking of *each* type as a proposition (see Sect. 9.5 above). Accordingly, h-semantics hardly supports the interpretation of all formulas as judgements in Frege's sense of the word. Consider a special instance of (Sect. 9.1) where all formulas are judgements of basic form (iii) (see Sect. 9.5 above):

$$\frac{a_1 : A_1, \ldots, a_n : A_n}{b : B} \tag{9.3}$$

If $A_1, \ldots, A_n, B$ are (-1)-types then the name of judgement used for these expressions of form $a : A$ can be understood in Frege's sense: here A stands for a proposition and a stands for its proof (witness, truth-maker). But when A is n-type A^n with $n > -1$ the h-semantics suggests to read the same expression $a : A$ as a mere *declaration* of token a of certain higher, i.e., *extra-logical* type A. This extra-logical object a does perform a logical function: it guaranties that (-1)-type A^{-1} obtained from A^n via the propositional truncation is non-empty and thus gives rise to valid judgement $a^* : A^{-1}$. So there is an indirect sense in which token a of type A also functions as a truth-maker even if A is not a proposition. This fact, however, hardly justifies using the name of judgement (in anything like its usual sense) for all expressions of form $a : A$ along with their h-semantics. In case of the other three basic forms of "judgements" the situation is similar. A possible name for such basic syntactic constructions, which we borrow from Computer Science and programming practice, is "declaration". We suggest to reserve the name of judgement only for those declarations, which involve only (-1)-types.

When $A_1, \ldots, A_n, B$ are higher types the above scheme Sect. 9.3 represents an extra-logical rule that justifies the construction of token b from given tokens $a_1, \ldots, a_n$. Once again the characterisation of this rule as extra-logical should be understood with a pinch of salt. The (-1)-truncated version of this rule is logical. However this truncated version represents only a simplified and impoverished version of the original rule. Thus we've got here a more precise answer to the question of how to represent extra-logical scientific methods with formal calculi (Sect. 9.4). Generally, assumptions in MLTT/HoTT-rules have a more complex form than in (9.3). We leave it for a further research to study these complex forms from the same viewpoint.

Declarations of form $a : A$ (and of other forms) can be always rendered as meta-propositions (including the case when type A is not propositional) (Tsementzis 2017); in English such a meta-proposition can be expressed by stating that a *is* a token of type A. Then scheme (9.2) can be used as usual. The obtained meta-theoretical propositional translation can be useful, in particular, for foundational purposes if one wants (as did Voevodsky) to certify HoTT and study its model-theoretic properties on an established foundational basis such as ZF. However this way of developing Model theory of HoTT is evidently not appropriate for our present purpose.

An available alternative consists of interpreting formal rules of form (9.2) in terms of extra-logical mathematical constructions (generally, not purely symbolic) which input and output certain non-propositional contents. In the special case (9.3) such assumptions are tokens of mathematical objects of certain types and the conclusion is a new token of certain new type. For a suggestive traditional example think of Euclid's geometrical rules that he calls *postulates*: the first of these rules justifies a construction of straight segment by its given endpoints. The first-order statement "For any pair of distinct points there exists a straight segment having these points as its endpoints" used by Hilbert in his axiomatic reconstruction of Euclidean geometry is a meta-theoretical reconstruction of Euclid's extra-logical rule rather than just a logically innocent linguistic paraphrase (Rodin 2018).

When an appropriate construction c that instantiates given formal rule R is exhibited, rule R is justified in the sense that it provably has a model, which is c itself. Since such a notion of modelling a rule does not involve a satisfaction relation defined in terms of truth-conditions, it marks a more significant departure from the conceptual foundations of Model theory laid down by Tarski back in 1950-ies. Whether or not this conceptual difference implies a significant technical difference remains unclear to the author of this paper. The Model theory of HoTT developed by Voevodsky, which we overview in the next Section, is intended to be formalizable both in ZFC (via a meta-propositional translation described above) and in UF. The latter option amounts to using HoTT/UF as its proper meta-theory; in this case preformal h-semantics turns into a form model-theoretic semantics. In this way h-semantics becomes formal.

9.8 Initiality: Theories as Generic Models

HoTT as presented in the HoTT Book (Univalent Foundations Program 2013) uses the syntax of MLTT extended with the Univalence Axiom; the semantic "homotopic" aspect of this theory, a central fragment of which has been briefly described above (Sect. 9.5), remains here wholly informal: the language of the Book deliberately and systematically confuses logical and geometrical terminology referring to types as "spaces", to terms as "points" and so on. This is what we call the preformal h-semantics throughout this paper.

A mathematical justification of such terminological liberty, which proves very useful in providing one's reasoning in and of MLTT with a helpful geometrical intuition, lies in the fact that "all of the constructions and axioms considered in this book have a model in the category of Kan complexes, due to Voevodsky"

(Univalent Foundations Program 2013, p. 11; Kapulkin and Lumsdain 2012–2016).[10]

A closer look at Kapulkin and Lumsdain (2012–2016) shows that the concept of model used by the authors is quite far from being standard. In particular, the authors of this paper use techniques of *Functorial Model theory* (FMT), which stems from the groundbreaking Ph.D. thesis of W.F. Lawvere accomplished in 1963 (Lawvere 1963).[11]

The idea of Functorial Model theory can be described in form of the following construction:

- a given theory $\mathbf{T}$ is presented as a category of syntactic nature (a canonical syntactic model of $\mathbf{T}$);
- models of $\mathbf{T}$ are construed as functors $m : \mathbf{T} \to S$ from $\mathbf{T}$ into an appropriate background category (such as the category of sets), which preserve the relevant structure;
- models of $\mathbf{T}$ form a functor category $C_{\mathbf{T}} = S^{\mathbf{T}}$;
- in the above context theory $\mathbf{T}$ is construed as a *generic model*, i.e., as an object (or a subcategory as in Lawvere 1963) of $C_{\mathbf{T}}$.

Voevodsky and his co-authors proceed according to a similar (albeit not quite the same) pattern:

- Construct a general model of given type theory $\mathbf{T}$ (MLTT or its variant) as a category $\mathcal{C}$ with additional structures which model $\mathbf{T}$-rules. For that purpose the authors use the notion of *contextual category* due to Cartmell (1978); in later works Voevodsky uses a modified version of this concept named by the author a *C-system*.
- Construct a particular contextual category (variant: a C-system) $\mathcal{C}(\mathbf{T})$ of syntactic character, which is called *term model*. Objects of $\mathcal{C}(\mathbf{T})$ are MLTT-contexts, i.e., expressions of form

$$[x_1 : A_1, \ldots, x_n : A_n]$$

taken up to the definitional equality and the renaming of free variables and its morphisms are substitutions (of the contexts into $\mathbf{T}$-rule schemata) also identified up to the definitional equality and the renaming of variables. More precisely, morphisms of $\mathcal{C}(T)$ are of form

$$f : [x_1 : A_1, \ldots, x_n : A_n] \to [y_1 : B_1, \ldots, y_m : B_m]$$

[10]Kapulkin and Lumsdain (2012–2016) is a systematic presentation of Voevodsky's results prepared by Kapulkin and Lumsdain on the basis of Voevodsky's talks and unpublished manuscripts. The Book (Univalent Foundations Program 2013) refers to the first 2012 version of this paper. In what follows we refer to its latest upgraded version that dates to 2016.

[11]For a modern presentation of FMT see Johnstone (2002, vol. 2, D1-4) and Nourani (2016).

where f is represented by a sequent of terms $f_1, \ldots, f_m$ such that

$$x_1 : A_1, \ldots, x_n : A_n \vdash f_1 : B_1$$
$$\vdots$$
$$x_1 : A_1, \ldots, x_n : A_n \vdash f_m : B_m(f_1, \ldots, f_m)$$

Thus morphisms of $\mathcal{C}(T)$ represent derivations in **T**.

- Define an appropriate notion of morphism between contextual categories (C-systems) and form category $CTXT$ of such categories.
- Show that $\mathcal{C}(\mathbf{T})$ is initial in $CTXT$, that is, that for any object $\mathcal{C}$ of $CTXT$ there is precisely one morphism (functor) of form $\mathcal{C}(\mathbf{T}) \to \mathcal{C}$.

The latter proposition is stated in Kapulkin and Lumsdain (2012–2016) as Theorem 1.2.9 without proof; the authors refer to Streicher (1991) where a special case of this theorem is proved and mention that "the fact that it holds for other selections from among the standard rules is well-known in folklore".

The authors state that the initiality property of $\mathcal{C}(\mathbf{T})$ justifies the qualification of $\mathcal{C}$ (a general contextual category/C-system) as a model of **T** (Definition 1.2.10 in Kapulkin and Lumsdain 2012–2016). Since the initiality condition does not belong to the conceptual background of the standard Model theory this statement calls for explanation. We offer here such an explanation in terms of our proposed *constructive* view of theories (Sect. 9.2 above). Think of generic term model $\mathcal{C}(\mathbf{T})$ as a special presentation of theory **T** in form of *instruction*, i.e., a system of rules presented symbolically. This instruction is a schematic syntactic construction; it is schematic in the sense that it is applicable to more than one context. Available contexts are objects of $CTXT$. The initiality property of $\mathcal{C}(\mathbf{T})$ in $CTXT$ guarantees that in each particular context $\mathcal{C}$ instruction$\mathcal{C}(\mathbf{T})$ is interpreted and applied unambiguously. A useful instruction can be schematic but it cannot be ambiguous.

The initiality property of contextual categories and other relevant mathematical structures was in the focus of Vladimir Voevodsky's research during the last several years of his life and work. He considered it as a genuine open mathematical problem and dubbed it the Initiality Conjecture. This conjecture is not a mathematical statement which waits to be proved or disproved but a problem of building a general formal semantic framework for type theories, which includes a syntactic object with the initiality property explained above. This conjecture still stands open to the date of writing the present paper.

9.9 Conclusion

We have seen that the epistemological argument between partisans of the so-called Syntactic View and the Semantic View of scientific theories, which took place in the twentieth century, was at a great extent determined by basic features of formal logical tools available at that time (Sects. 9.1 and 9.3.2). We have also seen that

Ernest Nagel and Morris Cohen, using the traditional philosophical prose, expressed epistemological views on science and on the role of logic in science, which could not be supported by formal logical means available to them in a ready-made form at this point of history (Sect. 9.3.1). This was a price payed by these people for their decision to develop epistemology and philosophy of science on a formal logical basis. This price may be or be not worth paying but one should keep in mind that formal logical tools in their turn are developed with certain intended purposes and with some ideas about the nature of logic and its epistemic role. Even if formal logic is more tightly connected to mathematics than to empirical sciences, important logical ideas may come from sciences. If a system of logic represented with a symbolic calculus does not support one's epistemological view on science this may be a reason to revise both one's epistemological ideas *and* one's logical ideas and logical techniques rather than only the former.

In this paper we defend a *constructive* view of scientific theories according to which *methods* (including empirical, mathematical and eventually logical methods) are essential constituents of scientific theories, which should not be left out as a part of "context of discovery" or in any other way (Sect. 9.2). This epistemological view is certainly not original and dates back at least to René Descartes and Francis Bacon. The brief word description of this view given in this paper is evidently very imprecise and allows for various specifications, many of which may be not compatible. However our aim in this paper is to support this general view with a new formal technique rather than elaborate on it using only philosophical prose. We call this general epistemological view "constructive" referring to an essential feature of formal framework, viz. Homotopy Type theory, that we use for supporting this view; the relevant sense of being constructive we have explained in some length elsewhere (Rodin 2018). In order to avoid a possible confusion let us mention here that our proposed constructive view of theories does not imply a form of epistemological constructivism that treats objects of science as mental or social constructions; in fact our constructivism is compatible with a version of scientific realism.

HoTT has emerged as an earlier unintended homotopical interpretation of Constructive Type theory due to Martin-Löf (MLTT); MLTT in its turn has been developed as a mathematical implementation of constructive notion of logic, which focuses on proofs and inferences rather than on the relation of logical consequence between statements, and which is concerned primarily with epistemic rather than ontological issues (Sect. 9.4.3). A significant part of this intended preformal logical semantics (albeit not all of this semantics) of MLTT is transferred to HoTT (Sect. 9.5); the semantic features of HoTT inherited from MLTT support a constructive logical reasoning in HoTT. However HoTT also brings about a new extra-logical semantics, which is an essential ingredient of our proposed concept of constructive theory. Just as in the case of standard set-theoretic semantics of first-order theories the extra-logical semantics of HoTT is tightly related to the logical semantics: h-hierarchy of types and the notion of propositional truncation (Sect. 9.5) provides a clear sense in which extra-logical terms have a logical impact (Sect. 9.7.3). However the relationships between logical and extra-logical elements are organised here differently; a major difference is that the h-semantics allows for making sense of the notion of extra-logical *rule* (or more precisely, an extra-logical application of formal rule), which is not available in the standard setting. This is a reason why we believe that a HoTT-based framework can be a better representational tool for science than the standard first-order framework.

Model theory of HoTT does not exist yet in a stable form. In this paper we discussed conceptual issues that arise when one attempts to use with HoTT and akin theories the standard notion of satisfaction due to Tarski (Sect. 9.7). Then we reviewed a work in progress started by Voevodsky and continued by his collaborators where the category-theoretic concept of initiality (Sect. 9.8) plays an important role. The purpose of this analysis is to provide a conceptual foundation for prospective applications of HoTT and akin theories in the representation of scientific theories along with the Model theory of such theories. Taking into account the Model theory of HoTT is an important upgrade of preformal h-semantics to its more advanced formal version.

Admittedly, a possibility of using HoTT or a similar theory as a standard representational tool in science at this point is a philosophical speculation rather than a concrete technical proposal. Nevertheless there are reasons to believe that this new approach can be more successful than its set-theoretic ancestor. Unlike the standard set-theoretic foundations of mathematics Univalent Foundations are effective and "practical" in the sense that they allow for an effective practical formalization of non-trivial mathematical proofs and checking these proofs with the computer. An effective UF formalization of the mathematical apparatus of today's Physics and other mathematically-laden sciences paths a way to a similar effective formalization and useful digital representation of scientific theories. The step from foundations of mathematics to representation of scientific theories is by no means trivial. This paper attempts to provide conceptual preliminaries for it.

References

Bourbaki, N. (1974). *Algebra I*. Paris: Hermann.

Cartmell, J. (1978). *Generalised algebraic theories and contextual categories*. Ph.D. thesis, Oxford University

Cohen, C., Coquand, Th., Huber, S., & Mörtberg, A. (2016). Cubical type theory: A constructive interpretation of the univalence axiom. *Leibniz International Proceedings in Informatics, 23*, 1–34.

Fantl, J. (2017). Knowledge how. In E. N. Zalta (Ed.), *The Stanford encyclopedia of philosophy*. Metaphysics Research Lab, Stanford University, fall 2017 edition.

Frege, G. (1892). Ueber Sinn und Bedeutung. Zeitschrift für Philosophie und Philsophische Kritik, NF 100, pp. 25–50. Translated in Beaney, M. (Ed.). (1997). *The Frege reader* (pp. 151–171). Oxford: Blackwell.

Frege, G. (1984). Logical investigations. Translated in: B. McGuiness, Trans. M. Black & V.H. Dudman, P. Geach, H. Kaal, E.-H. Kluge, B. McGuiness, & R. H. Stoothoff (Eds.), Collected papers on mathematics, logic, and philosophy. Oxford: Basil Blackwell. Blackwell (1918–1926).

Gentzen, G. (1969). Investigations into logical deduction. In M. E. Szabo (Ed.), The Collected Papers of Gerhard Gentzen (pp. 68–131). Amsterdam/London: North-Holland Publishing Company; original publication in German: Untersuchungen über das logische Schliessen, Mathematische Zeitschrift, 39 (1934–1935), S. 176–210, 405–431.

Glymour, C. (2013). Theoretical equivalence and the semantic view of theories. *Philosophy of Science, 80*(2), 286–297.

Halvorson, H. (2012). What scientific theories could not be. *Philosophy of Science, 79*(2), 183–206.

Halvorson, H. (2016). Scientific theories. In P. Humphreys (Ed.), *The Oxford handbook of philosophy of science*. New York: Oxford University Press.

Hilbert, D., & Bernays, P. (1939). *Grundlagen der Mathematik v. 2*. Berlin: Springer.

Hodges, W. (1997). *A shorter model theory*. Cambridge: Cambridge University Press.

Hodges, W. (2013). *First-order model theory*. Stanford Encyclopedia of Philosophy, https://plato.stanford.edu/entries/modeltheory-fo/. Accessed January 13, 2018.

Hofstra, P., & Warren, M. (2013). Combinatorial realizability models of type theory. *Annals of Pure and Applied Logic, 164*(10), 957–988.

Johnstone, P. (2002). *Sketches of an elephant: A topos theory compendium*. Oxford: Oxford University Press.

Kapulkin, Ch., & Lumsdain, P. L. (2012–2016). *The simplicial model of univalent foundations (after Voevodsky)*. arXiv:1211.2851.

Kolmogorov, A. N. (1991). On the interpretation of intuitionistic logic. In V. M. Tikhomirov (Ed.), *Selected works of A. N. Kolmogorov* (Vol. 1, pp. 151–158). Springer (Original publication in German: Zur Deutung der Intuitionistischen Logik. Math. Ztschr., 35, S. 58–65, 1932).

Krause, D., & Arenhart, J. R. B. (2016). *The logical foundations of scientific theories: Languages, structures, and models*. New York: Routledge.

Lawvere, F. W. (1963). *Functorial semantics of algebraic theories*. Ph.D., Columbia University, New York.

Martin-Löf, P. (1984). *Intuitionistic type theory* (Notes by Giovanni Sambin of a series of lectures given in Padua, June 1980). Napoli: BIBLIOPOLIS.

Martin-Löf, P. (1985). Constructive mathematics and computer programming. In *Proceeding of a Discussion Meeting of the Royal Society of London on Mathematical Logic and Programming Languages* (pp. 167–184).

Martin-Löf, P. (1996). On the meanings of the logical constants and the justifications of the logical laws. *Nordic Journal of Philosophical Logic, 1*(1), 11–60.

Martin-Löf, P. (1998). An intuitionistic theory of types. In G. Sambin & J. Smith (Eds.), *Twenty-Five Years of Constructive Type Theory: Proceedings of a Congress Held in Venice, October 1995, Oxford University Press* (pp. 127–172).

Muller, F. A. (2011). Reflections on the revolution at Stanford. *Synthese, 183*(1), 87–114.

Nagel, E., & Cohen, M. R. (1934). *An introduction to logic and scientific method*. London: Routledge.

Nourani, C. F. (2016). *A functorial model theory*. Palm Bay: Apple Academic Press.

Piecha, Th., & Schroeder-Heister, P. (Eds.). (2016). *Advances in proof-theoretic semantics* (Trends in logic, Vol. 43). New York: Springer.

Prawitz, D. (1974). On the idea of a general proof theory. *Synthese, 27*(1/2), 63–77.

Prawitz, D. (1986). Philosophical aspects of proof theory. *G. Fløistad, La philosophie contemporaine. Chroniques nouvelles, Philosophie du langage, 1*, 235–278.

Reus, B. (1999). Realizability models for type theories. *Electronic Notes in Theoretical Computer Science, 23*(1), 128–158.

Rodin, A. (2014). *Axiomatic method and category theory* (Synthese library, Vol. 364). Cham: Springer.

Rodin, A. (2017). Venus homotopically. *IfCoLog Journal of Logics and Their Applications, 4*(4), 1427–1446.

Rodin, A. (2018). On the constructive axiomatics method. *Logique et Analyse, 242*(61), 201–231.

Schreiber, U. (2013). *Classical field theory via cohesive homotopy types* (preprint). arXiv: 1311.1172.

Schroeder-Heister, P. (2016). Proof-theoretic semantics. In E. N. Zalta (Ed.), *The Stanford encyclopedia of philosophy* (Winter 2016 Edition). Metaphysics Research Lab, Stanford University. https://plato.stanford.edu/entries/proof-theoretic-semantics/.

Streicher, Th. (1991). *Semantics of type theory* (Progress in theoretical computer science). Boston: Birkhäuser Inc.

Sundholm, G. (1994). Existence, proof and truth-making: A perspective on the intuitionistic conception of truth. *Topoi, 13*, 117–126.

Suppe, F. (1989). *The semantic conception of theories and scientific realism*. Chicago: University of Illinois Press.

Suppes, P. (2002). *Representation and invariance*. London: College Publications

Tarski, A. (1941). *Introduction to logic and to the methodology of deductive sciences*. New York: Oxford University Press.

Tarski, A. (1956). On the concept of logical consequence. In *Logic, semantics, metamathematics: Papers from 1923 to 1938 (J. H. Woodger, translator), Oxford, Clarendon Press 1956* (pp. 409–420).

Tsementzis, D. (2017, Forthcoming). Meaning explanation for HoTT (preprint). *Synthese*. http://philsci-archive.pitt.edu/12824/.

Univalent Foundations Program. (2013). *Homotopy type theory: Univalent foundations of mathematics*. Princeton: Institute for Advanced Study. Available at http://homotopytypetheory.org/book/.

van Fraassen, B. C. (1980). *The scientific image*. Oxford: Oxford University Press.

Part III
Comparing Set Theory, Category Theory, and Type Theory

Chapter 10
Set Theory and Structures

Neil Barton and Sy-David Friedman

Abstract Set-theoretic and category-theoretic foundations represent different perspectives on mathematical subject matter. In particular, category-theoretic language focusses on properties that can be determined up to isomorphism within a category, whereas set theory admits of properties determined by the internal structure of the membership relation. Various objections have been raised against this aspect of set theory in the category-theoretic literature. In this article, we advocate a methodological pluralism concerning the two foundational languages, and provide a theory that fruitfully interrelates a 'structural' perspective to a set-theoretic one. We present a set-theoretic system that is able to talk about structures more naturally, and argue that it provides an important perspective on plausibly structural properties such as cardinality. We conclude the language of set theory can provide useful information about the notion of mathematical structure.

The authors wish to thank Ingo Blechschmidt, Andrew Brooke-Taylor, David Corfield, Patrik Eklund, Michael Ernst, Vera Flocke, Henning Heller, Deborah Kant, Cory Knapp, Maria Mannone, Jean-Pierre Marquis, Colin McLarty, Chris Scambler, Georg Schiemer, Stewart Sharpiro, Michael Shulman, Thomas Streicher, Oliver Tatton-Brown, Dimitris Tsementzis, Giorgio Venturi, Steve Vickers, Daniel Waxman, John Wigglesworth, and an anonymous reviewer for helpful discussion and comments, as well as an audience in Mussomeli, Italy at Filmat 2018. They would also like to thank the editors for putting together the volume. In addition, they are very grateful for the generous support of the FWF (Austrian Science Fund) through Project P 28420 (*The Hyperuniverse Programme*).

N. Barton (✉) · S.-D. Friedman
Kurt Gödel Research Center for Mathematical Logic (KGRC), Vienna, Austria
e-mail: neil.barton@univie.ac.at; sdf@logic.univie.ac.at

S. Centrone et al. (eds.), *Reflections on the Foundations of Mathematics*,
Synthese Library 407, https://doi.org/10.1007/978-3-030-15655-8_10

10.1 Introduction

Two approaches (and, as the current volume shows, maybe more) in current debates on foundations provide radically different perspectives on mathematical subject matter. The *set-theoretic*[1] perspective holds that all mathematical objects may be modelled in the sets, a formalism given in terms of a primitive membership-relation, and that this (in some sense) provides a *useful* foundation. The *category-theoretic* perspective, on the other hand, holds that all mathematical objects may be modelled by the kinds of *mapping* properties they have with respect to others. As we shall see below, the two provide somewhat different perspectives on mathematics. In light of this datum, a question which has sprung up in the literature is which foundation we should use for interpreting mathematics (assuming that a foundation is desirable at all).

One particular application to which category theory has been seen as especially suited to is elucidating the notion of *mathematical structure*.[2] A definition of mathematical structure is somewhat difficult to provide, but for the purposes of this paper we will take it that the existence of an isomorphism is sufficient for sameness of mathematical structure, and that this provides a useful way of getting at the notion (even if only partially).

This paper is directed at the following question:

Main Question. To what extent is (material) set theory a useful tool for discovering interesting facts about structures?

We will argue that set-theoretic language can be useful for conveying important structural information. In particular, we provide a theory of sets and classes which better respects isomorphism invariance, but nonetheless makes extensive use of the ambient material set theory. This is especially important if one holds that *cardinality* is a structural property; the theory we provide allows us to see how theories and cardinality interact (via a version of the Morley Categoricity Theorem).

Our strategy is as follows. First (Sect. 10.1) we briefly outline the set-theoretic and categorial approaches, and explain the tasks to which each is best suited. Next (Sect. 10.2) we examine the difficulties that some proponents of each foundation have seen for the other party, and provide some responses on behalf of each. We argue that they are not in themselves problematic, but raise a challenge for the advocate of set-theoretic foundations. We then (Sect. 10.3) present a theory of sets, urelements, and the classes that can be built over them. We argue that this language provides a modification of (material) set theory that better respects structural properties, whilst providing us with the resources to easily talk about notions like

---

[1] This term is slightly coarse since it is ambiguous between *material* and *categorial* set theories (we will distinguish these later). For those that know the difference between the two types of set theory, we mean "material set theory" by "set theory" until we make the distinction, and lump categorial set theories in with category-theoretic foundations for now.

[2] See, for example, Awodey (1996).

cardinality and how it interacts with structural notions. Finally (Sect. 10.4) we make some concluding remarks and present some open questions.

10.2 Two Perspectives on Foundations: Set-Theoretic and Categorial

In this section we'll explain the basic difference we see between category-theoretic and (material) set-theoretic foundations. We'll then refine both our set-theoretic and category-theoretic perspectives to give a better account of the subject matter they concern.

The distinction between the set-theoretic and category-theoretic perspective may be cast in different ways, however the most basic contrast is in how they approach the *representation* of garden-variety mathematical entities. The question is one of a perspective on which mathematics is about *objects* (and the internal membership-structure those objects exhibit), versus one on which mathematics is about particular kinds of *roles* a mathematical object can perform within a wider context. Under set-theoretic foundations, we focus on translating the language of a particular mathematical theory **T** into the language of set theory $\mathscr{L}_\in$ (consisting of only the language of first-order logic and a single non-logical symbol $\in$), and then find a model for **T** in the sets (given, of course, some antecedently accepted set theory). From the category-theoretic perspective, we would rather see what the essential *relationships* **T**-objects have to one another, and then try and capture these relationships through the notions of *arrow* and *composition*. To see this difference, a couple of examples are pertinent:

Example 1 (Singletons) In set theory the singleton of an object x is the one-element set $\{x\}$. Different singletons can look very different from the set-theoretic perspective; for example $\{\emptyset\}$ and $\{\beth_\omega\}$ are both singletons, but have very different properties (for example, their transitive closures look very different).

Conversely in category theory, we have the notion of a terminal object 1, where an object 1 is terminal in a category $\mathcal{C}$ iff there is one and only one arrow from any other $\mathcal{C}$-object to 1. Terminal objects can have a variety of different properties in different situations. For example, if we think of a partial order $\mathbb{P} = (P, \leq_\mathbb{P})$ as a kind of category (so there is an arrow from p_0 to p_1 just in case $p_0 \leq_\mathbb{P} p_1$), then if $\mathbb{P}$ has a maximal element it will be terminal. Interestingly, in the set-theoretic context, we can form a category **Set** consisting of all sets as objects and the functions between them as arrows. We then see that between any set A and any singleton $\{x\}$ there is exactly one function given by the rule $f(a) = x$ for every $a \in A$, and so the terminal objects of **Set** are exactly the singletons. Nonetheless, from the category-theoretic perspective it doesn't really matter which terminal object we talk about, since all terminal objects are isomorphic within any particular category. This contrasts sharply with the set-theoretic case where different singletons can have different interesting set-theoretic properties (such as identity of transitive closure).

Example 2 (Products) In set theory, we define the product $A \times B$ of two sets A and B by first picking a canonical choice of ordered pair, and then letting $A \times B = \{\langle a, b\rangle | a \in A \wedge b \in B\}$ (giving us the 'rectangle' of A and B). Products of more complicated objects are then defined component-wise. For example, the direct product of two groups $G = (D_G, *_G)$ and $H = (D_H, *_H)$ is defined as the following group:

$$G \times_{\text{Group}} H =_{df} (D_G \times D_H, *_{G\times H})$$

Where $*_{G\times H}$ is defined component-wise for $g \in G$ and $h \in H$:

$$\langle g_1, h_1\rangle *_{G\times H} \langle g_2, h_2\rangle =_{df} \langle g_1 *_G g_2, h_1 *_H h_2\rangle$$

Conversely, in category theory, a product of two $\mathcal{C}$-objects A and B is another $\mathcal{C}$-object $A \times B$ together with two $\mathcal{C}$-arrows $Pr_A : A \times B \to A$ and $Pr_B : A \times B \to B$, such that for any pair of arrows $f : C \to A$ and $g : C \to B$, there is exactly one arrow $\langle f, g\rangle : C \to A \times B$ making the following diagram commute:

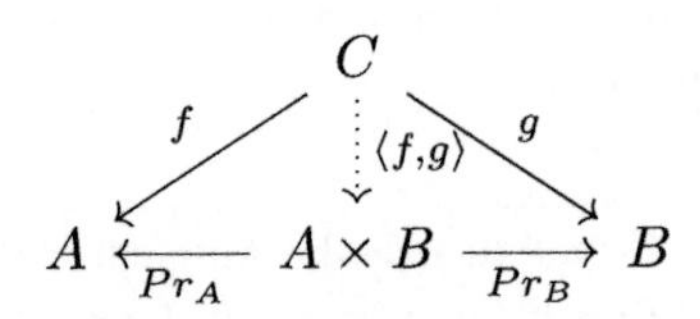

From the category-theoretic perspective, any objects that fulfil this role are a product (and, indeed, products are unique up to $\mathcal{C}$-isomorphism). In the case of sets and groups, the definition of set-theoretic product will (in **Set**) fulfil this role (using the relevant projections as Pr_A and Pr_B), as will the definition of direct product for groups (when we consider the category **Grp** consisting of all groups as objects and group homomorphisms as arrows). However, any other definition of product resulting from a different definition of ordered pair would equally well qualify as a category-theoretic product (and indeed, we could find a function between the sets involved in the two products, 'factoring' one product through the other).

The difference in the above cases is the following: In set-theoretic foundations, representations of mathematical objects are obtained by construction from the membership relation and a suitable coding. On the category-theoretic perspective, we simply state what arrow-theoretic properties an entity must have in order to fulfil the functions it does in normal mathematical reasoning.

The eagle-eyed and well-informed reader may regard the distinction between set-theoretic and category-theoretic as a false dichotomy, since one can give *categorial* theories of *sets* by axiomatising the external functional properties attaching to the objects in a (or maybe 'the') universe of sets. This is precisely what is done on many categorial[3] set theories such as on Lawvere's[4] *Elementary Theory of the Category of Sets* (**ETCS**), which we'll examine in a little more detail later. In this way, it seems like the term 'set theory' can be correctly applied to certain categorial theories. For this reason we make the following distinction:

Definition 3 (Informal) *Material* set theories are those that axiomatise a primitive notion of membership (denoted by '$\in$'), from which mathematical objects may be coded. *Categorial* set theory on the other hand provides a theory of sets formulated in the language of category theory, and on which objects are coded by systems of arrows resembling the usual properties we expect set-theoretic functions to have. Membership in categorial set theory is a defined relation, often[5] explained in terms of functions $x : 1 \to A$ (read: "x is a member of A"), since one can think of any such function (from a 'singleton') as 'picking' a member of A.[6]

To make our initial question more precise, we are interested in the extent to which *material* set theory tells us interesting information about structures. Where the term "set theory" occurs without a qualifier, we mean *material* set theory and take *categorial* set theory to be a part of *categorial* foundations.

Both set theory and category theory allow us to identify objects up to structural equivalence. Exactly how they do so is a tricky issue, and provides us with a third:

Example 4 (Isomorphisms) In set theory, working within first-order logic, we settle upon some relevant coding of *vocabulary* (i.e. function, constant, and relation symbols), of *structure* (usually as an ordered tuple), and *satisfaction on a structure* of formulas in this language (given by an interpretation function on a structure). We then say that two structures *in the same vocabulary* $\mathfrak{A}$ and $\mathfrak{B}$ are isomorphic iff there is a (coded) bijection between their domains such that for every relation symbol R of the vocabulary (respectively for function

[3]There is some dispute over the use of the term 'categorial' versus 'structural' when axiomatising sets in category theory. We use the term 'categorial' since we reserve structure-like terms for the philosophical notion of structure.

[4]See Lawvere (1965) for the original presentation, and Lawvere and McLarty (2005) for an updated version. A clean and concise presentation (with some informal commentary) is available in Leinster (2014).

[5]As with many notions in category theory, there are different arrow-theoretic ways of getting at the same idea. See, for example, Goldblatt (1984, Ch. 4) for some discussion.

[6]We are grateful to Michael Shulman and Dimitris Tsementzis for emphasising the importance of making this distinction.

and constant symbols) and for every finite sequence $\vec{a}$ from A, $R^{\mathfrak{A}}(\vec{a})$ iff $R^{\mathfrak{B}}(f(\vec{a}))$.

Importantly (an issue often glossed over in mathematics)[a] discussion of isomorphism only makes sense once the vocabulary (and some suitable coding thereof) has been chosen (on top of the coding-dependence of the set-theoretic analysis of bijection).

In category theory, however, the notion of isomorphism is dealt with by external arrow-theoretic properties. An arrow $f : X \to Y$ is an *isomorphism* (in a category $\mathcal{C}$) iff there is a $\mathcal{C}$-arrow $g : Y \to X$ in such that $g \circ f = Id_X$ and $f \circ g = Id_Y$ (i.e. composing the functions in either direction yields the identity morphism). Two objects are said to be isomorphic (in $\mathcal{C}$) iff there is an isomorphism between them. Importantly, the notion of isomorphism only makes sense *within a category*.

The treatment of isomorphism through a particular kind of arrow results in contexts in which the notion of set-theoretic and category-theoretic isomorphism come apart. For example, there are cases where we have category-theoretic isomorphisms that are not bijective homomorphisms (in the material set-theoretic sense). One such kind is when the relevant arrows are simply not functions, such as in the category of proofs which has sentences as objects and equivalence classes of proofs as arrows (so there is a single arrow $f : P \to Q$ when Q is derivable from P). Here, since isomorphisms are equivalence classes of proofs between equivalent sentences, we have isomorphisms that are not (properly speaking) bijections of any kind. In the context where there is a functorial relationship between the category and **Set**, however, there can be no non-bijective isomorphisms (since functors preserve isomorphisms).

However, in the case where there is no functor between the category and **Set**, this is possible. An interesting (yet complex) case is the homotopy category that has topological spaces as objects and homotopy classes of continuous functions as arrows. Here, the inclusion of the unit circle into the punctured plane is an isomorphism (its inverse is the radial projection map), which is not bijective. In fact, Freyd (1970) showed that this is not a concrete category (i.e. there is no nice faithful functor from this category to **Set**), which facilitates the consideration of non-bijective iso-arrows.[b,c]

---

[a]See Baldwin (2018, Ch. 1, esp. §1.2) for an argument that this is an often ignored distinction.

[b]We are grateful to Andrew Brooke-Taylor for bringing this example to our attention, and some further discussion of the issue. We would also like to thank Ingo Blechschmidt and Jean-Pierre Marquis for some further helpful conversations, in particular emphasising the pervasiveness of the non-concreteness phenomenon. For additional discussion see Marquis (2013), and for results showing how non-concreteness permeates see Di Liberti and Loregian (2018).

[c]A further simple (but somewhat silly) example is the following category which we define material-set-theoretically. The category has just one object $\{a, b\}$, and a single morphism defined by $f(a) = f(b) = b$. Here $f = Id_{\{a,b\}}$ (in the category), and so is trivially iso, but is nonetheless non-bijective.

The above example is important, since it shows that even the notion of *structural similarity* (as captured by the notion of isomorphism) is differently interpreted by the two perspectives. Thus, whether or not a property is 'isomorphism' invariant depends already on whether one holds one of the two perspectives to be privileged.

A second issue is whether or not there are notions of *sameness of structure* that are not underwritten by isomorphism. For example two categories $\mathcal{C}$ and $\mathcal{D}$ are said to be *equivalent* iff there are functors $F : \mathcal{C} \rightarrow \mathcal{D}$ and $G : \mathcal{D} \rightarrow \mathcal{C}$, such that there are natural isomorphisms $f : Id_{\mathcal{C}} \rightarrow G \circ F$ and $g : Id_{\mathcal{D}} \rightarrow F \circ G$. In other words, when composing the functors one does not get the *identity* back, but rather something *isomorphic*. This has significant consequences; for example the category of finite sets is equivalent to the category of finite ordinals, but the former has proper-class-many elements whereas the latter has only ω-many. Some authors (e.g. Marquis 2013) maintain that it is categorial equivalence, rather than isomorphism, that constitutes the 'correct' notion of sameness of structure for categories (a fact also supported by the development of the subject).

These cluster of issues show that there are plausibly notions of structure that are not underwritten by isomorphisms as understood through bijection. In order to focus discussion, we will focus on the notion of 'sameness of structure' as given by isomorphisms that are bijective (what (Marquis 2013) terms the 'extensional' perspective). It is an important open question (one we shall identify in Sect. 10.5) how the current work might be modified to handle different cases.

It should also be noted that talk of 'objects' is dispensable from the category-theoretic perspective. Really, by taking the notion of domain and co-domain as part-and-parcel of the arrow, we could just speak purely in terms of arrows and composition. Material set theory and category theory thus represent two different perspectives on the nature of mathematical subject matter; on the one hand, we might think of mathematical objects as constructed and determined by their internal membership-structure, and on the other we might think of them as determined (up to isomorphism) by their relationships to other entities, and the role they play in the category as a *whole*.

This underlying difference in perspective represents two sides of a long-standing[7] philosophical divide: Should we think of the subject matter of mathematics as given by individual objects with particular intrinsic relations determining their properties, or should we rather think of the subject matter of mathematics as concerned with purely structural properties (i.e. those invariant up to isomorphism)? The material set-theoretic and categorial perspectives are interesting representatives of different sides of this divide (though, as we shall see, issues are more subtle than they first appear).[8]

---

[7] At least since Benacerraf (1965).

[8] A salient third option (especially given the topic of the current volume) is Homotopy Type Theory. Here type theory endowed with a homotopy-theoretic interpretation is employed, providing a foundation that meshes elegantly with category-theoretic methods. See the excellent (The Univalent Foundations Program 2013) for discussion.

10.3 Objections: Refining the Perspectives

As it stands, however, there are puzzles for each conception of foundations. In this section, we explain some of the complaints that have been made about the different foundational viewpoints and argue that these are easily resolvable. We'll argue that this suggests a possible route of inquiry for the friend of set-theoretic foundations; to modify their language in order to better respect isomorphism invariance and structure.

10.3.1 Objections to Categorial Foundations and Schematic Types

One supposed 'problem' for the friend of category-theoretic foundations concerns its *subject matter*. What, exactly, is category theory *about*? For, as it stands, category theory merely defines particular kinds of *algebraic* structure. The discipline seems to be of a piece with algebraic enterprises such as group theory or other areas of abstract algebra. One begins by laying down conditions on what a system of arrows must satisfy in order to be a *category* (existence of identity and composition morphisms, and associativity of composition). This defines an algebraic structure much like that of group (in fact, there is a corresponding abstract algebraic structure for category that is slightly more general than that of *group*; namely then notion of being a *monoid*), which can then be made more specific with additional constraints. For example, insisting that particular diagrams exist and commute in a category yields the definition of a *topos*: a Cartesian closed category with a subobject classifier. This kind of category is very useful for studying the algebraic properties instantiated by various logical and mathematical systems, and while it is exceptionally rich in structure, it still (in the spirit of category theory) just corresponds to particular algebraic properties that a system of arrows can instantiate. Hellman sums up this thought:

> ... this theory [i.e. category theory] itself is presented *algebraically*, via first-order 'axioms' only in the sense of *defining conditions* telling us what a *category* is, together with further ones defining *topoi* of various sorts. As such these 'axioms' are like the conditions defining a group, a ring, a module, a field, etc. By themselves they *assert nothing. They merely tell us what it is for something to be a structure of a certain kind.* (Hellman 2006, p. 134)

Hellman's point (an important one) is that the axioms of category theory (even when expanded to isolate categories of more complex varieties) make no *existential claims* (this is what Hellman means by saying that the axioms "assert nothing") in that they specify what it is for a system of objects to satisfy certain axioms, without asserting that anything satisfying them actually exists.

Some (including Hellman 2006) have taken this as an *objection* to category-theoretic foundations. An appropriate foundation for mathematics should state that some objects exist and that mathematics can be interpreted within this structure, thereby laying a framework ontology on which mathematics can be built. Thus category theory appears to contrast with the usual set-theoretic foundations, where the axioms of Infinity, Power Set, and Replacement all make existential claims within **ZFC**, and many other axioms extending **ZFC** also make existential claims.[9]

This objection should not trouble the friend of category-theoretic foundations. One salient response (made by Mclarty 2004) is that no-one has ever proposed the axioms of *category theory* as a foundation, the proposal is rather to assert that some topos or other exists and mathematics either can or should be interpreted there. Good examples here are categorial theories of sets (such as **ETCS**) or attempts to axiomatise a category of all categories (e.g. **CCAF**).

Moreover, we might also think that Hellman's objection simply misses the mark. His remarks reveal something about the general practice of category theory: It *is* an algebraic discipline, no matter whether it *can* be modified to yield assertory content, as McLarty suggests. When practising category theory, we care only about whether we have the relevant relations between objects, and this *does* contrast with set theory where (largely speaking) we are interested in the properties of some specific structure (namely the cumulative hierarchy).[10] This feature of the two frameworks is further witnessed by attitudes to categoricity. In a categoricity proof, we aim to show that a certain theory has just one model (up to isomorphism). In the context of set theory (or indeed number theory and analysis) the project of providing a categoricity proof makes sense; we wish to see that our axiomatisation has been successful and we have (given the determinacy of the concept) pinned down a single structure up to isomorphism.[11] In the case of number theory and analysis we have proofs of full categoricity (by the work of Dedekind), and in the case of set theory we have quasi-categoricity: We can show that $\mathbf{ZFC}_2$ augmented with anti-large cardinal axioms can be fully categorical, and any two $\mathbf{ZFC}_2$ structures are either isomorphic or one is isomorphic to an initial segment of the other.[12] In the case of category theory though, to attempt such a proof for a categorial theory would be an absurd endeavour, the whole *point* of category theory is to isolate structural properties that can be shared by radically non-isomorphic structures.

Thus categories demand no single place to be interpreted, and the complaint that category theory fails to delimit a determinate range of objects is misplaced. The friend of set-theoretic foundations, for example, will regard it as of a piece with

---

[9]Good examples here are so called *large cardinal axioms*, as well as *forcing axioms*, and *inner model hypotheses*.

[10]Here we are playing slightly fast-and-loose with debates in the foundations of set theory; under a natural interpretation of Joel Hamkins' multiverse perspective, set theory also should be understood as purely algebraic. See Hamkins (2012) for the original presentation of this view and Barton (2016) for an argument to the effect that this results in a purely algebraic interpretation.

[11]The exact dialectic import of a categoricity proof is something of a vexed question, see Meadows (2013) for discussion. An argument that the quasi-categoricity of $\mathbf{ZFC}_2$ shows that our axiomatisation has been successful is available in Isaacson (2011).

[12]The original proof of this is available in Zermelo (1930), and is subsequently tidied up in Shepherdson (1951, 1952).

group theory; if category-theory has any subject matter at all, it is the category-theoretic structure that can be instantiated in various ways in the sets (and so she should countenance category theory as a useful foundational language, even if it is not her favourite foundation). There is no pressure to find '*the*' structure of '*the*' category-theoretic world; the discipline is designed to be flexible and resist such a characterisation.

A friend of category-theoretic foundations might thus regard the subject matter of mathematics as fundamentally algebraic, category theory as providing a good axiomatisation of this perspective, but nonetheless resisting the characterisation of a unique concrete subject matter. If category theory has a subject matter at all, then it is a *purely structural* one.

An immediate and difficult question is how we should think of this category-theoretic structural subject matter. Landry and Marquis provide the following interesting idea:

> A category, too, is neither a privileged abstract kind of system nor is it an abstract Fregean "structure" qua an "object": it is a Hilbertian style abstract structure qua a schematic type, to be used as a framework for expressing what we can say about the shared structure of the various abstract kinds of structured systems in terms of 'having' the same type of structure. (Landry and Marquis 2005, p. 35)

The thought here is to think of categories as providing structure as a 'schematic type', rather than a 'particular structure'. Of course, it bears explaining what a 'schematic type' is. One kind of schematic type is well known to mathematical logicians—the notion of first-order theory. These, if they can be instantiated in any infinite context, can be instantiated in many infinite contexts. We leave it open whether non-first-orderisable content can be expressed categorially. If higher-order content (with a version of the full semantics) can be encoded in a categorial language, then it is at least *possible* that we might determine a structure up to (set-theoretic) isomorphism categorially. Whether or not this is possible we leave open; for now, we note that the ability to systematise relationships across non-isomorphic contexts *just is* one of the main strengths of category theory, and many proponents of categorial foundations *do* see category theory in this light (e.g. Mac Lane 1986).

Thus we take the target of category theory to be showing the basic relationships objects have to have to one another to fulfil their functional roles. In order to understand better the notion of *schematic type*, it is useful to return to the analogy with group theory. While it makes sense to speak of 'the group-theoretic structure', there is not a *single way* the group-theoretic structure can be instantiated, rather it admits of satisfaction in multiple different ways and contexts (and indeed this is one of the reasons *why* abstract algebra has been so fruitful in contemporary mathematics). This is much the same for categories, which provide a useful framework for systematising these relationships. We thus provide the following:

Definition 5 (Informal and philosophical) A *schematic type* is a system of relationships that can be instantiated in many different non-isomorphic contexts.

Viewing category theory as the appropriate theory for formalising schematic types, we see that the problem of 'subject matter' is not really a problem at all. Rather, category-theoretic foundations provide a language and context in which to study algebraic relationships, and for this reason precisely *resist* the identification of a concrete subject matter.[13] Categories correspond to theories of mappings and can be multiply instantiated throughout concrete systems of objects, and there is no pressure to identify a unique subject matter or make it assertory in nature.[14]

10.3.2 Objections to Set Theory and the Combinatorial Perspective

In this subsection, we'll delve into some of the criticisms of material set theory. We'll show that when set theory is understood as applying to an appropriate subject matter (namely providing an analysis of possible combinations of objects) the objections fail to gain traction. We'll therefore suggest that a methodological pluralism is an attractive attitude to foundations. Before moving onto the final section, we'll identify that a possible line of inquiry for the friend of set-theoretic foundations is to provide a modification of her language that better respects isomorphism invariance.

The objections to set-theoretic foundations come in two broad kinds, as Feferman (speaking about Mac Lane 1971) explains:

> Two views are intermixed in (Mac Lane 1971) as to current set-theoretical foundations, namely that (i) they are inappropriate for mathematics as practised, and (ii) they are inadequate for the full needs of category theory. (Feferman 1977, p. 149)

Our strategy will be the following. We expand on Feferman's two dimensions, articulating the different objections we find in the literature. For reasons that will

---

[13]Moreover, one might think that category theory formalises these schematic types in a way that highlights privileged conceptual routes (such as when we know that a particular property is *universal*). Marquis, for example, writes:

> "The point I want to make here is extremely simple: category theory, and not just its language, provides us with the proper code to represent the map of mathematical concepts." (Marquis 2017a, p. 92)

[14]A second objection, one that we will not consider here, is the point raised by Mathias (2000, 2001) that category theory lacks the logical strength to discuss certain strong statements of analysis that relate to large cardinal axioms. While the objection merits a response, we set this aside for several reasons: (1) research is ongoing here, and it is unclear that category theory *cannot* do the job, (2) there are, in any case, logically strong category-theoretic statements (see below), and (3) the possible responses to the objection do not help us elucidate the philosophical role being played by category theory in terms of schematic types. See also Ernst (2017) for some discussion of these issues, as well as a general survey of the comparisons between categorial and set-theoretic foundations.

become apparent (we find the former dimension to be the more challenging of the two), we deal with these in reverse order, starting with inadequacy. As we go, we will provide responses from the set-theoretic standpoint. We do, however, have to be careful about dialectical strategy. In 'defending the set-theoretic viewpoint' we could be doing one of (at least) two things:

1. We could be arguing that, despite the category-theoreticians best efforts, set theory is still the best foundation for discussing mathematical structure.
2. Slightly weaker, we could contend that despite arguments to the contrary, set theory is still interesting as a foundation as it has *plenty to tell us* about mathematical structure.

We wish to emphasise that *it is this latter claim* we wish to support. We wish to claim that despite many criticisms in the literature, set theory can still provide interesting information about how certain structural properties are instantiated.

10.3.2.1 Inadequacy

The problem of inadequacy is roughly the following: Set theory does not provide enough of something (either objects or information). The key issue is raised by Mac Lane:

> Our fundamental observation is just this: There is an appreciable body of results about categories... but the received methods of defining categories in terms of sets do not provide any one single context (i.e. any one model of a standard set theory) within which one can comprehensively state these results. (Mac Lane 1971, p. 235)

The fundamental idea is the following: It is undeniable that the methods of category theory have provided a versatile method for modern mathematics. This raises the question for the friend of set-theoretic foundations: "Given that category theory provides structural information, what sets should we interpret category theory as about?".

Mac Lane's point is that there is no single context in which we can interpret category theory unrestrictedly. This is visible in two related but distinct dimensions: (i) Which *model*[15] we should take to found category theory, and (ii) Which *categories* we should expect set theory to found.[16]

[15]We use the term 'model' in a loose and informal way here, and intend it to apply to possibly proper-class-sized structures. For example, we will at least *allow* $(L, \in)$ as a model, even though it is proper-class-sized.

[16]In the quotation above, Mac Lane is specifically interested in the first point we consider. However, the intuition expressed transfers naturally to other objections he makes, as outlined below.

What model?

The first issue concerns exactly what the relevant model should satisfy. Simply put, the widespread independence phenomenon in set theory has challenged the classical idea that there is a single maximal universe of sets in which we may interpret all mathematical discourse. This is discussed by Mac Lane:

> These results, and others too numerous to mention, show that many interesting Mathematical questions cannot be settled on the basis of the Zermelo-Fraenkel axioms for set theory. Various additional axioms have been proposed, including axioms which insure the existence of some very large cardinal numbers and an axiom of determinacy (for certain games) which in its full form contradicts the axiom of choice. This variety and the undecidability results indicate that set theory is indeterminate in principle: There is no unique and definitive list of axioms for sets; the intuitive idea of a set as a collection can lead to wildly different and mutually inconsistent formulations. On the elementary level, there are options such as **ZFC**, **ZC**, **ZBQC** or intuitionistic set theory; on the higher level, the method of forcing provides many alternative models with divergent properties. The platonic notion that there is somewhere the ideal realm of sets, not yet fully described, is a glorious illusion. (Mac Lane 1986, p. 385)[17]

Since our expertise is primarily in higher-set theory (i.e. **ZFC** and its extensions) and independence, we approach the issue from that perspective. As is well-known, there are many set-theoretic sentences independent of our canonical set theory **ZFC** (e.g. CH). Mac Lane takes this to show that there is no one notion of 'set' and hence no one place that category theory can be interpreted. We have two responses to this argument:

Response 1 This explicitly takes a stand on the status of certain questions in the philosophy of set theory. In particular, it turns on how many universes of set there are (or at least how many distinct but legitimate concepts of set there are). While the independence phenomenon is certainly challenging, this does not mean that there are multiple 'meanings' to the word "set". Thus, for the theorist who simply rejects the claim that the independence phenomenon indicates semantic or ontological indeterminacy, the objection gains no traction (without further argument).

Response 2 Even if we allow the existence of different universes or concepts of set, Mac Lane's criticism is subject to a tu quoque. This is because, as we explained above, category theory by its nature *does not demand* a single context for interpretation (in fact quite the *reverse*). Rather, we argued, category theory should be understood as providing a uniform language in which to study algebraic properties and schematic types. Thus to insist on a single model or axiomatisation in which category theory should be interpreted is to impute content to it that is simply not there. Thus, insofar as this is a problem for set theory at all, it is also one for categorial foundations.

This allows a quick response to Mac Lane's objection: Even if there are multiple set-theoretic concepts or universes, and no overarching context, this does not matter.

[17]Similar remarks are made repeatedly in Mac Lane (1986), cf. pp. 359–360, 373.

Wherever we study category-theoretic properties set-theoretically (discussion of particular set-theoretic interpretations is provided below), we know that our results will transfer to the alternative cases via the schematic properties of category theory. It is enough for us to study category-theoretic structure set-theoretically to find *one* set-theoretic structure exemplifying the relevant schematic type. For the purposes of set-theoretic foundations, we do not need to find *the* set-theoretic subject matter corresponding to category theory.[18]

Which Categories?

The second problem of inadequacy concerns what one has *within* a particular context. Simply put, category theory seems to speak about structures that are proper-class-sized, and so do not have any set-theoretic representative. An obvious example here is **Set**, the category of all sets that has as arrows set-theoretic functions (this can be given direct category-theoretic axiomatisation by **ETCS** or its extensions).[19] There are, however, many such categories (e.g **Grp**, **Top**, **Fld**, etc.).

There are two main strategies for overcoming this problem. The first is to posit the existence of *Grothendieck universes* and interpret category theory there. More formally:

Definition 6 A *Grothendieck universe* is a set U with the following properties:

(i) U contains the set of all natural numbers.
(ii) U is transitive.
(iii) U is closed under the formation of pair sets.
(iv) U is closed under the formation of power sets.
(v) U is closed under set-theoretic union.
(vi) If $x \in U$, $f : x \to y$ is a surjection, and $y \subseteq U$, then $y \in U$.

We can then interpret category theory as concerned with any such universe. For instance, **Set** can be interpreted as concerned with the sets in some U (let's call this category $\mathbf{Set}_U$), and this (along with any functor categories) is a perfectly legitimate object formed in the stages of the cumulative hierarchy above U.

Grothendieck himself (in proposing a set-theoretic interpretation of category theory) suggested the axiom that there should be an unbounded sequence of these universes. In fact, being a universe is clearly equivalent to being V_κ where κ is an inaccessible cardinal, and so the proposal comes down to interpreting category theory within any one of an unbounded sequence of inaccessible universes.

The standard objection to this strategy is that it apparently 'prevents' considering perfectly acceptable categories, such as the *actual* category of *all* sets. Given any

[18] A similar point is made in Maddy (2017). For some other remarks on what we would like from set-theoretic foundations, see also Maddy's contribution to the present volume (Maddy).

[19] The topos axiomatised by **ETCS** is that of a well-pointed topos with a natural number object and satisfying the categorial version of the Axiom of Choice.

such interpretation of category theory, there are sets outside of that interpretation. But (so the objection goes) category theory is about any and all sets instantiating the relevant category-theoretic interpretation. Any such restriction seems ad hoc. Muller expresses the point as follows:

> Any stipulation to the effect that the category-theoretician is only allowed to grab at some fixed set whereas outside this set there are more sets, so that he is not permitted to grab at all of them, is artificial and barks at his explicit intentions. The category- theoretician has every right to refuse to dance to the cardinality tunes [of] the set-theoretician whistles. Category-theory is about form & structure, irrespective of how much & how many; it 'only' wants to have everything which is available. The category-theoretician means *all* sets when he makes the category **Set** of all sets, period. Set-theories which cannot accommodate *this* are flawed. (Muller 2001, pp. 11–12)

A different strategy then is to allow proper-class-sized categories (so-called 'large' categories), and adopt a two-sorted class-theoretic language and theory (such as **NBG** or **MK**) in providing a category-theoretic interpretation. The problem here is that often category-theorists will consider functor categories between two categories. Taking two categories $\mathcal{C}$ and $\mathcal{D}$, it is natural to consider the category $\mathcal{D}^{\mathcal{C}}$ consisting of all functors from $\mathcal{C}$ to $\mathcal{D}$ as objects, and natural transformations between functors as arrows. Such a category, however, is often one type higher than both $\mathcal{C}$ and $\mathcal{D}$. In the case then when both $\mathcal{C}$ and $\mathcal{D}$ are large, even *with* proper classes we are not guaranteed the existence of a class-theoretic representative concerning $\mathcal{C}^{\mathcal{D}}$ (normally proper classes cannot be members).[20] The issues concerning both interpretations are summed up as follows:

> Using universes, all the functor categories are there, but there is no category of all groups. Using Godel-Bernays, one has the category of all (small) groups, but only a few functor categories.[21] (Mac Lane 1971, p. 235)

What should we take from this? Again we hold that these objections fail to gain traction:

Response 1 A simple point, but one based on a non-trivial theorem, is that for certain categories (in particular the category of all reflexive graphs), it is not clear that the requirements Mac Lane wishes to place on interpretations of category theory are consistent. Specifically Ernst (2015) shows that there are restrictions on what categories one can have. The proof proceeds by considering a version of Cantor's Theorem in the category of all reflexive graphs and shows that certain desirable

[20] Of course, the material set-theorist might just accept the existence of proper-classes, hyper-classes, hyper-hyper-classes and so on. This is naturally interpretable in an ontology on which every universe can be extended in height, but there is also a question of whether the believer in one maximal unique universe of sets could also make use of nth-order hyper-classes. Normally it is assumed not, but this question remains philosophically open.

[21] Given the topic of the present volume, there is an interesting question as to the extent to which this difficulty is avoided in homotopy type theory. We thank Dimitris Tsementzis for the suggestion that this difficulty could possibly be overcome in this foundation.

conditions on the existence of categories are jointly inconsistent.[22] Thus we cannot simply naively insist on the existence of any category whatsoever without some restrictions.

Response 2 The problem implicitly takes a stand on issues in the philosophy of set theory, though in the opposite direction to the previous section. For, in the case where we think there is no maximal conception of the notion of *ordinal*, and rather that any universe of set theory can be extended in height, we are always implicitly restricted to a particular bounded universe of sets anyway. Thus, to the theorist who holds that there are no 'maximally high' universes, the objection fails to gain any traction.[23]

Response 3 Even if we *do* allow that there can be 'maximally high' universes of set theory, the objection again fails to account for the schematic nature of category theory. Given this algebraic character, when a category theorist asks us to consider "The category **Set** which has as objects all sets and arrows all functions" this should be understood as shorthand for communicating various category-theoretic properties. Some of these are captured by first-order axiomatisations such as **ETCS** and its extensions, but again, they are about a *schematic type* (in particular a *first-order* one) rather than a particular concrete subject matter. So for studying this schematic type, it is enough that we have *just one* structure exemplifying the schematic type. Results proved about this schematic type can then be exported to other contexts (and other structures instantiating the category-theoretic properties), the results are about the schematic type and *not* the concrete instantiation of it in the sets.

Consider the case where we have both the Grothendieck-universe and class-theoretic interpretations available to the set theorist. More precisely, suppose she believes the following: (1) There is a single unique universe of sets, (2) There are unboundedly many inaccessible cardinals, and (3) There is a satisfactory interpretation of Morse-Kelley class theory over V. Now, consider the category theorist's consideration of **Set**, and how this relates to the set theorist's universe. The category-theoretic structure of **Set** is multiply instantiated, both by each individual V_κ where κ is inaccessible, and also V (as well as many other structures besides, some of them countable[24]). Now suppose we consider some 'super-large' functor category $\mathbf{Set}^{\mathcal{C}}$ (for some category $\mathcal{C}$). The $\mathbf{Set}^{\mathcal{C}}$ schematic type will be instantiated by the various V_κ with the sets above them, but not by V (since there are no levels above V). But this does not matter, any properties proved about the schematic

[22]See Ernst (2017), Maddy (2017) and Maddy for further discussion of the significance of this result.

[23]One might even think, in the opposite direction, that some distinction between 'small' and 'large' is essential for making sense of certain category-theoretic results. See Shulman (2008) for arguments to this effect, and Ernst (2017) for some additional discussion.

[24]An example of such a structure would be the Skolemisation and Mostowski Collapse of any set-theoretic structure satisfying an appropriate amount of set theory.

type **Set** using the schematic type $\mathbf{Set}^{\mathcal{C}}$ can just be exported back to V, the proof concerning **Set** just depends on there being *some* set-theoretic counterpart in which $\mathbf{Set}^{\mathcal{C}}$ is instantiated, not that *every* instantiation of **Set** has a corresponding $\mathbf{Set}^{\mathcal{C}}$ representative. While it is the case that when a category theorist states a theorem of the form "**Set** has an extension to $\mathbf{Set}^{\mathcal{C}}$" can only be interpreted to concern the small instantiations, this does not harm the results about the schematic type **Set**, and the fact that any results proved on the basis of $\mathbf{Set}^{\mathcal{C}}$ about *that* structure can perfectly well be exported back to V. This is much like the case with first-order **ZFC**, it would be bizarre to claim that results proved about **ZFC** using extensions in the model theory of sets are inapplicable to V because V lacks extensions.[25]

A natural rejoinder is that occasionally category theorists will consider explicitly *large* categories like **SET** (i.e. the category of all sets) rather than just **Set** (interpreted as the category of small sets in the first universe). Nonetheless, exactly the same considerations apply concerning schematic types. Even if one insists that there is a distinction between 'small' and 'large' sets, one can still have **SET** instantiated in some small structure, it is just that structure does not *think* it is small. These considerations are familiar from the set-theoretic framework; one can easily have a particular V_κ satisfying the claim that there are proper-class-many cardinals of some kind Φ, without V_κ witnessing that there are *literally* (i.e. in V) proper-class-many such cardinals. One just requires that the Φ-cardinals are unbounded in V_κ for there to be a universe satisfying this property. Indeed, as above, any conclusions based on the theory **ZFC**+ "There is a proper class of Φ-cardinals", made by considering the extension of a model thereof can be exported back to V (on the assumption of course that V does contain such a proper class). Similarly, one just needs a universe containing an inaccessible (in fact if the schematic content of category theory is only first-order, one needs much less) for there to be set-sized set-theoretic contexts in which there is a meaningful distinction between small and large categories capable of instantiating the relevant algebraic content. To argue that set theory fails to provide an appropriate surrogate for $\mathbf{SET}^{\mathcal{C}}$ is to impute non-algebraic content to category theory which is quite simply not there.

10.3.2.2 Inappropriateness

The inappropriateness dimension of Feferman's taxonomy concerns set theory giving us *too much* of something (either objects or information). We'll see that while these problems are also resolvable, an additional line of inquiry is suggested by the complaint that set theory provides too much non-isomorphism-invariant information; namely to modify set theory so that the language respects isomorphism invariance.

[25]There may, nonetheless, be certain *philosophical* considerations here, as well as technical issues concerning how much *higher-order* reasoning we can capture using extensions. See Barton (2018) and Antos et al. for discussion.

Logical Strength

The first issue concerns the logical strength of set theory. For the practical purposes of founding mathematics, so the argument goes, we do not need anything like the strength of **ZFC** set theory. Landry and Marquis record this sentiment:

> Second, it is fair to say that category theorists and categorical logicians believe that mathematics does not require a unique, absolute, or definitive foundation and that, for most purposes, frameworks logically weaker than **ZF** are satisfactory. Categorical logic, for instance, is taken to provide the tools required to perform an analysis of the shared logical structure, in a categorical sense of that expression, involved in any mathematical discipline. (Landry and Marquis 2005, p. 19)

We do not wish to disagree that large portions of mathematics do not require the logical strength of **ZFC** set theory. However, we do wish to make two rejoinders:

Response 1 First, set theory does not aim at being 'minimal' in any sense. Rather, we wish to provide the most generous theory possible (often understood through maximising consistency strength) to be used in interpreting any conceivable mathematics we might consider. So, while the objection might be convincing to a theorist who has a penchant for minimising logical strength, it fails to be convincing to the friend of set-theoretic foundations.

Response 2 Second, we have another tu quoque here: There are interesting category-theoretic principles that turn our to have significant large cardinal strength. Bagaria and Brooke-Taylor, for example, note the following (in an article on colimits and elementary embeddings):

> Many problems in category theory, homological algebra, and homotopy theory have been shown to be set-theoretical, involving the existence of large cardinals. For example, the problem of the existence of rigid classes in categories such as graphs, or metric spaces or compact Hausdorff spaces with continuous maps, which was studied by the Prague school in the 1960's turned out to be equivalent to the large cardinal principle now known as Vopěnka's Principle... Another early example is John Isbell's 1960 result that $\mathbf{Set}^{\mathbf{Op}}$ is bounded if and only if there is no proper class of measurable cardinals. (Bagaria and Brooke-Taylor 2013, p. 1)

A key point to attend to in the above is that these are not category-theoretic principles that were dreamt up by set theorists. These are principles that were naturally studied by category theorists that turned out to not just be independent from **ZFC**, but also have substantial large cardinal strength (Vopěnka's Principle is quite high in the large cardinal hierarchy). Moreover, with certain additional assumptions one can find models of material set theories like **ZFC** in categorial set theories like **ETCS**.[26] In virtue of this, the claim that set theory is somehow

[26]For instance one way to do this is to find arrow-theoretic trees in a model of strengthened **ETCS** that correspond to the relevant membership trees required to build a model of **ZFC**. We are grateful to Michael Shulman for discussion here, and directing us to his useful Shulman (2010) and Osius (1974).

unfavourably distinguished by its logical strength when compared to category theory (in all its guises) seems dubious.

Isomorphism Invariance

The second problem of inappropriateness concerns the earlier discussed fact that set theory makes decisions about non-isomorphism invariant facts. The key issue is that a central practice in set-theoretic foundations involves specific choices of 'canonical' representatives.

For example, in the earlier discussion of products, the exact object that is eventually selected as 'the' product will vary depending on numerous conventional stipulations. We might, for example, represent the ordered pair $\langle a, b\rangle$ along the lines of Hausdorff as $\{\{a, 1\}, \{b, 2\}\}$ (rather than the usual Kuratowski definition: $\langle a, b\rangle =_{df} \{\{a\}, \{a, b\}\}$) resulting in a different choice of product. This then raises the following question: If set theory is meant to tell us *what* mathematical objects exist, then what is the fact the matter about *which* definition of product is the *actual* product?

The problem was noticed in the philosophical literature at least as early as the seminal (Benacerraf 1965). There, he presses this problem of non-isomorphism invariant choices to be made via the existence of 'junk' theorems, where a theorem is 'junk' when it concerns non-isomorphism invariant properties. For example, is it true that $5 \in 7$? A natural answer is "Yes"; the canonical choice of ordinals is the von Neumann definition, and under that definition it is the case that $5 \in 7$. However, under the Zermelo definition, it is not the case that $5 \in 7$. So, if we think that mathematical objects 'are' sets (or, as Benacerraf points out, any objects) what is the fact of the matter concerning whether or not $5 \in 7$? Since the truths of number theory are invariant under domain (we just need some things that have the standard natural number structure under the relevant relations of the required vocabulary for arithmetic to be true in that context) there seems to be no good response; whatever objects we pick as 'the' natural numbers, they satisfy the same arithmetic sentences.

The simple response is that the many (if not most) set theorists do not subscribe to the heavy-duty set-theoretic reductionism required to generate the problem. Rather, most friends of set-theoretic foundations (or, at least, this is the line we shall present here), take set theory to be a device of *representation*. Using the membership relation and axioms of first-order logic we are able to *interpret* mathematical claims as ones about sets, with axioms telling us how these objects can be *combined* to yield other mathematical properties, providing a context in which different mathematical claims can be interrelated.

Claims like "$5 \in 7$" just *do not make sense* for the friend of set-theoretic foundations until we have settled on a particular interpretation of number theory. If we pick the von Neumann ordinals, then the question of whether $5 \in 7$ corresponds via our chosen interpretation to whether $5 < 7$ in number-theoretic terms, and so is obviously true (in fact trivially so). If we had picked a different representation, however, it may have been false (as is the case on the Zermelian conception of finite ordinals).

This has implications for the purposes to which set theory is put in foundations. The language is excellent for studying what kinds of mathematical properties are compossible, and what kinds of objects are required to find an interpretation of a piece of mathematics. Hence, the default context for studying independence results (and their implications) is models of set theory,[27] and the indexing of consistency strength is accomplished using set-theoretic principles. If one wishes to know whether it is possible to have one mathematical property with or without another, one studies the models of set theory in which they have interpretations. If one wishes to know how much logical strength is implied by the adoption of a particular mathematical system, a standard technique is to find a model of the principle in the sets. However, once a model has been found, few set theorists (if any) would suggest that the mathematician should change their practice and adopt the (highly baroque) set-theoretic language. The interpretation has been given, and this acts as a certificate that the original (and probably more fluid) language is in good working order.[28]

This has implications for the kind of applications that we see for set theory to mathematical structure. Rather than seeing it as a language and framework for the *working* mathematician, it should rather be used in finding interpretations of the working languages and comparing them. Structurally, this might have important consequences. One might, for example, see *cardinality* as an important structural property. Set theory then provides us with a useful framework in which schematic types of certain kinds and cardinality interact, to yield helpful structural information. A good example here is the celebrated:

Theorem 7 (The Morley Categoricity Theorem) *Suppose that a countable first-order theory* **T** *has exactly one model up to isomorphism in a single uncountable cardinal. Then it has one model (up to isomorphism) in every uncountable cardinal.*

The standard proof of this proceeds against a background of material set theory. Assuming that one does hold that cardinality is a structural property, it yields information about how first-order schematic types and cardinality interact,

[27] This said, there are category-theoretic options here. See Bell (2011).

[28] Vladimir Voevodky himself was clear about this role for **ZFC** with respect to Homotopy Type Theory. See, for example, his abstract for the 2013 North American Meeting of the Association of Symbolic Logic, where he says:

> Univalent foundations provide a new approach to the formal reasoning about mathematical objects. The languages which arise in this approach are much more convenient for doing serious mathematics than **ZFC** at the cost of being much more complex. In particular, the consistency issues for these languages are not intuitively clear. Thus **ZFC** retains its key role as the theory which is used to ensure that the more and more complex languages of the univalent approach are consistent. (Voevodsky 2014, p. 108)

We are grateful to Penelope Maddy for bringing this to our attention.

specifically if one has a first-order schematic type **T** (this could even be given categorially) it provides conditions that tell us when there is only one way (up to isomorphism) that **T** could be instantiated in a structure of a particular size.

We thus make the preliminary conclusion; though material set theory is baroque and choices must be made about canonical representatives in a fashion that is not isomorphism invariant from the perspective of certain vocabularies (e.g. number theory), it nonetheless provides a useful perspective for stating how schematic types interact with other structural properties (namely cardinality). We thus subscribe to a methodological pluralism in foundations; category theory is the appropriate theory for explaining how schematic types interact with one another, but set theory is the appropriate language for explaining how schematic types interact with concrete systems of objects.[29]

This response is satisfactory as far as it goes. However, a challenge remains for the set theorist: Could we possibly factor out the use of canonical representatives to yield a conception of set on which we are able to use the combinatorial power of set theory whilst considering isomorphism invariant properties?

The point is brought into especially clear focus when contrasting material set theory with *categorial* theories of sets (like **ETCS**). We should note that meaningful statements of categorial set theories like **ETCS** are not fully isomorphism invariant. Take, for example, the claim "f has co-domain B" for some f and B. This interpretable as a perfectly good formula in the language of category theory, but is obviously not isomorphism invariant; there might be objects isomorphic to B which differ as to whether they are the co-domain of f or not.[30] However, something that

[29]The following analogy may be helpful. Viewing mathematics as describing a kind of quasi-computational enterprise, set theory is something like a theory of machine-code: It tells us what kinds of things can be built, and what we need to build them. Category theory on the other hand is like a high-level programming language, codifying what effects can be achieved by different structural relationships in different contexts. Both the set theorist as computational engineer and the category theorist as programmer have important roles to play in mathematics. This analogy (or, at least, something similar) was originally communicated to the first-author by David Corfield after a talk at the LSE in November 2013. He is also grateful to Dr. Corfield for subsequent discussion of categorial foundations.

[30]Strictly speaking, we have used a formulation of category theory here on which we have variables for objects as well as arrows. In a purely arrow-theoretic framework (where equality is only defined between parallel morphisms) one has the result that any two equivalent categories satisfy the same sentences. (We are grateful to Ingo Blechschmidt for helpfully forcing us to be precise about this issue.) However, if one wants to use parameters, for instance if one is developing the theory of modules in **ETCS**, or using more than one variable, then the result no longer holds. (Thanks here are due to Jean-Pierre Marquis for this useful comment.) There are some theories that aim to make it impossible to state non-isomorphism-invariant properties in their language. Two candidates here are Makkai's **FOLDS** (see Makkai 1998; Marquis 2017b) and Univalent Foundations (see The Univalent Foundations Program 2013; Tsementzis 2017). We are grateful to Dimitris Tsementzis for stressing this use of Univalent Foundations and making us aware of the non-isomorphism invariance of category theory (in conversation and his Tsementzis 2017), as well as directing us to Theorem 8. We are also grateful (again) to Jean-Pierre Marquis for emphasising to us the interest of **FOLDS**.

ETCS *does* offer is a way of 'modding out' this non-isomorphism invariant noise. For instance, one can prove:

Theorem 8 (ETCS) *Let $\phi(X)$ be a formula in the language of* **ETCS** *with no constants and no free variables except the set variable X. Then if X and Y are isomorphic, then $\phi(X)$ iff $\phi(Y)$.*[31]

Thus **ETCS** provides us with a clean-cut class of formulas for which we have isomorphism invariance. Our task for the rest of this paper is to make a preliminary step in providing a material set theory that does the same for *structure*. We will do so by proposing a theory of sets and classes built over *structures* composed of *urelements*, and end up with a theory that better respects isomorphsim invariance whilst facilitating the consideration of how schematic types and systems of concrete objects interact (as in the Morley Categoricity Theorem).

10.4 How Set Theory Helps the Structural Perspective

We now have a challenge for the friend of set-theoretic foundations; find a use of set-theoretic language that better respects isomorphism invariance. In this section, we do just that. The broad idea is to think of material set theory as *built over* some mathematical subject matter, conceived of as composed of structures with urelements and the functions and relations between them. Our point is the following; we can factor out the arbitrary choices of coding to find a material set theory that by design respects isomorphism invariance.

10.4.1 Set Theory with Structures (ZFCS)

In taking inspiration from structuralism,[32] we will consider structures as composed of featureless points, and the functions and relations between them. Treating a featureless point as a kind of *urelement*, we will build a theory of sets and classes over these urelements, and show how by doing so we can develop a more structurally respectful theory in the language of sets and classes. However, this will also be a framework in which it is possible to use the rich combinatorial power afforded by material set theory in discussing notions of cardinality and structure.

[31] See here McLarty (1993, p. 495).

[32] See here, for example, Shapiro (1991).

Definition 9 The theory of *Set Theory with Structures* (or **ZFCS**) is defined as follows:

(i) Symbols:

(a) We have *three sorts* of variables: $u_0, u_1, \ldots, u_n, \ldots$ will range over *urelements* (to be featureless points in the domains of structures), $s_0, s_1, \ldots, s_n, \ldots$ will range over *structures*, and $x_0, x_1, \ldots, x_n, \ldots$ will range over *sets*.
(b) The usual logical symbols (so one's favourite connectives and quantifier(s)), and one non-logical symbol '$\in$' (to denote material set membership).
(c) Symbols: A single symbol U (for the universe), $f_{m,n}$ (for m-ary functions), $R_{m,n}$ (for m-ary relations) and c_n (for constants), where m, n are natural numbers and $m > 0$. These will be used to describe structures.

(ii) Atomic formulas:

(a) $a = b$ where a, b are variables of the same sort.
(b) $a \in b$ where a is a variable and b is a set-variable.
(c) $U(s, a)$ where s is a structure-variable and a is an urelement-variable. (*Intended meaning:* a belongs to the universe (or domain) of the structure s.)
(d) $f_{m,n}(s, u_1, \ldots, u_m) = u$ where s is a structure-variable, the u_i and u are urelement variables. (*Intended meaning:* the u_i and u belong to the universe of the structure s and the interpretation of the m-ary function symbol $f_{m,n}$ in s sends $(u_1, \ldots, u_m)$ to u.)
(e) $R_{m,n}(s, u_1, \ldots, u_m)$ where s is a structure-variable and the u_i are urelement variables. (*Intended meaning:* The m-tuple $(u_1, \ldots, u_m)$ belongs to the interpretation of the m-ary predicate symbol by the structure s.)
(f) $c_n(s) = u$ where s is a structure-variable and u is an urelement variable. (*Intended meaning:* The interpretation of the constant symbol c_n by s is u.)

(iii) Compound formulas: Obtained from atomic formulas by closing under connectives and quantifiers in the usual way. (Though, since the language is 3-sorted, there will be three kinds of quantifier; one for urelements, one for structures, and one for sets.)

(iv) Axioms:

(a) Extensionality for sets.
(b) Formula-Foundation for Sets: If a formula holds of some set then it holds of some set which is disjoint from all other sets for which the formula holds.
(c) The Axiom of Infinity: Usually rendered as concerning the existence of an inductive pure set.
(d) Pairing, Union, Powerset, Separation and Collection for sets.
(e) Axiom of Choice for sets.
(f) The domain of every structure is a set: i.e. $\forall s \exists x \forall a (U(s, a) \leftrightarrow a \in x)$.
(g) The Anti-Urelement Set Axiom: No set contains all of the urelements.

Some remarks concerning the definition are in order:

Remark 10 First, whilst the Anti-Urelement Set Axiom merits philosophical discussion,[33] consideration in detail would take us too far afield here. We make this assumption simply to avoid bounding the sizes of the structures we have available, and settle for the pragmatic justification that we are trying to show that material set theory *can* convey important structural information, not that it *must* in every situation.[34]

Remark 11 Second, we are taking inspiration from the structuralist literature in the following sense: Structures are to be understood as composed of featureless points (given by the urelemente) and the ways they may correspond (with functions and relations). Effectively, we layer sets *on top* of antecedently given structures conceived of in this sense. As we'll see, this facilitates cardinality comparisons whilst allowing for a theory that respects isomorphism invariance.

10.4.2 Class Theory with Structures (NBGS)

We are now in a position where we have a theory of structures and the sets that can be built over them. In order to arrive at a language that respects isomorphism invariance we now augment with class variables. As we shall see, this allows us to latch onto a range of isomorphism invariant classes.

Definition 12 Our *Class Theory with Structures* (or **NBGS**) comprises the following:

1. Symbols: All the symbols of **ZFCS**, with an additional kind of variables $X_0, X_1, \ldots, X_n, \ldots$ for classes.
2. Atomic formulas: In addition to the well-formed formulas of **ZFCS**, we admit $X_n = X_m$ for class variables X_n and X_m as well-formed, as well as $v_0 \in X_n$ for class-variable X_n and v_0 is either a set, structure, or urelement variable.
3. Compound formulas: Obtained inductively from the connectives, $\in$, urelemente quantifiers, structure quantifiers, set quantifiers, and class quantifiers.
4. Axioms:
   (a) All axioms of **ZFCS**.
   (b) Extensionality for classes (i.e. X_n and X_m are equal iff they have the same members).
   (c) Predicative Class Comprehension:
   $\exists X \forall u \forall s \forall x [(\phi(u) \leftrightarrow u \in X) \wedge (\psi(s) \leftrightarrow s \in X) \wedge (\chi(x) \leftrightarrow x \in X)]$

[33]For two examples of such discussion, see McGee (1997) and Rumfitt (2015).

[34]An alternative would be to take the 'wide sets' approach of Menzel (2014) and modify Replacement.

(Where u is a urelement variable, s is a structure variable, and x is a set variable, there are no class quantifiers[35] in ϕ, ψ, and χ, and each of ϕ, ψ, and χ is free for u, s, and x respectively.)

Effectively we allow extensional classes composed of objects of mixed types. The intuition behind the theory is this; once we have built our sets and structures out of sets and urelements, we can then talk definably about them (much in the same way as it is legitimate to talk about definable classes in the **ZFC** context). As we'll shortly see, we can then restrict to certain classes in using our set and class-theoretic language to latch onto isomorphism invariant properties.

One technical issue is how to treat isomorphism within this framework. We really have two notions available:

Definition 13 Two structures s_0 and s_1 are *structure-theoretically isomorphic* iff there is a third structure s within which there is a binary relation between the universes of s_0 and s_1 satisfying the usual rules of isomorphism.

Definition 14 Two structures s_0 and s_1 are *set-theoretically isomorphic* iff there is a set-theoretic bijection between the domains if s_0 and s_1 satisfying the usual rules of isomorphism.

Remark 15 Effectively the set-theoretic notion of isomorphism is the usual one, whilst the structure-theoretic notion pulls talk of structures and isomorphism into the prior given theory of structure.

What this prior given theory of structure satisfies will be important for our results and arguments, since we require the closure of our **NBGS** world under isomorphism. What we show below will always hold for the *set-theoretic* notion of isomorphism, but there is a chance that if the antecedent theory of structures over which we layer **NBGS** is too impoverished, then set-theoretic isomorphisms might not be mirrored by structure-theoretic ones. For example, consider a world which has **NBGS** layered over a theory of structure on which there is only one structure s_0 composed of an isolated point u_0 with no relations on it. This is be perfectly legitimate as a **NBGS** structure. However, while there are set-theoretic isomorphisms in this world (specifically f defined by the rule $f(u_0) = u_0$), the underlying theory of structure lacks the resources to even see that s_0 is isomorphic to itself.

One fix would be to introduce the following axiom:

Axiom 16 (The Structural Richness Axiom) *Any set-theoretic isomorphism has a corresponding extensionally equivalent structure-theoretic isomorphism.*[36]

[35]We make this assumption merely for technical hygiene since **NBGS** will do the job we want neatly. One could also drop this restriction, and use an impredicative comprehension scheme yielding a structural form of Morse-Kelley (call it **MKS**). This may well have interesting additional properties, such as the ability to define a satisfaction predicate for the universe and first-order formulas.

[36]In fuller formalism: If f is a set-theoretic isomorphism between s_0 and s_1, then there is an s such that s maps u_α to u'_α iff f does.

This would guarantee the existence of the isomorphisms we need for our results both set-theoretically and structure-theoretically. Indeed, we might derive the Structural Richness Axiom from the idea that any set-theoretic structure should be mirrored by a structure-theoretic one, postulating the following:

Axiom 17 (The Structural Radical Richness Axiom) *Any set-theoretic structure is mirrored by a corresponding structure-theoretic structure.*[37]

The extent to which our theory of structure should mirror what we have going on in the sets is an interesting one, but we shall set it aside from our technical result. Since we wish to leave it entirely open what the underlying theory of structure should satisfy (maybe, for example, we just want to build **NBGS** over the structural groups and nothing else) we shall simply concern ourselves for the technical work with the set-theoretic notion of isomorphism. Later, we will reconsider structural richness axioms in discussing how set theory has interesting things to say about the interaction of schematic types with structures.

We now have the resources to factor out non-isomorphism invariant properties from our theory whilst having combinatorial sets around to facilitate proof. This is shown by the following analogue of the earlier mentioned **ETCS** theorem concerning isomorphism invariance:

Theorem 18 (**NBGS**) *Suppose that $\phi(v)$ is a formula without parameters in the language of* **NBGS** *and v is a variable ranging over structures. Suppose that $\mathfrak{M}$ is a model for* **NBGS** *and s_0 and s_1 are structures in $\mathfrak{M}$ which are isomorphic in $\mathfrak{M}$. Then $\mathfrak{M} \models \phi(s_0) \leftrightarrow \phi(s_1)$.*

Proof The idea of the proof is simply to take an isomorphism π between s_0 and s_1 and then use it to build an automorphism π' from $\mathfrak{M}$ to $\mathfrak{M}$, moving $\mathfrak{M}$'s satisfaction of $\phi(s_0)$ to $\phi(s_1)$ (dual reasoning obtains the converse implication). So, let π be the witnessing structure-theoretic isomorphism. We define π' as follows:

(i) $\pi'(u) = \pi(u)$ if u is an urelement in $dom(\pi)$.
(ii) $\pi'(s)$ is obtained from s by replacing each urelement u in the universe of s by $\pi(u)$.
(iii) We can then (by induction and the well-foundedness of membership) replace a set x by defining $\pi'(x)$ as $\{\pi'(y)|y \in x\}$.
(iv) We similarly replace a class X by $\pi'(X) =_{df} \{\pi'(y)|y \in X\}$.
(v) This π' yields an automorphism of $\mathfrak{M}$ sending s_0 to s_1. Thus $\mathfrak{M}$ satisfies "$\phi(s_0)$ iff $\phi(s_1)$".[38]

[37] In fuller formalism: For any set X of urelements and set-theoretic functions $f^X_{m,n}$, relations $R^X_{m,n}$ on X, and constants c^X_n in X, there is an s such that $U(s, u_\alpha)$ (for each $u_\alpha \in X$), and structural relations $f^s_{m,n}$, $R^s_{m,n}$, and c^s_n equivalent to $f^X_{m,n}$, $R^X_{m,n}$, and c^X_n in the obvious way.

[38] This can be proved by the usual tedious induction on the complexity of ϕ.

Remark 19 A very similar theorem holds for formulas with more free variables, providing that the domains of the structures are non-overlapping. In this way, **NBGS** can provide a kind of isomorphism invariance *stronger* than the one for **ETCS** as given in Theorem 8.

Thus, while there are many non-isomorphism invariant facts we can state within **NBGS** (e.g. for an urelement or structure v, "$v \in X$"), we can factor out this non-isomorphism invariant 'noise' in a precise way. Thus, for ϕ of the appropriate form, if we use the ambient material set-theoretic properties to prove ϕ about s, we can transfer ϕ to any structure isomorphic to s.

If we then allow ourselves some axiom of structural richness (e.g. either Axiom 16 or Axiom 17) we can talk about inter-structural relationships such as embeddability and cardinality of structures using set-theoretic resources, whilst factoring out the structural content. For example, concerning cardinality, every cardinal number exists, both with respect to the relevant set-theoretic code living in the universe of **NBGS** under consideration, but also there is (given an assumption of structural richness) a structure of pure cardinality (i.e. of featureless points with no relationships between them). In this way the theory provides a way of combinatorially linking bona fide structures with their set-theoretic codes.

This has ramifications for how we treat theorems best suited for material set theory. The Morley Categoricity Theorem, for example, can be recast as the claim that for any countable **T** in $\mathscr{L}_{\omega,\omega}$ (i.e. **T** is a particular kind of first-order schematic type) if **T** is satisfied by exactly one structure (up to isomorphism) for structures with domain of some particular uncountable size, then given any two structures satisfying **T**, if their domains are uncountable and the same size, then they are isomorphic.

This version of the Morley Categoricity Theorem facilitates extraction of structural information: All the above can be recast as talk *directly* about theories and structures. However, the ambient material set theory provided by **NBGS** has a role to play: The proof could be formalised exactly as is usually done in the material set-theoretic case, since we have the relevant set-theoretic resources, set-theoretic codes, and model theory lying around.[39] It is just that *in the end* we can easily extract the *purely structural* content from Morley's result using Theorem 18 and assumptions of structural richness, and then talk directly and transparently about structures conceived of in a sense independent of set-theoretic coding.

Moreover, not only do we now have a way to extract the purely structural information from the material set theory, but we can also find non-arbitrary representatives for structures:

Definition 20 In **NBGS**, we say that a class X consisting of structures is (set- or structure-theoretically) *invariant* if X is closed under isomorphism between structures. If in addition any two structures in X are isomorphic we refer to X as a (set- or structure-theoretic) *isomorphism type*.

[39] See, for example, Tent and Ziegler (2012) for a presentation of the usual proof.

Thus, in **NBGS** we can equate specific mathematical structures with isomorphism types. This then gives us a non-arbitrary, non-coding dependent representative for the structure in question (since isomorphism types include all and only the structures isomorphic to one another). However, all this occurs in a framework where we have the full resources of material set theory available to speak about their relationships.

10.5 Conclusions and Open Questions

How does this perspective provide philosophical progress? Our main claim (as specified in the introduction) was to show that material set theory seems to be interesting from the structural perspective. In particular, it still represents our best theory of cardinality, and how cardinality interacts with properties of syntactic theories. In this respect it is of interest to friends of both set-theoretic and category-theoretic foundations.

Of course one might simply reject that cardinality is a structural property, and that cardinality considerations fade away when one takes seriously a structural perspective as coded categorially.[40] This raises an important first question for future research:

Question 21 To what extent should cardinality be viewed as a structural property?

If we accept that cardinality is structural, however, it is interesting that we can, by taking much structural talk at face value, come up with a theory (namely **NBGS**) that allows us to easily extract structural content from set-theoretic claims. In this way, we have both the combinatorial power of set theory but also a way of factoring out non-isomorphism invariant content.

The theory represents a small first step on a wider research programme in the foundations of mathematics: How can we *fuse* different perspectives to yield new mathematical and philosophical perspectives and results? In particular, we have left open the following questions:

Question 22 There is the possibility of looking at categorial structure of theories like **NBGS**. Some, are easy: The category of isomorphism invariant classes that has as objects isomorphism types and equivalence classes of embeddings as arrows is an obvious choice. However, there is also the possibility of looking at the categorial structure of the classes of mixed type. What sorts of categorial relationships does this world support?

One remaining challenge is the following: In the introduction it was noted that certain categories have non-function-like relationships either for trivial reasons (such as the category of proofs of a logical system) or deeper ones concerning the

[40] We are grateful to Steve Vickers for pressing this point.

nature of the category itself, such as with non-concrete categories like the homotopy category. There is also the question of how to understand non-bijective notions of structural similarity like categorial equivalence. The framework we have provided depends explicitly on a notion of structure on which isomorphism is understood as a functorial (in fact bijective) relationship between structures. A question then is:

Question 23 Can a similar perspective be found that brings set-theoretic combinatorial power to bear on non-concrete categories? What about for more relaxed notions of sameness of structure such as categorial equivalence?

This would further help to inform the following project:

Question 24 Aside from the normal interpretations of category theory in set theory (either through universes or classes) or set theory in category theory (say via the use of objects and arrows coding membership-trees), what further ways are there of combining the different perspectives? We have shown how this can be done with respect to 'structure' and 'cardinality', but can we have further philosophically and mathematically fruitful fusions of these two perspectives?

For the moment, we hope to have shown that there is at least the possibility of combining different perspectives to yield interesting fusions of foundational theories.

References

Antos, C., Barton, N., & Friedman, S.-D. Universism and extensions of V. Submitted. https://arxiv.org/abs/1708.05751.

Awodey, S. (1996). Structure in mathematics and logic: A categorical perspective. *Philosophia Mathematica, 4*(3), 209–237.

Bagaria, J., & Brooke-Taylor, A. (2013). On colimits and elementary embeddings. *The Journal of Symbolic Logic, 78*(2), 562–578.

Baldwin, J. T. (2018). *Model theory and the philosophy of mathematical practice: Formalization without foundationalism*. Cambridge: Cambridge University Press.

Barton, N. (2016). *Multiversism and concepts of set: How much relativism is acceptable?* (pp. 189–209). Cham: Springer International Publishing.

Barton, N. (2018). Forcing and the universe of sets: Must we lose insight? Forthcoming in the *Journal of Philosophical Logic*.

Bell, J. (2011). *Set theory: Boolean-valued models and independence proofs*. Oxford: Oxford University Press.

Benacerraf, P. (1965). What numbers could not be. *Philosophical Review, 74*(1), 47–73.

Caicedo, A. E., Cummings, J., Koellner, P., & Larson, P. B. (Eds.). (2017). *Foundations of Mathematics: Logic at Harvard Essays in Honor of W. Hugh Woodin's 60th Birthday* (Volume 690 of Contemporary Mathematics). Providence: American Mathematical Society.

Di Liberti, I., & Loregian, F. (2018). Homotopical algebra is not concrete. *Journal of Homotopy and Related Structures, 13*, 673–687.

Ernst, M. (2015). The prospects of unlimited category theory: Doing what remains to be done. *The Review of Symbolic Logic, 8*(2), 306–327.

Ernst, M. (2017). Category theory and foundations. In E. Landry (Ed.), Landry (2017) (pp. 69–89). Oxford: Oxford University Press.

Ewald, W. B. (Ed.). (1996). *From Kant to Hilbert. A source book in the foundations of mathematics* (Vol. I). Oxford: Oxford University Press.
Feferman, S. (1977). Categorical foundations and foundations of category theory. In R. E. Butts & J. Hintikka (Eds.), *Logic, foundations of mathematics, and computability theory* (pp. 149–169). Dordrecht: Springer.
Freyd, P. (1970). *Homotopy is not concrete* (pp. 25–34). Berlin/Heidelberg: Springer.
Goldblatt, R. (1984). *Topoi: The categorial analysis of logic*. Amsterdam: Dover Publications.
Hamkins, J. D. (2012). The set-theoretic multiverse. *The Review of Symbolic Logic, 5*(3), 416–449.
Hellman, G. (2006). Against 'absolutely everything'! In A. Rayo & G. Uzquiano (Eds.), *Absolute generality*. New York: Clarendon Press.
Isaacson, D. (2011). The reality of mathematics and the case of set theory. In Z. Noviak & A. Simonyi (Eds.), *Truth, reference, and realism* (pp. 1–75). Budapest/New York: Central European University Press.
Landry, E. (Ed.). (2017). *Categories for the working philosopher*. Oxford: Oxford University Press.
Landry, E., & Marquis, J.-P. (2005). Categories in context: Historical, foundational, and philosophical†. *Philosophia Mathematica, 13*(1), 1.
Lawvere, W. (1965). An elementary theory of the category of sets. *Proceedings of the National Academy of Science of the U.S.A., 52*, 1506–1511.
Lawvere, W., & McLarty, C. (2005). An elementary theory of the category of sets (long version) with commentary. In *Reprints in Theory and Applications of Categories 11* (pp. 1–35). TAC.
Leinster, T. (2014). Rethinking set theory. *The American Mathematical Monthly, 121*(5), 403–415.
Mac Lane, S. (1971). Categorical algebra and set-theoretical foundations. In Scott and Jech (1971) (pp. 231–240). Providence: American Mathematical Society.
Mac Lane, S. (1986). *Mathematics: Form and function*. New York: Springer.
Maddy, P. (2017). Set-theoretic foundations. In Caicedo et al. (2017) (pp. 289–322). Providence: American Mathematical Society.
Maddy, P. (2019). What do we want a foundation to do? In this volume.
Makkai, M. (1998). Towards a categorical foundation of mathematics. In *Logic Colloquium'95 (Haifa)* (Lecture Notes in Logic, Vol. 11, pp. 153–190). Berlin: Springer.
Marquis, J. (2013). Mathematical forms and forms of mathematics: Leaving the shores of extensional mathematics. *Synthese, 190*(12), 2141–2164.
Marquis, J.-P. (2017a). Canonical maps. In E. Landry (Ed.), Landry (2017) (pp. 90–112). Oxford: Oxford University Press.
Marquis, J.-P. (2017b). Unfolding FOLDS: A foundational framework for abstract mathematical concepts. In E. Landry (Ed.), Landry (2017) (pp. 136–162). Oxford: Oxford University Press.
Mathias, A. R. D. (2000). Strong statements of analysis. *Bulletin of the London Mathematical Society, 32*(5), 513–526.
Mathias, A. (2001). The strength of mac lane set theory. *Annals of Pure and Applied Logic, 110*(1), 107–234.
McGee, V. (1997). How we learn mathematical language. *The Philosophical Review, 106*(1), 35–68.
McLarty, C. (1993). Numbers can be just what they have to. *Noûs, 27*(4), 487–498.
Mclarty, C. (2004). Exploring categorical structuralism. *Philosophia Mathematica, 12*(1), 37–53.
Meadows, T. (2013). What can a categoricity theorem tell us? *The Review of Symbolic Logic, 6*, 524–544.
Menzel, C. (2014). $ZFCU$, wide sets, and the iterative conception. *Journal of Philosophy, 111*(2), 57–83.
Muller, F. A. (2001). Sets, classes and categories. *British Journal for the Philosophy of Science, 52*, 539–573.
Osius, G. (1974). Categorical set theory: A characterization of the category of sets. *Journal of Pure and Applied Algebra, 4*(1), 79–119.
Rumfitt, I. (2015). *The boundary stones of thought: An essay in the philosophy of logic*. Oxford: Oxford University Press.

Scott, D., & Jech, T. (1971). *Axiomatic set theory* (Axiomatic Set Theory, Number pt. 1). Providence: American Mathematical Society.

Shapiro, S. (1991). *Foundations without foundationalism: A case for second-order logic*. New York: Oxford University Press.

Shepherdson, J. C. (1951). Inner models for set theory–Part I. *Journal of Symbolic Logic, 16*(3), 161–190.

Shepherdson, J. C. (1952). Inner models for set theory–Part II. *Journal of Symbolic Logic, 17*(4), 225–237.

Shulman, M. A. (2008). Set theory for category theory. https://arxiv.org/abs/0810.1279

Shulman, M. A. (2010). Stack semantics and the comparison of material and structural set theories. https://arxiv.org/abs/1004.3802

Tent, K., & Ziegler, M. (2012). *A course in model theory* (Lecture Notes in Logic). Cambridge: Cambridge University Press.

The Univalent Foundations Program. (2013). *Homotopy type theory: Univalent foundations of mathematics*. http://homotopytypetheory.org/book

Tsementzis, D. (2017). Univalent foundations as structuralist foundations. *Synthese, 194*(9), 3583–3617. https://doi.org/10.1007/s11229-016-1109-x

Voevodsky, V. (2014). *2013 North American Annual Meeting of the Association for Symbolic Logic: University of Waterloo, Waterloo, ON, Canada May 8–11, 2013. The Bulletin of Symbolic Logic, 20*(1), 105–133.

Zermelo, E. (1930). On boundary numbers and domains of sets. In Ewald (1996) (Vol. 2, pp. 1208–1233). Oxford University Press.

Chapter 11
A New Foundational Crisis in Mathematics, Is It Really Happening?

Mirna Džamonja

Abstract The article reconsiders the position of the foundations of mathematics after the discovery of the homotopy type theory HoTT. Discussion that this discovery has generated in the community of mathematicians, philosophers and computer scientists might indicate a new crisis in the foundation of mathematics. By examining the mathematical facts behind HoTT and their relation with the existing foundations, we conclude that the present crisis is not one. We reiterate a pluralist vision of the foundations of mathematics.

The article contains a short survey of the mathematical and historical background needed to understand the main tenets of the foundational issues.

11.1 Introduction

It seems by all evidence about the differences between set theory and homotopy type theory HoTT that there is a crisis in the foundations of mathematics. However, neither mathematicians nor even the philosophers have spoken enough about it. The preceeding crisis, happening at the beginning of the twentieth century and sparked by the Russell paradox, created many animated discussions and engaged a large number of mathematicians- not necessarily 'logicians' (as mathematical logic as a separate field of mathematics had not yet existed, properly speaking). All the mathematicians were interested in foundations and so were many philosophers. That was the time when mathematics formed a tameable, unified, body of research, and mathematicians could easily understand each other. The folklore has it that Hilbert was the last mathematician who knew all of mathematics, but he was not the only one: let us mention Poincaré, Picard and Von Neumann, for example. The discussion about foundation of mathematics engaged a wide audience of mathematicians. Today, logicians tend to hold foundational discussions with very little success in

M. Džamonja (✉)
School of Mathematics, University of East Anglia, Norwich, UK
IHPST, Université Panthéon-Sorbonne, Paris, France
e-mail: m.dzamonja@uea.ac.uk

S. Centrone et al. (eds.), *Reflections on the Foundations of Mathematics*,
Synthese Library 407, https://doi.org/10.1007/978-3-030-15655-8_11

engaging the mathematical community. For example, mathematicians spent 358 years solving Fermat's last problem. The solution, due to Wiles (1995) could not have possibly been the one that Fermat meant to write in his famous margin in 1637, as it involves mathematical techniques that did not exist even in a trace form at the time of Fermat. Could Fermat have found a different proof? If we prove that Fermat's last problem does not have a solution in Peano Arithmetic PA, which encompasses all the tools available to Fermat, then it is pretty clear that Fermat could not have solved it. Some FOM (Foundations of Mathematics) discussion groups seem convinced that this is the case, yet Angus Macintyre seems to have proved that Wiles' proof actually can be carried out in PA (Macintyre). However the general mathematical audience has not yet noticed this 'rumination by the logicians', as much as everybody would really love to know if Fermat had actually proved it.

The truth is that mathematical logic has become a very specialised subject, considered by mathematicians as an esoteric subject of little interest to their daily life. Therefore, even the mathematical foundations, since they are immersed in logic, are considered as esoteric. (It must be admitted that this feeling in the mathematical community is not completely decoupled from the rather dry writing style that logic papers tend to have). The purpose of this article is, firstly, to recall that just exactly the opposite is true, thinking about foundations is not esoteric and, if there is indeed a crisis in the foundation of mathematics, then *everybody* should be concerned. Not only the logicians and the mathematicians but also all the scientists, who base their reasoning upon logic and mathematics. One cannot build a high-rise without knowing how stable are the pillars.[1] Secondly, in the main body of the article we shall describe the foundational question under consideration in such terms that every mathematician, philosopher of mathematics or computer scientist can follow the exposition, with the purpose of engaging a wider community in the foundational discussion. Lastly, we shall give our own opinion, calming the fire of panic and arguing that the crisis is relative and solvable and that it brings us more of the positive than of the negative information. We shall argue for a vision of foundations that envelops this information into a new stage of the development of the existing foundations of mathematics.

11.2 For a Mathematician

Many of our glorious ancestors in set theory in the 1960s and later put an enormous effort to show to mathematicians that the fact that there were simple statements, such as the continuum hypothesis,[2] about simple objects, such as the set of the real numbers, that turned out to be independent of the axioms of set theory – then almost

[1]This metaphor might be overused. But I still like it.

[2]Stating that every infinite subset of the set of the real numbers is bijective either with the set of the natural numbers or with the set of the real numbers.

universally considered as *the* foundation of mathematics – should be of interest to every mathematician on the planet. This effort does not seem to have been very successful, even though the independence can appear almost everywhere, see David H. Fremlin (1984) for some early examples in various areas of mathematics. Well, the point is that it is *almost* everywhere, and many mathematicians have had the reaction to withdraw to those areas of mathematics that were immune to independence and to relegate any independence to the inferior rang of 'pathological' objects. Infinite combinatorics- pathological, uncountable measures- pathological, non-separable Banach spaces- pathological, independence- "can't possibly happen to me".

Fine, but what if somebody tells you that your art of mathematics is going to go the way of the games chess and go, it is going to be replaced by formal arguments doable by a machine which can do it better than you or any other human? That is how the program of formalisation of mathematics through univalent foundations teamed with automatic theorem proving, presented by a winning mathematician who invented it, looks to a naked eye.

Ironically, it is exactly those areas of mathematics most remote from independence, that are protected from independence by their inherent definability and constructibility, which are in danger of this new formalisation and computerisation. Formalisation of mathematics uses constructive mathematics.'Pathological' objects in general are not constructive, they often rely on the axiom of choice and further set-theoretic machinery, with no specific witness for existential properties. This is what makes them prone to independence, but rather safe from formalisation.[3] So, one way or another, foundations have not been kind to mathematicians, and it looks like one should start paying attention and ask the question if the mathematics we are producing today is safe from a foundational and existential crisis.

Ever since the nineteenth century most (or just many?) people have considered set theory as the foundations of mathematics. This is in spite of the great crisis in the foundations of mathematics brought about by Russell's paradox in 1902 and the subsequent introduction to mathematics not only of several different axiomatisations of set theory[4] and various very successful alternative approaches to foundations of mathematics, such as category theory, type theory and the theory of toposes. While it is clear, for example, that it is much more reasonable to do certain parts of mathematics by using category theory, it has been known that category theory and set theory with large cardinals are to some extent bi-interpretable (see e.g. Feferman 1977) and apart from philosophical reasons, there is no reason for an 'ordinary mathematician' to take his time from his mathematical considerations in order to think if one or another foundation fits better with the purpose at hand.

In 2006 a new kid appeared on the block. It is the homotopy type theory including univalent foundations. These foundations are different from set-theoretic foundations. In some respect these foundations appear to contradict the set-theoretic

[3] Although see later on Isabelle and see Mizar's project website http://mizar.org/project/.

[4] In this paper we take Zermelo-Fraenkel set theory with choice (ZFC) for our basis and we freely dispose of a sufficient amount of large cardinals.

foundations. A great disquiet, or perhaps excitement about the newly proposed foundations can be seen by the following quote by Michael Harris[5] in (2017):

> It's impossible to overstate the consequences for philosophy, especially the philosophy of Mathematics, if Voevodsky's proposed new Foundations were adopted.

Our thesis is that there is no contradiction in sight, moreover, these foundations complement the set-theoretic foundations. We feel that any suggestion of a contradiction between the two or the idea that either should or could be replaced by the other comes more from a lack of understanding than anything else. We attempt in this article to give a simple account of the most relevant developments in order to demonstrate our thesis. We shall lead the reader through a presentation of the basics of several different parts of mathematics and logic that have gone into the developments in question and sometimes we shall have to simplify to a great extent in order to make ourselves understandable. The reader interested in the details will easily be able to find them, either in the other chapters of this book or elsewhere in the literature and references mentioned here.

11.3 On Set Theory and Type Theories

Set theory is based on the classical first order logic. Basic entities are sets and they are completely determined by their elements: the Axiom of Extensionality states that for any sets A and B,

$$[\forall x(x \in A \iff x \in B)] \iff A = B.$$ [6]

The Law of Excluded Middle holds and the Axiom of Choice is assumed (as we are working with ZFC). This type of set theory was formalised by Ernst Zermelo (1908) and ramified in the work of Zermelo and Fraenkel, as well as Skolem. A detailed history of the development of ZFC and some other set theories can be found in Fraenkel et al. (1973).

Univalent foundations are a vision of Vladimir Voevodsky of a foundational system for mathematics in which the basic objects are homotopy types. They obey a type theory satisfying the univalence axiom. They are formalisable in a computer proof assistant and are constructivist. To introduce univalent foundations, let us start by discussing type theory. We present the types together with the idea of a proof, which is not how these concepts were developed initially.

[5]A mathematician who does care about the foundations of mathematics.

[6]The intended meaning of the symbol $\in$ is that is stands for the actual real' membership relation. A common misunderstanding is that this is the case with all models of set theory. However, the setting of set theory in the logic of first order, does not allow to rule out models in which the symbol is interpreted by other means. They have to be dealt by separate methods, including the Mostowski collapse theorem.

In set theory, basic variables are interpreted as sets and atomic formulas are of the form $x = y$ and $x \in y$. Naive set theory in which every collection of the form $\{x : \varphi(x)\}$ forms a set, leads to paradoxes, the most celebrated of which is the one developed by Russell in 1902.[7] Presuming that $B = \{x : x \notin x\}$ is a set, we see that the question if $B \in B$ does not have a unique answer. Set theory resolved this issue by various axiomatisations, in our case of ZFC by introduction of Restricted Comprehension by Zermelo in 1908. This type of Comprehension only allows formation of sets of the form $\{x \in y : \varphi(x)\}$, that is, the comprehension of all elements of a given set with a given property, into a set. Russell's own solution to the paradox was to introduce type theory. There are many versions of type theory, but the following version from Fraenkel et al. (1973) is equivalent to the one used by Russell and Whitehead in Whitehead and Russell (1990) in which they succeeded to give foundations of the entire extent of the mathematics then known in terms of type theory.

In this context, each variable has a type, which is a natural number, and writing x^i means that x is a variable of type i. Atomic formulas are of the form $x^i \in y^{i+1}$ and $x^i = z^i$. That is, a variable can only be equal to a variable of the same type and can only belong to a variable of type exactly one higher. The Axiom of Infinity is added to level 0 and Axiom of Choice at every level.

The advantageous formalisation power of type theory is to some extent balanced off by the fact that is rather cumbersome to write mathematics in this way. The reader may find it entertaining to contemplate the notions of complement or cardinality in this setting.

For the homotopy type theory it is more useful to think of the Dependent Type Theory, developed by Martin-Löf starting from the 1960s.[8] Athough the original idea of dependent types evolves from Martin-Löf's study of randomness in probability, we can see dependent type theory in a computational context, like a computer language (this idea goes back to de Bruijn). It is particularly well suited for the theory of proofs, as we shall now see.

Type theory does not have semantics, only syntax. The basic expressions are of the from **term:Type**, for example $x : \mathbb{N}$ and $y : \mathbb{R}$, which mean that x is a variable of the non-negative integer type and y a variable of the real type. Such expressions, variable typings, are called *contexts*. From this we can advance to *proofs* or *judgements* using some predetermined rules. For example

$$x : \mathbb{N}, y : \mathbb{R} \vdash x \times y : \mathbb{N} \times \mathbb{R}.$$

[7] Zermelo seems to have discovered the paradox about at the same time, although he did not publish it.

[8] In our explanation, we shall privilege intuition to exactness. A much more precise, yet still rather compact, survey of Type Theory and Univalent Foundations can be found in Thierry Coquand's article for the Stanford Encyclopedia of Philosophy (Coquand 2018). A longer survey of these matters is in the recent article by Daniel Grayson (Grayson 2018). Another recent article, by Martin Escardo (2018) gives an entirely self-sufficient presentation of the Axiom of Univalence. We have partially followed an article by Michael Shulman and its notation, Shulman (2017).

The symbol $\vdash$ means 'proves'. In this judgement we can see a typical operation on types, which in this case is the product of two. A list of all basic types and the operations on them can be found on the slides by Coquand (2014). It is useful to think of this as a category, in which objects are the contexts and the morphisms are the judgements. This is called the *Classifying Category* and denoted **Ctx**. So, when we write

$$\Gamma \vdash a : A$$

we consider a as a morphism from Γ to A. Contexts with variables are 'types' and those without variables, where we substituted a variable by a constant, are morphisms. It is exactly the way to handle the idea of substitution which makes this into the *dependent type theory*. For example, we allow families of contexts, such as B **type**, where we interpret

$$\Gamma \vdash B \textbf{ type}$$

as a judgement allowing to conclude from Γ that there is an object of type B (or B is a type in Γ). The intuition is that B 'depends' on Γ. The dependent types also can be combined, for example

$$n : \mathbb{N} \vdash C_n \textbf{ type}$$

corresponds to "for all n we have an object of type C_n".[9] We can think of the dependencies as factor or slice categories, where for example the statement $\Gamma \vdash B$ **type** corresponds to the existence of type B in the category $\mathbf{Ctx}/\Gamma$, the category of all contexts in which Γ holds, or precisely to the statement that B is a type in that category. Shulman (2017) states: "The point is that substitution into a dependent type presents the pullback functor between slice categories". Intuitively speaking this means that substitutions behaves as we are used to having it, where 'free' variables can be replaced by other free variables of the same type.

These developments lead to a proof system by allowing deductions, which are the axioms of judgements. These axioms come in two kinds: ones that allow us to construct a type and others which allow to eliminate in order to prove equivalence. This resembles, for example, the natural deduction, but in this context it is more complex. An example of a deduction rule looks like this:

$$\frac{\Gamma \vdash A \textbf{ type}; \Gamma; n : A \vdash B \textbf{ type}}{\Gamma \vdash \prod_{n:A} B \textbf{ type}}$$

[9]Of course, type theory has no internal logic in the sense of semantics, but the operations of product and join correspond to the existential and universal quantifiers, under the Curry-Howard correspondence, described below. See Escardó (2018). The intuitionistic higher-order logic interpretable in MLTT is often referred to as the internal logic of (dependent) type theory.

where the premise is written in the numerator and the conclusion in the denominator. This model of reasoning is very formal and well suited for implementation on a computer, hence proof assistants such as Coq. Choosing a particular collection of type constructors specifies the rules of a particular type theory, that is, dependent type theory is not a unique theory but rather a template for producing proof theories.

It is possible to interpret propositional logic in type theory, namely we interpret propositions P as types with at most one element. P is 'true' if there is an element in it (these are types of level 1). From the point of view of proofs, saying that type is inhabited amounts to saying that there is a proof of that type. We can also interpret sets, as properties (types of level 2).[10] By a property of elements of a type A we mean a judgement of the form

$$x : A \vdash P : \Omega.$$

Under this interpretation, we obtain the Curry-Howard correspondence (Howard 1980) between the operations of intuitionistic propositional logic and (some of) the constructors of type theory. So theorems of intuitionistic propositional logic are provable in the deduction system. The Law of Excluded Middle LEM does not fall into this correspondence. LEM is addable artificially as a judgement, but then it tends to cause difficulties in the execution of the proof assistants, although it seems relatively harmless in Coq.[11] There are proof assistants based on higher order logic, such as Isabelle/HOL and Isabelle/ZF which are able to deal with AC and LEM. For example, Lawrence C. Paulson in (2002) mechanised Gödel's proof that the consistency of ZF implies that of ZFC using the proof assistant Isabelle. The HoTT libraries of computer proofs are based on Coq.

We note that the full Axiom of Choice is inamicable to the intuitionistic logic in the sense that by proofs of Diaconescu (1975) and Goodman and Myhill (1974), the Axiom of Choice implies the Law of Excluded Middle. In a model of dependent type theory, the Axiom of Choice can be enforced on the level of sets, at the likely price of losing the connection with the proof assistants. A very interesting article about the relation of type theory and the Axiom of Choice is Martin-Löf's (2009), in which it is shown that when a certain amount of choice is added to type theory the resulting theory interprets ZFC.

[10]This may look like type theory is a generalisation of set theory, as mentioned in some references, but this is not true since set theory is not only made of objects but also of the underlying logic and the axioms.

[11]Andrej Bauer in private correspondence (27 February 2018) says "Coq does not loop if we postulate excluded middle, it just gets stuck on some computations, and then requires human intervention to get unstuck."

11.4 An Identity Crisis and Univalence

To explain what the univalence is, let us first discuss some intuitions. Frege in the celebrated Appendix of his Frege (1903) (in reaction to Russell's paradox and attempts to solve it) stated that

> Identity is a relation given to us in such a specific form that it is inconceivable that various kinds of it should occur.

Indeed, in set theory, there is only one kind of equality, given by the Axiom of Extensionality and in model theory it is well known that we can always assume that the equality relation in any model is interpreted by the true equality.

Martin Löf's type theory MLTT builds up on Dependent Type Theory by introducing some new types, including universes $U_0, U_1, \ldots$ such that $U_n : U_{n+1}$ and the identity type $Id_A(x, y)$ for propositions x, y. This type is the type of proofs of equality between x and y, which are both of type A. So if this type is inhabited, it means that the elements x and y of type A are provably equivalent. Another type, called refl proves that any x is equal to itself. This setting gives rise to *two* different kinds of identity: definitional $A = B$ (for types) and $x = y : A$ for elements of a type, and the propositional one $Id_A(x, y)$. Furthermore, building on $Id_A(x, y)$ we can also declare that any two proofs for this identity are equivalent etc., extending the propositional equivalence to all levels of types. Then it is clear (in fact it follows from the rule of identity introduction) that the definitional identity implies the propositional one, but it is not at all clear that the converse should be true.

This problem, which is to some sense the completeness problem for proof theory, was noticed early on by Martin-Löf and others, who tried to resolve it by inserting explicit judgements to this extent. However, in the full generality, this leads to paradoxes. The sameness of the two notions of equality became a known problem and search was open for more subtle models of dependent type theory where this sameness is true.[12]

Intuitively speaking, Voyevodsky's univalence fulfils this dream. It says that the identity type $\mathrm{Id}(X, Y)$ is precisely the type $\mathrm{Eq}(X, Y)$ of equivalences (which we have not defined), in the sense of being in one- to-one correspondence with it. To quote (Grayson 2018), *it offers a language for mathematics invariant under 'equivalence'*. This intuitive understanding of the axiom, present in most presentations of univalence for general audience (hence this one), is somewhat misleading. Much is to be gained by understanding a more exact definition of the univalence, which is possible to do by following the short and exactly-to-the point (Escardó 2018). The conclusion is that univalence is a property of Martin-Löf's identity type of a universe of types. Without univalence, refl is the only way to construct elements of the identity type, with univalence this is also possible by other

[12] This leads to intensional versus extensional type theory and is a long and interesting story, see Hofmann (1995) for more.

means. The *univalence axiom* states that a certain type is inhabited, the univalence type. It takes a number of steps to build the univalence type, for these steps the reader may consult §3 of Escardó (2018).

Voyevodsky's HoTT is an extension of MLTT with the Univalence Axiom, using topological insights which we shall describe in §5. It should be noted that the Univalence Axiom is not a consequence of MLTT as there is also a model of MLTT where it fails, namely any model of Streicher's Axiom K (see Hofmann and Thomas 1998).

11.5 The Topological View

Vladimir Voevodsky came into the field of logic in a rather round-about way, and one may say, somewhat naively. Paraphrasing his talk in the 2013 Logic Colloquium in Evora, Portugal, we learn that after obtaining a Fields medal in 2002, he was feeling increasingly isolated in the position to publish any proof he wanted (due to his reputation) and to have a shrinking audience of real readers (due to the complexity of the proofs in question). He realised that he was not alone in the situation and judged that mathematics badly needed – computerised proof verification. Only then did he understand that there already existed a vibrant community of researchers working not only on proof verification but also on proof assistants. At learning of the relevant issues, Voevodsky realised that the univalency question had appeared in a different part of mathematics under a different name and that it had been successfully solved. The field was homotopy theory and the key to the solution is the notion of an ∞-groupoid (a notion that was already used by some other researchers in the logic community, as it happens, see below for historical remarks). The main idea of Voevodsky was to interpret the category **Ctx** as the homotopy category, so that types become spaces and morphisms. Propositional equality then becomes the homotopic equivalence and the structure obtained becomes an ∞-groupoid. Finally, Voyevodsky's vision become the foundation of mathematics expressed in the language of type theory in which the objects are represented in an ∞-groupoid. This vision is succinctly represented in Coquand's Bourbaki seminar presentation (Coquand 2014).

For us here, let us briefly recall the relevant topological notions and their history, discovered and developed in 1950s by Daniel Kan, building on an even earlier work by Henry Whitehead. We start with the *simplex category* Δ, whose objects are linear orders on a natural number $n = \{0, 1, 2, \ldots, n-1\}$ and whose morphisms are (non-strictly) order-preserving functions. We fix some large enough infinite cardinal number (may be a large cardinal) and consider all sets that have cardinality smaller than that cardinal. This category is called **Set**. A *simplicial set* X is a contravariant functor $X : \Delta \to \mathbf{Set}$. Simplicial sets form a category under natural (the word natural here has a technical meaning) transformations and this category is called **sSet**.

The point is that it is a classical result in algebraic topology that a certain fibration **sSet**$/W$ of **sSet** called Kan complexes is equivalent to the category of 'nice' topological spaces and homotopy classes of maps. Because of this fact one can, when working up to homotopy, think of simplicial sets as of combinatorial representations of shapes or spaces, of simplicial paths as paths on these spaces etc. Kan complexes form a model of an ∞-groupoid, the formal definition of which requires higher category theory, but which in essence deals with a category with objects of every type n for $n < \omega$ and n-morphisms. The homotopy types of the geometric realizations of Kan complexes give models for every homotopy type of spaces with only one non-trivial homotopy group, the so called Eilenberg-Mac Lane spaces. It is conjectured that there are many different equivalent models for ∞-groupoids all which can be realized as homotopy types, some of which are known.

Grothendick's *homotopy hypothesis* states that the ∞-groupoids are equivalent as in the equivalence of categories, to topological spaces.

A central theorem of the homotopy type theory with univalence, and one which gave it its name is the following:

Theorem 1 (Voevodsky, Kapulkin and LeFanu Lumsdaine 2012) *Modulo the existence of two inaccessible cardinals, it is consistent that* **sSet**$/W$ *forms a model of Martin Löf's type theory with the Univalence Axiom.*

In the proof one can also assume LEM on the level of propositions and the Axiom of Choice on the level of sets. Working in this particular model is what became known as the *univalent foundations*, although some other models were found later, as we shall mention.

The revolutionary result of Voevodsky had some precursors and parallels, notably in the work of Michael Makkai (1995), Hofmann and Streicher (1998) and Steve Awodey (see Awodey 2010); Martin Hofmann's Ph.D. thesis Hofmann (1995) already mentioned above, for example, contains a study of the groupoid of intensional equivalences. See Wikipedia (20xx) for a detailed historical description. It also has descendants, which are well worth mentioning. Another model of Univalence Axiom and with less consistency strength, the cubical model, was found by Bezem et al. (2017). Work of Peter Aczel (1999) shows that dependent type theory for constructive sets has a much weaker proof theoretic strength than that of ZF. Namely, this model can be done in constructive set theory CZF with universes. This set theory is proof theoretically strictly weaker than ZFC, and has the same strength as MLTT/dependent type theory (without the Univalence Axiom). A corollary of the work by Bezem et al. (2017) is that adding the Univalence Axiom does not add any proof theoretic power to dependent type theory. The story of the proof-theoretic strength of these systems is rather involved. A lot of results and the history can be found in Michael Rathjen's paper (Rathjen 1994) and, specifically on the strength of the univalence axiom, in the paper (Rathjen 2017).

Furthermore, the relation between this and the classical logic and set theory is also very intricate. Work of Crosilla and Rathjen (2002) shows that the proof-theoretic strength of CZF+ LEM + universes is at least ZFC+ unboundedly many inaccessibles. So it is LEM that adds strength to univalent foundations.

11.6 Conclusion, Crisis or No Crisis

Much discussion has been generated by the advent of the univalent foundations and this volume is certainly a forum for it, as is the HoTT book (The Univalent Foundations Program 2013). To certain fans of HoTT, some of them present at the conference "Foundations of Mathematics: Univalent Foundations and Set Theory" in Bielefeld in 2016 which initiated this volume, this subject means the end of classical mathematics as we know it, as we can just move to a model of univalent foundations and let the ever-developing libraries of computer proofs do all our mathematics for us. A mathematician of the future becomes a new incarnation of a chess player, not capable to come up with anything weakly comparable to what a computer can do on the same subject. The less ardent fans have called to a measured reflexion on the choice between this foundation and the classical ones, implying perhaps the eventual adoption of one or the other. The HoTT book states in the Introduction: "This suggests a new conception of foundations of mathematics, with intrinsic homotopical content, an invariant conception of the objects of mathematics and convenient machine implementations, which can serve as a practical aid to the working mathematician. This is the Univalent Foundations program.... we therefore believe that univalent foundations will eventually become a viable alternative to set theory as the implicit foundation for the unformalized mathematics done by most mathematicians." Viewed from this angle, the situation looks like quite a serious crisis in foundations, putting two foundational programmes in a competition.

Our view is that this crisis is not one and that, as ever in mathematics, mathematical results speak for themselves. First, there is no obvious reason of why one should choose to limit oneself to a specific model of mathematics, be it the simplicial set model or the cubical one or, for that matter, the constructible universe L or some other set-theoretic universe; the prevalence of consistency results in mathematics has rendered us very careful on this issue. But there are also more tangible observations following from the above results. If a theory has the consistency strength less or equal ZFC+ some amount between 1 and 2 of inaccessible cardinals, then this theory cannot resolve any of the questions for which we know that a larger consistency strength is required. This includes many questions in modern logic, such as projective determinacy, the existence of a measurable cardinal, the failure of the singular cardinal hypothesis and so forth. If we take the view that large cardinals represent the ultimate concept of the 'actual infinity' or the 'higher infinite', then these notions cannot be formalised by univalent foundations.[13] On another topic, if a theory has a much weaker strength than that same theory when accompanied by the Law of Excluded Middle, then that theory cannot answer all questions which need this law and cannot reproduce all proofs where this law is used. Some proofs by contradiction have a translation which does not require this law, some can be formalised, but the results quoted above show that this cannot

[13]By the 'higher infinite' we do not mean the uncountable, which can be formalised in type theory. Rather, we follow the usage the reference book on large cardinals by Kanamori (2003).

be the case with all such proofs. And much of modern mathematics uses proofs by contradiction. So, there is no way of formalising into a computerised proof in HoTT all that is known or will be known to mathematicians, even those who are not interested in the actual infinity but simply in the use of classical logic.

Of course, there are and have been many mathematicians (Kronecker, Poincaré to some extent and Brouwer to some extent, for example) who have rejected either the actual infinity or the Law of Excluded Middle or both. The fact is that much of what they, in that vision, considered as mathematics is now within the reach of computer formalisation. Is this a bad or a good fact? It is neither bad or good. Mathematics is not about judgement but about furthering human knowledge and it recognises facts as steps to further advancement. Computer formalisation is a big step forward for advancing certain parts of mathematics and a big philosophical advance which works together with, and not against, classical mathematics. The fact that human beings invented the printing press did not stop the development of writing, it enhanced it, and the same is true of any invention (weapons excluded) of our various civilisations. We should not fear that the computers are going to replace us- because they are not- we should welcome the new insights that they will make us have.

We have already suggested elsewhere (Džamonja 2017) that the solution of the crisis in the foundations might be to recognise the illusion of that ontological *the* in "the foundations". The foundations are there to serve mathematics and not to bound its growth. If we need different foundations for different parts of mathematics, so be it, as long as we do not run into obvious contradictions. The foundations offered by set theory and category theory (which in our own view are different even if bi-interpretable) remain as important as ever, the univalent foundations have their place.[14] Mathematics is also about intuition and it should be done by everybody in a way that maximises their own intuition, it may be ZFC for one and constructive mathematics for another. For example, with our ZFC eyes, we may pick from the article (Bezem et al. 2017) the fact that there are Π^0_1-statements that cannot be proved in MLTT + Univalence and can be proved in ZFC. But, to a mathematician like Voevodsky it may be more striking that there are statements about abstract category theory and abstract homotopy theory that are intuitively constructive (and should not require strong proof theoretic means) but that become

[14]This view is easy to take for a platonist, which the author happens to be, who anyway considers that all our mathematics is only a development towards the unreachable but existent world of real Mathematics and that the creation consists in being able to reach further towards that goal. How to do it is the essence of creativity. However, it does not seem to us that any of the known mathematical philosophy positions contradicts the evidence for pluralists foundations as represented by the above results.

A 'natural hero to logical pluralist', in the words of Greg Restall (2002) is Carnap, since, still citing (Restall 2002), 'for him logic is of fundamental importance to philosophical inquiry, but nothing about this inquiry narrows the field of play to just one logic.'

apparently non effective, needing the Axiom of Choice, when formulated in ZFC.[15] When expressed in set theory, the proof requires the Axiom of Choice, which may be counter intuitive. It would then be natural to look for a formalism where such mathematical fact can be stated and proved in a more direct way, an intuition that it confirmed by the result about constructivity of univalence in Bezem et al. (2017), which is the actual point of the paper from the point of view of its authors.

In fact, a more careful reading of the history of logic than is within the scope of this article shows that various encodings of the foundations have always coexisted.[16]

Let us finish by a quote from the masters who wisely predict the fate of foundations of mathematics, in words that entirely represent our own view. Bourbaki writes in the introduction to the first volume of Bourbaki (1966)[17]:

> We now know that, logically speaking, it is possible to derive almost all present-day mathematics from a unique source, set theory. By doing this we do not pretend to write a law in stone; maybe one day mathematicians will establish different reasoning which is not formalisable in the language that we adopt here and, according to some, recent progress in homology suggests that this day is not too far away. In this case one shall have to, if not, totally change the language, at least enlarge the syntax. It is the future of mathematics that will decide this.

Acknowledgements Many thanks to the organisers of the FOMUS conference in July 2016 for their invitation to give a talk and to participate in the panel discussion. I would also like to thank Peter Aczel, Andrej Bauer, Mark Bezem, Thierry Coquand, Laura Crossila, Deborah Kant, Angus Mcintyre, Marco Panza, Duško Pavlović, Michael Rathjen, Christian Rosendal and Andrés Villaveces, as well as to the anonymous referee, for very useful and interesting discussions about various parts of this paper. My thanks equally go to the audiences in Paris, Nancy, Oxford, Teheran and Mexico City who have listened and contributed by their comments to the talk that accompanied the development of this paper.

---

[15] We thank Thierry Coquand for the following example of such a statement: a functor which is essentially surjective and full and faithful is an equivalence, as well as for the above comments.

[16] Several readers of the first draft of the paper have commented that one should more didactically stress as a conclusion something that emerges as evidence from all stated above. So I will take the invitation to state what seems to me as an obvious conclusion, or more positively, a good beginning for future reflection. Pluralism is the only reasonable and honest choice in the foundations of mathematics today. Machines, including computers, bring forward human knowledge but do not replace it and it is by embracing what they have to offer that we are going to advance, not by positioning ourselves in camps in which ones fight with chalk and the others with computer chips.

Of course, we could explain and explain some more, but to quote Duško Pavlović, from his comments on this paper, which are so interesting and detailed that they could in themselves form another paper:

> There are so many misunderstandings between good people for bad reasons, which it should be possible to clear. But the conundrum is: the more we try to explain, the more words we write, there is more chance for further misunderstandings. I think the only way to resolve such a conundrum is honesty and trust.

[17] The translation here is by the present author.

References

Aczel, P. (1999). On relating type theories and set theories. Available on author's web site.

Awodey, S. (2010). Type theory and homotopy. arXiv:1010.1810 [math.CT].

Bezem, M., Coquand, T., & Huber, S. (2017). The univalence axiom in cubical sets. arXiv:1710.10941 [math.LO].

Bourbaki, N. (1966). Éléments de mathématique. Fasc. XVII. Livre I: Théorie des ensembles. Chapitre I: Description de la mathématique formelle. Chapitre II: Théorie des ensembles. Actualités Scientifiques et Industrielles, No. 1212. Troisième édition revue et corrigée. Hermann, Paris.

Coquand, T. (2014). *Théorie des types dépendants et axiome d'univalence.* Presentation at the Bourbaki Seminar, Juin 2014, author's web site.

Coquand, T. (2018). Type theory. https://plato.stanford.edu/entries/type-theory/#7

Crosilla, L., & Rathjen, M. (2002). Inaccessible set axioms may have little consistency strength. *Annals of Pure and Applied Logic, 115*(1–3), 33–70.

Diaconescu, R. (1975). Axiom of choice and complementation. *Proceedings of AMS, 51*, 176–178.

Džamonja, M. (2017). Set theory and its place in the foundations of mathematics-a new look at an old question. *Journal of Indian Council of Philosophical Research, 34*, 415–424.

Escardó, M. H. (2018). A self-contained, brief and complete formulation of Voevodsky's univalence axiom. arXiv:1803.02294v2 [math.LO], March 2018.

Feferman, S. (1977). Categorical foundations and foundations of category theory. In *Logic, foundations of mathematics and computability theory. Proceedings of the Fifth International Congress of Logic, Methodology and Philosophy of Science, University of Western Ontario, London, ON, 1975, Part I* (University of Western Ontario Series in Philosophy of Science, Vol. 9, pp. 149–169). Dordrecht: Reidel.

Fraenkel, A. A., Bar-Hillel, Y., & Lévy, A. (1973). *Foundations of set theory* (2nd revised ed.). Amsterdam: North-Holland.

Frege, G. (1903). *Grundgesetze der Arithmetik. Band I, II* (2009 reprint ed.). Paderborn: Mentis Verlag GmbH. Begriffsschriftlich abgeleitet. [Derived in conceptual notation], Transcribed into modern formula notation and with a detailed subject index provided by Thomas Müller, Bernhard Schröder and Rainer Stuhlmann-Laeisz.

Fremlin, D. H. (1984). *Consequences of Martin's axiom. Cambridge tracts in mathematics.* Cambridge: Cambridge University Press.

Girard, J.-Y. (1972). *Interprétation fonctionelle et élimination des coupures de l'arithmétique d'ordre supérior*. Ph.D. thesis, Université Paris VII.

Goodman, N. D., & Myhill, J. (1974). Choice implies excluded middle. *Zeitschrift fur Mathematische Logik und Grundlagen der Mathematik, 24*, 461.

Grayson, D. R. (2018). An introduction to univalent foundations for mathematicians. arXiv:1711.01477v3 [math.LO], February 2018.

Harris, M. (2017). *Mathematics without apology: Portrait of a problematic vocation.* Princeton: Princeton University Press (2017)

Hofmann, M. (1995). *Extensional concepts in intensional type theory.* Ph.D. thesis, University of Edinburgh.

Hofmann, M., & Streicher, T. (1998). The grupoid interpretation of type theory. In G. Sambin & J. M. Smith (Eds.), *Twenty five years of constructive type theory* (pp. 83–112). Oxford: Clarendon Press.

Howard, W. A. (1980). The formulae-as-types notion of construction. In J. P. Seldin & H. J. Roger (Eds.), *To H.B. Curry: Essays on combinatory logic, lambda calculus and formalism* (pp. 479–490). London: Academic. Original manuscript (1969).

Hurkens, A. J. C. (1995). A simplification of Girard's paradox. In M. Dezani-Ciancaglini & G. Plotkin (Eds.), *Typed lambda calculi and applications* (pp. 266–278). Berlin/Heidelberg: Springer.

Kanamori, A. (2003). The higher infinite. Springer monographs in mathematics. In *Large cardinals in set theory from their beginnings* (Second ed.). Berlin: Springer.

Kapulkin, C., & LeFanu Lumsdaine, P. (2012). The simplicial model of univalent foundations (after Voevodsky). arxiv:1211.2851.

Macintyre, A. Wiles' proof only needs PA. Private communicaton on an announced result.

Makkai, M. (1995). *First order logic with dependent sorts, with applications to category theory.* Authors's webpage.

Martin-Löf, P. (2009). 100 years of Zermelo's axiom of choice: What was the problem with it? In *Logicism, intuitionism, and formalism* (Volume 341 of Synthese Library, pp. 209–219). Dordrecht: Springer.

Paulson, L. C. (2002). *The relative consistency of the axiom of choice – Mechanized using Isabelle/ZF* (Technical report UCAM-CL-TR-551). Cambridge: Cambridge University.

The Univalent Foundations Program. (2013). *Homotopy type theory: Univalent foundations of mathematics*. Princeton: Institute for Advanced Study.

Rathjen, M. (1994). The strength of some Martin–Löf type theories. *Archive for Mathematical Logic, 33*, 347–385.

Rathjen, M. (2017). Proof theory of constructive systems: Inductive types and univalence. In G. Jäger & W. Sieg (Eds.), *Feferman on foundations – logic, mathematics, philosophy*. Cham: Springer.

Restall, G. (2002). Carnap's tolerance, language change and logical pluralism. *Journal of Philosophy, 99*, 426–443.

Shulman, M. (2017). Homotopy type theory: The logic of space homotopy type theory: The logic of space homotopy type theory: The logic of space homotopy type theory: Logic of space. arXiv:1703.03007v1, March 8, 2017.

Whitehead, A. N., & Russell, B. (1990). *Principia mathematica* (2nd ed.). Cambridge: Cambridge University Press.

Wikipedia. (20xx). *Homotopy type theory*. https://en.wikipedia.org/wiki/Homotopy_type_theory

Wiles, A. (1995). Modular elliptic curves and Fermat's last theorem. *Annals of Mathematics (2), 141*(3), 443–551.

Zermelo, E. (1908). Untersuchungen über die Grundlagen der Mengenlehre. I. *Mathematische Annalen, 65*(2), 261–281.

Chapter 12
A Comparison of Type Theory with Set Theory

Ansten Klev

Abstract This paper discusses some of the ways in which Martin-Löf type theory differs from set theory. The discussion concentrates on conceptual, rather than technical, differences. It revolves around four topics: sets versus types; syntax; functions; and identity. The difference between sets and types is spelt out as the difference between unified pluralities and kinds, or sorts. A detailed comparison is then offered of the syntax of the two languages. Emphasis is put on the distinction between proposition and judgement, drawn by type theory, but not by set theory. Unlike set theory, type theory treats the notion of function as primitive. It is shown that certain inconveniences pertaining to function application that afflicts the set-theoretical account of functions are thus avoided. Finally, the distinction, drawn in type theory, between judgemental and propositional identity is discussed. It is argued that the criterion of identity for a domain cannot be formulated in terms of propositional identity. It follows that the axiom of extensionality cannot be taken as a statement of the criterion of identity for sets.

12.1 Introduction

Set theory and type theory are both said in many ways. Here I shall understand by set theory standard axiomatic set theory, thus ZFC or some variation thereof. By type theory I shall understand Martin-Löf type theory. There are in fact several versions of Martin-Löf type theory, but the differences between these will not play any significant role for us. The only feature I shall assume that is not common to all versions of type theory is a hierarchy of higher types, also called a logical framework. Homotopy type theory is an extension of Martin-Löf type theory obtained by adding the univalence axiom and rules allowing the formation of so-called higher inductive types. Most of what I shall say regarding type theory should

A. Klev (✉)
Institute of Philosophy, Czech Academy of Sciences, Prague 1, Czechia
e-mail: klev@flu.cas.cz

S. Centrone et al. (eds.), *Reflections on the Foundations of Mathematics*,
Synthese Library 407, https://doi.org/10.1007/978-3-030-15655-8_12

apply equally well to this extension (which, of course, is not to say that everything one can say regarding Martin-Löf type theory applies equally well to homotopy type theory).

The aim of this contribution is to note some of the ways in which type theory differs from set theory and the predicate logic that it is written in. I hope to make it plausible that the somewhat complex surface structure of type theory hides a very rich conceptual architecture that may be of considerable value to those interested in the philosophical analysis of logical and mathematical notions. I will thus concentrate on the more conceptual aspects of type theory, rather than its more technical aspects. Philosophically inclined logicians and logically inclined philosophers make up the intended audience. It is not presupposed that the reader is an expert in set theory or type theory.

The topics I shall deal with are: sets versus types (Sect. 12.2); syntax (Sect. 12.3); functions (Sect. 12.4); and identity (Sect. 12.5). Each of these topics provides a lens through which fundamental differences between type theory and set theory may be observed. For the purposes of a comparison of the kind intended here other topics could also have been included. One that springs to mind is axioms (as in set theory) versus rules (as in type theory); but, for reasons of space and time, a study of such further topics will have to wait for another occasion.

12.2 Sets and Types

Set theory is a theory of sets and type theory a theory of types. Both the notion of set and the notion of type may be described as the notion of a class, group, sum, totality, plurality, manifold, etc. of things. The way in which sets are handled in set theory is, moreover, quite similar to the way in which types are handled in type theory. For instance, in set theory we have a set of natural numbers, whose existence is guaranteed by the axiom of infinity; and in type theory a type of natural numbers, whose existence is guaranteed by the rules of natural-numbers formation, introduction, and elimination. In set theory we have an empty set, $\emptyset$, whose existence is either postulated outright or taken to follow from the existence of any other set by application of the axiom of separation; and in type theory a type, often written $\mathbf{0}$, that by definition is empty, namely by its not having any introduction rule. And just as new sets may be formed from old by various set-theoretical operations (union, intersection, pairing, separation, power set), so new types may be formed from old by various type-theoretical operations, indeed operations that often are similar to ones available in set theory (disjoint union of two types, disjoint union of a family of types, generalized Cartesian product).

The fundamental notion of set theory would thus appear to be quite similar to the fundamental notion of type theory. This appearance fades on closer scrutiny: sets and types are in fact objects of quite different natures. The difference is implicit in Cantor's very first definition of the notion of set (my translation)[1]:

> A manifold (a sum, a set) of elements belonging to some conceptual sphere is called well-defined if, on the basis of its definition and in accordance with the logical principle of the excluded third, it must be regarded as internally determined, both whether any object of that conceptual sphere belongs as an element to the mentioned set, and also whether two objects belonging to the set, in spite of formal differences in the mode of givenness, are equal to each other or not.

Cantor here presupposes a distinction between a set and the conceptual sphere to which the elements of the set belongs. A type can very well be conceived of as a conceptual sphere: a type is a kind, or sort, of thing, whence the things that fall under, or belong to, a type have a certain homogeneity and may therefore be said to constitute a conceptual sphere. Implicit in Cantor's definition is thus the distinction between, on the one hand, a set as a collection of things and, on the other, a type, that is, a kind, or sort, of things.

A similar distinction is quite explicit in another foundational document of set theory. Zermelo (1930), namely, draws a distinction between sets (*Mengen*) and domains of sets (*Mengenbereiche*). A domain of sets in Zermelo's sense is what one today more commonly calls a model of set theory, or a universe of sets. It is this notion of a domain, or universe, of sets rather than that of an individual set that most closely resembles the notion of a type. This understanding of the relation of sets and types is nicely illustrated by Aczel (1978), who provides a model of the constructive set theory CZF—which by the addition of the law of excluded middle yields classical ZF—inside a particular type **V** described by Aczel (p. 61) as "a type theoretic reformulation of the classical conception of the cumulative hierarchy of sets". Thus a universe of sets in the sense of CZF is taken to be a certain type, and an individual set to be an object of this type. De Bruijn (1975) assumes the same understanding of the relation between sets and types in his discussion of how to implement set theory in type theory: he suggests doing so by the introduction of a separate type, SET, and a relation, $\in$, on this type subject to one's favourite axioms of set theory.

Zermelo is careful to distinguish ordinals from sets representing ordinals inside a domain of sets. Thus he seems to assume that besides domains of sets there is also a domain of ordinals. The same assumption is upheld in type theory, where there is in fact an indefinite range of domains of ordinals, namely types usually written $\mathbf{W}(A, B)$, where A is a type of individuals and B a family of types of individuals over A. This illustrates the general point that in type theory each kind of object has its own domain. For instance, a type of ordinals $\mathbf{W}(A, B)$ is not a "subtype" of any

[1] Cantor (1882, pp. 114–115): "Eine Mannichfaltigkeit (ein Inbegriff, eine Menge) von Elementen, die irgend welcher Begriffssphäre angehören, nenne ich *wohldefinirt*, wenn auf Grund ihrer Definition und in Folge des logischen Princips vom ausgeschlossenen Dritten es als *intern bestimmt* angesehen werden muss, *sowohl* ob irgend ein derselben Begriffssphäre angehöriges Object zu der gedachten Mannichfaltigkeit als Element gehört oder nicht, *wie auch* ob zwei zur Menge gehörige Objecte, trotz formaler Unterschiede in der Art des Gegebenseins einander gleich sind oder nicht."

other type, in particular not of Aczel's **V**; nor is the type of natural numbers **N** a "subtype" of any other type, in particular not of any type of ordinals $\mathbf{W}(A, B)$; nor, finally, is the type **n** that by definition has n elements a "subtype" of any other type, in particular not of the type **N**.

According to another of Cantor's definitions, a set is "any plurality that can be thought as a one".[2] The pair of notions of many and one, or plurality and unity, has a long history in philosophy. For example, Parmenides held that what there is is one; his pupil Zeno argued that there is no plurality; and Anaxagoras countered that what there is is a plurality.[3] An example closer to our concerns here is the definition, found in Aristotle (*Metaphysics* XI.9) and Euclid (*Elements* VII Def 2), of a number as a plurality of units or indivisibles. It is natural to think of numbers thus understood as sets of units, as Cantor himself defines numbers.[4] Indeed, it is natural to regard the pair of notions of plurality and unity as the ancient predecessor of the notion of set: a set is a unified plurality. Such is the understanding of sets assumed not only by the early pioneers of set theory, viz., Cantor and Dedekind (cf. his 1888, art. 2), but also by the, nowadays widely accepted, iterative conception of set (cf. e.g. Linnebo 2010).

The notion in traditional philosophy that most appropriately may be called a predecessor of the notion of type as understood in type theory is, to my mind, not the notion of plurality, but rather that of genus. According to Aristotle's definition (*Topics* I.5), the term P is a genus of the term S if P reveals *what* S is, in other words, if P is an appropriate answer to the question of what S is. Aristotle adds a qualification to the effect that S cannot be an individual term. Thus he regards living being as a genus of animal and colour as a genus of white, but not man as a genus of Socrates or white as a genus of a trope of whiteness, such as that found in a piece of paper on my desk. The tradition, especially following Porphyry's *Isagoge* (cf. Barnes 2003), speaks of the predicate in these latter cases as a species of its subject. But also a species is taken to be a predicate that reveals a *what* of its subject; hence, for our purposes—which are not exegetical—it seems justified to say that P is a genus of S also when S is an individual term, provided, of course, that P reveals what S is.

It is indeed quite natural to think of the type of an object in type theory as the *what* of that object. What is 5? A natural number. What is cos? A function from $\mathbb{R}$ to $\mathbb{R}$. Any mathematical object is an object of a certain type, and not just an object full stop. In taking this view of mathematical ontology, Martin-Löf type theory differs not only from set theory, but also from type theories with a single type of individuals, usually called ι. Such are, for instance, the simple type theories of Frege (1893) and

---

[2]Cantor (1883, p. 587): "Unter einer Mannichfaltigkeit oder Menge verstehe ich nämlich allgemein jedes Viele, welches sich als Eines denken lässt…"

[3]For these claims and arguments see Kirk et al. (1983, esp. pp. 249–251, 266, 358). For a general overview of the topics of plurality and unity in ancient and mediaeval philosophy, see Meier-Oeser (2001).

[4]See Cantor (1895, p. 482) and the discussion in Hallett (1984, pp. 119–141).

Church (1940) as well as the ramified type theory of the *Principia Mathematica* (Russell and Whitehead 1910). Here all individuals are regarded on a par, and natural numbers, for instance, are coded as objects in some already constituted type rather than taken to constitute their own type of individuals. In this respect, therefore, such type theories are more closely related to set theory than to Martin-Löf type theory.[5]

Both "kind" and "sort" are natural English translations of the Greek *genos*; and both have been used in the sense of "type" in logic. Thus, the various sorts, or domains of individuals, in many-sorted logic are just types of individuals[6]; and in the so-called logical framework of Harper et al. (1993) higher types are called "kinds". Contemporary philosophy also knows the notion of a sortal concept, which moreover plays a role in philosophical logic (cf. e.g. Lowe 1989). Russell (1903, § 497), when introducing the term "type" in the relevant sense into logic, could well have chosen "kind", "sort" or "category"; but perhaps he wanted a term less tainted by the philosophical tradition. The Greek *tupos*, from the verb *tuptō*, to beat, or strike, originally meant the effect of a blow, thus an impression or engraving or the like; but it also meant figure as well as the form or shape of something.[7] In English, "type" was first used to mean symbol, so that a river might be called a type of human life; but it now also means, among other things: (i) a general form or pattern of a class or kind of things; (ii) such a kind or class itself; (iii) an example—sc., a typical example, an archetype or prototype—of such a class.[8] Of these different meanings, a type in Russell's sense, that is, in the sense of type theory, comes closest to (ii). When philosophers contrast types with tokens, and when mathematicians speak of order-types and isomorphism-types, they rather follow sense (i).

Cantor's first definition of set may be regarded as too restrictive in its requirement that the elements of a set should all be taken from one conceptual sphere. It is, namely, a natural thought that a set can collect together things of arbitrary kinds. There might, for instance, be a set containing the author of this paper, the Mona Lisa, the boolean value t, and the empty set. The notion of set may thus be said to be formal in the sense of topic-neutral: a set is not tied to any one topic, any one conceptual sphere, but can have elements from arbitrary conceptual spheres. Types, by contrast, may be thought of as being just these topics or conceptual spheres themselves. In the terminology of Husserl (1913) we may call the notion of set a formal category, and types may be called regions (for more discussion, see Klev 2017a). This is then yet another articulation of the basic difference between the notion of type and the notion of set that has been emphasized here and that consists

---

[5]The relation is closer still if the simple hierarchy of types is taken to be a hierarchy of classes; see Quine (1956) and Linnebo and Rayo (2012).

[6]The idea of a many-sorted language is implicit in Hilbert (1899). An early, perhaps the first, use of the term "many-sorted logic" (or, its German cousin), as well as a definition of the thing itself, can be found in Schmidt (1938). Schmidt (p. 485) emphasizes that he uses "sort" instead of "type" or "kind" so as to avoid the association of levels characteristic of type theory.

[7]Cf. Liddell & Scott's *Greek–English Lexicon*, s.v. "τύπως".

[8]Cf. *Oxford English Dictionary*, s.v. "type".

in a type's being a kind, or sort, and a set's being a collection, namely a collection of objects each of which is of some kind.

12.3 Syntax

The syntax of a formal language may seem like a rather technical topic, of more interest to the metamathematician or the programmer than to the philosopher. But the syntax of a formal language is the skeleton of its conceptual architecture, so if we want to understand the latter, we should also need to understand the former. In particular, if we wish to compare conceptual aspects of set theory and type theory, we shall do well in comparing the syntax of the languages they are written in.

12.3.1 Type Structure

Some remarks on type structure are necessary to begin with. I will assume the presence of a hierarchy of higher types in Martin-Löf type theory (cf. Nordström et al. 1990, chs. 19–20). The types in this hierarchy may be divided into three kinds: types of individuals, or individual domains; function types; and a "supertype" **dom** of types, or domains, of individuals. This latter type is often called **set** in the literature, but since we are here using "set" in the sense of set theory, another name seems advisable.[9]

Examples of types of individuals are the type **N** of natural numbers and the type **bool** of boolean values, or truth-values. Being types of individuals, both **N** and **bool** are objects of type **dom**.

It is characteristic of function types in Martin-Löf type theory that they are *dependent*. That a function f is of dependent type means that the type of $f(a)$ depends on the argument a. Thus, for different arguments, a and b, the type of $f(a)$ may differ from the type of $f(b)$.[10] For our purposes here it is, however, enough to consider the simpler case of non-dependent function types. Thus we say that $(A)B$ is a type, namely the type of functions from A to B, whenever A and B are types. The notion of function will be discussed in more detail in Sect. 12.4 below.

For the development of logic inside type theory a type of propositions is required. In the simple type theory of Church (1940) the type o of propositions is, in the presence of an axiom of extensionality, just the type of truth-values. Truth-values

[9]The use of the term "set" for types of individuals, perhaps inspired by Bishop (1967, p. 13), goes back to Martin-Löf (1984). "Domain" was used for the same purpose in Diller and Troelstra (1984). One reason why this, rather natural, terminology was not taken up by Martin-Löf himself seems to have been the use of the term "domain" for a very different notion in Dana Scott's so-called theory of domains (e.g. Scott 1982).

[10]For a general discussion of dependent types, see Aspinall and Hofmann (2005).

were in fact present already in the logic of the *Grundgesetze* (Frege 1893), where they served as the meanings (*Bedeutung*) of sentences. For Frege, however, truth-values did not constitute a separate type, but were part of the one grand universe of individuals.[11] In Martin-Löf type theory propositions are identified with types of individuals, in line with the so-called propositions-as-types principle, or Curry–Howard isomorphism.[12] On the one hand, a proposition A may be regarded as a type of individuals, namely as the type of its proofs, or truthmakers. On the other hand, a type of individuals A may be regarded as the proposition that A is inhabited. We may therefore identify the type **dom** of types of individuals with the type **prop** of propositions:

$$\mathbf{prop} = \mathbf{dom}$$

A type of propositions is thus accommodated within the hierarchy of higher types by being identified with the "supertype" **dom** of individual domains.[13] The resulting conception of proposition is, of course, very different from that assumed in the simple theory of types and predicate logic.

We may also think of the language of set theory as presupposing a certain type structure. Namely, there is a type V of individuals called sets; there is a type **prop** of propositions; and for any types A and B, there is a type $(A)B$ of functions from A to B. How propositions are conceived in this context is of little consequence for us; what matters only is that there is a type of propositions.

The function types introduced above are types of unary functions. Functions of greater arity may be explained by the technique of "Currying".[14] Note first that if A, B, and C are types, then so is $(B)C$ and therefore also $(A)(B)C$. But a function f of type $(A)(B)C$ may be regarded as a binary function whose first argument is of type A and whose second argument is of type B. Thus, for suitable arguments a and b we may write $f(a, b)$ instead of $f(a)(b)$. Conversely, if f is such a binary function, then for any a of type A, we have a unary function $f(a, \_)$ from B to C. Thus we may regard f as a function of type $(A)(B)C$. Similar considerations apply to functions of arity greater than two.

[11] In the reconstruction of Frege's type hierarchy given by Dummett (1973, pp. 44–45) there is a separate type of propositions.

[12] See Howard (1980), Martin-Löf (1984, pp. 5–6, 11–13), de Bruijn (1995), or Wadler (2015).

[13] In homotopy type theory there is also a more specific conception of propositions. A so-called *mere proposition* is a type of individuals that, intuitively, has at most one element; see The Univalent Foundations Program (2013, ch. 3).

[14] Systematic employment of this technique stems from Schönfinkel (1924, § 2), but the idea is present already in Frege (1893, § 36).

12.3.2 Terms and Types

In the presentation of the syntax of a predicate-logical language it is usual to distinguish terms and formulae. The corresponding distinction in a type theory adhering to the propositions-as-types principle is the distinction between terms and types. Indeed, the propositions-as-types principle is sometimes called the formulae-as-types principle. When comparing the syntax of set theory with the syntax of type theory it is therefore natural to begin by comparing the way in which terms and formulae, respectively types, are formed.

In the language of set theory there are no primitive term-forming operators, nor any primitive individual constants; hence, variables are the only terms formable by the primitive vocabulary. The rules for forming formulae follow the usual pattern of predicate logic, with the two basic schemes: (i) function applied to a suitable number of arguments; (ii) variable-binding operator applied to a suitable number of operands.[15] Thus we are thinking of $\in$ and $=$ as binary functions on V, namely of type $(V)(V)\mathbf{prop}$, and of the propositional connectives as functions on **prop**. The quantifiers are the variable-binding operators of set theory.

It is well known that the application of a variable-binding operator to an operand may be analyzed as a combination of pure variable-binding and function application. For instance, the application of the quantifier $\forall x$ to a formula $A(x)$ may be analyzed as

$$A(x) \dashrightarrow x.A(x) \dashrightarrow \forall(x.A(x)),$$

where $x.A(x)$ is $A(x)$ with the variable x bound. Since we have assumed the presence of a hierarchy of higher types, we may, moreover, identify variable-binding with abstraction, sc., lambda-abstraction, or functional abstraction. What is here written $x.A(x)$ may thus be thought of as a function from V into **prop**, thus of type $(V)\mathbf{prop}$. A more familiar notation for $x.A(x)$ would be $\lambda x.A(x)$, but we shall employ the lambda for a slightly different purpose below. The universal quantifier, finally, may then be regarded as a higher-order function, of type $((V)\mathbf{prop})\mathbf{prop}$, for it takes a function, say $x.A(x)$, and yields a proposition $\forall(x.A(x))$.

Both the terms and the types of type theory are formed by the same two schemes (i) and (ii). In particular, in the version of type theory assumed here there is only one variable-binding operator, namely abstraction.[16] The modes of construction in set theory and type theory, both endowed with a hierarchy of higher types, are thus the same: function applied to a suitable number of arguments and variable-binding in the form of abstraction. The primitive vocabulary available for forming terms and types in Martin-Löf type theory is, however, much richer than the primitive vocabulary of set theory. Indeed, whereas the primitive vocabulary of set theory consists of

[15]For the distinction between function and operator, see Church (1956, §§ 3, 6).

[16]Dependent function types introduce a further variable-binding operator; thus with dependent function types present, there are two variable-binding operators.

$\in$ and $=$ together with the logical connectives and the quantifiers, the range of primitive constants in Martin-Löf type theory is indefinite. Martin-Löf type theory is, namely, open in the sense that the objects of type **dom** are not given once and for all, in a list or inductive definition, say. Rather, objects of type **dom** (i.e., types of individuals) and functions with co-domain **dom** may, in effect, be introduced at will, subject only to the requirement that the types thus introduced or formable satisfy the general description of what a type of individuals is (cf. Martin-Löf 1984, pp. 7–8). In general, this requirement is met by providing any newly postulated type of individuals A or function Φ forming such types with one or more introduction rules. Once introduction rules for A or Φ are provided, an elimination rule may be formulated,[17] whereby further pieces of primitive vocabulary are introduced.

The openness of the primitive vocabulary of type theory may be illustrated by the following, somewhat perverse, example. In order to do arithmetic in type theory one usually introduces a type of natural numbers, **N**, and three associated constants: the individual constant 0, the successor function, s, and a recursor R, which makes possible definition by recursion and proof by induction on **N** (e.g. Martin-Löf 1984, pp. 71–76). Assuming all of this to have been introduced, there is, however, nothing stopping one from introducing further primitive vocabulary, $\mathbf{N}^*$, 0^*, s^*, and R^*, satisfying rules structurally identical to those governing the corresponding non-starred constants. In other words, there is nothing stopping one from introducing a duplicate of the type of natural numbers and its associated constants. There might not be much use in such duplication; but the question of usefulness is besides the point here, where we just want to illustrate the openness of the primitive vocabulary of type theory. One can of course exploit this openness in more inventive ways than mere duplication of well-known types and their associated constants. In fact, inductively defined types quite generally are to be formalizable in type theory; and any such formalization introduces new primitive vocabulary.

12.3.3 Judgements

Besides term and type, there is in type theory a further logico-grammatical category: judgement.[18] At the most general level, there are two forms of (categorical) judgement in type theory:

$$a : \mathscr{C}$$
$$a = b : \mathscr{C}$$

[17] Dybjer (1994, 2000) provides a general scheme for the formulation of such rules.

[18] The term "judgement" in this sense was introduced by Martin-Löf (1982). The term had of course been used in logic before, but it had fallen out of fashion during, say, the first decades of the twentieth century; see e.g. Carnap (1934, § 1). For the early history of the use of the term in logic, see Martin-Löf (2011); for aspects of the later history, see Sundholm (2009).

The first may be read as saying that a is an object of category $\mathscr{C}$ and the second as saying that a and b are identical objects of category $\mathscr{C}$. Here "category" is used not quite as in "logico-grammatical category", but in a more ontological sense. This use of the term will seem quite natural if (i) we analyze $a : \mathscr{C}$ as being of the form subject–copula–predicate, and (ii) keep in mind that one of the meanings of the Greek *katēgoria* is "predicate". The specific forms of judgement assumed in any particular version of type theory depends on which categories are assumed. In the version of type theory assumed in this paper the categories are **type** and A, for any **type** A. Thus we get the following forms of (categorical) judgement:

$$A : \textbf{type}$$
$$A = B : \textbf{type}$$

and for any **type** A,

$$a : A$$
$$a = b : A$$

Examples of judgements are thus

$$\textbf{N} : \textbf{type}$$
$$\mathsf{t} : \textbf{bool}$$
$$\mathsf{s} : (\textbf{N})\textbf{N}$$
$$\mathsf{s}(2+1) = 2 + \mathsf{s}(1) : \textbf{N}$$

It is essential for anyone wishing to understand the relation between the language of set theory and the language of type theory to appreciate the difference between the type theorist's $a : A$ and the set theorist's $a \in b$. The first says that a is an object of the type, or kind, or sort A, whereas the latter says that, within the universe, or type, of sets, the relation $\in$ obtains between a and b. Instances of the form of judgement $a : A$ in the type structure of set theory are

$$\emptyset : V$$
$$\{\emptyset, \{\emptyset\}\} : V$$
$$\in : (V)(V)\textbf{prop}$$
$$Ord : (V)\textbf{prop},$$

where $Ord(a)$, for $a : V$, is the proposition that a is a Von Neumann ordinal. The first two of these judgements says that $\emptyset$, respectively $\{\emptyset, \{\emptyset\}\}$, is a set; the third that $\in$ is a binary propositional function; and the fourth that Ord is a unary propositional function on the universe of sets V. These are things that cannot be expressed in the language of set theory, in particular not by a formula of the form $a \in b$. That $\emptyset$ is a set, for instance, cannot be expressed by a set-theoretical formula. In fact, reflection on type predications—assertions that such and such is an object of some type—

shows that they cannot be taken to have the form of predicate-logical formulae (cf. Klev 2018a). We are not forced to conclude, as Wittgenstein (1922, esp. 4.126) seems to have done, that type predications cannot be expressed at all, that the type of an object a is something that cannot be said, but only shown in our use of a symbol for a. One can instead introduce—and explain the meaning of—a new form of statement for the expression of type predications. The judgement $a : A$ is such a novel form of statement.

When a type A is regarded as a proposition, an object a of type A may be regarded as a proof, or a truthmaker, of A.[19] The judgement $a : A$ is thus to be read as saying that a is a proof of the proposition A.[20] It should be emphasized that $a : A$ is not itself a proposition. In fact, the logico-grammatical categories of judgement and proposition are disjoint: no judgement is a proposition, nor is any proposition a judgement. Thus, iterated colons, as in "$b : (a : A)$", make no sense in type theory.[21] Nor is there any need for such iterated colons, since the relation between a term and its type is taken to be internal and not in need of proof: if a is of type A, then its being so can be seen either immediately from the form of a or else from the form of what results after a sequence of simple computation steps. This point is closely related to "the basic intuitionistic idealization that we can recognize a proof when we see one" (Kreisel 1962, p. 202) and to the requirement on a type theory that the relation between a term and its type be decidable in the recursion-theoretic sense.

The theorems of type theory are judgements and not propositions. Thus it is not

$$A$$

but

$$a : A$$

that is the form of a theorem of type theory (albeit not the only form). The logical form of a theorem must be the form of something that is assertion-apt, in the sense of being suitable as the object of an act of assertion. Propositions—be they conceived as types of proofs, as truth-values, or as sets of possible worlds, say—are not assertion-apt in this sense. The judgement that a is a proof of A, by contrast, is assertion-apt and thus has the logical form appropriate for a theorem.

If we can demonstrate that a proposition A has a proof, or a truthmaker, then we are certainly justified in asserting that A is true. Thus it makes good sense to

[19]For the relation between, on the one hand, recent discussions on truthmakers and truthmaker semantics and, on the other, Martin-Löf type theory, see Sundholm (1994) and Klev (2017b).

[20]In the literature, also the term "proof-object" is used, following Diller and Troelstra (1984).

[21]This contrasts with so-called justification logic (cf. Artemov and Fitting 2016), where $a : A$, understood as "a is a justification of A", is a proposition. Thus, for instance $b : (a : A)$ and $(a : A) \supset A$, are well-formed formulae there.

introduce a new form of judgement A *true* governed by the rule

$$\frac{a : A}{A\ \mathit{true}}$$

It must be emphasized that the *true* occurring here is not a truth-predicate in the ordinary sense. A predicate in the usual sense always yields a proposition (or formula) when applied to a suitable number of arguments, but A *true* is a judgement and not a proposition. This *true* should rather be compared to the sign $\vdash$ of judgement, or assertion, in Frege's and Russell's systems:

$$\vdash A$$

This assertion sign, the so-called turnstile, is not used in forming new propositions, but in forming a judgement from a proposition. Thus, unlike the propositional connectives and the quantifiers, the assertion sign turns a proposition into something assertion-apt. Employing grammatical terminology, we may say that the assertion sign supplies the mood of the assertion $\vdash A$, whereas the proposition A is the content. A content by itself is not assertion-apt; what is required for assertion is a composite of content and mood, more specifically, of content and assertoric mood.

In the language of set theory, as in predicate-logical languages more generally, no distinction similar to that between proposition and judgement is drawn. In particular, the theorems of set theory are not regarded as being of a logico-grammatical category different from that of formulae. The string of symbols $\vdash A$ can of course be found in presentations of set theory, but it is then regarded as expressing a metamathematical property of A, namely that there exists a sequence of formulae coding a derivation of the formula A. We may regard this metamathematical "$\vdash A$" as a judgement, but only as a judgement in the metalanguage of set theory (or any other formal theory) and not as a judgement of set theory proper.[22] Inside set theory we must instead regard the formulae themselves as assertion-apt; no additional element such as the assertion sign is required in order to turn such a formula into something of the form of a theorem. If we take assertions to have content–mood structure, we must therefore regard the formulae of set theory as coming equipped with the assertoric mood. But then we shall be forced to hold that the antecedent A of an implication $A \supset B$ as well as any of the disjuncts of a disjunction $A \vee B$ have assertoric mood, which they clearly do not have.[23] For the purposes of conceptual clarity it is therefore quite useful to draw a distinction such as that between proposition and judgement.

[22]See Curry (1963, ch. 2.C) for such an account of the metamathematical "$\vdash A$".

[23]This is sometimes called the "Frege point" after Geach (1972, pp. 254–270); see e.g. Frege (1983, pp. 201–203).

12.3.4 Hypothetical Judgements

In the simple theory of types, as well as in higher-order logic and many-sorted logic, variables are typed globally, in the sense that the type of a variable is specified once and for all by the formation rules of the language. In Martin-Löf type theory, by contrast, variables are typed locally, namely in a so-called context

$$x_1 : A_1, \ldots, x_n : A_n,$$

where the A_i's are types. If there are free variables in a judgement $a : \mathscr{C}$ or $a = b : \mathscr{C}$, then such a judgement therefore does not have a determinate meaning unless a context is provided in which the types of these variables are declared. The general form of judgement in type theory must therefore also list the context. A common notation uses the turnstile, as follows:

$$x_1 : A_1, \ldots, x_n : A_n \vdash a : \mathscr{C}$$
$$x_1 : A_1, \ldots, x_n : A_n \vdash a = b : \mathscr{C}$$

These judgements should be read as saying that a is an object of category $\mathscr{C}$, respectively that a and b are identical objects of category $\mathscr{C}$, in the context $x_1 : A_1, \ldots, x_n : A_n$. This in turn may be recast as a *hypothetical* judgement: a is an object of category $\mathscr{C}$, or a and b are identical objects of category $\mathscr{C}$, provided $x_1 : A_1, \ldots, x_n : A_n$. We get a more familiar form of hypothetical judgement by letting $\mathscr{C}$ be a proposition A and by thinking of the type assignment $x_i : A_i$ as the assumption that A_i is true:

$$A_1 \; true, \ldots, A_n \; true \vdash A \; true$$

This is the judgement that A is true provided, or on the assumption, or condition, that $A_1, \ldots, A_n$ are all true.

Hypothetical judgement is another novelty of type theory in comparison with set theory and ordinary predicate logic. In particular, hypothetical judgements should not be confused with implications, that is, propositions of the form $A \supset B$, for a hypothetical judgement is not a proposition (cf. Sundholm 2012). A hypothetical judgement should rather be compared to a sequent in the sense of Gentzen (1936, § 5), namely a sequent

$$A_1, \ldots, A_n \longrightarrow A$$

to be read as saying that A is true on the assumption that $A_1, \ldots, A_n$ are all true and governed by introduction and elimination rules together with structural rules (these sequents are therefore not quite those of ordinary sequent calculi, which have so-called left and right rules rather than introduction and elimination rules). This, in turn, may be understood as a gloss on the assertion that the relation of consequence holds between the propositions $A_1, \ldots, A_n$ and A.

12.4 Functions

Among the fundamental notions of mathematics one finds not only the notion of a domain or collection of objects, but also that of an operation on, or transformation, or mapping, of objects of such a domain, what is usually called a function. We have already come across (non-dependent) function types: if A and B are types, then so is $(A)B$. In Martin-Löf type theory we take this type to be explained by the following two elimination-like rules:

$$\frac{f : (A)B \qquad a : A}{f(a) : B} \qquad \frac{f : (A)B \qquad a = a' : A}{f(a) = f(a') : B.}$$

The first rule says that whenever $f : (A)B$ and $a : A$, then there is an object $f(a)$ of type B, namely the result of applying f to a. With the function type $(A)B$ there is thus associated a primitive notion of application to an argument. The second rule says that a function applied to identical arguments yield identical results.

In set theory it is not common to regard the notion of function as a primitive notion. On the contrary, one there usually takes the notion of function to be explained in terms of the primitive notions of set theory. More precisely, a function f is there defined to be a set f of ordered pairs satisfying the condition that if $\langle x, y\rangle \in f$ and $\langle x, y'\rangle \in f$ are both true, then so is $y = y'$.[24] Inside set theory, there might thus be no need to talk about functions in addition to sets. If we are to analyze the language and conceptual architecture of set theory, we do, however, still need to invoke functions, as we have already done when sketching a type structure for set theory. In particular, the relations of elementhood, $\in$, and identity, $=$, will then be analyzed as binary propositional functions on the universe V of sets, as expressed by the following judgements:

$$\in : (V)(V)\mathbf{prop}$$
$$= : (V)(V)\mathbf{prop}$$

Yet another function is presupposed by the definition of the application of a function *qua* set of ordered pairs to an argument. Namely, in order to define the term $f(a)$, where both f and a are sets, the syntax of predicate logic requires that there be some function connecting f and a. Thus, we must think of $\__1(\__2)$ as a

[24] Hausdorff (1914, p. 33) appears to have been the first to define the notion of function in this way solely in terms of set theory. On Hausdorff's definition in a historical context, see Ulrich Felgner's article in Hausdorff (2002, pp. 621–633).

binary function on the universe V, as follows:

$$\_\_{1}(\_\_{2}) : (V)(V)V$$

Of course, one could also use some other notation here, such as a left quotation mark, thus writing f'a, as in Shoenfield (1967, p. 245).

Since the primitive language of set theory contains no term-forming operators, the definition of any n-ary function symbol F must there take the form

$$Fa_1 \dots a_n = b \leftrightarrow_{\text{def}} \varphi[a_1, \dots, a_n, b] \qquad \text{(Def-}F\text{)}$$

Here $\varphi[a_1, \dots, a_n, b]$ is a formula whose free variables are among $a_1, \dots, a_n, b$ such that

$$\forall x_1 \dots x_n \exists! y\, \varphi(x_1, \dots, x_n, y)$$

is a theorem. The application function, in particular, may be introduced by the definition

$$f(a) = b \leftrightarrow_{\text{def}} (\psi[f, a] \wedge \langle a, b\rangle \in f) \vee (\neg\psi[f, a] \wedge b = \emptyset),$$

where $\psi[f, a]$ is a formula expressing that f is a function in the sense of set theory and that there is x such that $\langle a, x\rangle \in f$. Notice, however, that the *definiendum* here is not $f(a)$, but rather $f(a) = b$. More generally, the *definiendum* in the definition (Def-F) is not the term $Fa_1 \dots a_n$, but rather the formula $Fa_1 \dots a_n = b$. What is explained by such a definition is, therefore, not the term, but the formula. The term $Fa_1 \dots a_n$ is thus rendered an improper symbol in the sense of Russell and Whitehead (1910, p. 66), viz., a symbol that has no meaning in isolation, but only in certain contexts. The term $f(a)$, in particular, has no meaning outside the context of formulae of the form $f(a) = b$ or formulae defined from such formulae. The proposition that $2+2$ is an even number, for instance, cannot therefore be analyzed simply as $Even(2+2)$, but must be analyzed as $\exists n(2+2 = n \wedge Even(n))$.

Functions *qua* sets of ordered pairs should thus not be compared to objects of function type $(A)B$. They may, however, be compared to objects of type $A \to B$. In Martin-Löf type theory $A \to B$ is a type of objects that, just as the relevant sets of ordered pairs, are individuals coding functions. It can be formed whenever A and B are types of individuals:

$$\frac{A, B : \mathbf{dom}}{A \to B : \mathbf{dom}}$$

There is a more general form of type, written $\Pi(A, B)$, of individuals coding dependent functions; but, as elsewhere in this article, we may concentrate on the simpler non-dependent case, $A \to B$. The introduction rule for $A \to B$ allows us to form an individual $\lambda(f)$ for any $f : (A)B$ when A and B are types of individuals:

$$\frac{A, B : \mathbf{dom} \qquad f : (A)B}{\lambda(f) : A \to B}$$

Just as in set theory, we need an application function to account for the application of an element $c : A \to B$ to an argument. Such an application function is provided by the elimination rule for $A \to B$:

$$\frac{c : A \to B \qquad a : A}{\mathsf{ap}(c, a) : B}$$

The application function ap is defined by laying down that it is to satisfy the following equation:

$$\mathsf{ap}(\lambda(f), a) = f(a) : B$$

From this equation it is clear that $\lambda(f)$ may indeed be thought of as a code of $f : (A)B$.

Those familiar with Frege's logical system in the *Grundgesetze* may recognize the similarity between the terms $\lambda(f)$ and Frege's courses-of-values (*Wertverläufe*; Frege 1893, § 9). A course-of-values $\grave{\alpha}.f(\alpha)$ can be formed whenever f is a function from individuals to individuals (or, from objects to objects, in Frege's terminology). Moreover, there is a binary application function, $-_1\frown-_2$, defined on individuals, satisfying

$$\grave{\alpha}.f(\alpha)\frown a = f(a)$$

The rules governing courses-of-values are thus quite similar to those governing $\lambda(f)$. An essential difference, however, is that a Fregean course-of-values is of the same type as any argument to which it may be applied via $-_1\frown-_2$. Self-application therefore makes sense, which is one of the reasons (though not the only reason) for the inconsistency of Frege's system. In Martin-Löf type theory, by contrast, the object $\lambda(f)$ is of type $A \to B$, whereas any object to which $\lambda(f)$ may be applied via ap is of type A.

We should emphasize the difference between the primitive notion of application associated with the function type $(A)B$ and application via an application function such as ap, $-_1\frown-_2$ or $\__1(\__2)$. Consider, for instance, the term $\mathsf{ap}(c, a)$. It involves both notions of application: the primitive application of ap to c and a and the application of c to a via the function ap. It is clear that some primitive notion of function application is needed. As Frege (1892, p. 205) and Bradley (1893, ch. 3) in effect noted, an infinite regress soon arises if we try to get rid of the primitive notion of function application in favour of application via an application function. For instance, if we invoke an application function ap^* to account for the application of ap to c and a, as in $\mathsf{ap}^*(\mathsf{ap}, c, a)$, then we either need to say that ap^* is applied primitively to its arguments or else invoke yet a further application function and thus be on the brink of regress.

12.5 Identity

Recall that in type theory we have two basic forms of (categorical) judgement: $a : \mathscr{C}$ and $a = b : \mathscr{C}$. Just as there is a fundamental difference between the formula $a : \mathscr{C}$ of type theory and the formula $a \in b$ of set theory, so there is a fundamental difference between the formula $a = b : \mathscr{C}$ of type theory and the formula $a = b$ of set theory or of predicate logic quite generally. The most fundamental difference is in logico-grammatical category: whereas the former is a judgement, the latter is a proposition. In particular, a formula $a = b$ of predicate logic is to be analyzed as the result of applying a certain binary propositional function, $= : (A)(A)\mathbf{prop}$, to objects $a, b : A$ of the underlying domain A. The judgement $a = b : \mathscr{C}$, by contrast, does not have function–argument structure, nor is it a proposition. As a consequence, the propositional connectives and the quantifiers are not applicable to identity judgements as they are to identity propositions.

The wish to do predicate logic in type theory very nearly forces one to introduce a propositional function **Id** such that for any type of individuals A and any $a, b : A$, the result $\mathbf{Id}(A, a, b)$ of applying **Id** to A, a, and b is the proposition that a and b are identical. Thus, we are to have the following formation rule:

$$\frac{A : \mathbf{dom} \qquad a : A \qquad b : A}{\mathbf{Id}(A, a, b) : \mathbf{prop}}$$

That is, whenever A is a type of individuals and a and b are elements of A, then there is a proposition $\mathbf{Id}(A, a, b)$. Because of the identification $\mathbf{prop} = \mathbf{dom}$, we could also have written $\mathbf{Id}(A, a, b) : \mathbf{dom}$ in the conclusion, but **prop** is more suggestive in this case. The formation rule for **Id** does of course not by itself make $\mathbf{Id}(A, a, b)$ into an identity proposition. But by supplying the function **Id** thus postulated with the introduction rule

$$\frac{a : A}{\mathsf{refl}(a) : \mathbf{Id}(A, a, a)}$$

and a corresponding elimination rule, it can be seen that $\mathbf{Id}(A)$ indeed behaves as identity on A. In particular, it follows easily from these rules that $\mathbf{Id}(A)$ is an equivalence relation such that if $\mathbf{Id}(A, a, b)$ is true, then a and b are indiscernible by unary propositional functions over A (cf. e.g. Nordström et al. 1990, pp. 57–59). Type theory thus recognizes both judgemental identity, $a = b : A$, and propositional identity, $\mathbf{Id}(A, a, b)$.

Identity types, $\mathbf{Id}(A, a, b)$, are central in homotopy type theory and have been the subject of many important metamathematical studies.[25] The relation between judgemental and propositional identity, and the treatment of identity more generally, is, moreover, one of the more intriguing philosophical aspects of Martin-Löf type theory. Here I will restrict myself to one topic, namely the conceptual priority of judgemental identity over propositional identity. Again we shall see how type theory allows a much deeper analysis of logical notions—in this case the notion of identity—than anything set theory and predicate logic can offer.

In order to have the right to take f to be a function on some domain A, we must presuppose that A has been equipped with a notion of identity. For we may take f to be a function on A only if we know that f is well-defined, that is, only if we know that the results $f(a)$ and $f(b)$ of applying f to identical arguments $a, b : A$ are identical; or, in symbols:

$$a = b \implies f(a) = f(b) \qquad \text{(Well-def)}$$

The propositional function of identity on A, namely the function $= \, : (A)(A)\mathbf{prop}$, is a binary function on A. Being such a function, it therefore presupposes that A comes equipped with a notion of identity. On pain of circularity, this notion of identity, presupposed by the propositional function of identity, cannot itself be propositional identity. In particular, the identity, $=$, featuring in (Well-def) cannot be propositional identity. Thus the notion of a function f on a domain A presupposes some notion of identity other than propositional identity.

Judgemental identity, since it is not a propositional function on A, may well serve as the notion of identity presupposed by the notion of a function on A. It does serve this role in Martin-Löf type theory. The formulation there of (Well-def) is, indeed, one of the primitive, meaning-giving rules governing functions, and invokes only judgemental identity (see Sect. 12.4):

$$\frac{f : (A)B \qquad a = b : A}{f(a) = f(b) : B}$$

The formulation of (Well-def) in terms of propositional identity **Id**, by contrast, is a theorem, easily established by the rule of **Id**-elimination. We may conclude that judgemental identity is conceptually prior to propositional identity, since the functionhood of propositional identity presupposes judgemental identity. It is worthwhile noting that this argument relies only on the difference in logico-grammatical category between judgemental and propositional identity, and not on the rules that govern these two notions of identity (which tend to vary between various versions of type theory).

Another way of seeing the priority of judgemental identity over propositional identity follows the idea that the criterion of identity for a type A cannot be

[25] Examples are Streicher (1993), Hofmann (1997), and Hofmann and Streicher (1998).

expressed by propositional identity, though it may be expressed by judgemental identity. It is a general requirement on a category $\mathscr{C}$ that it be determinate what it is for objects $a, b : \mathscr{C}$ to be identical. A good example is afforded by the rational numbers. Someone who does not know that different fractions may be the same rational number, such as are $\frac{1}{2}$ and $\frac{2}{4}$, cannot be said to have understood the concept of a rational number. Or, in a more ontological turn of phrase: it is part of the essence of rational numbers that fractions $\frac{m}{n}$ and $\frac{m'}{n'}$ such that $mn' = m'n$ are the same rational number, namely they are the same "in spite of formal differences in the mode of givenness", to quote Cantor (see the quotation in Sect. 12.2).[26] The explanation of a type A, being a category, must therefore include an explanation of what it is for objects $a, b : A$ to be the same. Or, again in a more ontological terminology: the essence of a type A must include a criterion of identity, namely a criterion that determines what it is for objects $a, b : A$ to be identical. It can then be seen by reasoning similar to that above that the criterion of identity for A cannot be expressed in terms of the propositional function of identity on A (cf. Klev 2018b). This function, namely, presupposes that A has already been equipped with a notion of identity; but it is the criterion of identity for A that is to determine what it is for objects $a, b : A$ to be identical.

It follows that we cannot regard the axiom of extensionality in set theory,

$$a = b \leftrightarrow \forall x(x \in a \leftrightarrow x \in b),$$

as a formulation of the criterion of identity for sets. Just as all other axioms of set theory, this is an ordinary predicate-logical formula, so the identity sign occurring in it must be taken to signify a propositional function on the universe V of sets, and therefore to presuppose a criterion of identity for V. Or, again, the well-formedness of the axiom of extensionality presupposes that V has been equipped with some notion of identity; hence it cannot be by means of this axiom that V is so equipped.

The identity judgement $a = b : V$, by contrast, lends itself to the expression of the identity criterion for V, since it does not presuppose any notion of identity on V. Indeed, it is by means of judgements $a = b : \mathscr{C}$ that the identity criterion for a category $\mathscr{C}$ in Martin-Löf type theory is stated. For instance, the criterion of identity of the function type $(A)B$ may be taken to be contained in the following two rules:

$$\frac{f = g : (A)B \qquad a : A}{f(a) = g(a) : B} \qquad \frac{f, g : (A)B \qquad x : A \vdash f(x) = g(x) : B}{f = g : (A)B}$$

Thus, we are to have $f = g : (A)B$ if and only if $f(a) = g(a) : B$ for arbitrary $a : A$. The criterion of identity for the function type $(A)B$ is thus in a sense reduced to the criterion of identity for B. The criterion of identity for a type A of individuals follows from a stipulation, stated in terms of judgemental identity, of when two so-

[26]Cantor's use of the term "mode of givenness" (*Art des Gegebenseins*) might have inspired Frege's use of the same term in his elucidation of the notion of sense; see Sundholm (2001, pp. 65–67).

called canonical objects of A are identical; we shall not go into the details here (cf. Martin-Löf 1984, pp. 7–10).

12.6 Concluding Remarks

A theory that is put to service as a foundation for mathematics provides mathematics not only with a theoretical foundation, that is, a set of basic principles (axioms or rules, say) on the basis of which the theorems of mathematics are to be demonstrable: it also provides it with a certain conceptual architecture, or an "ideology", an articulation of its basic concepts and their interrelations. In the foregoing pages I have contrasted the two ideologies thus provided by set theory and type theory. We have seen, for instance, how the two primitive notions of set and type differ from each other, as well as how type theory differs from set theory in taking the notion of function to be primitive. Besides outlining such differences, I have tried to convey how rich and nicely detailed the conceptual architecture of type theory is. On first sight such intricacy may seem superfluous. Do we, for instance, need a distinction between judgements and propositions or a distinction between judgemental and propositional identity? Set theory draws no such distinctions, but it has nonetheless been quite successful as a foundation for mathematics. Indeed, the ideological simplicity of set theory is certainly part of the explanation of its success. I have tried to show how the subtle distinctions offered by type theory can be of use in the philosophical analysis of logical and mathematical notions, namely as tools that allow one to dig deeper where set-theoretical ideology gives in.

Acknowledgements I am grateful to Deborah Kant for inviting me to contribute to this volume. The critical comments of two anonymous readers on an earlier draft helped me in the preparation of the final version of the paper. While writing the paper I have been supported by grant nr. 17-18344Y from the Czech Science Foundation, GAČR.

References

Aczel, P. (1978). The type theoretic interpretation of constructive set theory. In A. Macintyre, L. Pacholski, & J. Paris (Eds.), *Logic colloquium 77* (pp. 55–66). Amsterdam: North-Holland.

Artemov, S., & Fitting, M. (2016). Justification logic. In E. N. Zalta (Ed.), *The Stanford encyclopedia of philosophy*. Metaphysics Research Lab, Stanford University. https://plato.stanford.edu/archives/win2016/entries/logic-justification/

Aspinall, D., & Hofmann, M. (2005). Dependent types. In B. C. Pierce (Ed.), *Advanced topics in types and programming languages* (pp. 45–86). Cambridge: MIT Press.

Barnes, J. (2003). *Porphyry. Introduction. Translated with an introduction and commentary.* Oxford: Oxford University Press.

Bishop, E. (1967). *Foundations of constructive analysis*. New York: McGraw-Hill.

Bradley, F. H. (1893). *Appearance and reality*. Oxford: Clarendon Press. Cited from the Ninth Impression (1930).

Cantor, G. (1882). Ueber unendliche, lineare Punktmannichfaltigkeiten. Nummer 3. *Mathematische Annalen, 20*, 113–121.
Cantor, G. (1883). Ueber unendliche, lineare Punktmannichfaltigkeiten. Nummer 5. *Mathematische Annalen, 21*, 545–591.
Cantor, G. (1895). Beiträge zur Begründung der transfiniten Mengenlehre (Erster Artikel). *Mathematische Annalen, 46*, 481–512.
Carnap, R. (1934). *Logische Syntax der Sprache*. Vienna: Julius Springer.
Church, A. (1940). A formulation of the simple theory of types. *Journal of Symbolic Logic, 5*, 56–68.
Church, A. (1956). *Introduction to mathematical logic*. Princeton: Princeton University Press.
Curry, H. B. (1963). *Foundations of mathematical logic*. New York: McGraw-Hill.
de Bruijn, N. G. (1975). Set theory with type restrictions. In A. Hajnal, R. Rado, & T. Sós (Eds.), *Infinite and finite sets: to Paul Erdős on his 60th birthday* (pp. 205–214). North-Holland.
de Bruijn, N. G. (1995). Types in mathematics. *Cahiers du Centre de Logique, 8*, 27–54.
Dedekind, R. (1888). *Was sind und was sollen die Zahlen?* Braunschweig: Vieweg und Sohn.
Diller, J., & Troelstra, A. (1984). Realizability and intuitionistic logic. *Synthese, 60*, 253–282.
Dummett, M. (1973). *Frege. Philosophy of language*. London: Duckworth. Cited from the second edition (1981).
Dybjer, P. (1994). Inductive families. *Formal Aspects of Computing, 6*, 440–465.
Dybjer, P. (2000). A general formulation of simultaneous inductive-recursive definitions in type theory. *Journal of Symbolic Logic, 65*, 525–549.
Frege, G. (1892). Über Begriff und Gegenstand. *Vierteljahrsschrift für wissenschaftliche Philosophie, 16*, 192–205.
Frege, G. (1893). *Grundgesetze der Arithmetik I*. Jena: Hermann Pohle.
Frege, G. (1983). *Nachgelassene Schriften*. Hamburg: Felix Meiner Verlag.
Geach, P. T. (1972). *Logic matters*. Oxford: Blackwell.
Gentzen, G. (1936). Die Widerspruchsfreiheit der reinen Zahlentheorie. *Mathematische Annalen, 112*, 493–565.
Hallett, M. (1984). *Cantorian set theory and limitation of size*. Oxford: Clarendon Press.
Harper, R., Honsell, F., & Plotkin, G. (1993). A framework for defining logics. *Journal of the ACM, 40*, 143–184.
Hausdorff, F. (1914). *Grundzüge der Mengenlehre*. Leipzig: Veit & Comp.
Hausdorff, F. (2002). *Gesammelte Werke. Band II*. Heidelberg: Springer.
Hilbert, D. (1899). *Grundlagen der Geometrie*. Leipzig: Teubner.
Hofmann, M. (1997). *Extensional constructs in intensional type theory*. London: Springer. Reprint of Ph.D. thesis, University of Edinburgh (1995).
Hofmann, M., & Streicher, T. (1998). The groupoid interpretation of type theory. In G. Sambin & J. M. Smith (Eds.), *Twenty-five years of constructive type theory* (pp. 83–111). Oxford: Oxford University Press.
Howard, W. A. (1980). The formulae-as-types notion of construction. In J. P. Seldin & J. R. Hindley (Eds.), *To H.B. Curry: Essays on combinatory logic, lambda calculus and formalism* (pp. 479–490). London: Academic Press.
Husserl, E. (1913). *Ideen zu einer reinen Phänomenologie und phänomenologischen Philosophie. Erstes Buch*. Halle: Max Niemeyer.
Kirk, G. S., Raven, J. E., & Schofield, M. (1983). *The presocratic philosophers* (2nd ed.). Cambridge: Cambridge University Press.
Klev, A. (2017a). Husserl and Carnap on regions and formal categories. In S. Centrone (Ed.), *Essays on Husserl's logic and philosophy of mathematics* (pp. 409–429). Dordrecht: Springer.
Klev, A. (2017b). Truthmaker semantics: Fine versus Martin-Löf. In P. Arazim & T. Lávička (Eds.), *The logica yearbook 2016* (pp. 87–108). London: College Publications.
Klev, A. (2018a). The concept horse is a concept. *Review of Symbolic Logic, 11*, 547–572.
Klev, A. (2018b). The logical form of identity criteria. In P. Arazim & T. Lávička (Eds.), *The logica yearbook 2017* (pp. 181–196). London: College Publications.

Kreisel, G. (1962). Foundations of intuitionistic logic. In E. Nagel, P. Suppes, & A. Tarski (Eds.), *Logic, methodology and the philosophy of science* (pp. 198–210). Stanford: Stanford University Press.

Linnebo, O. (2010). Pluralities and sets. *Journal of Philosophy, 107*, 144–164.

Linnebo, O., & Rayo, A. (2012). Hierarchies ontological and ideological. *Mind, 121*, 269–308.

Lowe, E. J. (1989). *Kinds of being*. Oxford: Basil Blackwell.

Martin-Löf, P. (1982). Constructive mathematics and computer programming. In J. L. Cohen, J. Łoś, et al. (Eds.), *Logic, methodology and philosophy of science VI, 1979* (pp. 153–175). Amsterdam: North-Holland.

Martin-Löf, P. (1984). *Intuitionistic type theory*. Naples: Bibliopolis.

Martin-Löf, P. (2011). When did 'judgement' come to be a term of logic? Lecture held at *École Normale Supérieure* on 14 October 2011. Video recording available at http://savoirs.ens.fr//expose.php?id=481

Meier-Oeser, S. (2001). Vielheit. In J. Ritter, K. Gründer, & G. Gabriel (Eds.), *Historisches Wörterbuch der Philosophie* (Vol. 11, pp. 1041–1050). Basel: Schwabe.

Nordström, B., Petersson, K., & Smith, J. (1990). *Programming in Martin-Löf's type theory*. Oxford: Oxford University Press.

Quine, W. V. (1956). Unification of universes in set theory. *Journal of Symbolic Logic, 21*, 267–279.

Russell, B. (1903). *Principles of mathematics*. Cambridge: Cambridge University Press.

Russell, B., & Whitehead, A. N. (1910). *Principia mathematica* (Vol. 1). Cambridge: Cambridge University Press.

Schmidt, A. (1938). Über deduktive Theorien mit mehreren Sorten von Grunddingen. *Mathematische Annalen, 115*, 485–506.

Schönfinkel, M. (1924). Bausteine der mathematischen Logik. *Mathematische Annalen, 92*, 305–316.

Scott, D. (1982). Domains for denotational semantics. In M. Nielsen & E. M. Schmidt (Eds.), *Automata, languages and programming* (pp. 577–613). Heidelberg: Springer.

Shoenfield, J. R. (1967). *Mathematical logic*. Reading: Addison-Wesley.

Streicher, T. (1993). *Investigations into intensional type theory*. Habilitation thesis, Ludwig Maximilian University Munich.

Sundholm, B. G. (1994). Existence, proof and truth-making: A perspective on the intuitionistic conception of truth. *Topoi, 13*, 117–126.

Sundholm, B. G. (2001). Frege, August Bebel and the return of Alsace-Lorraine: The dating of the distinction between *Sinn* and *Bedeutung*. *History and Philosophy of Logic, 22*, 57–73.

Sundholm, B. G. (2009). A century of judgement and inference, 1837–1936: Some strands in the development of logic. In L. Haaparanta (Ed.), *The development of modern logic* (pp. 263–317). Oxford: Oxford University Press.

Sundholm, B. G. (2012). "Inference versus consequence" revisited: Inference, consequence, conditional, implication. *Synthese, 187*, 943–956.

The Univalent Foundations Program (2013). *Homotopy type theory: Univalent foundations of mathematics*. Princeton: Institute for Advanced Study. http://homotopytypetheory.org/book

Wadler, P. (2015). Propositions as types. *Communications of the ACM, 58*, 75–84. Extended version available at http://homepages.inf.ed.ac.uk/wadler/papers/propositions-as-types/propositions-as-types.pdf

Wittgenstein, L. (1922). *Tractatus logico-philosophicus*. London: Routledge & Kegan Paul.

Zermelo, E. (1930). Über Grenzzahlen und Mengenbereiche. *Fundamenta Mathematicae, 16*, 29–47.

Chapter 13
What Do We Want a Foundation to Do?

Comparing Set-Theoretic, Category-Theoretic, and Univalent Approaches

Penelope Maddy

Abstract It's often said that set theory provides a foundation for classical mathematics because every classical mathematical object can be modeled as a set and every classical mathematical theorem can be proved from the axioms of set theory. This is obviously a remarkable mathematical fact, but it isn't obvious what makes it 'foundational'. This paper begins with a taxonomy of the jobs set theory does that might reasonably be regarded as foundational. It then moves on to category-theoretic and univalent foundations, exploring to what extent they do these same jobs, and to what extent they might do other jobs also reasonably regarded as foundational.

Mainstream orthodoxy holds that set theory provides a foundation for contemporary pure mathematics. Critics of this view argue that category theory, or more recently univalent foundations, is better suited to this role. Some observers of this controversy suggest that it might be resolved by a better understanding of what a foundation is. Despite considerable sympathy to this line of thought, I'm skeptical of the unspoken assumption that there's an underlying concept of a 'foundation' up for analysis, that this analysis would properly guide our assessment of the various candidates. In contrast, it seems to me that the considerations the combatants offer against opponents and for their preferred candidates, as well as the roles each candidate actually or potentially succeeds in playing, reveal quite a number of different jobs that mathematicians want done. What matters is these jobs we want our theories to do and how well they do them. Whether any or all of them, jobs or theories, deserves to be called 'foundational' is really beside the point.

The forces behind the rise of set-theoretic foundations (in the late nineteenth and early twentieth centuries) and its subsequent accomplishments (as of the early twenty-first) are explored and assessed in §13.1. §13.2 turns to the criticisms lodged against set-theoretic foundations as category theory developed (in the 1940s) and the

---

P. Maddy (✉)
University of California, Irvine, Irvine, CA, USA
e-mail: pjmaddy@uci.edu

S. Centrone et al. (eds.), *Reflections on the Foundations of Mathematics*,
Synthese Library 407, https://doi.org/10.1007/978-3-030-15655-8_13

subsequent case for category-theoretic foundations (beginning in the 1960s). The current proposal for univalent foundations is examined in the concluding §13.3.

13.1 Set-Theoretic Foundations[1]

It's commonplace to note that all standard mathematical objects and structures can be modeled as sets and all standard mathematical theorems proved from the axioms of set theory[2] – indeed, familiarity may well have dulled our sense of just how remarkable this fact is. For our purposes, though, let me draw attention to another commonplace: when called upon to characterize the foundational role of set theory, many observers are content merely to remind us that mathematics can be embedded in set theory in this way. But simply repeating that this is so leaves our present questions untouched: what's the point of this embedding?, what need does it serve?, what foundational job does it do?

To answer these questions, we should look back at the conditions under which set theory arose in the first place. Over the course of the nineteenth century, mathematics had expanded dramatically in an extraordinary variety of directions. This great torrent of new mathematics brought with it a pair of epistemological losses, as the subject outstripped available thinking on what ultimately justifies mathematical claims. Early efforts to make good on those losses eventually needed support of their own, and it was at this point that set theory entered the picture. A quick sketch of these developments should help explain what jobs set theory was at least partly designed to do.

Consider first the case of geometry. From the diagrams of Euclidean times to Kant's late eighteenth century theory of spatial intuition, geometry was generally understood as grounded in some variety of visualization or intuition. That changed in the nineteenth century with the introduction of 'points at infinity' (where parallel lines meet) and 'imaginary points' (with complex numbers as coordinates). There was no denying the fruitfulness of regarding geometry from this new perspective, but the imaginary points at which two disjoint circles 'intersect' can hardly be visualized or intuited! So this is the first epistemic loss: visualization and intuition were no longer adequate measures of correctness for the brave new geometry. What justification, then, could be given for admitting these new, invisible points, what guarantee that they wouldn't severely compromise the whole subject? Geometers were understandably queasy about this expansion of the proper domain of their inquiry.

---

[1]Many themes of this section are explored in more detail, with sources, in §I of [2017].

[2]Items like the category of all groups or the category of all categories are exceptions. There is no set of all groups or set of all categories for the same reason that there's no set of all sets: sets are formed in a transfinite series of stages, and there's no stage at which all of them (or all of them that are groups or all of them that are categories) are available to be collected. The category-theoretic cases are explored in the next section.

The second epistemic loss came with the rise of pure mathematics during this same period.[3] Up through the eighteenth century, there was no distinction between pure and applied; mathematics was considered the study of the mathematical structure literally present in the physical world. Galileo, Newton, Euler, Fourier, and others took the goal of natural science to be the isolation of purely mathematical laws governing the behavior of observable phenomena (e.g., planetary motions or the distribution of heat in a body) without appeal to hidden causes (e.g., Cartesian vortices or caloric fluids). This strategy was a tremendously successful at the time, encouraging precise mathematization and eschewing dubious mechanical explanations. The ongoing clash between mathematization of observable behavior and causal explanation re-emerged in the late nineteenth century in the study of thermodynamics: descendants of the purely descriptive, mathematical tradition, proposed the experientially exception-less second law, that entropy can only increase, while descendants of the causal, explanatory tradition developed the kinetic theory, according to which a decrease in entropy is just highly unlikely. In the early years of the twentieth century, the tables of history turned: kinetic theory with its atoms and molecules in random motion was experimentally confirmed. This meant that the laws of classical thermodynamics were revealed to be merely probable, and more generally, that the many hard-won differential equations of the eighteenth and nineteenth centuries were highly effective approximations, smoothed-out versions of a more complex, discrete microstructure.

By the end of the nineteenth and beginning of the twentieth centuries, as pure mathematics proliferated and applied mathematics lost its claim to literal truth, it became clear that mathematics isn't actually in the business of discerning the precise formal structure of the physical world. Rather, it provides an array of abstract models for the scientist to choose from for any particular descriptive job. Most of these mathematical descriptions involve explicit idealizations and approximations, and applied mathematicians expend considerable effort on explaining how and why these non-literal representations are nonetheless effective, often within a limited range of cases. Some such descriptions are effective despite our inability to explain exactly what worldly features they're tracking (e.g., in quantum mechanics). These are crucial morals for the philosophy of science, but our concern here is the epistemic loss suffered by mathematics itself: in the new, pure mathematics, there was no place for physical interpretation or physical insight to guide developments. In Euler's day, a scientist's feel for the physical situation could help shape the mathematics, could constrain it to effective paths even when rigor was in short supply. Now that mathematicians had declared their independence – their freedom to pursue whatever paths caught their purely mathematical interest – physical results and physical intuition could no longer serve to support or justify mathematical work. Without this guidance, how were mathematicians to tell which among the proliferation of new, purely mathematical abstract structures under were trustworthy, legitimate, worthwhile?

[3]For more on this development, see [2008] or chapter 1 of [2011].

In the face of these epistemic set-backs, two types of remedies were offered. The first, in response to queasiness about the new, un-intuitable geometric points, came in the mid-nineteenth century, when Karl von Staudt managed to build surrogates for the suspicious entities out of straightforwardly acceptable materials. So, for example, a point at infinity where two given parallel lines meet can be identified with the collection ('pencil') of all lines parallel to those two, and this point at infinity taken to be on a given line if the line is in that collection.[4] In this way, a previously suspicious mathematical development is domesticated by building it up from previously justified mathematics. This method was widely used, but eventually a new question has to be faced: which means of building new from old are trustworthy, and why?

The second remedy, in response to the proliferation of abstract structures, came later in the nineteenth century, with Hilbert's axiomatic method: each proposed structure should be rigorously axiomatized; if that axiomatization is coherent, the structure is legitimate. Though careful isolation of appropriate axioms might reveal unnoticed incoherence in some cases, for most others a new worry is immediate: how do we tell which axiomatizations are coherent?[5] A second concern is less obvious, but also serious. Suppose I've devised an axiom system for the natural numbers, another for the real numbers, another for analysis, another for point-set topology, another for computable functions, and so on – and I prove theorems in each of these. Can I use the theorems I've proved in one area to prove theorems in another? John Burgess illustrates how centrally modern mathematics replies on the ability to move easily between its branches:

> There is the interconnectedness of the different branches of mathematics, a phenomenon evident since the seventeenth century in the use of coordinate methods, but vastly expanded in the nineteenth century. With the group concept, an idea originating in algebra is applied to geometry. With 'functional analysis', ideas originating in geometry or topology are applied to analysis, as functions come to be considered 'points' in an auxiliary space, and operations like differentiation and integration come to be considered 'transformations' of that space. (Footnote: One reason one needs to allow in pathological functions like the Riemann-Weierstrass examples is in order to achieve a certain 'completeness', analogous to the completeness of the real number-line, in the 'space' of functions.) And so on across the whole of mathematics.
>
> Interconnectedness implies that *it will no longer be sufficient to put each individual branch of mathematics separately on a rigorous basis.* (Burgess 2015, pp. 59–60, emphasis in the original)

Today it's hard to see how Wiles could have proved Fermat's Last Theorem if he'd been confined to one or another of the individual axiom systems!

[4]Readers of Frege (1884) will recognize this as 'the direction of a line' and recall how it serves as Frege's model for identifying a natural number with a collection of equinumerous collections.

[5]This was before the development of formal languages and deductive systems, before a clear understanding of consistency, satisfiability, and of course, before Gödel's completeness and incompleteness theorems.

The epistemic and methodological questions raised by these nineteenth-century losses and their partial remedies permeated the climate in which set theory first arose. Though much of the initial motivation for introducing sets was purely mathematical – Cantor, for example, was just trying to generalize his theorem on the uniqueness of trigonometric representations – it soon became clear that the various building methods of von Staudt and the rest were all essentially set-theoretic in character, all derivable from a few elementary set-theoretic operations (like taking subsets, intersections and unions, cross-products, and power sets). Indeed it turned out that all the various items previously axiomatized in separate systems (natural numbers, real numbers,[6] analysis, etc.) could be constructed set-theoretically – the beginnings, in other words, of the famous embedding of mathematics in set theory.

In this way, set theory made progress on our first two questions: the trustworthy building methods are those of set theory; the coherent axiom systems are those that can be modeled by sets. Of course this is cold comfort unless we know that set theory itself is reliable – a particularly dubious proposition at the time, given both the paradoxes and wide-spread debates over fundamentals (the well-ordering principle, the axiom of choice, the continuum hypothesis, etc.). Working in the Hilbertian tradition, Zermelo set out to axiomatize the subject, successfully isolating the basic assumptions underlying the informal practice while forestalling the known routes to paradox. Though he hoped to include a consistency proof in his original presentation, Hilbert encouraged him to publish the axioms first and follow with the consistency proof when it was ready. Years later it became clear what good advice this was, when Gödel showed that only an stronger system could prove the consistency of Zermelo's axioms (assuming they are consistent).

So, much as we might like to have an iron-clad guarantee of the consistency of set theory, and thus of the trustworthiness of the methods embedded therein, this is a forlorn hope; all we get is the assurance that the embedded methods are no more dangerous than Zermelo's set theory. Most estimates of that danger have decreased substantially over the intervening century, with the development of a compelling intuitive picture of the universe of sets (the iterative conception), a finely-articulated model within set theory (the constructible universe), and a vast, intricate and far-reaching mathematical theory with no sign of contradiction. Meanwhile, various levels of consistency strength have been delineated and explored – from relatively weak subsystems of second-order arithmetic to ever-larger large cardinal axioms –

[6]It's worth noting that Dedekind's set-theoretic construction of the reals was different in character from von Staudt's construction of imaginary points. Von Staudt was faced with a practice in good working order, but with questionable posits. Dedekind was faced with a defective practice (basic theorems of the calculus couldn't be proved). So von Staudt's challenge was to remove queasiness about the posits by domesticating them, while Dedekind's was to produce a more precise replacement that would both conform to previous practice and extend it (proving those basic theorems). Thus Dedekind's construction had a different, plausibly 'foundational' function (called **Elucidation** in [2017]). As both category-theoretic and univalent foundations are content to relegate **Elucidation** to ETCS, a weak category-theoretic theory of collections (see [2017], §II, and UFP (2013), p. 8, respectively), it won't figure in the comparative analysis here.

yielding a hierarchy that's now routinely used to calibrate the level of danger a proposed theory presents. Presumably the ability to assess these risks is something mathematicians value. Insofar as we're inclined to regard **Risk Assessment** as a 'foundational' virtue, this is one foundational job that contemporary set theory does quite well.

Finally, the other question raised by the axiomatic response to the loss of older forms of justification (intuition/visualization, physical interpretation/insight) concerned the interrelations between the various branches of pure mathematics: if each branch is characterized by its own separate list of axioms, how can work in one branch be brought to bear in another?

> To guarantee that rigor is not compromised in the process of transferring material from one branch of mathematics to another, it is essential that the starting points of the branches being connected should ... be compatible. ... The only obvious way to ensure compatibility of the starting points ... is ultimately to derive all branches from a common, unified starting point. (Burgess 2015, pp. 61–62)

This 'common, unified starting point' emerges when the various branches are all embedded in a single theory of sets, when all theorems are treated as theorems in the same system. In this way, set theory provides a **Generous Arena** where all of modern mathematics takes place side-by-side and a **Shared Standard** of what counts as a legitimate construction or proof. These are the striking achievements of the well-known embedding of mathematics in set theory. Insofar as they fairly count as 'foundational', set theory is playing two more crucial foundational roles.

Let's pause a moment to notice that everything claimed so far on set theory's behalf has been at the level of straightforward mathematical benefits: the embedding of mathematics in set theory allows us to assess the risk of our theories, to bring results and techniques one branch of mathematics to bear on concepts and problems in another, and to agree on standards of construction and proof. Some observers, especially philosophers, have been tempted to draw – in addition – strong metaphysical or epistemological morals: we've discovered that all mathematical entities were really sets all along, or that our knowledge of mathematics is reducible to our knowledge of sets.[7] These further claims might rightly be called 'foundational', too, but they're also controversial, to say the least. For mathematical purposes, the metaphysical claim is beside the point: it doesn't matter whether we say the von Neumann ordinals are the numbers or the von Neumann ordinals can serve as fully effective mathematical surrogates for the numbers. As for the epistemological claim, it's just false: however it is that we know the things we know in the various, far-flung branches of mathematics, it isn't by deriving them from the axioms of set theory. Most of the time, it's our conviction that the mathematics is correct that makes us think there must be a formal proof from those axioms!

While dubious philosophical claims like these are unlikely to affect practice, other intrusions of irrelevant ontological thinking might come uncomfortably close.

---

[7] These are the spurious foundational virtues called **Metaphysical Insight** and **Epistemic Source** in [2017].

Quite generally, if we take the claim that set theory determines the ontology of mathematics too seriously, we might be tempted to think of it as 'the final court of appeal', charged with passing stern judgement on new mathematical avenues. In fact, I think this gets the situation backwards: casting set theory as the **Generous Arena** isn't intended to limit mathematics; rather it places a heavy responsibility on set theory to be as generous as possible in the types of structure whose existence it implies.[8] This admonition to maximize is one of the most fundamental and powerful methodological principles guiding the development of set theory. If we imagine, in our overly philosophical mood, that set theory has some kind of special access to the metaphysical facts about what abstracta exist, then we might be tempted to put the onus on mathematical practice to conform to the dictates of set theory, to raise that special access above informed judgements of mathematical advantage. I trust we can all agree that this would be a grave mistake.

A more subtle danger in the same general direction arises from the fact that our embedding of mathematics in set theory is more like von Staudt's approach than Hilbert's: a surrogate for the mathematical item in question is constructed by set-theoretic means, as an item in V, the set-theoretic universe; it's not enough, as the Hilbertian would have it, that there's a model somewhere in V that thinks there is such an item. A simple example would be a proof of $1 = 0$ from the axioms of (first-order) Peano Arithmetic: PA + not-Con(PA) is consistent (assuming PA is), so it has a model that thinks there's a proof of $1 = 0$ from PA; but viewed set-theoretically, that model is benighted, the thing it takes for a proof of $1 = 0$ has non-standard length, isn't really a proof. For a more interesting example, consider a definable[9] well-ordering of the real numbers. There is such an ordering in Gödel's inner model, the constructible universe L, but if we add large cardinal axioms to our list, as many set theorists these days do, then that model is benighted: the thing it takes for a well-ordering of the reals only orders the reals present in L; in fact, there is no definable well-ordering of all the reals.

Speaking loosely, we might express this by saying that the inconsistency proof and the well-ordering exist on the Hilbertian standard, while on the von Staudtian set-theoretic standard, they don't. This way of talking is expressive and largely benign, but it can lead us astray if we forget that it's figurative, if we fall into taking it too literally. We need to bear in mind that the cash value of 'these things exist in V' is just 'the existence of (surrogates for) these things can be proved from the axioms of set theory' – a straightforward manifestation of set theory's role as **Shared Standard** of proof. To say that 'the universe of sets is the ontology of mathematics' amounts to claiming that the axioms of set theory imply the existence of (surrogates for) all the entities of classical mathematics – a simple affirmation of set theory's role as **Generous Arena**.

[8]The underlying methodological maxim here is to prefer non-restrictive, maximizing theories. [1997] concludes with an early attempt to formalize this notion. Various developments of this idea and alternatives to it have been suggested, but the problem remains open.

[9]That is, projectively definable.

The danger in taking figurative ontological talk too seriously is that it can lead to a sort of rigidity in practice. Consider that definable well-ordering of the reals. Suppose a pure mathematician has a clever and fruitful approach to a certain problem, or an applied mathematician has a way to effectively model some physical situation, by means of such an ordering. If we believe that set theory is the 'ontology' of mathematics, 'the final court of ontological appeal', we'll be tempted to say 'tough luck, it might be nice if there were such a thing, but there isn't'. But this seems wrong. Both mathematicians' activities can be carried out inside L – by which we mean, in set theory with $V = L$ as an additional axiom. Since that theory includes the standard axioms, it provides a fairly **Generous Arena** all by itself: the usual constructions and techniques are ready to hand; to speak in the figurative idiom, L is a pretty good place to do mathematics. The disadvantage is that results proved using $V = L$ can't automatically be exported to other areas of mathematics, and results from other areas that depend on large cardinals can't automatically be imported. But as long as these import/export restrictions are observed, as long as the use of axioms beyond the standard ones is carefully flagged, there's no reason to rule out these developments. The pure mathematician's work on her problem is simply part of the investigation of L, a particularly important part of V; the applied mathematician has determined that it's most effective to model his physical situation in L rather than V.

This leaves us with a tempered version of the von Staudian 'final court of ontological appeal': the axioms for our **Generous Arena**, which constitute our **Shared Standard** of proof, include the usual axioms – plus some others, beginning with large cardinals, that add to their generosity – but these can be temporarily adjusted for mathematical or scientific purposes with suitable import/export restrictions. Once we reject the idea that the choice of a fundamental theory to do these foundational jobs is a matter of determining the 'true mathematical ontology', once we focus instead on the literal mathematical content of our decisions, we come to see that we can and should allow some wiggle room for both pure and applied mathematicians to work in well-motivated variants of the fundamental theory. I won't attempt to explicate what counts as 'well-motivated' – this requires the sound judgment of insightful practitioners – but one clear qualifier is the existence of an attractive, well-understood model inside V,[10] as in the case of L and $V = L$.[11] Though this marks a slight adjustment to strict von-Staudism, it's still very far from full Hilbertism, where any consistent theory as good as any other – precious few such theories can deliver a pure mathematical theorem worth proving or an applied mathematical model amenable to actual use.[12]

[10]This is, the existence of such a model can be proved from the fundamental axioms.

[11]Another well-known example is the theory $ZF + V = L(R) + AD$. Again, separating the 'mathematically worthy' from the unworthy no doubt requires keen mathematical discernment and well-informed good judgement.

[12]For successful application, it's not enough that our theory prove the existence of a suitable structure; it must exist in a context with enough mathematical tools to study and manipulate that structure. See [2011], pp. 90–96, for a related discussion.

One last point. Returning once more to the historical development of set theory, Zermelo's axioms were soon supplemented with replacement and foundation, and his imprecise notion of 'definite property' was sharpened to 'formula in the first-order language of set theory'. This generated what we now know as the formal theory ZFC. At that point, the embedding of mathematics in set theory came to serve yet another purpose: once mathematics was successfully encoded in a list of formal sentences, meta-mathematical tools could be brought to bear to prove theorems about its general features. Among the greatest of these results were those of Gödel – classical mathematics, if consistent, can't prove its own consistency or the negation of the Continuum Hypothesis – and Cohen – or the Continuum Hypothesis itself. Here set theory provides a **Meta-mathematical Corral**, tracing the vast reaches of mathematics to a set of axioms so simple that they can then be studied formally with remarkable success. Perhaps this accomplishment, too, has some claim to the honorific 'foundational'.

So my suggestion is that we replace the claim that set theory is a (or 'the') foundation for mathematics with a handful of more precise observations: set theory provides **Risk Assessment** for mathematical theories, a **Generous Arena** where the branches of mathematics can be pursued in a unified setting with a **Shared Standard** of proof, and a **Meta-mathematical Corral** so that formal techniques can be applied to all of mathematics at once. I haven't offered any argument that these accomplishments must be understood to be 'foundational', but it seems to me consistent with the ordinary use of the term to so apply it. I take it for granted that these accomplishments are of obvious mathematical value, whatever we decide about the proper use of the term 'foundational'.

Let's now turn to two of set theory's purported rivals: first category-theoretic foundations, then univalent foundations.

13.2 Category-Theoretic Foundations[13]

By the end of the 1930s, ZFC had been codified in its first-order form and its role as **Generous Arena, Shared Standard, Meta-mathematical Corral**, and in **Risk Assessment** were widely accepted. Soon thereafter, mathematical pressures in abstract algebra gave rise to category theory, and category theorists began to criticize set theory as a 'foundation'. By the 1960s, category theory was being proposed as alternative to set theory that could overcome these weaknesses. A look at the objections raised and the solutions offered should help us determine what jobs the critics thought a 'foundation' was supposed to do.

So, what was wrong with set-theoretic foundations? The first objection is that category theory deals with unlimited categories, like the category of all groups or the category of all categories or the category of all mathematical X's, but nothing

[13]For more on many themes of this section, with sources, see §II of [2017].

like this can be found in the universe of sets.[14] Grothendieck overcame this problem by positing an ever-increasing sequence of 'local universes' and noting that any category-theoretic activity can be carried out in a large enough one of these. In set-theoretic terms, this is to add inaccessible cardinals, the smallest of the large cardinals beyond ZFC. In other words, the risk of Grothendieck's category theory is no greater than that of ZFC + Inaccessibles. If **Risk Assessment** is the foundational goal in question, set theory is still doing fine. Likewise, surrogates for the categories are available in set theory's **Generous Arena**, so **Shared Standard**, and **Metamathematical Corral** also seem to be in order.

Given that the foundational goals of set theory were still being met, it's difficult to see what the category theorists took to be missing. The objection seems to be that in any particular category-theoretic context, the item serving as a set-theoretic surrogate for the category of all X's doesn't include all the set-theoretic X's:

> Categorical algebra ... uses notions such as that of the category **G** of all groups. ... To realize the intent of this construction it is vital that this collection **G** contain *all* groups; however, if 'collection' is to mean 'set' ... this intent cannot be realized. (Mac Lane 1971, p. 231)

This is true, but it doesn't keep set theory from meeting any of the identified foundational goals. Of course it doesn't tell us what the category of X's *really* is, any more than the use of the von Neumann ordinals as surrogates for the natural numbers tells us what they really are, but this dubious metaphysical goal has easily been seen to be mathematically irrelevant.[15]

I'm not entirely confident that this is what left the category theorists dissatisfied with the Grothendieck-style move, but fortunately, subsequent developments reveal that this isn't a point that needs to be settled. In hope of overcoming this purported shortcoming of set-theoretic foundations, it was proposed that category theory itself could provide a proper 'foundation' for unlimited categories – where this was taken to mean a fundamental theory in which the category of all X's can be formed for any mathematical notion X, and the usual operations and constructions of category theory can be carried out. Progress toward this goal was incremental until just recently: Ernst (2015) shows that any such theory is actually inconsistent.[16] This means that set theory is not to blame for its inability to form categories of all X's – no consistent theory, including category theory itself, can do that. So whatever the problem was supposed to be, it's a problem that can't be solved by any 'foundation'.

The second early criticism lodged by category theorists against set-theoretic foundations concerns the nature of the embedding. Though surrogates for various mathematical items can be found in the universe of sets, that universe as a whole

[14]Because new groups or categories or mathematical X's keep being formed at higher and higher ranks in the iterative hierarchy of sets, there's never a rank at which they all be collected together. Recall footnote 2.

[15]Recall footnote 7 and surrounding text.

[16]Ernst shows, in particular, that a contradiction arises from the assumption that such a theory can form the category of all graphs.

is vast, its construction techniques wildly indiscriminate, so it includes hordes of useless structures and – this is the important point – no way of telling the mathematically promising structures from the rest. Furthermore, the set-theoretic surrogates have lots of extraneous structure, artifacts of the way they're constructed. Here the hope was to find a foundation that would guide mathematicians toward the important structures and characterize them strictly in terms of their mathematically essential features. Such a foundation would actually be useful to mainstream mathematicians in their day-to-day work, not remote, largely irrelevant, like set theory; it would provide **Essential Guidance**. Proponents held that this is precisely what category theory had done for algebraic geometry and algebraic topology.

Now it could be that some over-zealous partisan of set-theoretic foundations at one time or another claimed that mathematics would be better off if all mathematicians thought like set theorists, but as far as I can tell, this was never one of the foundational jobs that set theory was seriously proposed to do. No reasonable observer would suggest that an algebraic geometer or algebraic topologist would do better to think in set-theoretic rather than category-theoretic terms. But it seems equally unreasonable to suggest that an analyst, or for that matter a set theorist, would do better to think in category-theoretic terms.[17] What's intriguing here is that proponents of category-theoretic 'foundations' would apparently agree. Mac Lane, for example, writes:

> Categories and functors are everywhere in topology and in parts of algebra, but they do not yet relate very well to most of analysis.
>
> We conclude that there is as yet no simple and adequate way of conceptually organizing all of Mathematics. (Mac Lane 1986, p. 407)

If a 'foundation' is to reveal the underlying essence, the conceptual core, omit all irrelevancies, and guide productive research, then it's unlikely that it can encompass all areas of mathematics. Faced with this tension between **Essential Guidance** and **Generous Arena**, Mac Lane seems willing to forego **Generous Arena**, and with it presumably **Shared Standard** and **Meta-Mathematical Corral**.

This preference is more-or-less explicit in the theory of categories that's proposed as our fundamental foundation. The 'Category of Categories as a Foundation' (CCAF) was introduced by Lawvere in the 1960s and subsequently improved by McLarty in the 1990s. CCAF is a actually a minimal background theory which is then supplemented as needed to guarantee the existence of particular categories for this or that area of mathematics. One such special category is 'The Elementary Theory of the Category of Sets' (ETCS), which codifies a relatively weak theory of collections (ZC with bounded separation). Collections in this sense are understood in a natural way in terms of their arrows rather than their elements, but to gain a category-theoretic set theory with sufficient strength for, say, **Risk Assessment**,

[17] See, e.g., the work of Mathias discussed in [2017].

more characteristically set-theoretic notions have to be translated in from outside.[18] A category for synthetic differential geometry is another example that could be added with a suitable axiom. As might be expected from the Hilbertian flavor of this approach, it isn't conducive to **Generous Arena**.

So despite the rhetoric – pitting category theory against set theory, proposing to replace set-theoretic foundations with category-theoretic foundations – the two schools are aimed at quite different goals. Set theory provides **Risk Assessment**, **Generous Arena**, **Shared Standard**, and **Meta-mathematical Corral**, and it apparently continues to do these jobs even in the context of category-theoretic foundations. What category theory offers is **Essential Guidance**, but only for those branches of mathematics of roughly algebraic character. I have no objection to calling this a 'foundational' achievement, so long as it isn't taken to supersede the other foundational goals explored here. What category theory has accomplished – however this achievement is labeled – is a way of thinking about a large part of mathematics, of organizing and understanding it, that's been immensely fruitful in practice. Proponents of set-theoretic foundations should have nothing but admiration for this achievement. It raises deep and important methodological questions about which 'ways of thinking' are effective for which areas of mathematics, about how they differ, about what makes them so effective where they are and ineffective where they aren't, and so on.

So, should we regard set theory's range of accomplishments for mathematics in general as more 'foundational' than category-theory's conceptual achievements across several important areas of the subject, or vice versa? I confess that this doesn't strike me as a productive debate. In contrast, a concerted study of the methodological questions raised by category theory's focus on providing a fruitful 'way of thinking' would almost certainly increase our fundamental understanding of mathematics itself. I vote for that.

13.3 Univalent Foundations

With these nineteenth and twentieth century developments in the background, the turn of the 21st brought a new critique of set-theoretic foundations and a new proposal for its replacement. Like set theory and category theory, this more recent effort also arose out of ongoing mathematical practice. The mathematics involved this time is homotopy theory, which, like category theory, has its roots in abstract algebra; proponents of the subject describe it as 'an outgrowth of algebraic topology and homological algebra, with relationships to higher category theory' (UFP 2013, p. 1). The program of univalent foundations involves using homotopy theory to interpret Martin-Löf's type theory, then adding the so-called 'Univalence Axiom' –

[18]Of course set theory also translates notions from outside when locating their surrogates, but set theory isn't claiming to provide **Essential Guidance**.

which has the effect, understood roughly, of identifying isomorphic structures.[19] The result is declared to be 'incompatible with conventional [presumably, set-theoretic and category-theoretic] foundations' (Awodey (2014), p. 1) and to provide 'a completely new foundation' (Voevodsky 2014a, b, p. 9).

We've seen that set-theoretic foundations arose in response to the serious practical questions in the wake of the profound shift from mathematics as a theory of the world to mathematics as a pure subject in its own right. In contrast, category-theoretic practice was functioning well enough with Grothendieck's understanding; the impetus this time came from the hope for truly unlimited categories (misconstrued at the time as a shortcoming of set-theoretic foundations) and the promise that category theory could do a new and different foundational job (**Essential Guidance**). Univalent foundations takes a page from each book: there was a real practical problem to be addressed, and addressing it introduced a new foundational goal. Let me explain.

Grothendieck's work in category theory was already so complex that 'the intellectual faculties are being strained to their uttermost limit' (Burgess 2015, p. 176), and as younger mathematicians pushed these ideas further, there was some evidence those limits had been breached. Vladimir Voevodsky, one of the leaders in this development and the originator of univalent foundations, describes how the troubles began:

> The groundbreaking 1986 paper 'Algebraic Cycles and Higher K-theory' by Spencer Bloch was soon after publication found by Andrei Suslin to contain a mistake in the proof of Lemma 1.1. The proof could not be fixed, and almost all the claims of the paper were left unsubstantiated.
>
> The new proof, which replaced one paragraph from the original paper by thirty pages of complex arguments, was not made public until 1993, and it took many more years for it to be accepted as correct. (Voevodsky 2014a, p. 8)

Soon, a similar problem hit closer to home. In 1999–2000, Voevodsky lectured at Princeton's Institute for Advanced Study on an approach to motivic cohomology that he, Suslin, and Eric Friedlander had developed, an approach based on earlier work of Voevodsky. That earlier work was written while the jury was still out on Bloch's lemma, so necessarily did without it. As the lectures progressed, the details were carefully scrutinized.

> Only then did I discover that the proof of a key lemma in my [earlier] paper contained a mistake and that the lemma, as stated, could not be salvaged. Fortunately, I was able to prove a weaker and more complicated lemma, which turned out to be sufficient for all applications. A corrected sequence of arguments was published in 2006. (ibid.)

Perhaps even worse, in 1998 a counterexample was reported to a 1989 paper of Michael Kaparonov and Voevodsky, but because of the complexities involved, Voevodsky reports that he didn't believe it himself until 2013!

---

[19]See Awodey (2014, p. 1).

It's easy to sympathize with the cumulative effect of these mishaps on Voevodsky: 'This ... got me scared' (ibid.). It became hard to ignore the fact that proofs in this area were so complex as to be prone to hidden glitches, a worry exacerbated by the further fact that correcting these glitches made the proofs even more complex. To top off the anxiety, at this point Voevodsky was hoping to push even further, into something new he called '2-theories'.

> But to do the work at the level of rigor and precision I felt necessary would take an enormous amount of effort and would produce a text that would be very hard to read. And who would ensure that I did not forget something and did not make a mistake, if even the mistakes in much more simple [!!] arguments take years to uncover? (Voevodsky 2014a, p. 8)

This, then, is the pressing new problem faced by mathematical practioners in this field: how can we be confident that our proofs are correct? To this point, various sociological checks had been enough – proofs were carefully examined by the community; mathematicians of high reputation were generally reliable; and so on – but those checks had apparently been outstripped.

The need to address this problem gives rise to a new goal – a systematic method for **Proof Checking** – and it seems reasonable to classify this goal, too, as 'foundational'. As we've seen, set-theoretic foundations originated in the embedding of standard mathematics in set theory. For this purpose, as Voevodsky puts it, all we need is to

> ... learn how to translate propositions about a few basic mathematical concepts into formulates of ZFC, and then learn to believe, through examples, that the rest of mathematics can be reduced to these few basic concepts. (Voevodsky 2014a, p. 9)

Here we have the embedding expressed in formal terms. Despite its metamathematical virtues, this formal system isn't one in which any mathematician would actually want to prove anything; in fact (as noted earlier), our confidence that there is a formal proof is usually based on our confidence in the informal proof, combined with our informed belief that all informal proofs can be formalized in this way. The demands of **Proof Checking** are quite different: we need a system that can represent actual proofs, 'a tool that can be employed in everyday mathematical work' (Voevodsky 2014a, p. 8).[20]

[20]Awodey traces the roots of univalent foundations in traditional foundational work to Frege rather than Zermelo: 'this new kind of ... formalization could become a practical tool for the working mathematician – just as originally envisaged by Frege, who compared the invention of his Begriffsschrift with that of the microscope, (Awodey 2016a, p. 8, see also Awodey and Coquand (2013, p. 6). While Frege does make this comparison, it involves a contrast between the microscope and the eye: 'because of the range of its possible uses and the versatility with which it can adapt to the most diverse circumstances, the eye is far superior to the microscope' (Frege 1879, p. 6). Frege's formal system 'is a device invented for certain scientific purposes, and one must not condemn it because it is not suited to others' (ibid.). The 'scientific purpose' in question is to determine whether arithmetic can be derived by pure logic; the Begriffsschrift was needed 'to prevent anything intuitive from penetrating here unnoticed ... to keep the chain of inferences free of gaps' (ibid., p. 5). It seems to me likely that Awodey's 'practical tool for the working

Now there are actually several proof checking technologies on offer these days, some even based on set theory. In his contribution to this volume, Paulson touches on a range of options and remarks that 'every formal calculus ... will do some things well, other things badly and many other things not at all' (Paulson 2019, Chap. 20, pp. 437–453). The proponents of univalent foundations have their own preferred system, combining ideas from Martin-Löf's type theory with insights from the study of computer languages – a system called 'the calculus of inductive constructions' (CIC). The project is to express ordinary mathematical reasoning in these terms – a process that might 'become as natural as typesetting ... papers in TeX' (UFP 2013, p. 10) – and to apply the associated proof assistant (Coq) to mechanically check the validity of those arguments.

Obviously this is a heady undertaking, still in its early stages,[21] but the ambitions of these theorists go beyond the original goal of testing the complex arguments of homotopy theory: Voevodsky holds that univalent foundations, 'like ZFC-based foundations and unlike category theory, is a complete foundational system' (Voevodsky 2014a, p. 9).[22] By this he means that both set-theoretic and univalent foundations aim to provide three things:

(1) a formal language and rules of deduction: first-order logic with the axioms of set theory, on the one hand; the aforementioned deductive system CIC, on the other.
(2) an intuitive interpretation of this deductive system: the iterative hierarchy, on the one hand; homotopy types, on the other.[23]
(3) a method for encoding mathematics: the well-known embedding of mathematics in set theory, on the one hand; an encoding in homotopy types on the other.

The presence of (3) indicates that **Generous Arena** and **Shared Standard** are goals of univalent foundations, though Voevodsky admits that 'this is ... the least understood part of the story' (Voevodsky 2014a, p. 9).

The question that needs answering is whether this encoding in homotopy types is like set theory's proof that there is a set-theoretic surrogate or like category theory's postulation of a category with the desired features – recalling von Staudt vs. Hilbert – as only the former serves to unite the encodings in a single **Generous Arena**. There's probably an easy answer to this question, but if so, it's unknown to me. Voevodsky's strong analogy between set-theoretic and univalent foundations, summarized above, suggests the former; while some of Awodey's remarks appear to lean toward to latter. The move to univalent foundations, Awodey writes,

---

mathematician' would be analogous to the eye, not the microscope, that serving as such a practical tool is one of those purposes for which the microscope and Frege's formal system are 'not suited'.

[21]Cf. UFP 2013, p. 2: 'univalent foundations is very much a work in progress'.

[22]I don't know what Voevodsky finds lacking in category-theoretic foundations – perhaps that it fails to provide a **Generous Arena**?

[23]Interestingly, Voevodsky (2014a) observes that 'our intuition about types of higher levels comes mostly from their connection with multidimensional shapes, which was studied by ZFC-based mathematics for several decades' (p. 9).

> ... has the practical effect of simplifying and shortening many proofs by taking advantage of a more axiomatic approach, as opposed to more laborious analytic [e.g., set-theoretic] constructions. (Awodey 2016b, p. 3)

In a footnote, Awodey alludes to Russell's famous remark about 'the advantages of theft over honest toil' (ibid.).

In broad outline, it appears that the foundational theory into which mathematics is to be embedded begins by postulating a countable hierarchy of 'universes' (UFP 2013, p. 549) that obey a series of 'rules' (ibid., pp. 549–552). To this 'type theory', we add three axioms of homotopy theory: function extensionality, univalence, and higher inductive types (ibid., §A.3). Set theory, for example, is encoded as the category of all the 0-types in one or another of these universes, together with the maps between them (ibid., pp. 398, 438). So far, this looks more like honest toil than like theft. But to get even to ETCS, we have to add the axiom of choice, which incidentally brings with it the law of excluded middle (ibid., §10.1.5). If we simply assert that there is such a category, our procedure begins to look more like the axiomatic method of category-theoretic foundations – start with CCAF and add axioms as needed, asserting the existence of individual categories with the desired features for the various areas of mathematics – and we've seen that this sort of approach doesn't even aim for a **Generous Arena**. I'm in no position to assess how far univalent foundations extends in this direction – whether these are minor variations that can be handled with careful import/export restrictions or something more Hilbert-like – so I leave this as a question to its proponents: is your theory intended to provide a **Generous Arena** for all branches of mathematics and a **Shared Standard** of proof – and if so, how?

Whatever the answer to this question may be, further doubts on the viability of univalent foundations for **Generous Arena** and **Shared Standard** arise when we consider **Essential Guidance**, the key new foundational goal of category-theoretic foundations. Following the category theorists, Voevodsky seems to endorse this goal: he holds that 'the main organizational ideas of mathematics of the second half of the 20th century were based on category theory' (Voevodsky 2014a, p. 9); seeks 'a tool that can be employed in everyday mathematical work' (ibid., p. 8); and counts set theory's failure in these areas against its suitability as a foundation.[24] So, for example, it isn't enough that we find a way to embed set theory in the theory of homotopy types; we need to find a way that reveals the true nature of the subject, unlike ZFC:

> The notion of set ... is fundamental for mathematics. ... However, the theory of sets [has] never been successfully formalized. ... The formal theory ZFC ... is not an adequate

[24] Similarly, Awodey bemoans the 'serious mismatch between the everyday practice of mathematics and the official foundations of mathematics in ZFC' (Awodey 2016a, p. 2) and connects univalent foundations with structuralist tendencies in the philosophy of mathematics that frown on the extraneous features of set-theoretic surrogates.

> formalization of the set theory which is used in mathematics. (Voevodsky 2014b, lecture 2, slides 21–22)[25]

Voevodsky takes this to be accomplished in the new foundation:

> As part of Univalent Foundations we now have a formalization of set theory in the form of the theory of types of h-level 2 in MLTT [i.e., Martin-Löf type theory].[26] I believe that this is the first adequate formalization of the set theory that is used in pure mathematics. (Ibid, lecture 3, slide 11)[27]

Set theorists would most likely dispute this claim,[28] but for our purposes, what matters is that the goal of **Essential Guidance** is more or less explicit. And as we've seen, it seems unlikely that any one way of thinking is best for all areas of mathematics, so aiming for **Essential Guidance** tends to undercut **Generous Arena** and **Shared Standard**.

So, given that **Generous Arena** and **Shared Standard** are once again threatened by **Essential Guidance**, likely to return to the province of set theory, what of the other foundational goals? Speaking of the new formal system, Voevodsky remarks

> Currently we are developing new type theories more complicated than the standard Martin-Löf type theory and at the same time more convenient for practical formalization of complex mathematics. Such type theories may easily have over a hundred derivation rules. (Voevodsky 2013, slide 18)

Notice again the contrast with formalized ZFC. The first-order logic used there is designed to be a simple as possible, with as few formation and inference rules as possible, facilitating meta-mathematical study of theories expressed therein. Because the system of univalent foundations is designed to be as natural as possible a format for actual mathematical reasoning, it ends up being considerably more complex, so the goal of **Metamathematical Corral** presumably also remains with set theory. Furthermore, the complexity of univalent foundations leaves the question of consistency unsettled, much as in the early days of pure mathematics, and the solution is the same:

> Thus a careful and formalizable approach is needed to show that the newly constructed type theory is at least as consistent as ZFC with a given structure of universes [that is, with inaccessibles]. (Voevodsky, ibid.)

In other words, the role of 'a foundational system ... as a standard of consistency' (Voevodsky 2014a, p. 8) – **Risk Assessment** – also falls to set theory.[29]

---

[25]Cf. Awodey and Coquand 2013, p. 6: 'the fundamental notion of a set ... in univalent foundations turns out to be definable in more primitive terms'.

[26]Colin Mclarty was kind enough to explain to me that 'types of h-level 2' is just a different terminology for the '0-types in one or another of these universes' in the previous paragraph.

[27]Cf. UFP (2013, p. 9).

[28]I'm not sure what these thinkers take to be wrong with ZFC, but it could be something akin to the category-theorist's conviction that a neutral notion of 'collection' is better understood in top-down function-based terms (as in ETCS) rather than bottom-up element-based terms (as in ZFC).

[29]See also UFP (2013, p. 15).

To sum up, then, **Risk Assessment**, **Metamathematical Corral**, **Generous Arena**, and **Shared Standard** all appear to continue as the province of set-theoretic foundations. We're left with **Proof Checking**, the new goal introduced by univalent foundations. The promise is that ordinary mathematical reasoning will be easily and directly expressed in CIC and the validity of proofs then checked automatically in COQ, and thus that homotopy type theory will provide a framework for reliable **Proof Checking**:

> I now do my mathematics with a proof assistant. I have lots of wishes in terms of getting this proof assistant to work better, but at least I don't have to go home and worry about having made a mistake in my work. I know that if I did something, I did it, and I don't have to come back to it nor do I have to worry about my arguments being too complicated or about how to convince others that my arguments are correct. I can just trust the computer. (Voevodsky 2014a, p. 9)

I think we can all agreed that this is a very attractive picture, even if it would only apply to areas of mathematics amenable to this sort of conceptualization.

13.4 Conclusion

The upshot of all this, I submit, is that there wasn't and still isn't any need to replace set theory with a new 'foundation'. There isn't a unified concept of 'foundation'; there are only mathematical jobs reasonably classified as 'foundational'. Since its early days, set theory has performed a number of these important mathematical roles – **Risk Assessment**, **Generous Arena**, **Shared Standard**, **Meta-mathematical Corral** – and it continues to do so. Demands for replacement of set theory by category theory were driven by the doomed hope of founding unlimited categories and the desire for a foundation that would provide **Essential Guidance**. Unfortunately, **Essential Guidance** is in serious tension with **Generous Arena** and **Shared Standard**; long experience suggests that ways of thinking beneficial in one area of mathematics are unlikely to be beneficial in all areas of mathematics. Still, the isolation of **Essential Guidance** as a desideratum, also reasonably regarded as 'foundational', points the way to the methodological project of characterizing what ways of thinking work best where, and why.

More recent calls for a foundational revolution from the perspective of homotopy type theory are of interest, not because univalent foundations would replace set theory in any of its important foundational roles, but because it promises something new: **Proof Checking**. If it can deliver on that promise – even if only for some, not all, areas of mathematics – that would be an important achievement. Time will tell. But the salient moral is that there's no conflict between set theory continuing to do its traditional foundational jobs while these newer theories explore the possibility of doing others.[30]

[30]Many thanks to Colin McLarty, Lawrence Paulson, and an anonymous referee for very helpful explanations, discussions, and comments.

References

Awodey, S. (2014). Structuralism, invariance, and univalence. *Philosophia Mathematica, 22*, 1–11.

Awodey, S. (2016a). *Univalence as a principle of logic*. Unpublished. Available at https://www.andrew.cmu.edu/user/awodey/

Awodey, S. (2016b). *A proposition is the (homotopy) type of its proofs* (Unpublished). Available at https://www.andrew.cmu.edu/user/awodey/

Awodey, S., & Coquand, T. (2013, Summer). Univalent foundations and the large-scale formalization of mathematics. *Princeton Institute for Advanced Study Letter*. https://www.ias.edu/ideas/2013/awodey-coquand-univalent-foundations

Burgess, J. (2015). *Rigor and structure*. Oxford: Oxford University Press.

Ernst, M. (2015). The prospects of unlimited category theory. *Review of Symbolic Logic, 8*, 306–327.

Frege, G. (1879). *Begriffsschrift* (S. Bauer-Mengelberg, Trans., reprinted in von Heijenoort, Ed.). From Frege to Gödel, (Cambridge, MA: Harvard University Press, 1967), pp. 5–82.

Frege, G. (1884) *Foundations of arithmetic* (J. L. Austin, Trans.). (Oxford: Blackwell, 1980

Mac Lane, S. (1971). Categorical algebra and set-theoretic foundations. In D. Scott & T. Jech (Eds.), *Axiomatic set theory, proceedings of the symposium in pure mathematics of the AMS, UCLA 1967* (pp. 231–240). Providence: AMS.

Mac Lane, S. (1986). *Mathematics: Form and function*. New York: Springer.

Maddy, P. (1997). *Naturalism in mathematics*. Oxford: Oxford University Press.

Maddy, P. (2008). How applied mathematics became pure. *Review of Symbolic Logic, 1*, 16–41.

Maddy, P. (2011). *Defending the axioms*. Oxford: Oxford University Press.

Maddy, P. (2017). 'Set-theoretic foundations', to appear in A. Caicedo et al, *Foundations of mathematics* (Contemporary mathematics, Vol. 609). Providence: AMS.

Paulson, L. (2019). *Formalizing mathematics in simple type theory* (this volume).

Univalent Foundations Program, group author (UFP). (2013). *Homotopy type theory: Univalent foundations of mathematics*. Princeton: Institute for Advanced Study.

Voevodsky, V. (2013, May 8). *Slides for a plenary talk to the Association for Symbolic Logic*. Available at https://www.math.ias.edu/vladimir/lectures

Voevodsky, V. (2014a, Summer). The origins and motivations of univalent foundations. *Princeton Institute for Advanced Study Newsletter*, pp. 8–9. https://www.ias.edu/ideas/2014/voevodsky-origins

Voevodsky, V. (2014b, September) '*Foundations of mathematics: Their past, present and future*', the Bernays lectures I-III, ETH Zurich.

Part IV
Philosophical Thoughts on the Foundations of Mathematics

Chapter 14
Formal and Natural Proof: A Phenomenological Approach

Merlin Carl

Abstract In this section, we apply the notions obtained above to a famous historical example of a false proof. Our goal is to demonstrate that this proof shows a sufficient degree of distinctiveness for a formalization in a Naproche-like system and hence that automatic checking could indeed have contributed in this case to the development of mathematics. This example further demonstrates that even incomplete distinctivication can be sufficient for automatic checking and that actual mistakes may occur already in the margin between the degree of distinctiveness necessary for formalization and complete distinctiveness.

14.1 Introduction

It is a striking consequence of Gödel's completeness theorem (see e.g. chapter 3 of Rautenberg 2006) that, whenever there is a correct mathematical proof of a certain sentence ϕ from any fixed set S of axioms stated in first-order predicate calculus, there is also a formal derivation of ϕ in the sense of a system of formal deduction.[1] It is this force of the completeness theorem that makes the study of formal proofs relevant to mathematical practice, as it demonstrates a certain kind of adequacy of formal proofs as a model of normal mathematical arguments. This adequacy, however, is rather weak: The only guarantee is that the set of correctly provable assertions coincides with the set of formally derivable assertions. No claim is made on the relation between mathematical arguments and formal derivations.

[1]The reasoning here is roughly this: If ϕ is not formally derivable from S, then there is a model of $S + \neg\phi$, i.e. a way to interprete the occurring notions in such a way that S becomes true, but ϕ becomes false. In the presence of such an interpretation, no argument claiming to deduce ϕ from S can be conclusive. A similar point is made by Kreisel in Kreisel (1965).

M. Carl (✉)
Europa-Universität Flensburg, Flensburg, Germany
e-mail: merlin.carl@uni-konstanz.de; Merlin.Carl@uni-flensburg.de

S. Centrone et al. (eds.), *Reflections on the Foundations of Mathematics*, Synthese Library 407, https://doi.org/10.1007/978-3-030-15655-8_14

There are in fact various reasons to assume that this relation cannot be too close: For example, Boolos (1987) constructed an example of a statement ϕ that is easily seen to be derivable in first-order logic from a set S of premises, but all such derivations are provably of such a vast length that they cannot possibly be actually 'written down' in any sense of the word. On an even more fundamental level, as most interesting theories like Peano arithmetic or ZFC set theory are incomplete and there are statements rather canonically associated with them (i.e. consistency) that might be used in a natural argument, one might even doubt how far carrying out natural mathematics or even merely particular areas as number theory in a fixed axiomatic system can work in principle.

On the other hand, there is the program of formal mathematics, where mathematics is actually carried out in a strictly formal framework. This has now been done for a huge variety of important and non-trivial theorems, including e.g. the Gödel completeness theorem (see Koepke and Schlöder 2012), Brouwer's fixed point theorem (Pak 2011) and the prime number theorem (Avigad et al. 2006). There is even a 'Journal of Formalized Mathematics', entirely dedicated to completely formal proofs. Such proofs are usually done by formalization, i.e. a translation or re-formulation of mathematical arguments in a formal system. The success of this approach suggests that, in spite of the objections mentioned above,[2] formal proofs can be adequate to mathematical arguments in a stronger sense; namely, that correct arguments can be translated into derivations. However, this process of translation is often highly nontrivial and in many cases, the essence of an argument seems to get lost in translation. For any non-formalist view of mathematics, this seems inevitable: If mathematical arguments have content and are about 'objects', then this essential relation to 'objects' must be lost when one passes to formal derivations, which are void of content. It is hence a crucial question for the philosophy of mathematics to determine the relation between arguments and derivations.

In Azzouni (2005a, 2009), Jody Azzouni considers this question while he searches for an explanation for what he calls the 'benign fixation of mathematical practice', namely the fact that mathematics, considered as a social practice, is remarkably stable when compared to other social practices such as art, religion, politics, philosophy and even the natural sciences. The notion of a mathematical proof appears to be particularly invariant: While the standards for what can count as evidence in, say, physics or biology have considerably changed over the last 2000 years, we can still evaluate an argument from e.g. Euclidean geometry and agree on its correctness. Even where the practice splits, for example into a classical and an intuitionistic branch, this agreement is not lost: For the intuitionist mathematician, a valid classical argument may seem invalid from his standpoint, yet he will usually

[2]For example, Boolos' proof has recently been formalized, see Benzmüller and Brown (2007). The matter appears to be more a question of the choice of the formal system than one of formal vs. informal proof or of first-order vs. higher-order logic: When a proof is 'possible, but far too long' in a certain first-order axiomatic system, there are usually natural stronger systems that allow for the necessary abbreviations.

be able to distinguish it from a classically invalid argument. Similarly, one doesn't need to become an intuitionist to see whether an argument is intuitionistically valid.

Azzouni's explanation, which he labels the 'derivation indicator view', or DI-view, of mathematical practice, goes roughly as follows: There is a notion of proof, namely formal proofs in one or another setup, that allows for a purely mechanical proof-check. That is, the correctness of a proof given in this form can be evaluated by simply processing the symbols of which it consists according to a certain algorithm. Since any two persons (and, in fact, a trained monkey or even a computer) applying this algorithm will obtain the same result, this explains the broad agreement at least for formal proofs.

But proofs as they appear in mathematics are virtually never formal proofs in a certain proof system. In fact, formal proofs but for the most trivial facts tend to become incomprehensible to the human reader. What we find in textbooks are arguments presented in natural language, mixed with formal expressions, diagrams, pictures etc. Checking those is not a mechanical procedure; rather, it requires careful concentration in carrying out the indicated mental steps in one's mind, while questioning every step, sustaining it if possible and rejecting it otherwise. The question hence arises how we account for the broad agreement on proofs presented in this manner.

Azzouni's answer is that such proofs, while not formal themselves, 'indicate', 'point to' formal proofs. They are to be considered as recipes for producing a fully formal version of an argument. This indication is clear to us in the same way it is, e.g., clear to us how a cooking recipe is to be transformed into a series of muscle movements in our kitchen. The notion of a formal proof is here independent from the choice of one or another concrete system of representation; rather, it is a form of proof in which every step is a single inference according to some valid deduction rule. This concept is prior to the development of actual representations for formal proofs and could well have been intended by mathematicians in the era before formal logic systems were introduced. In particular, in such a proof, every reference to an imagination of the concepts used can be put aside. We can see that 'If all zunks are zonks, and Jeff is a zunk, then Jeff is a zonk.' is true without knowing what zunks and zonks are or who Jeff is. The subjective component of the argument is hence eliminated as far as possible (all that remains is observing finite sign configurations) and this is the reason for the wide agreement.

In Rav (2007), Yehuda Rav objects to this view with an argument that I want to summarize as follows: Formal proofs cannot provide a basis for the explanation of our agreement on the correctness of proofs. This agreement is based on understanding. Once a proof is transformed into a form in which it is algorithmically checkable, it must be void of content: all contributions of our understanding must have entered the formalization as additional symbol strings. To do this, the argument must have been clarified to the last extent. Hence, at the moment where an algorithmically checkable proof is obtained, the 'battle is over', i.e. the checking is already finished as far as human understanding is concerned: The interesting work is done exactly along the way of formalizing the argument, and this process is non-algorithmical. It is based on an understanding of the occurring concepts, it has an 'irreducible semantic content'.[3] Therefore, carrying out the algorithmic checking for the (formalized) argument will not result in any epistemic gain concerning the (original informal) argument. In particular, it does not strengthen the position that the (original) argument is valid. It might show us that we made some mistake in the 'exercise'[4] of rewriting the proof in a formal system, but that tells us nothing about the proof itself, just as, in programming, an implementation mistake tells us nothing about the correctness of the algorithm we had in mind. Concerning the derivation-indicator view, this implies that it fails to explain the consensus about mathematics: For the consensus about formal proofs via algorithmic checking procedures is of no help unless we explain consensus about the relation between the natural argument and its 'formalization', for which no algorithmic checking procedure is at hand.

This argument has certainly a good degree of persuasive power. We want to evaluate this criticism closer. First, an epistemic gain through automatic proof checking is indeed possible when the latter is used as a means of communication: Even if the inventor of a proof would not learn anything new about the proof by having it automatically checked, the automatic checking can serve as a certificate for the correctness of the result for others. An attempt to communicate an otherwise hardly accessible proof by this means is the work on the formalization of Thomas Hales' proof of the Kepler conjecture in the Flyspeck project, which was recently announced to be completed (see e.g. Bauer and Nipkow (2006)). Focusing on the inventor of a proof himself, it is clear that indeed a lot can be learned about a proof through the process of formalization; being forced to work according to the outermost standards of precision, one is more likely to spot mistakes that otherwise evaded one's attention. But neither of these ways to obtain an epistemic gain from automatic proof checking concerns the point made by Rav. The question is then, more precisely: Can the actual process of automatic checking itself (in contrast to the production of an automatically checkable format) lead to an epistemic gain about a proof that one already knows (e.g. by being its originator)? This is what Rav's criticism is about.

On the surface, we claim that the image of an 'algorithmic system' underlying Rav's argument is too narrow: It falls short of taking into account e.g. methods for automatic language processing or the possibility of using an automatic theorem prover for bridging gaps. But our main intention is deeper: We want to examine when and how an algorithmic system may lead to an epistemic gain about a natural mathematical argument.

When we talk about gaining trust in a proof, we have obviously left the realm where one can consider a proof as mere text or string of symbols; we have to take into account our attitude towards the proof, the way it is given to us or it presents itself to us. The question concerning the epistemic gain should then be reformulated as follows: 'Is there a state of mind towards a proof that allows the construction of

[3]Cf. Rav (2007, p. 301).

[4]Rav (2007, p. 306).

an automatically checkable write-up, but is undecided about the correctness of the proof?'

This formulation makes it obvious that the question can't be decided by merely considering mathematical texts of different degree of formalization. Rather, the representation of a proof in its reader's consciousness has to be taken into account: It will e.g. be relevant whether the reader only briefly skimmed through it or studied it thoroughly, whether he worked the missing steps out or merely granted them, whether a cited result is applied with understanding or merely as a 'black box' etc. Such differences are idealized away in most approaches to logic. One of the rare approaches to logic which seriously takes into account such aspects is found in Husserl's 'Formal and transcendental logic' (Husserl 1900). This motivates us to choose this work as a starting point for our investigation.

Our approach here is hence to analyze to a certain (humble) extent the phenomenology of proof understanding. We will distinguish two qualities of the way how a proof can be mentally represented, applying the approach of Husserl's analysis of judgements, in particular his notions of 'distinctiveness' and 'clarity', to proofs. We will argue that, if a mental representation of a proof has both of these qualities to a maximal extent, then Rav is right in claiming that the 'battle is over'[5] and an algorithmic proof check cannot lead to an epistemic gain. On the other hand, we claim that only distinctiveness is necessary for putting an argument into a form that can be subjected to an automatic proof check. Therefore, we obtain a margin in which automatic proof checking can indeed give substantial information on the correctness of an argument: namely if the proof is mentally present in a distinct, but unclear manner. Considering several examples from the history and the folklore of mathematics, we demonstrate that this tends to occur frequently in mathematical practice.

To sustain our claims and make them more concrete, we will, in the course of this paper, refer to Naproche, a system for the automatic checking of natural mathematical arguments. Its aim is exactly, as in Rav's words, 'to do the work of automatic checking even an informal proof' and already in its current form it gives a vivid picture of the surprisingly natural form an automatically checkable proof can take. We will therefore start by shortly introducing the Naproche system in the next section.

In Sect. 14.3, we explain the distinction between distinctiveness and clarity, using several examples. In Sect. 14.4, we demonstrate that automatic checking requires the former, but not the latter quality, again giving examples. In Sect. 14.5, we argue that clarity and distinctiveness correspond in a certain way to a 'complete', 'gapless' derivation as they are represented in formats like natural deduction or the sequent calculus. The goal is to show which features of a mental representation of a proof are expressed in such a derivation. In Sect. 14.6, we analyze a famous historical example of a false proof in these terms, considering whether or not and how a Naproche-like system might have helped to spot the mistake. Section 14.7 contains a critical

[5]Feferman (1979), quoted after Rav (2007).

review of our account, suggesting various ways in which the use of automatic proof checking is limited. Finally, we give in Sect. 14.8 our conclusions and plans for future considerations on the topic.

14.2 Naproche

Naproche is an acronym for NAtural language PROof CHEcking, a joint project of mathematical logicians from Bonn, formal linguists from Duisburg-Essen and computer scientists from Cologne. It is a study of natural mathematical language with the goal to bridge the gap between formal derivations and the form in which proofs are usually presented. For this, the expressions of natural mathematical language are interpreted as indicators for certain operations, like introducing or retracting an assumption, starting a case distinction, citing a prior result, making a statement etc. We will describe only very roughly how this system works, as the details are irrelevant for our purpose. The interested reader may e.g. consult Cramer et al. (2009) for a detailed description. Also, more information and a web interface are available at Naproche Web Interface. A brief account on the philosophical significance of Naproche can be found in Carl and Koepke (2010).

In the course of the project, a controlled natural language (CNL) for mathematics is developed, which resembles natural mathematical language and is constantly expanded to greater resemblance. This Naproche CNL contains linguistic triggers for common thought figures of mathematical proofs. Texts written in the Naproche CNL are hence easy to write and usually immediately understandable for a human reader. If one was presented with a typical Naproche text without further explanation, one would see a mathematical text, though one in a somewhat tedious style.

Here is an excerpt from a short text about number theory in the Naproche CNL, due to Marcos Cramer and myself, accepted by the current Naproche version 0.47:

Definition 29: Define m to divide n iff there is an l such that $n = m \cdot l$.

Definition 30: Define $m|n$ iff m divides n.

Lemma DivMin: Let $l|m$ and $l|m+n$. Then $l|n$.
Proof: Assume that l and n are nonzero. There is an i such that $m = l \cdot i$. Furthermore, there is a j such that $m + n = l \cdot j$.
Assume for a contradiction that $j < i$. Then $m + n = l \cdot j < l \cdot i = m$. So $m \leq m + n$. It follows that $m = m + n$. Hence $n = 0$, a contradiction. Thus $i \leq j$.
Define k to be $j - i$. Then we have $(l \cdot i) + (l \cdot k) = (l \cdot i) + n$. Hence $n = l \cdot k$. Qed.

Via techniques from formal linguistics, namely an adapted version of discourse representation theory (see Kamp and Reyle 2008), the content of such texts can be formally represented in a format that mirrors its linguistical and logical structure. This format is called a proof representation structure (PRS). In particular, whenever a statement is made, it can be computed from the PRS whether this is supposed to be an assumption or a claim and, in the latter case, under what assumptions this claim is made. In this way, the text is converted into a series of proof goals, each asking to deduce the current claim made in the proof from the available assumptions. The Naproche system then uses automatic theorem provers to test whether the claim indeed follows in an obvious way from the available assumptions. This allows the system to close the gaps that typically appear in natural proofs, one of the crucial features in which natural proofs differ from formal derivations. In this way, every claim is checked and either deduced (and accepted) or not, in which case the checking fails and returns an error message indicating the first claim where the deduction could not be processed.

14.3 Intentions, Fulfillment, Clarity and Distinctiveness

In this section, we introduce Husserl's notions of the distinctiveness and clarity of a judgement as a starting point for our transfer of these concepts to proofs. Crucial for the difference between clarity and distinctiveness is the notion of 'fulfillment' of an intention. We therefore start with a brief introduction to this notion, see e.g. Husserl (2009, 1900) for Husserl's accounts of these notions.

14.3.1 Fulfilled and Unfulfilled Intentions

A central notion of phenomenology is the notion of intention, i.e. the directedness towards something. Whatever this something is, it must, according to Husserl, correspond to a possible way of presenting itself in some kind of experience. Here, 'experience' is taken in a very broad sense, including sensual experience in the usual sense of seeing, hearing etc., but not limited to it: E.g. remembering, imagining, reading a mathematical proof etc. count as legitimate forms of experience. The intention towards X is hence associated with a system of experiences in which X appears. What kind of experience is relevant for a certain intention depends on – or rather strictly corresponds to – the type of the intended object: A piece of music will present itself in hearing, a phantasm will present itself in imagination, a sensual object – say, a tree – will present itself in organized visual, tactile etc. perception. Fulfillment of an intention now simply means that experiences presenting the intended object are made: The piece is heard, the phantasm imagined, the tree seen and felt. Fulfillment may be partial, and in fact, for many types of objects, it will be necessarily so: For example, the intention towards a tree includes anticipations that it can be seen from all sides, including rough anticipations what it will look like. Thus, seeing only one side of a tree, the fulfillment of the intention is only partial: There are anticipations corresponding to the object type that remain unfulfilled. Even if one has walked completely around the tree, this does not change, as now the backside is not given in visual or tactile experience, but only in memory thereof. Proceeding along these lines, it is not hard to see that intentions towards physical objects are necessarily only partially fulfilled.

In the case of a judgement, the intention is directed towards a categorical object, i.e. the state of affairs expressed by the judgement, i.e. in the case 'The rose is red' the fact that the rose is red. In this case, some of the appearing partial intentions from which the judgement is built up (e.g. that towards a rose or towards redness) may be fulfilled while others remain unfulfilled: We may e.g. experience a rose, but not its redness (say under bad lighting). In the case of a work of fiction, we may form a vivid imagination of some of the described objects, but merely skip over the others, leaving the intention 'signitive', merely indicated by a symbol.

We want to apply the concept of fulfillment in the context of mathematical proofs. Our main concern, then, are intentions towards mathematical objects and their fulfillment. One might worry at this point that this will force us to accept some mysterious supernatural faculty for seeing abstract objects, but this worry falls short of taking into account the flexibility of the phenomenological treatment described above: Fulfillment of an intention means having the corresponding experiences. What these experiences are would be the topic of a phenomenological investigation of mathematical objects. Luckily, our treatment does not require the prior execution of such a monumental task, as we will only be concerned with mathematical objects in a very special context. Still, we indicate here two examples of possible interpretations that are hardly 'mysterious': The first would be the intuitionistic standpoint, taken e.g. by Becker and Heyting, according to which the intention towards a mathematical object is fulfilled by a construction of that object in the sense of intuitionism. Another interpretation is that of mathematical objects as 'inference packages': The intention towards a mathematical object is directed towards a system of techniques how to deal with inferences that contain this notion. The fulfillment of such an intention – e.g. in the course of a proof – would hence be the application of these techniques for the full explication of a proof step. An account of mathematical objects in this spirit, though apparently not with Husserl in mind, can be found in Azzouni (2005b); the treatment on p. 106–111 of Lohmar (1989) is of a similar spirit.[6]

[6] Another related perspective is that of Martin-Löf given in Martin-Löf (1980) (in particular on p. 7): He distinguishes 'canonical' from 'noncanonical' or 'indirect' proofs, where a 'noncanonical proof' is a 'method' or a 'program' for producing a 'canonical' proof; as an example, an indirect proof that $123^5 + 5^{123} = 5^{123} + 123^5$ would consist in first proving the general law of commutativity for addition and then instantiating it accordingly rather than carrying out the constructions described by both sides of the equation and checking that they actually lead to the same result. In a similar way, we may view an informal high-level argument as a recipe for obtaining a proof in which every formerly implicit inferential step is actually carried out. Of course, we don't need to go along with Martin-Löf's constructivist approach concerning mathematics here: The checking of non-constructivistic proofs is – regardless of how one views their epistemological value – a cognitive act which is, along with the underlying notion of correctness and its relation to automatization, accessible to a phenomenological analysis. For a perspective on the fulfillment of mathematical intentions based on the notion of construction, see van Atten (2010) and van Atten and Kennedy (2003).

This indicates how the notion of fulfillment can be applied to mathematical proofs. Consider the inferential snippet 'As $1 < a < p$ and p is prime, a does not divide p'. We can skim through it, leaving the empty intention to use the primeness of p to see why the result holds. Fulfilling the intention would amount to actually seeing it, i.e. completing the proof.

14.3.2 Husserl's Notions of Distinctiveness and Clarity

In Husserl (1900, 1999), Husserl offers a phenomenological analysis and foundation for logic. As the notions in question have their systematic place in this analysis, we give a short recapitulation.

At the beginning, logic is taken in the traditional sense as the study of the forms of true judgements. It soon becomes apparent that, in the way this is traditionally done, numerous implicit idealizations are presupposed concerning the judgement and the modus in which it is given. These idealizations are made explicit. In the course of this explication, logic quite naturally splits into several subsections depending on the stage of idealization assumed. It turns out that most of these subsections are not considered by traditional logic, which is concerned with what finally turn out to be distinctly given judgements to which we are directed with epistemic interest. Furthermore, in the study of the abstract forms of judgements, the extra assumption is made that the referents occurring in the judgement forms considered are to be interpreted in a way making the statement meaningful.[7,8]

The first subsection is what Husserl calls the purely logical grammar ('rein logische Grammatik'), i.e. the mere study of forms that can possibly be a judgement at all in contrast to arbitrary word sequences like 'and or write write', ignoring all connections with truth. This part will not concern us further.

A terminological distinction made in various places in Husserl's work is that between 'distinctiveness' and 'clarity'. We find it e.g. in chapter 20 of Husserl (1952) with respect to concepts: While 'distinctiveness' means having explicated what one means with a certain concept, 'clarity' brings about an intuition of the intended object. This is explained on page 101 of Husserl (1952):

> Die Verdeutlichung des Begriffs, des mit dem Wort Gemeinten als solchen, ist eine Prozedur, die sich innerhalb der bloßen Denksphäre abspielt. Ehe der mindeste Schritt zur Klärung vollzogen ist, während keine oder eine ganz unpassende und indirekte Anschauung mit dem Worte eins ist, kann überlegt werden, was in der Meinung liegt, z.B. in 'Dekaeder': ein Körper, ein Polyeder, regelmäßig, mit zehn kongruenten Seitenflächen. (…) Bei der Klärung überschreiten wir die Sphäre der bloßen Wortbedeutungen und des

[7] E.g. the statement 'The theory of relativity is green' is arguably neither true nor false, which nevertheless doesn't contradict the principle of the excluded middle.

[8] See Lohmar (1989) for a further discussion of this point.

Bedeutungsdenkens, wir bringen die Bedeutungen zur Deckung mit dem Noematischen der Anschauung (...)[9]

A parallel distinction concerning judgements instead of concepts is then introduced in Husserl (1900):

A judgement is given in a `distinct` manner when its parts and their references to each other are made explicit. The intentions indicated by its parts may remain unfulfilled, but the compositional structure of the partial intentions is apparent. Given a judgement of the form 'S is p', we can, going along with the formulation, 'carry out' the judgement by explicitly setting S and applying p to it without fulfilling the intentions corresponding to S or p. This explication shows that certain intentions such as those of the form $P \wedge \neg P$ are inexecutable in principle. The subdiscipline of logic concerned with this kind of givenness is 'consequence logic' which considers the executability of a judgement in principle, based on its mere structure, without regard to 'facts'.

`Clarity`, on the other hand, is obtained when the indicated intentions are 'fulfilled', e.g. the objects under consideration are brought to intuition. This may still lead to falsity and absurdity, but these are then of a semantic nature, not apparent from the mere form of the judgement. Of course, both clarity and distinctiveness come in degrees and can be present for certain parts of a judgement, but not for others.

A crucial point of the analysis is that the inexecutability of intentions indicated by certain distinctively given judgements already makes certain assumptions on the objects under consideration which are tacitly presupposed in logical considerations (see above).

Our aim is to apply this classification from single judgements to arguments, particularly mathematical proofs. For example, like single judgements, arguments have a hierarchical intentional structure, which can be given in a vague or in a distinct way and also can be partly or completely fulfilled or unfulfilled. The everyday experience with the process of understanding mathematical arguments suggests that something corresponds to these notions in the realm of such arguments. In particular, the difference between grasping the mere meaning ('Vermeinung') of an argument or actually mentally following it is probably well-known to readers of mathematical texts.

14.3.3 Proofs, Arguments and Understanding: A Clarification

Is it possible to understand a false proof? Certainly. We can be convinced by it, explain it to others (and convince them), translate it to another language, reformulate

[9]Making a concept distinct, i.e. making distinct that which is meant by the word by itself, is a procedure which takes place within the sphere of pure thinking. Before the least step of clarification is performed, while no or a completely inadequate intuition is associated with the word, we can reflect upon that which is meant, i.e. in 'decahedron': a solid, a polyhedron, regular, with ten congruent faces. (...) With a clarification, we transcend the sphere of mere meanings and thinking concerning meanings, we match the meanings with the noematic content of intuition. [Translation by the author]

it, recognize it in its reformulations etc. Even if we know it is false, this does not necessarily hinder our understanding, and it is even often possible (and sometimes takes some effort to avoid) to re-enter a state of mind in which it is still convincing. This for example seems for some people to be the case with the 'goat problem' (Monty Hall Problem (1990)).

Of course, in the usual understanding and despite common manners of speech, a 'false proof' is not a proof. It merely shares some features with a proof on the surface. Anyway, the word is often used in such a way that a proof can be false. This use seems to resemble closer the way we internally think of proofs. We could replace the word 'proof' e.g. by 'argument' to avoid this ambiguity, but we prefer to keep it. Hence, we use the word 'proof' in the sense of a proof attempt. Otherwise, we could never know if something is a proof, for in principle, we could always have been mistaken in checking it.

In this section, we make a humble approach to the study of the ways how a proof can appear to us. We take a phenomenological viewpoint: Hence, instead of asking what proofs might be in themselves – like platonic ideas, patterns of brain activity, mere sequences of tokens or of thoughts – we focus on the question how they give themselves when we encounter them in our mental activity. Mental activities directed towards proofs are e.g. creating it, searching for it, explaining it, remembering it, checking it etc. In such acts, we can experience a proof in different qualities. It is these qualities of proof experience that we consider here, focusing on two, namely clarity and distinctiveness. These are hence not properties of proof texts, but of our perception of proofs. The only way to point to such qualities is hence to create the corresponding experience and then naming it. This is what the following is about. Importantly, we will consider examples of proofs that are likely to lead to an experience with the quality in question, yet one must keep in mind that it is the experience, not the proof text, we are talking about, and that the same text may well be perceived in different ways. The point of this is to find out what is needed for our perception of a proof to make it checkable and compare it to what we need to formalize it. By Rav's claim, the qualities necessary for formalization presuppose those necessary for checking. We aim at demonstrating the contrary.

We now proceed to apply distinctiveness and clarity to proofs (in the sense above). In analogy to the case of judgements, a proof is distinctively given when its parts and their relations are made explicit. At this stage, we hence pay no attention to the correctness of the proof, only to its 'structure'. Distinctiveness about a proof means consciousness of what exactly is claimed and assumed at each point, from what a claim is supposed to follow, where assumptions are needed, which objects are currently relevant, which of the objects appearing are identical, how they are claimed to relate to and depend on each other etc. In analogy with distinct judgements, a distinct proof does not need to be correct. Not even its logical structure must be sound: A distinctively given argument can well be circular. However, from a distinctive perception of the argument, it will be apparent that it is. What we have with a distinct perception of a proof can be seen as a 'proof plan', a description of its logical architecture. In particular, distinctiveness includes consciousness of the sequential structure of the argument. To some extent, it is necessary whenever we even attempt to formulate it.[10] Naturally, distinctiveness comes in degrees. An argument can be distinct in certain parts but not in others. We frequently experience aquaintance with a proof without being able to state exactly where each assumption enters the argument, where each auxiliar lemma is used etc. Also, quite often in understanding natural proofs, we encounter some mixture of distinct deductive steps and imaginative thought experiments.

Concerning clarity, consider the following well-known 'proof' that $2 = 1$[11]:

Let $a, b \in \mathbb{R}$, $a = b$. *As* $a = b$, *we have* $a^2 = ab$, *hence* $a^2 - b^2 = ab - b^2$. *Dividing by* $(a - b)$, *we get* $a + b = b$. *With* $a = b$, *it follows that* $2b = b$. *Dividing by* b, *we obtain* $2 = 1$.

Is this proof – seen as a train of thought – lacking distinctiveness? Not at all. It is completely apparent which of these few steps is supposed to follow from which assumption or fact earlier obtained. In fact, it is in a form that closely resembles a formal derivation (in particular, it could easily be processed by Naproche), and not much would be necessary to make it completely formal.

Anyway, it is of course invalid, yet many people, including clever ones, at first don't see why. The problem here is obviously not that one does not really know what is stated in each step, or that one doesn't know what is supposed to follow or how; the problem is a misperception of division. Following the habit that 'you may cancel out equal terms', the semantic layer is left for the sake of a symbolic manipulation. On this level of consideration, one easily forgets about the condition imposed on such a step. If one takes the effort of really going back to what division is and why the rule that is supposedly applied here works, i.e. if one sharpens the underlying intuition of division, and if one additionally goes back to the meaning of the syntactical object '$a - b$', the mistake – division by 0 – is easily discovered. What is now added and was missing in the beginning is hence a more precise, adequate perception of the objects and operations appearing in the argument, in other words, a step towards the fulfillment of the intention given by the expression $a - b$. We call this degree of adequacy to which the notions are perceived the 'clarity' of the proof perception.

The above proof is hence distinct, yet not clear. Another famous example is the following:

$$1 = \sqrt{1} = \sqrt{(-1)(-1)} = \sqrt{-1}\sqrt{-1} = i \cdot i = i^2 = -1$$

Here, the mistake is obviously a misperception of the complex square root that probably comes from a prior intuition about square roots in the positive reals.

[10] Indeed, as a working mathematician, one occasionally experiences the perception of a vague proof idea which seems quite plausible until one attempts to actually write it down. When one finally does, it becomes apparent that the argument has serious structural issues, e.g. being circular. This particularly happens when one deals with arguments and definitions using involved recursions or inductions.

[11] Singh (2000), appendix, or see e.g. http://www.quickanddirtytips.com/education/math/how-to-prove-that-1-2

Already these primitive examples show that distinctiveness can be present without clarity.

Of course, we can have both distinctiveness and clarity. Every well-understood proof from a thorough textbook is an example. We can also have neither, and if one teaches mathematics, one will occasionally find examples in homework and exams. Further examples are most of the many supposed constructions for squaring the circle, most attempts at an elementary proof of Fermat's Last Theorem (see Fleck et al. 1909), disproofs for Gödel's Incompleteness Theorem, 'proofs' for the countability of $\mathbb{R}$ etc.

Clarity of a proof hence means fulfillment of the occurring intentions. These intentions involve those directed to mathematical objects as well as those directed towards logical steps, i.e. inferential claims: For a distinct understanding of a proof, it suffices to understand that, at a certain point, A is supposed to follow. This supposed inference is an empty intention directed at a derivation of A from the information given at this point. The fulfillment of this intention consists in carrying out this derivation. If a proof is given in perfect clarity, one knows exactly how to carry out each claimed inferential step, using the inferences associated with the occurring objects.

Let us briefly discuss the relation of the preceeding account to the view that proofs serve as fulfillments of mathematical intentions, as e.g. advocated by P. Martin-Löf or R. Tieszen (see Chapter 13 of Tieszen 2005). At first sight, our claim that proofs can be present in an unclear manner seems to conflict with the claim that proofs themselves are 'fulfillments' – for that would make them 'unfulfilled fulfillments'. This apparent conflict is, however, merely terminological: In Tieszen's account of Martin-Löf's position, proofs are cognitive processes of 'engaging in mental acts in which we come to 'see' or 'intuit' something' serving as evidence for mathematical claims. The intention towards a mathematical object would then be fulfilled by evidence for its existence, i.e. a – possibly preferably constructive – proof of its existence. From the point of view of a proof checker that we are interested in, on the other hand, a purported proof is, at first, a linguistic object: It can only serve as evidence once it has been worked through and understood, a process that may require considerable amounts of time, patience and cognitive involvement. In particular, it must be correct in order to provide evidence; in contrast, in our use of the word 'proof' as discussed at the start of the current section, we follow the usual manner of speech to allow for 'false proofs'.[12] The notion of proof that Tieszen seems to aim at is the way a proof is present after this process has been carried out. This is explicated on page 277 of Tieszen (2005):

> 'One can get some sense of the concept of 'evidence' that I have in mind by reflecting on what is involved when one does not just mechanically step through a 'proof' with little or no understanding, but when one 'sees', given a (possibly empty) set of assumptions, that

[12]This point is also discussed in Martin-Löf (1987), p. 418: 'And it is because of the fact that we make mistakes that the notion of validity of a proof is necessary: If proofs were always right, then of course the very notion of rightness or rectitude would not be needed.'

a certain proposition must be true. Anyone who has written or read proofs has, no doubt, at one time or another experienced the phenomenon of working through a proof in such a merely mechanical way and knows that the experience is distinct from the experience in which one sees or understands a proof. (...) To give a rough description, one might say that some form of 'insight' or 'realization' is involved, as is, in some sense, the fact that the proof acquires 'meaning' or semantic content for us upon being understood.'

It is exactly this process of acquiring 'meaning or semantic content for us' that we are concerned with. In the first sentence of the quoted passage by Tieszen, our notion of proof is present as 'proof' in quotation marks.

14.4 Distinctiveness, Clarity and Automatic Proof Checking

Having established the two qualities relevant for our approach, we now want to link them to natural and automatic proof checking. Our thesis is that, while a full formalization requires a distinct and clear presentation of a proof, which means that there's nothing left to do for an automatic checker, distinctiveness is sufficient for producing an automatically checkable text, but not for checking the proof 'by hand'.

14.4.1 Distinctiveness Is Sufficient for Automatic Checkability

We start by comparing a short and basic natural (though very thoroughly written) mathematical argument written up for human readers to its counterparts in the Naproche language. The following is a passage from the English translation of Edmund Landau's 'Grundlagen der Analysis' (Landau (2004)) (all variables denote natural numbers):

Theorem 9: For given x and y, exactly one of the following must be the case:

1) $x = y$.
2) There exists a u (exactly one, by Theorem 8) such that $x = y + u$
3) There exists a v (exactly one, by Theorem 8) such that $y = x + v$.

Proof: A) By Theorem 7, cases 1) and 2) are incompatible. Similarly, 1) and 3) are incompatible. The incompatibility of 2) and 3) also follows from Theorem 7; for otherwise, we would have $x = y + u = (x + v) + u = x + (v + u) = (v + u) + x$. Therefore we can have at most one of the cases 1), 2) and 3).
B) Let x be fixed, and let M be the set of all y for which one (hence, by A), exactly one) of the cases 1), 2) and 3) obtains.
I) For $y = 1$, we have by Theorem 3 that either $x = 1 = y$ (case 1) or $x = u' = 1 + u = y + u$ (case 2). Hence 1 belongs to M.
II) Let y belong to M. Then either (case 1) for y) $x = y$, hence $y' = y + 1 = x + 1$ (case 3) for y'); or
(case 2) for y) $x = y + u$, hence if $u = 1$ then $x = y + 1 = y'$ (case 1) for y'); but if $u \neq 1$, then, by Theorem 3, $u = w' = 1 + w$, $x = y + (1 + w) = (y + 1) + w = y' + w$ (case 2)

for y'); or (case 3) for y) $y = x + v$, hence $y' = (x + v)' = x + v'$ (case 3) for y'). In any case, y' belongs to M. Therefore, we always have one of the cases 1), 2) and 3).

Now compare this to the following variant, a fragment of a text which is accepted by the current version of Naproche (taken from Carl et al. (2009)):

Theorem 9: Fix x, y. Then precisely one of the following cases holds:

Case 1: $x = y$.
Case 2: There is a u such that $x = y + u$.
Case 3: There is a v such that $y = x + v$.

Proof: Case 1 and case 2 are inconsistent and case 1 and case 3 are inconsistent. Suppose case 2 and case 3 hold. Then $x = y + u = (x + v) + u = x + (v + u) = (v + u) + x$. Contradiction. Thus case 2 and case 3 are inconsistent. So at most one of case 1, case 2 and case 3 holds.
Now fix x. Define $M(y)$ iff case 1 or case 2 or case 3 holds.
Let y such that $y = 1$ be given. $x = 1 = y$ or $x = u' = 1 + u = y + u$. Thus $M(1)$.
Let y such that $M(y)$ be given. Then there are three cases:
Case 1: $x = y$. Then $y' = y + 1 = x + 1$. So $M(y')$.
Case 2: $x = y+u$. If $u = 1$, then $x = y+1 = y'$, i.e. $M(y')$. If $u \neq 1$, then $u = w' = 1+w$, i.e. $x = y + (1 + w) = (y + 1) + w = y' + w$, i.e. $M(y')$.
Case 3: $y = x + v$. Then $y' = (x + v)' = x + v'$, i.e. $M(y')$.
So in all cases $M(y')$. Thus by induction, for all y $M(y)$. So case 1 or case 2 or case 3 holds.
Qed.

We see some extra complications arise due to the fact that the Naproche language is a controlled language, so that formulations like 'For given x and y' or 'incompatible' need to be replaced by their counterparts 'Fix x, y' and 'inconsistent' in the Naproche language. This, of course, can easily be overcome by amending the language accordingly. The passage is given exactly in the way Naproche can currently read it to avoid the criticism of being speculative, but we are safe to assume that a line like 'exactly one of the following must be the case' can be processed by a slightly improved version. Now, what does it take to go from the natural to the Naproche version? Do we need to understand the proof in some depth or see its correctness? Certainly not. Rather, we reformulate the proof according to some linguistic restrictions. Hence, while a certain difference in the wording is obvious, these two texts are very similar in content and structure. Given a knowledge of the current Naproche language, passing from the first version to the second is trivial: One merely changes some formulations, permanently working along the original. One can do this with virtually no understanding of the original text, as long as one keeps the indicators for assuming, deducing and closing assumptions and uses the same symbol where the same object is meant. One does not even need to know the meaning of the symbols used. It seems that any state of mind allowing one to write the first text also allows one to write the second. In fact, even a faint memory of a vague understanding might suffice.

This basic example indicates what is necessary for producing an automatically checkable version of a proof: The argument must be given to us as a sequence of steps in such a way that we can see what is currently claimed and assumed, which objects are considered, when a new object is introduced and when something new is claimed about an object introduced earlier (so we will e.g. use the same symbol). A mere image of some mental movement, which is indeed often the way one remembers or invents an argument, is not sufficient. One needs an explicit consciousness of the way primitive intentions are built together to form judgements and then how these complex intentions are used to build up the argument. On the other hand, it is not necessary at all to reduce everything to formal statements and simple syllogisms. Whether or not a concrete checker will succeed in a particular case depends of course on how well the checker captures the semantics of natural mathematical language, but in principle, arguments at this stage of understanding are open to an automatic checking process. Hence, to produce an automatically checkable format, it suffices to have a distinct understanding of the proof.

We consider another example, which deserves special interest, as it is explicitly given by Rav in (2007) as an example of an 'ordinary' proof, which he comments thus:

> The proof of this theorem as given above is fully rigorous (by currently accepted standards). It requires, however, on the part of the reader a certain familiarity with standard mathematical reasoning (...). No formal logic is involved here; it is a typical reasoning that Aberdein has aptly characterized as the `informal logic` of mathematical proof. [Rav 2007, p. 305]

He then goes on to describe what to do for obtaining a version of this proof that can be automatically checked, claiming that this needs, 'as must have been noted, a logician's know-how'. Here is the proof as given in Rav (2007, pp. 304), including some preliminaries about group theory (we omit Rav's bracketed observations on the proof):

> Recall that a group is defined as a set G endowed with a binary operation (to be written as juxtaposition) having a distinguished identity element, denoted by 'e' and satisfying the following first-order axioms:
>
> - (i) $(\forall x)(\forall y)(\forall z)[x(yz) = (xy)z]$; (associativity)
> - (ii) $(\forall x)(ex = x)$; ('e' is a left-identity)
> - (iii) $(\forall x)(\exists y)(yx = e)$. (every element has a left-inverse)
>
> On the basis of these axioms, one proves that the postulated left-identity is also a right-identity – in symbols, $(\forall x)(xe = x)$ – from which it will follow that the identity element is unique. Here is the proof:
>
> Let u be an arbitrary element of G. By axiom (iii), there exists t such that (1) $tu = e$; once more by axiom (iii), for the t just obtained, there exists $s \in G$ such that (2) $st = e$. Hence: (3) $ue = e(ue) = (eu)e = [(st)u]e = [s(tu)]e = (se)e = s(ee) = se = s(tu) = (st)u = eu = u$. (...) Since u is an arbitrary element of G and by (3) $ue = u$, we conclude that $(\forall x)(xe = x)$.
>
> Now if there were a second element e' with the property that $(\forall x)(e'x = x)$, we would conclude from what has just been proven that $(\forall x)(xe' = x)$ also, and hence $e = ee' = e'$. Thus, the identity element is unique. Hence, we have proven the following:
>
> Theorem: (a) $(\forall x)(xe = x)$ (b) $(\forall y)(\forall x)[yx = x \implies y = e]$.

The following Naproche version of this proof, due to Marcos Cramer,[13] is accepted by the current version of the system:

Suppose that there is a function $*$ and an object e satisfying the following axioms:

Axiom 1. $\forall x \forall y \forall z(x * (y * z) = (x * y) * z)$.
Axiom 2. $\forall x(e * x = x)$.
Axiom 3. $\forall x \exists y(y * x = e)$.

Theorem A: For all x, we have $x * e = x$.
Proof: Let u be given. By Axiom 3, there is a t such that $t * u = e$. By Axiom 3, there is an s such that $s * t = e$. Then $u * e = e * (u * e) = (e * u) * e = ((s * t) * u) * e = (s * (t * u)) * e = (s * e) * e$ and $(s * e) * e = s * (e * e) = s * e = s * (t * u) = (s * t) * u = e * u = u$. Thus $\forall x(x * e = x)$. Qed.
Theorem B: If $\forall x \forall y(x * y = x)$, then $y = e$.
Proof: Assume that e_1 is such that $\forall x(e_1 * x = x)$. Then by Theorem A, $\forall x(x * e_1 = x)$. Hence $e = e * e_1 = e_1$. Thus $e_1 = e$. Qed.

Again, the transition from the original text to the Naproche text is quite elementary and needs neither a logician's expertise nor knowledge of group theory. It is particularly interesting to note that the equivalence of these texts can be observed even without a knowledge of the meaning of the corresponding terms. This demonstrates again that it is distinctiveness, not clarity, which is required for the transition as well as for judging the adequacy of such a translation.

Let us briefly recall at this point how Naproche proceeds to check a text like the above: From the text, it builds a representation of the intended logical structure of the proof. These proof representation structures (PRSs) contain the occurring statements along with their intended relations, as A is to be assumed, B is to be deduced, C is to be used in the step from A to B etc. Other than a formal derivation which, by its definition, cannot be false, a PRS is completely neutral to the correctness or soundness of the represented argument: What it represents is the distinctively given intentional structure of a proof. Notably, presuppositions hidden in certain formulations (e.g. uniqueness in the use of definite articles as in 'Let n be the smallest natural number such that...') are made explicit. In the next step, an attempt is made to construct a fully formal proof from the PRS: In particular, when B is, according to the PRS, claimed to follow at a certain step (possibly with the help of an earlier derived statement A), then an automatic theorem prover is used to attempt to prove B from the available assumptions at this point, where indicators like explicit citations of earlier statements help to direct the proof search. In particular, the automatic prover will attempt to apply definitions of the occurring notions as well as results in which they appear: It will, when a notion T occurs, quite naturally try to use the 'inference package' associated with T for creating the actual deductions claimed to exist. It will, in other words, attempt to do something

[13] We thank Marcos Cramer for the kind permission to use the Naproche version of this text in our paper.

analogous to fulfilling the intentions given in the distinctive representation of the proof.

14.4.2 Distinctiveness Is Not Sufficient for Natural Checking

The degree of understanding obtained by distinctly disclosing the structure of an argument is not sufficient for performing a proof check. In fact, we can have perfect distinctiveness and still be completely agnostic concerning correctness. This is already indicated by our examples above. One reason for this is that, in a natural argument, we do not have a fixed, manageable supply of inference rules justifying each step. When checking a step, we often use some mental representation ('image') of the objects under consideration. This representation is different from a formal definition and usually precedes it.[14] However, these images are imperfect in directing us towards the objects we mean and may carry false preconceptions concerning these objects. If, for example, concept B is a generalization of concept A, there is a certain tendency of assuming properties of A for B. There is a vast amount of frequent mistakes compatible with a distinct presentation. A strong source of mistakes is some kind of a closed world assumption that excludes objects we can't really imagine. This danger remains even after we come to know about counterexamples. Imagining e.g. a continuous function from $\mathbb{R}$ to $\mathbb{R}$ as a 'drawable line' is often very helpful, but it also misdirects us in many cases. For another example, in spite of strong and repeated efforts, some students in set theory courses never acknowledge the existence of infinite ordinals and keep subtracting 1 from arbitrary ordinals. The idea of a non-zero 'number' without a predecessor is apparently hard to accept.

Such preconceptions derived from a misinterpretation of mental images are a common source of mistakes even in actual mathematical research practice. Math Overflow Discussion (2010a,b) contain long and occasionally amusing lists of common misperceptions in mathematics, most of which are instances of such misinterpretations. We will get back to this below when we consider classical examples of false proofs.

Of course, this kind of perception of mathematical objects is all but a dispensible source of mistakes: In fact, it is exactly this ability that steers the process of proving and creating mathematics, thereby making the human mathematician so vastly superiour to any existing automatic prover.

Let us now briefly reconstruct this example-driven account in the more general terminology set up above for this purpose. Given a distinct presentation, we only see what is claimed to happen, a 'proof intention'. We can at this point already

[14]Historically, at least. The deductive style dominant in mathematical textbooks confronts the student with the inverse problem: Namely making sense of a seemingly unmotivated given formal definition.

spot some mistakes, e.g. circularity, or that the statement actually proved is not the statement claimed; but we cannot check the proof: A 'proof map', as helpful as it is for finding one's way through a more involved argument, does not require for its creation an understanding or checking of the proof. To check a proof, it is necessary to fulfill these inferential intentions; but for this, clarity is needed.

14.5 Distinctiveness, Clarity and Formalization

In this section, we consider the question what kind of understanding is necessary for carrying out a full formalization of a proof in a common system of first-order logic like, say, natural deduction. We argue that Rav is indeed right in claiming that such an understanding allows checking and that, in fact, the checking is almost inevitably carried out in the process of obtaining such an understanding.

To do this, let us reflect on the process of formalization. A formal proof is one in which the manipulation of symbols is justified without any reference to a meaning of these symbols. It is clear how a certain symbol may be treated without knowing what it means, without even taking into account that it might mean anything. This is achieved by replacing semantic reference by formal definitions. For instance, the meaning of the word 'ball', representing a certain geometric shape, will be replaced by rules that allow certain syntactical operations once a string of the form $ball(x)$ shows up. Still, the formal definition must capture the natural meaning if the formal proof is to be of any semantic relevance, not just a symbolic game: The formal definitions have to be "adequate" in a certain sense. How do we arrive at adequate formal definitions? Obviously by observing the role a certain object plays in proofs and then formulating precisely what about this object is used. Then, the notion of the object is replaced by the statements used about it. (See Tieszen (2005) and Tieszen (2011) for thorough discussions of phenomenological aspects involved in the forming and clarification of mathematical concepts.)

If we replace an informal by a purely formal proof, we have to make all implicit references to the content explicit to eliminate them in the formalization. This means that the role of the object in the argument must be clear. Consequently, to obtain a fully formal derivation from an informal proof, we must have distinctiveness **and** clarity. But when all hidden information is made explicit as part of a complete understanding of the argument, i.e. if all intentions are fulfilled, mistakes will inevitably become apparent. The only questionable part remaining is then the connection between the original semantic references and the formal definitions.[15] But in established areas of mathematics, these definitions have stood the test of time,

[15]An excellent example of the delicate dialectics involved in forming definitions of intuitive concepts is the notion of polyhedron in Euler's polyhedron formula as discussed in Landau (2004). Sometimes, this is the really hard part in creating new mathematics. Another prominent example is the way how the intuitive notion of computability was formalized by the concept of the Turing machine.

and even though, particularly in new areas, there are debates about the adequacy of definitions and though the focus occasionally shifts from one definition to another providing a deeper understanding of the subject (often indicated by amending the original notion with expressions like 'normal', 'acceptable', 'good' etc.), this issue virtually doesn't come up in mathematical practice. Even if it does, it is usually considered to affect the degree to which the result is interesting, not the correctness of the proof.[16]

Considering a distinct proof presentation, a good automatic prover will be able to draw from formal definitions what we draw from correct intuition. Note that we make no claim on the question whether formal definitions can exhaust semantic content, nor do we need such a claim. The process of replacing steps referring to understanding and perception of abstract objects by derivations from formal definitions is what corresponds to the activity of fulfilling intentions involved in the course of a clarification. Conceptually and mentally, this may well be a very different operation. However, as explained above, Gödel's completeness theorem ensures that, whenever an argument can be brought to clarity, there will be a derivation from the definitions. This is the reason why an automatic proof checker, using an enhanced formalism as described above, can give us information on the possibility of clarification.

For a more concrete picture of what information this kind of automatic checking can give us, let us suppose that we have given a certain proof B to our proof-checker (e.g. Naproche) and got a negative feedback, i.e. that our assumptions were found to be contradictory or that some supposed consequence could not be reproduced by the system. We are thereby made aware of a particular proof step that needs further explanation, and it is here that we may become aware of an actual mistake in our intended proof. This can already be seen in the most basic examples above, e.g. the 'proof' that $2 = 1$: Here, the demand to divide by $(a - b)$ will trigger the presupposition that $(a - b) \neq 0$ as an intermediate proof goal which will then be given to the automatic theorem prover. It is then a matter of seconds that we will be informed that this proof goal was found to contradict the assumptions which will make it very obvious what goes wrong in the intended proof. Here, we thus have a clear example how automatic proof checking can lead to an epistemic gain.

It is not so obvious what information there is to be drawn from the opposite scenario, e.g. a positive feedback from the system. Certainly, we are informed that the proof goal is actually formally provable as it should be, but that doesn't imply that our proof was correct. Getting information about our proof from a positive feedback would need a close connection between our natural way to think about missing proof steps and the automatic theorem prover, which is certainly a

[16]Suppose, for example, that someone came up with a non-recursive function that one can evaluate without investing original thought so that one is inclined to accept the evaluation of this function as an instance of 'calculation', thus disproving the Church-Turing thesis. As a consequence, recursiveness would lose its status as an exact formulation of the intuitive concept of calculation. But this would not affect the correctness of recursion theory.

fascinating subject for further study, but currently far remote from reality. This point will be discussed in Sect. 14.7.

14.6 An Historical Example

In this section, we apply the notions obtained above to a famous historical example of a false proof. Our goal is to demonstrate that this proof shows a sufficient degree of distinctiveness for a formalization in a Naproche-like system and hence that automatic checking could indeed have contributed in this case to the development of mathematics. This example further demonstrates that even incomplete distinctivication can be sufficient for automatic checking and that actual mistakes may occur already in the margin between the degree of distinctiveness necessary for formalization and complete distinctiveness.

Example (Cauchy 1821)[Cf. Rickey][17]
Claim: Let $(f_i|i \in \mathbb{N})$ be a convergent sequence of continuous functions from $\mathbb{R}$ to $\mathbb{R}$, and let s be its limit. Then s is continuous.
Proof: Define $s_n(x) := \Sigma_{i=1}^{n} f_i(x)$, $r_n(x) := \Sigma_{i=n+1}^{\infty} f_n(x)$. Also, let $\varepsilon > 0$. Then, as each f_i is continuous and finite sums of continuous functions are continuous, we have $\exists\delta\forall a(|a| < \delta \implies |s_n(x+a) - s_n(x)| < \varepsilon)$.
As the series $(f_i|i \in \mathbb{N})$ converges at x, there is $N \in \mathbb{N}$ such that, for all $n > N$, we have $|r_n(x)| < \varepsilon$.
Also, the series converges at $x + a$, so there is N such that, for all $n > N$, we have $|r_n(x + a)| < \varepsilon$.
So we get: $|s(x+a) - s(x)| = |s_n(x+a) + r_n(x+a) - s_n(x) - r_n(x)| \leq |s_n(x+a) - s_n(x)| + |r_n(x)| + |r_n(x+a)| \leq 3\varepsilon$.
Hence s is continuous.

This example is taken from Rickey and closer analyzed in the appendix of Landau (2004). The mistake becomes obvious when one focuses on the dependencies between the occurring quantities: The δ shown to exist in line 3 of the proof depends on ε, x and n. The N from line 4, on the other hand, only depends on ε and x. However, the N used in line 6 obviously also depends on a. Hence N is in a subtle way used in two different meanings. The dependence on a can only be eliminated if there is some M bigger than $N(\varepsilon, x + a)$ for all $|a| < \delta(\varepsilon, x, n)$. This property means that $(f_i|i \in \mathbb{N})$ is uniformly convergent, which is much stronger than mere convergence.

Simple as this mistake may seem, it has a long success story (see again Landau 2004): The (wrong) statement it supposedly proves was considered trivially true for quite a while by eminent mathematicians, and when the first counterexamples

[17]This formulation is sometimes disputed as not correctly capturing the argument Cauchy had in mind. Some claim that Cauchy meant the variables implicit in his text to not only range over what is now known as the set of reals, but also over infinitesimals. However, the formulation we offer captures the way the proof was and still is understood and at first sight considered correct by many readers, so we will not pursue this historical question further.

occurred, they were considered either as pathologies that shouldn't be taken seriously as functions or violently re-interpreted as examples. It was no other then Cauchy who first felt the urge to give a proof and published the above argument in his monograph (Cauchy 1821). It took several decades before the mistake was spotted and the statement was corrected by strengthening the assumption to uniform convergence.

Reproducing the understanding of this argument shows what is going on: In the argument for the existence of N, one gets the imagination of a 'sufficiently large number' and then reuses the object in a new context in an inappropriate way because hidden properties of the object – its dependencies on others in its construction – are ignored. That is, while the train of thought described here gives distinct intentions to certain objects N and N' which are then identified, a fulfillment of these intentions is not possible.

Now, suppressing the arguments on which an object depends is quite common in mathematical writings. A formalizer, of course, must reconstruct this information. The way a Naproche-like system models a text can easily allow for such a convention. Apart from that, the text is certainly not lacking distinctiveness. It also uses only very little natural language and not in any complicated way. It would be quite feasible to enrich the vocabulary of e.g. Naproche to process it in the precise form given here. But when the formalization is carried out, the proof breaks down. It will be very interesting to actually carry this out on concrete systems once they are sufficiently developed.

14.7 Discussion

We have explained above how automatic checking with Naproche-like systems can lead to an epistemic gain concerning a proof. We will now take a close look at the assumptions on which our account relies, thereby sharpening the picture when such gains can be expected and when not.

Under what circumstances, then, do we get new information from Naproche and what is that information? Certainly, it is informative if Naproche spots a non-intended contradiction – in this case, the proof contains a mistake. The false proofs of $1 = 2$ and $-1 = 1$ above are examples where a tacit assumption contradicts the information given, which an automatic proof checker can spot and report. But what does it mean in general when Naproche fails to confirm a proof? And what does it mean for the correctness of the proof if it succeeds? An informal proof, as found in a mathematical journal, a math exam or a math olympiad is roughly seen as correct when it convinces the critical expert: I.e. when it provides a person with the right background with the information needed to construct a complete, detailed argument. This notion of 'right background' is highly context-dependent: An original research paper at the frontier of some area of core mathematics may leave proof steps to the reader as 'clear' that would require from the average beginner student several years of study and concentrated work to complete; on the other hand, proof exercises for beginner students of mathematics often require details even – occasionally especially – for steps that are supposedly immediately clear. This is only one respect in which the correctness of proof texts is a delicate notion, involving sociological aspects like what background knowledge and what heuristic power is to be expected by the audience addressed. The criterion realized in Naproche in its current implementation is comparably weak, namely whether each claim can be formally deduced from the information given at that point with limited resources (typically within 3 s processing time). This is a very rough model of the inferences regarded as admissible for bridging proof steps by human readers to whom formally extremely complex inferences may be clear based on spatial intuition, analogy with previous arguments etc. Consequently, a failure of Naproche to confirm a proof does not necessarily mean that this proof is not a rational and sensible way of convincing the reader of the correctness of the conclusion. Conversely, at least in principle, we could have false positive results. If the checking succeeds, then a formal proof of the conjecture in question was generated, so a correct proof is confirmed to exist; but this does not necessarily mean that the original informal proof was correct by the standards for informal proofs. Suppose e.g. that, in a proof of a statement A, the end-result A is deduced as an intermediate step in the process of bridging a gap in that proof; in this case, the formalization should certainly not be seen as a confirmation that the original argument was sound. Similarly, if the work needed to confirm a certain inference in a proof is much more complex than the whole proof itself, one has a reason to doubt the proof. This becomes particularly obvious when one thinks of tutorial contexts: When asked to prove something, some students try to trick the corrector by making some deductions from the information given forwards, some steps backwards from the goal and then simply writing the results next to each other, claiming that the latter follows from the former. Even though this claim may be right, these texts fail to show that the author knew how to do it, and usually, the step left out is as complicated as the original problem itself. It is an interesting field of further study to see what kind of inferences may be 'left to the reader' and build formal models of those. More generally, the notion of correctness for informal proofs, in contrast to that for formal proofs, certainly deserves further attention.

Still, granting these objections and as decent as our currently existing software may be, the experience with Naproche and our analysis thereof above should suffice to demonstrate that there is more to get from automatic proof checking than Rav (and many others) might expect. However, one should not forget that this kind of checking relies on various convenient circumstances; for example, it assumes a stable formal framework which in particular allows to replace the understanding of a term by a formal definition for all purposes of deduction. Where such a framework is missing, the emulation of fulfillment of mathematical intentions by automatic proving will not work. Such a framework is, of course, not always present; rather, it typically occurs as a rather late stage in the development of a mathematical theory. The clarification and development of notions is an important part of mathematics, and proofs play a role as a part of this process (as e.g. Lakatos has impressively demonstrated in Lakatos 1976). A similar caveat holds with respect to the use of axiomatic systems: Not always is the content of a theory canonically codified in an axiomatic system; furthermore, axiomatic systems codifying a theory can be complemented when defects become apparent in its usage. (A well-known examples is Zermelo's proof of the well-ordering principle introducing the axiom of choice.)

A vast majority of contemporary mathematical work, however, does indeed work in such stable environments where we have formal definitions and a universally accepted axiomatic background. Still, there are parts of mathematicians' proof practice (arguably the philosophically most interesting ones) that evade automatic checking in the sense explained above. For those, there may indeed be fundamental reasons to expect that automatic checking will not be helpful. Let us look at one example: Namely Turing's work (Turing 1937) on the Entscheidungsproblem, which was widely accepted as settling this problem. It depends on a technical part (the Entscheidungsproblem is not solvable by a Turing machine) as well as a concept analysis (Turing machine computability captures the informal notion of computability, as intended in the formulation of the Entscheidungsproblem). The first part is, technical difficulties aside, open to automatic checking. But the second part seems to be of an entirely different nature: Can a machine even in principle help us to decide whether some formalization of a certain intuitive notion is adequate? Apparently, we can't have told the machine what such a notion means without knowing it ourselves. If, by some other means (like automatic learning) we had a machine answering 'yes' or 'no' to such questions, how could we know it is right? The machine would have to take part in the discussion arising, providing experiences, thought experiments exploring the borders of the informal notion etc. In some cases, the underlying informal understanding of such notions will be rooted far outside of mathematics in what may be called the life-world. Is the set of reals an adequate model for a 'line' or for time? Does Turing computability indeed capture computability in the intuitive sense? Does ZFC provide a reasonable understanding of what it means to be a set and are the set-theoretical formalizations of mathematical notions adequate to them? These questions play a crucial role in the acceptance of various formal proofs as answers to mathematical questions, and they are only non-mathematical when one arbitrarily limits the scope of mathematics to formalism. In all of these cases, one can easily imagine how the discussion about them will touch on aspects of human thinking and experience far outside of mathematics. These notions are human notions, made by humans for humans. Human understanding is the ultimate criterion for their adequacy, so that no outer authority like a computer can tell us what they mean. It is hence quite plausible that we cannot write a program of which we are justified to believe that it gives correct answers to questions concerning the adequacy of a concept analysis, but there is nothing mysterious about this 'non-mechanical nature of the mind': It is a mere consequence of the specific evidence type of (human) concept clarification.[18]

[18]Even more, taking the speculation a bit further, being able to meaningfully and convincingly participate in this kind of discourse is a plausible criterion for not calling something a 'machine' any more. (Moreover, the definiteness of a machine's response, which is a main motivation for striving for automatization in the first place, is lost when a machine becomes merely another participant of a discourse.) Of course, the rules of a discourse are made by its participants; so in the

To briefly summarize our discussion: There are indeed important aspects and parts of mathematicians' proof practice that are likely beyond automatic checking in principle, and definitely in the current state or any state to be expected in the foreseeable future. The great consensus on the Church-Turing-Thesis and the general acceptance of Turing's work as a solution to the Entscheidungsproblem (or Matiyasevich's work on diophantine equations as a solution to the 10th problem, see Matiyasevich) is something that Azzouni's derivation-indicator-view can hardly explain. But many great mathematicians have done their work in stable frameworks and a great deal of mathematics takes part in those: Here, automatic proof checking can lead to epistemic gains. And here, Naproche-like systems work, demonstrating that the derivation-indicator view is not as easily discarded by arguments like those of Rav.

14.8 Conclusions and Further Work

We hope to have made it plausible that phenomenological considerations and the corresponding shift of focus can be fruitfully applied to questions concerning the philosophy of mathematical practice with a relevance to mathematical research itself. Namely, we have argued that, in spite of the claims against it, automatic proof checking can lead to an epistemic gain about an argument in providing evidence that the indicated intentional acts can be carried out in a distinct and clear manner. The reason for this was that human proof checking needs clarity about a proof, while automatic checking can be performed once a certain degree of distinctiveness is obtained. For this argument, we crucially used the phenomenological turn from proofs in the way they are usually considered to the ways in which they occur.

A phenomenological theory of proof perception has, to our knowledge, not yet been given.[19]

end, the possibility of computers becoming influential even in the conceptual part of mathematics might boil down to the question whether we are willing to accept a computer as a participant in such a debate on equal terms. However, we now have reached a level of speculation at which it is better to stop.

[19]There are, however, several phenomenological accounts of mathematical proofs focusing on other aspects. We are thankful to the anonymous referee of an earlier version of this paper for pointing out to us four very worthwhile works on the phenomenology of proofs, and we take the opportunity to briefly discuss their relation to our work:

In Tragesser (1992), R. Tragesser points out that, in the philosophy of mathematics, actual proof practice is often ignored in favour of or replaced with formal proofs and remarks that this 'idealizes away' many elements that are important for mathematics. To remedy this, he proposes phenomenology as a better approach, where phenomenology means focusing on 'the thing itself' and striving for a deep an understanding as possible rather than being satisfied with pragmatical considerations. In particular, he emphasizes that possible proofs are as good as actual proofs and that this is a sense in which mathematics can be said to be a priori. His approach does not build on

It would certainly be interesting in its own right. As one consequence, it would contain a thorough study of proof mistakes, which, on the one hand, might become relevant in pedagogical considerations, but would also sharpen our understanding of what automatic proof checkers can add to our trust in a proof and how they can do this.

A concrete application of such considerations would be the development of proving tools suitable for Naproche-like systems. Such a prover is supposed to bridge steps in natural proofs which are assumed to be supplied by the reader. In a sense, these proofs are hence 'easy' and 'short'. Of course, such steps often take place in e.g. spatial or temporal intuition rather than formal reasoning. There is therefore no obvious relation between a 'simple, short argument' and the number of lines in a corresponding derivation.[20] A next step is hence to consider common elementary operations that are performed in supplying such proof steps and give formal background theories to replace them. The goal of this would be to make the automatic prover's activity more resemblant to an actual human reader. (Suppose

---

Husserl's phenomology and also does not apply, as ours does, to the analysis of mistakes, as for Tragesser, a proof is always a correct proof.

A similar critique of focusing on formal representations of proofs is given by G.-C. Rota in Rota (1997), who also presents a similar understanding of phenomenological analysis, which should consider as close as possible actual proofs without ignoring important aspects as belonging to other fields – such as psychology – or due to confusing description with norms. Rota in particular points out that 'proof' has aspects beyond 'verification', such as understanding, degrees of understanding, giving 'reasons', different kinds of proof and the heuristical value that proofs and theorems play for the further development of mathematics. He also gives historical examples to substantiate his thesis that mathematics is really a strive for understanding in which theorems and proofs both play equal roles.

In Sundholm (1993), G. Sundholm notes that there is a 'tension' between various plausible claims put forward concerning proofs, which leads him to distinguish between proof-acts (series of mental acts), proof objects (the result obtained at the end of a proof) and proof-traces (the written presentation, that allows others to reproduce the series of acts). His work does not build on Husserl's analysis of logic, but is related to our work as the present paper can be viewed as a closer analysis of the process of reproducing a proof-act on the basis of a proof-trace, distinguishing different qualities that such an reproduction can have and attempting to determine which of these needs to be achieved in order to check the proof for correctness or to produce an automatically verifiable version.

In Hartimo and Okada (2016), M. Hartimo and M. Okada compare Hankel's approach to formal arithmetic and mathematics to Husserl's and demonstrate strong parallels between Husserl's treatment of arithmetic and term-rewriting systems commonplace in computer science. As an example given on p. 955 of Hartimo and Okada (2016) that goes back to Husserl, an arithmetical expression like 18+48 is here considered to be 'fulfilled' by a normal form expression, in this case 66, and the possibility of fulfilling the intentions signified by such expressions is demonstrated by showing how to reduce them to such a unique normal form by a certain algorithm.

[20]To appreciate this difference, one might consider the equation $((a+b)+c)+d = ((d+c)+b)+a$ over the reals, which is obviously true for a human reader who thinks of addition as taking the union of two quantities. In our experience with number-theoretical texts in Naproche (Carl), the automatic prover, having to derive this from commutativity and associativity of addition, often got lost in the countless alternative possibilities which rule to apply. This is a striking example for the pragmatical difference between formal definitions and intuitive concepts.

e.g. that the automatic prover proves an auxiliary lemma in a proof in a very complicated way, obtaining the final theorem as an intermediate step. We would certainly not call this a valid reconstruction of the argument.) This could help to considerably increase the contribution of natural-language oriented automatic proof-checkers: In areas like elementary number theory, where crucial appeal to intuition is rare and proofs can be translated rather naturally, a Naproche reconstruction of an informal proof will usually correspond well to the proof intended. Even if it doesn't, we gain trust in the theorem from a positive checking, as we obtain a formal proof, whether it adequately captures the original proof or not. But of course, the goal of a proof checker is not just to check whether the theorem claimed to be proved is provable, but whether the purported proof actually is one. For succeeding at this task, the checker would have to become 'pragmatically closer' to the intended human reader.

Once sufficient background theories are built up, one should actually carry out the examples given above and others to see what Naproche does with them. Will it find the 'right' mistake? This asks for a systematic study of wrong proofs in e.g. flawed research papers, wrong students' solutions etc. Such a reconsideration of well-known mistakes can serve both as a source of inspiration for the development of natural proof-checkers and as powerful demonstration of what has been achieved.

Acknowledgements We thank Marcos Cramer for the kind permission to use his Naproche version of Rav's proof Sect. 14.4.1. We thank Dominik Klein and an anonymous referee for various valuable comments on former versions of this work that led to considerable improvements. We also thank Heike Carl for her thorough proofreading.

References

Avigad, J., Donnelly, K., Gray, D., & Raff, P. (2006). A formally verified proof of the prime number theorem. *ACM Transactions on Computational Logic, 9*(1),

Azzouni, J. (2005a). *Tracking reason. Proof, consequence, and truth.* New York: Oxford University Press.

Azzouni, J. (2005b). Is there still a sense in which mathematics can have foundations? In G. Sica (Ed.), *Essays on the foundations of mathematics and logic* (pp. 9–47). Monza: Polimetrica.

Azzouni, J. (2009). Why do informal proofs confirm to formal norms? *Foundations of Science, 14*, 9–26.

Bauer, G., & Nipkow, T. (2006). Flyspeck I: Tame graphs. *Archive of Formal Proofs*. Available online at http://afp.sourceforge.net/devel-entries/Flyspeck-Tame.shtml

Benzmüller, C. E., & Brown, C. E. (2007). The curious inference of Boolos in Mizar and OMEGA*. *Studies in Logic, Grammar and Rhetoric, 10*(23), 299–386.

Boolos, G. (1987). A curious inference. *Journal of Philosophical Logic, 16*(1), 1–12.

Carl, M. An Introduction to elementary number theory for humans and machines. Work in progress.

Carl, M., & Koepke, P. (2010). Interpreting Naproche – An algorithmic approach to the derivation-indicator view. Paper for the International Symposium on Mathematical Practice and Cognition at the AISB 2010.

Carl, M., Cramer, M., & Kühlwein, D. (2009). Chapter 1 of Landau in Naproche, the first chapter of our Landau translation. Available online: http://www.naproche.net/inc/downloads.php

Cauchy, A. (1821). Cours d'analyse, p. 120.

Cramer, M., Koepke, P., Kühlwein, D., & Schröder, B. (2009). The Naproche System, paper for the Calculemus 2009.
Feferman, S. (1979). What does logic have to tell us about mathematical proofs? *The Mathematical Intelligencer, 2*(1), 20–24.
Fleck, A., Maennschen, Ph., & Perron, O. (1909). Vermeintliche Beweise des Fermatschen Satzes. *Archiv der Mathematik und Physik, 14*, 284–286.
Hartimo, M., & Okada, M. (2016). Syntactic reduction in Husser's early phenomenology of arithmetic. *Synthese, 193*(3), 937–969.
Husserl, E. (1900). *Formale und transzendentale Logik. Versuch einer Kritik der logischen Vernunft*. Tübingen: Niemeyer.
Husserl, E. (1952). *Ideen zu einer reinen Phänomenologie und phänomenologischen Philosophie. Drittes Buch*. Haag: Matrinus Nijhoff.
Husserl, E. (1999). *Erfahrung und Urteil. Untersuchungen zur Genealogie der Logik*. Hamburg: Felix Meiner.
Husserl, E. (2009). *Ideen zu einer reinen Phänomenologie und phänomenologischen Philosophie*. Hamburg: Meiner.
Kamp, H., & Reyle, U. (2008). *From discourse to logic: Introduction to model-theoretic semantics of natural language, formal logic and discourse representation theory*. Netherlands: Springer.
Koepke, P., & Schlöder, J. (2012). The Gödel completeness theorem for uncountable languages. *Journal of Formalized Mathematics, 30*(3), 199–203.
Kreisel, G. (1965). Informal rigor and completeness proofs. In I. Lakatos (Ed.), *Problems in the Philosophy of Mathematics. Proceedings of the International Collquium in the Philosophy of Science* (Vol. 1).
Lakatos, I. (1976). *Proof and refutation*. Cambridge: Cambridge University Press.
Landau, E. (2004). Grundlagen der Analysis. Heldermann, N.
Lohmar, D. (1989). Phänomenologie der Mathematik: Elemente einer phänomenologischen Aufklärung der mathematischen Erkenntnis nach Husserl. Kluwer Academic Publishers.
Martin-Löf, P. (1980). *Intuitionistic type theory* (Notes of Giovanni Sambin on a series of lectures given in Padova). Available online: http://www.cip.ifi.lmu.de/~langeh/test/1984%20-%20Loef %20-%20Intuitionistic%20Type%20Theory.pdf
Martin-Löf, P. (1987). *Truth of a proposition, evidence of a judgement, validity of a proof. Synthese, 73*, 407–420.
MathOverflow-Discussion: Examples of common false beliefs in mathematics. (2010a). http:// mathoverflow.net/questions/23478/examples-of-common-false-beliefs-in-mathematics-closed
MathOverflow-Discussion: Widely accepted mathematical results that were later shown wrong. (2010b). http://mathoverflow.net/questions/35468/widely-accepted-mathematical-results-that-were-later-shown-wrong
Matiyasevich, Y. (1993). *Hilbert's 10th problem* (MIT Press Series in the Foundations of Computing. Foreword by Martin Davis and Hilary Putnam). Cambridge: MIT Press.
Monty Hall Problem. (1990). Wikipedia-article available at http://en.wikipedia.org/wiki/Monty_Hall_problem
Naproche Web Interface. Available at http://www.naproche.net/inc/webinterface.php
Pak, K. (2011). Brouwer fixed point theorem in the general case. *Journal Formalized Mathematics, 19*(3), 151–153.
Rautenberg, W. (2006). *A concise introduction to mathematical logic*. New York: Springer.
Rav, Y. (2007). A critique of a formalist-mechanist version of the justification of arguments in mathematicians' proof practices. *Philosophia Mathematica (III), 15*, 291–320.
Rickey, V. F. Cauchy's famous wrong proof. Available online: http://www.math.usma.edu/people/ rickey/hm/CalcNotes/CauchyWrgPr.pdf
Rota, G.-C. (1997). The phenomenology of mathematical proof. *Synthese, 111*(2), 183–196.
Singh, S. (2000). Fermats letzter Satz. Die abenteuerliche Geschichte eines mathematischen Rätsels. dtv.
Sundholm, G. (1993). Questions of proof. *Manuscrito, 16*, 47–70.

Tieszen, R. (2005). *Phenomenology, logic, and the philosophy of mathematics*. Cambridge: Cambridge University Press.

Tieszen, R. (2011). *After Gödel. Platonism and rationalism in mathematics and logic*. Oxford: Oxford University Press.

Tragesser, R. (1992). Three insufficiently attended to aspects of most mathematical proofs: Phenomenological studies. In M. Detlefsen (Ed.), *Proof, logic and formalization* (pp. 71–87). London: Routledge.

Turing, A. (1937). On computable numbers, with an application to the Entscheidungsproblem. *Proceedings of the London Mathematical Society, Ser. 2, 42*, 230–265.

van Atten, M. (2010). Construction and constitution in mathematics. *The New Yearbook for Phenomenology and Phenomenological Philosophy, 10*, 43–90.

van Atten, M., & Kennedy, J. (2003). On the philosophical development of Kurt Gödel. *The Bulletin of Symbolic Logic, 9*(4), 425–476.

Chapter 15
Varieties of Pluralism and Objectivity in Mathematics

Michèle Friend

Abstract The phrase 'mathematical foundation' has shifted in meaning since the end of the nineteenth century. It *used to mean* a consistent general theory in mathematics, based on basic principles and ideas (later axioms) to which the rest of mathematics could be reduced. There was supposed to be only one foundational theory and it was to carry the philosophical weight of giving the ultimate ontology and truth of mathematics. Under this conception of 'foundation' pluralism in foundations of mathematics is a contradiction.

More recently, the phrase *has come to mean* a perspective from which we can see, or in which we can interpret, much of mathematics; it has lost the realist-type metaphysical, essentialist importance. The latter has been replaced with an emphasis on epistemology. The more recent use of the phrase shows a lack of concern for absolute ontology, truth, uniqueness and sometimes even consistency. It is only under the more modern conception of 'foundation' that pluralism in mathematical foundations is conceptually possible.

Several problems beset the pluralist in mathematical foundations. The problems include, at least: paradox, rampant relativism, loss of meaning, insurmountable complexity and a rising suspicion that we can say anything meaningful about truth and objectivity in mathematics. Many of these are related to each other, and many can be overcome, explained, accounted for and dissolved by concentrating on crosschecking, fixtures and rigour of proof. Moreover, apart from being a defensible position, there are a lot of advantages to pluralism in foundations. These include: a sensitivity to the practice of mathematics, a more faithful account of the objectivity of mathematics and a deeper understanding of mathematics.

The claim I defend in the paper is that we stand to learn more, not less, by adopting a pluralist attitude. I defend the claim by looking at the examples of set theory and homotopy type theory, as alternative viewpoints from which we can learn about mathematics. As the claim is defended, it will become apparent that 'pluralism in mathematical foundations' is neither an oxymoron, nor a contradiction, at least

---

M. Friend (✉)
Department of Philosophy, George Washington University, Washington, DC, USA
e-mail: Michele@gwu.edu

S. Centrone et al. (eds.), *Reflections on the Foundations of Mathematics*,
Synthese Library 407, https://doi.org/10.1007/978-3-030-15655-8_15

not in any threatening sense. On the contrary, it is the tension between different foundations that spurs new developments in mathematics. The tension might be called 'a fruitful meta-contradiction'.

I take my prompts from Kauffman's idea of eigenform, Hersh's idea of thinking of mathematical theories as models and from my own philosophical position: pluralism in mathematics. I also take some hints from the literature on philosophy of chemistry, especially the pluralism of Chang and Schummer.

15.1 Introduction[1]

There is a deep difficulty for the pluralist. The pluralist is not just pluralist in one respect, as we shall shortly see. He is pluralist in some of, or at least: epistemology, foundations, methodology, ontology and truth. I'll work through the definitions and some easy examples in the next section. For the pluralist who is pluralist in most, or all, respects, the difficult task is to account for the apparent robustness and objectivity of mathematics. Is the robustness and objectivity an illusion, if not, what accounts for it?

15.2 Pluralism in General and Pluralism in Mathematics

What is pluralism? Let us start with some definitions.

Pluralism is a philosophical position where the trumping characteristic is a tolerance towards other points of view, theories, methodologies, values and so on. The tolerance is not an act of kindness. It is motivated by scepticism and honesty.

The *mathematical pluralist* I canvass here[2] is someone who occupies a philosophical position concerning *mathematics*. The trumping characteristic of is that he shows a tolerance towards different approaches in mathematics concerning: epistemology, foundations, methodology, objectivity and truth. He is a principled sceptic.

A *principled sceptic* is not just someone who doubts out of habit or obduracy. Rather, he is someone who has been brought to doubt in light of (1) being aware of several epistemologies, etc. in mathematical practice, and (2) lacking a strong conviction that one particular of these is 'correct'.

[1]I should like to thank Amita Chatterjee for her excellent comments as referee for improving the paper. The paper was originally published in 2017 in: *Mathematical Pluralism*, Special Issue of the Journal of the Indian Council of Philosophical Research, Mihir Chakraborty and Michèle Friend (Guest Editors) Springer. JICPR Vol. 34.2 pp. 425–442. DOI https://doi.org/10.1007/s40961-061-0085-3. ISSN: 0970–7794. pp. 425–442. I should also like to thank the Instituto de Investigaciones Filosoficas at the Universidad Autonoma de México for support.

[2]There are other aspects in which one can be a pluralist. Since I shall be focusing on the question of objectivity, these aspects are sufficient to draw out the problem.

Thus, the *principled sceptic* can be thought of as modest and honest. The modesty is such that it is not even necessarily a *stable* position, since the principled sceptic does not rule out the possibility that in the future he might acquire the relevant conviction, but he is also open to the idea that there might simply never be a unique correct epistemology, foundation etc. for him or for anyone else, although he recognises the hope or conviction in others.

In my experience, many mathematicians refer to themselves as pluralist. I suspect that what they mean by this is that they are pluralist in one respect or other, possibly one not discussed here, and they do not specify in which sense.

15.3 Varieties of Pluralism

Let us work through the varieties I propose to look at here, in order to understand the problem of objectivity.

Pluralism in epistemology: The pluralist in epistemology believes that there are different methods of knowing mathematical concepts and that it is far from obvious, given the present state of play, that there is anything like ultimate explanations, justifications or complete knowledge in (not *of*) mathematics,[3] or unique convergence on 'best' explanations, justifications or knowledge in mathematics.

First, let us distinguish between, on the one hand, learning and teaching mathematics, and knowing (parts of) mathematics. When we talk of epistemology we are talking of knowledge, not pedagogy. In epistemology we are interested in how we can be quite certain of a concept, where 'certainty' is not merely phenomenological, but justified; or how can we be sure that there is no ambiguity in a concept.[4] These are different questions from those that surround 'learning styles' or 'teaching styles'. About these, we might ask which are most efficient, or which teaching styles suit which kinds of students, and how can we detect this in advance? The difference is that in teaching mathematics at the lower levels, it is usually sufficient for the purposes at hand to give *some* grasp of a mathematical concept. Pluralism in learning or teaching styles, I take it, is rarely controversial.[5] That is, we tolerate different learning and teaching styles, since some part of the population

[3] I think that no one would claim a complete knowledge of mathematics, but some people might well claim a complete knowledge of some parts of mathematics, say, the theorem '$2 + 2 = 4$' in Peano arithmetic.

[4] Arguably, there is always some ambiguity in any concept. For example, there is always ambiguity in context. However, the ambiguity might not be important for our concerns or purposes, or it might not be recognised yet, since we do not yet make the subtle distinctions necessary to notice the ambiguity.

[5] The exceptions are, for example, when a new way of teaching mathematics is adopted nation-wide. Experiments of this sort were introduced in France under the 'Code Napoleon', in the USA as the 'New Math' (*Sic!*) and might very well be the case in some countries with a language not known to many people outside the country – for lack of textbooks, schools of mathematical

learns or teaches better in one style rather than another. Pluralism in epistemology is, I think, a more controversial claim.

Let me tell a more-or-less standard story. We shall then have occasion to refine it.

Proofs are one example of an epistemological route to mathematics. Others are: contemplation of an axiom or theorem, or application of a theorem or result in another area of mathematics or outside mathematics, experimenting: by drawing or calculating values within parameters so one watches the numbers change through the calculations.

We all know that some mathematicians (and people) think linearly and others think geometrically. Some mathematicians (and people) think generally and some think in a more detailed way. More controversially, some philosophers, logicians and mathematicians think that we can *only know* mathematics if we reason in a way that can be represented using a constructive logic. According to such constructivists, we construct mathematics step-by-step starting from ideas or principles or truths it would be impossible to doubt or the doubting of which would make thinking impossible (in some sense) or from immediate truths,[6] that is, truths we cannot further justify. Immediate truths are first principles of mathematics or logic that we *know*. We proceed from principles we know in this very strong sense of 'know' through inference rules that preserve knowledge. Those who reason otherwise are not entitled to claim to *know* mathematics! Of course, this raises the question as to which are the immediate truths, and what do we do if someone disagrees with our selection?

Continuing with the standard story, some philosophers, logicians and mathematicians on the other end of the epistemological spectrum think that we grasp, discover or see, mathematical truths, ideas and so on through contemplation and reasoning, not step-by-step, but by having insights into the reality beyond our sense experience. This is the more Platonist view. The privilege of being human is to be able to have these insights and to communicate them to others. Of course, we might be wrong etc.

With a few refinements, these two epistemologies are the only options, according to the standard story. The epistemological pluralist is (at least) agnostic between the constructive understanding of epistemology and the Platonist understanding.

The epistemological pluralist might also be more than agnostic. As an epistemological pluralist, I do not think that either of the above stories is quite right because, first, they do not preclude each other; we might use a proof *and* contemplation of an axiom or theorem to gain knowledge; second, knowledge is partly phenomenological. What is right about the story is that mathematicians make knowledge claims. They (rarely bother to, but could) claim to know that $2 + 8 = 10$

---

learning etc. All of these are geographical, historical and politically situated. The subject is not without interest.

[6]'Immediate' is not meant in the temporal sense of 'right away', or in the sense of 'obvious'. Some constructively 'immediate truths' might take a long time to appreciate. It remains, however, that there is no further direct *justification* we can add to bring about appreciation. At best, we can give a number of examples, and expose someone to them, hoping they will 'sink in'.

in Peano arithmetic, for example. As a pluralist, the way I interpret this is to think that what they mean is that that they can give a proof, from the axioms of arithmetic, and that they *feel* it to be true. So, there is a justification accompanied by the phenomenology of conviction. The conviction "that $2 + 8 = 10$ in Peano arithmetic" *comes from* a rigorous proof, or from knowing that a rigorous proof can be constructed, together with a pretty good idea that one could work out how to do it pretty quickly. However, the right sort of phenomenology even accompanied by *a* justification is not enough to claim knowledge, for, we still have to (meta-)justify our conviction *in* the particular justification.

It is far from clear that '$2 + 8 = 10$' (not *in Peano arithmetic* but *tout court*) is 'known'! It is not a justified *true* belief, since it is false in arithmetic mod 8.[7] In arithmetic mod 8, $2 + 8 = 2$.

Maybe this is disingenuous. After all, we rarely use arithmetic mod 8, so we tend to specify it if we are, so there is something inductively safe (or by default appropriate) in saying that one knows that $2 + 8 = 10$, without adding the qualifier 'in Peano arithmetic'. But $2 + 8 = 10$ is not just a *default truth* either, insofar as one can make sense of the notion of a default truth at all. Rather, it is a statement confirmed by use over and over again, in all sorts of contexts. That is, there is a good (meta) justification behind taking to be true by default. It is a highly *confirmed* default.

We can already begin to develop the thesis about objectivity. Following Wright (1992), we can call this "wide cosmological role" for the statement $2 + 8 = 10$. A 'truth' has a wide cosmological role when it can be imported to other discourses (Wright p. 196) and is treated as true in each. $2 + 8 = 10$ is very useful in all sorts of discourses: about pieces of wood, about loaves of bread, about river basins in a country and so on. This gives the 'truth' that $2 + 8 = 10$ a robustness, since it is counter-checked and confirmed widely. But there is *still* something more, and this is what we shall see in sections 4, 5 and 6, when we develop the notion of objectivity further in the form of what I shall refer to as borrowing, crosschecking and fixtures within mathematics.

For now, let us consider the next sort of pluralism.

> *Pluralism in foundations* consists in a tolerance towards the idea that there might be different *foundations* in mathematics. More moderately: it is an agnosticism concerning which foundation is 'the correct one', if any. Less moderately, it is the conviction that there is no reason to believe, on present evidence, that there is a unique foundation, together with agnosticism as to whether or not this situation might *ever* change.

Pluralism in foundations can be thought of in at least three ways: as epistemic, ontological or alethic. The epistemic version of pluralism in foundations is that we do not know which, if any, foundation is 'correct'. The ontological version is that there is no unique ontology referred to by the several purportedly foundational

[7] This is a sort-of 'clock arithmetic', where the numbers go up to 8 and then start again at 1, going in a circle.

theories. The alethic version is that there is no truth of the matter whether any particular proposed foundation is the truth-conferring foundation for mathematics.

These sorts of pluralism are convincing if we take seriously the practice of working mathematicians. Few of them work only in one of these foundations. Most do not even care to make explicit reference to one of these, they might make reference to several of these or to none of them. This is, of course, just a report about the practice, and the practice could be by-and-large wrong, but this remains to be shown and the burden of proof lies with the monist who advocates a particular unique foundation for mathematics.

Moreover, we should be alerted to the fact that the word 'foundation' has changed its use and meaning in the context of mathematics. Evidence of this can be found in this volume. See the chapters by Claudio Ternullo, Andrei Rodin, Penelope Maddy and Lawrence C. Paulson. It was used to mean a reducing discipline that would take on the philosophical charge of providing the ontology and ultimate truth of all of mathematics. Now, it tends to mean 'is a perspective from which we can see a lot of mathematics, and so unite a lot of mathematics into a coherent whole'. This is the modern, more pluralist, sense in which 'foundation' is used by Shapiro-type structuralism, by group theorists, algebraists, by most modern category theorists and set theorists and, more recently, by homotopy type theorists.[8] These modern

[8]I'll explain the terms. Shapiro's structuralism is a philosophy of mathematics characterised roughly as follows: mathematical theories can all be thought of as structures. A structure consists in a domain of objects together with some predicates, relations and functions that bear between the objects, and give structure to the objects, together with operations we can perform on formulas in a second-order set theoretic language. We use standard classical model theory to compare structures to each other. Group theory studies objects, such as the positive integers, which must include an identity element and operations that correspond closely to addition and multiplication. Group theorists impose their idea of groups, and findings about groups on other areas of mathematics, bringing new insights to those areas. Algebra is also a very basic way of looking at other areas of mathematics. Algebra is about working out which formulas are equivalent to which other formulas. Category theory was proposed as an alternative to set theory. In category theory, we have categories. They are made up of objects and 'arrows'. The claim of category theorists is that (almost) any part of mathematics can be thought of as a category. Moreover, category theorists compare categories and work out meta-functions (called 'functors' in category theory) which take us from one category to another. This, too, tells us about connections between different areas of mathematics. Set theory was proposed as a real foundation for mathematics at the beginning of the twentieth century. It was so proposed because so much of mathematics was found to be reducible to set theory. That is, it was a mathematically comprehensive theory. Alternative 'foundations', such as Whitehead and Russell's type theory, were proposed at the time or since the development of set theory. Homotopy type theory is more recent. Some mathematicians claim that it, too, can be thought of as a foundation for mathematics. From what I understand, it is based on the constructive dependent type-theory of Martin-Löf. What the homotopy type theorists bring to the table is not only the very sophisticated individuation of statements in mathematics in terms of type, but also the homomorphisms, that is, roughly, the relations and functions, between the types that are normally glossed over as equality. Each of the above areas of mathematics is foundational in the sense that they can make the claim that they can 'see' or 'interpret' much of what is counted as 'mathematics' today. Moreover, each theory brings its own way of looking and of seeing mathematics. We learn something from each, and we would deprive ourselves of understanding and insight if we were to ignore some.

'foundationalists' defend their favoured foundation with claims about making sense of, understanding parts of, or giving insight into, particular results in mathematics more easily or lucidly. This more sober use of 'foundation' is arguably pluralist in the sense above. Taking the working mathematician seriously we also observe a pluralism in methodology.

What is pluralism in methodology?

Pluralism in methodology is a tolerance towards different *methodologies*. In mathematics, we might see this in the form of borrowing techniques developed in one area of mathematics in an area otherwise foreign to it. We might do this in order to generate a proof of a theorem.

What we quite often see in modern mathematics is the borrowing of results and proof techniques from one area of mathematics to another. For example, proof techniques, such as proof by contradiction or diagonalisation techniques, are freely shared between areas of mathematics. Diagrams or tables developed in one area of mathematics might be imported to another for reasons of suggestion or demonstration. As we shall see in later sections when we further develop the notion objectivity available to the pluralist, this is not necessarily a problem because of the fixtures and crosschecking and the method of 'chunk and permeate' that organises our reasoning into compartments. All of these will be explained in due course. The surprising thesis here is that it is *borrowing* that *accounts for* the objectivity of the discipline of mathematics. Before I develop this properly, let us turn to the varieties of pluralism that most directly lead us to the problem of objectivity for the pluralist.

The problem of accounting for objectivity is made quite stark for the philosopher of realist *inclination*, but not *conviction*, when we observe pluralism in ontology and truth, since together they disrupt the philosopher's *usual* account of objectivity. The usual account runs as follows: there is an ontology, studied by mathematicians, and the ontology is what makes the statements in mathematics true or false. This is the general realist picture, and it is one of long standing in philosophy of mathematics.

To already cast doubt on the realist account we can disentangle realism in ontology from realism in truth in the following way. Realists in ontology think that the ontology of mathematics is independent of us, but that truth is, for example, a feature tied to language and concept formation, so we might get the truths wrong, in the sense of not quite referring to the underlying ontology. In contrast, a realist in truth but not in ontology, might think that a lot of mathematical statements are true without there being any ontology, since, for example, ontology has to be causal or physically detected by us. So, for example, someone might say that gauge theory in physics is true (since it gives us all the right measurements and predictions), but there is no 'underlying reality' in the sense of a causal ontology. In terms of predictions concerning matter or forces, gauge theory has *extra* mathematical and epistemological structure; structure that does not bear on, ontology and properties of objects. It might, nevertheless be true.[9] But, in this paper, it is not enough to

[9] James Weatherall "Understanding gauge". Paper presented at the conference: Logic, Relativity and Beyond 2015, Budapest, 9–13 August 2015.

disrupt the realist account with disentangling ontology from truth, since what we are contemplating is a position that is pluralist in *both* ontology and truth. So, as philosophers we are robbed of our traditional account of objectivity in mathematics.

A *pluralist in ontology* denies that there is a unique, absolute, consistent[10] well-defined (or well-definable) ontology which accounts for the whole of mathematics as it is practiced.

A *pluralist in truth* believes that truth in mathematics is not an absolute term, or at least is not a well-defined concept. Truth-in-a-theory is perfectly understood, except maybe in some half-developed theories; but truth *of* a theory is not. Truth is, then, always relative to a context, namely, a theory of mathematics, and a meta-mode in which we (implicitly and temporarily) take the theory to be true.

The question we shall now address is: is it possible for there to be, and if so in what way can there be, objectivity in mathematics without thinking that there is one epistemology, foundation, methodology, ontology or truth in mathematics. The task seems daunting. So, we might give up and think that we are quite delusional about the objectivity of mathematics. It is all a mere fiction[11] after all. But if we give up like this, then we are faced with the equally difficult task of accounting for the delusion! Mathematics certainly seems to be objective. The indicator for this is that we are often surprised by results, and when we are surprised we adjust our understanding and adopt the new information as fact.

15.4 Objectivity Within Proofs: Chunk and Permeate

For the pluralist, there are at least two separate sources of objectivity in mathematics. One is rigour of proof or demonstration. The other is borrowing, crosschecking and fixtures. We shall address the latter in the next section. Now, we concentrate on rigour of proof and preservation of 'objectivity' within a proof.

It often happens, especially in modern mathematics, that in a particular proof, results or methods are borrowed from different areas of mathematics. If we were to put all of the theories together we would have inconsistent or trivial theories. A theory might be inconsistent but not trivial, just in case it is paraconsistent or relevant.[12] But in such borrowing proofs, we do not use all of the resources of all of the theories. So, maybe the parts we do borrow are all consistent with each other? Unfortunately, things are not always so simple.

---

[10]For the paraconsistent logicians: substitute 'coherent' or 'non-trivial' for 'consistent'.

[11]I deliberately use the qualifier 'mere' to distance this attitude of giving up on the objectivity of mathematics from the Field's factionalism which is a positive account of mathematics as fiction.

[12]A theory is consistent if and only if it has a consistent model, or less impredicatively, a theory is consistent just in case the underlying logic, or formal representations of reasoning is/are not paraconsistent or relevant, and there are no contradictions. A theory is trivial if and only if every well-formed formula of in the language of the theory can be derived in the theory. A theory is paraconsistent, or relevant, if and only if inconsistency in the theory does not engender triviality.

There are particular proofs in mathematics that (1) have inconsistent sets of premises. That is, from them we could derive a contradiction. Moreover, (2) the underlying logic is assumed to be neither paraconsistent nor relevant, and therefore the theory that would result would be trivial, were we to take the premises as axioms and close the theorems under other-than-paraconsistent or relevant rules of inference. When both (1) and (2) occur in a proof, we have to be very careful. Otherwise we could be accused of working within a trivial theory and this is not acceptable mathematical practice.[13] Paraconsistent and relevant mathematical theories are uncommon in mathematics, so we shall ignore this area of mathematics forthwith. To get a flavour of how such work is done, see Graham Priest's chapter in this volume. Paraconsistent treatments of mathematics are very interesting. The question then is, can a proof where (1) and (2) occur still be coherent, objective or still make sense in mathematics?

One way of helping with such situations is to draw on the proof strategy called 'chunk and permeate'. (Brown and Priest 2004).

More precisely, we shall develop a very weak sort of objectivity: 'objectivity-modulo chunk and permeate'. *Informally*, and *overlooking many subtleties*, 'chunk and permeate' is a method of re-constructing proofs that use together inconsistent premises. Chunk and permeate, arranges such proofs into a series of chunks of reasoning. Within each chunk we have perfectly acceptable formal reasoning. Between chunks, we only allow *some* information to permeate, or flow – just enough to come to the conclusion of the chunk. In considering proofs, we are interested in preservation of objectivity, where 'objectivity' is not contrasted to subjectivity, but to context sensitivity, or context-relativity. One hypothesis of this paper[14] is that:

> a chunk and permeate re-arrangement of proofs is enough to preserve objectivity from premises to conclusion since each chunk belongs to a separate and objective context (background theory).

The notion of 'objective context' is what will be fleshed out in the next two sections. Chunk and permeate proofs are objectivity *preserving* but are not *guarantors* of objectivity by themselves, just as validity of a proof is truth preserving but not a guarantee of truth. That is, if the various premises are objective relative to their context, then so is the conclusion; but there is no guarantee that the premises are objective *tout court*. Nevertheless, if the hypothesis is correct, then we can accept the following definition:

> A proof is *objectivity preserving* if it *has been* or *can be* successfully arranged in a chunk and permeate proof.[15]

---

[13]The question whether intentionally working within a trivial theory *should* be acceptable, or under what circumstances it should be acceptable is under investigation by Luis Estrada Gonzalez.

[14]This has not been proved.

[15]It is an interesting and open question whether any correct proof can be arranged in a chunk-and-permeate fashion. It is an interesting and open problem whether any *in*correct proof can also be so arranged, and if so what this means. These questions are under investigation by the author and others, especially my colleagues in Mexico. The questions are quite delicate especially if we

To see this better: the philosophical problem with objectivity preservation within a proof is that the conclusion is true within the context of the premises (on condition that we accept the premises, that they are objective in the ways to be specified in the next sections) and possibly that we conducted the proof in the right order (for reasons of chunk and permeate) but there is no guarantee that the conclusion is true in general, i.e., globally true in mathematics (in the absence of one epistemology, foundation, methodology, ontology and truth). There is no guarantee because if we were to combine the whole theories or contexts from which we imported the theorems/ premises then we would end up with a trivial theory.

To solve the problem we turn to the other source of objectivity in mathematics: fixtures and crosschecking. We need the latter to account for the objectivity of mathematics taken as a whole in the light of global inconsistency, that is, inconsistencies between mathematical theories.

15.5 Objectivity in Mathematics: The Thesis of Borrowing, Crosschecking and Fixtures

Before I turn to the account of borrowing, crosschecking and fixtures. In case it is useful for situating the following sort of enquiry, I should note that inspiration and courage to undertake this account of objectivity in light of pluralism comes from Wright's treatment of objectivity in his book Truth and Objectivity (Wright 1992). There we find that there is not one account of truth or one account of objectivity, but several that depend on some properties of the discourse. Here I tentatively explore some ideas about objectivity in mathematics. The objectivity, in light of pluralism, takes the form of borrowing, crosschecking and fixtures.

Borrowing occurs when an axiom, rule, lemma, result, theorem, proof technique, diagram-representation, operation *and so on* are imported from one theory of mathematics to another.

Crosschecking occurs when we prove that two theories share features that are 'the same'.

Of course, the judgement that it is 'the same' axiom, rule etc. that is imported can only rigorously be made within a theory (or class of theories), and depends on our translation. See for example the chapter by Benedikt Ahrens and Paige North and the chapter by Neil Barton and Sy Freidman in this volume. We have to make a proof about the translation being correct or loyal. Such proofs of 'sameness', if they exist, are usually informal and carried out at the meta-level. Thus a *particular* theory in which such an informal proof is made is rarely stipulated. The informal proofs are made in a class of theories. This is because when we do not stipulate a particular theory for our informal proof, it follows that once we have the informal proof, all we can do is rule out some theories. We do this when we notice that

are allowed to use paraconsistent reasoning or even trivial representations of reasoning within a chunk, but then if we have to use such representations then this might reveal too high a price for acceptance or objectivity of the conclusion. These are all open questions.

some presuppositions or inference moves were made that are forbidden by some other theories, so the proof cannot belong to those theories. Similar remarks apply to crosschecking. Examples of crosschecking are: equi-consistency proofs, embeddings, reductions to a 'foundation' and to some extent borrowings and inter-translations.

> *Fixtures* are notions which stay fixed across some mathematical theories.[16] They are preconditions for crosschecking in mathematics, and items that are extensively borrowed become fixtures.

Modern mathematics, more than any other discipline, involves borrowing and crosschecking.[17] Let me give an example where borrowing, crosschecking and fixtures come together. For another example see the chapter by Philip Welch in this volume.

If I can show that proofs in a formal language can be expressed in a normal form, and then that they can be arranged in a tree (partial order), and that the tree (partial order) is infinite, then it follows by *reductio* and induction that either at least one node has an infinite number of branches, or that there is at least one branch that is infinitely long. An informal proof like this is ostensibly about proofs. It borrows from developments in logic (formal languages and normal form), partial orders and theories about infinity. None of these ideas are exotic in mathematics. They are each central. They each turn up in all sorts of areas of mathematics. They are *confirmed* by use and borrowing. Through borrowing, more exotic parts of mathematics are gradually incorporated into the cannon of mathematics. As we do so, as we borrow or crosscheck, we have to keep some things fixed, such as working within a theory or by translating and keeping language fixed or by using logical constants, and we might be surprised. If we cannot keep such features stable throughout the proof, then we would have to use the strategy of chunk and permeate. For example, we might find that there is an underlying ambiguity in a 'constant' or a definition, that only surfaces when we change the context.

> I advance the thesis that *objectivity preserving proofs* on the one hand and *borrowing, crosschecking and fixtures* on the other hand, together, *supplant* the need for absolute truth, absolute and independent ontology, a traditional foundation (to which all of mathematics can be reduced) or for a single epistemology or methodology.

This thesis overhauls *many* preconceptions in the philosophy of mathematics, and I think that it offers a better account of the present day practice of mathematics. More mildly, the thesis is enough to show that the pluralist canvassed here can give an account of the objectivity or apparent objectivity of mathematics. Before the thesis can be accepted, we should delve more deeply into the pair of notions: crosschecking and fixtures.

---

[16]The word 'fixtures' is supposed to be suggestive of the notion of a fixed point in mathematics, but it is a little looser than that of fixed point.

[17]This was less the case in the past, especially when geometry and arithmetic were kept quite separate. The later interaction between the two was also important. Today, we do not so much use arithmetic to check geometry and geometry to check arithmetic, but rather, we use set theory, and the ultimate tool: model theory, to do this. Or we use a combination of the two.

15.6 Objectivity in Mathematics: Crosschecking and Fixtures, Using Model Theory and Structures

A good start to understanding the notion of a fixture is to consider Shapiro's structuralism. A structuralist uses model theory in order to compare theories, or structures, using models. This way of comparing theories comes with a warning. The whole model theory approach assumes the language of set theory and is wedded to classical logic. These are the limitations of the model theoretic approach, at least as it has been developed by Shapiro. Nevertheless, it is useful, provided we bear these limitations in mind. The limitation is why model theory is *not the only way* of detecting fixtures for the pluralist. Let us begin with a few definitions.

> A *structure* is an n-tuple consisting in operations, a domain of objects, a set of object constants, a set of predicate, relation and function constants that pick out members of the powerset of the domain of objects. To our n-tuple, we add an assignment function which maps object constants to objects in a domain, first-order predicates to sets of objects in the domain, binary first-order relations to ordered pairs of objects in the domain, and so on for n-tuple relations and functions.

With Shapiro, we are assuming a second-order language with object, predicate, relation and function variables. We do not need to make this assumption, but it helps us to get a lot more fixtures and crosschecks using one mathematical tool: model theory.

> A *theory* is a set of formulas, possibly including axioms or base formulas closed under some rules or operations.

In model theory, the interpretation is the semantics of the theory. The semantics is the models. Models satisfy, or fail to satisfy sets of sentences or whole theories. Sets of sentences are consistent if and only if they have a model. Consistent structures have models. We attribute properties to theories by investigating the relations (mappings), which bear between structures. We also compare theories, by comparing their models. The comparison is made formally, using functions defined in the language of set theory. Using model theory, we learn the limitations of theories: we prove 'limitative results' about a theory or language. For example, we use model theory to study: embeddings, completeness, compactness, the Löwenheim-Skolem properties, categoricity, decidability and so on.[18] Some or all of these are what I shall call 'crosschecks'. Crosschecks re-enforce objectivity.

[18]Explaining the vocabulary: an embedding of one structure into another is a demonstration that the embedded structure is a part of a greater structure. A theory is complete if and only if the semantics and the syntax match in what they consider to be theorems (on the syntactic side) or be true (on the semantic side). That is, the syntax will prove all and only the truths of the theory. A theory is compact if and only if when an infinite set of theorems of the theory has a model, every finite set of theorems also has a model. The upward Lôwenheim-Skolem property states that if a countable theory has a countable model then it has a model for 'every' cardinal that is greater than countable. The downward Löwenheim-Skolem theorem states that if a countable theory has a model of countable size, then it has a model of every cardinality less than countable (i.e., of finite size). A theory is categorical if and only if all of its models have the same cardinality. A theory

Let us look at a simple example. A model theorist will think of a group as a structure. The structure contains an identity element, 1, two binary function symbols, +, x, which name a group addition and product operations and one unary function symbol, $^{-1}$, naming the inverse operation. (Hodges 1997, 3). The theory (the set of formulas that are true in the theory) is closed under some version of addition and multiplication. This structure can be *embedded in* a larger structure which includes everything in a group plus the operation of division, for example.

Embedding is a type of crosscheck. Say we show we can embed one theory/ structure, A, into another, B. Since there is a common structure to both, namely the structural elements of A, it follows that these are fixed between the structures. They serve as a basis for comparison of B to A and A to B. A is a proper sub-structure of B. Every formula true in A is also true in B, every formula provable in A is also provable in B and so on.

A and B check the objectivity of each other. That is, there is a meta-theory, or context, in which both are equally legitimate. Here our meta-theory is model theory. We might think of this as a pair-wise equi-objectivity-crosscheck, but then the language becomes ugly, so we shorten this to the, marginally less ugly, crosscheck objectivity.

> *Crosscheck objectivity* occurs when we have crosschecked one theory against another using a more-or-less formal theory such as model theory. The objectivity is relative to the theory used to make the check and up to the formality and objectivity of the proof of check.

The collection of objectivity crosschecks is part of what objectivity across mathematics consists in.

Equi-consistency,[19] compactness and the Löwenheim-Skolem results are also forms of objectivity crosscheck, all are studied in model theory. Remember the warning about the limitations of model theory and structuralism? There are other mathematical theories that provide crosschecks. We can find these in many of the still so-called 'foundational' theories, since they are all means of doing some meta-mathematics. Let us look at another source of objectivity crosscheck, not belonging to a foundational theory.

15.7 'Logical Notions' and Invariance: More Crosschecks and Fixtures

Tarski gives a very interesting answer to the question: what are the logical notions? The answer is ambiguous; and I shall return to this. First, let me outline some of Tarski's discoveries. Tarski extends 'logic' following what Klein had done in

is decidable if and only if the syntax of the theory can decide in every case whether a conclusion follows from a set of premises or not (Shoenfield 1967).

[19]Equi-consistency is the proved result that if one theory is consistent then so is the other. That is, they are either both consistent, or both inconsistent.

geometry (Tarski 1986 146). It is evident that this technique and conception can be applied more or less widely, and encounters some interesting limitations even within the application Tarski made.

Klein tried to find a unified approach to geometry by means of the study of space invariances with respect to a group of transformations. This idea was pressing because of the presence, and increasing acceptance of, non-Euclidean geometries. Not only did these force us to revise our compliance and faith in our geometrical intuitions, but the non-Euclidean geometries introduced new considerations on groups (therefore, introducing new algebraic ideas). The algebraic ideas then fed back to the geometries by way of finding invariances across geometries. The 'finding of the invariances' was essentially an analysis of the common 'logical structure' of geometry. Following Klein's insight, Tarski used the notion of *invariance under a transformation of a domain* of objects (on to itself) to identify *invariant*, and so, *logical* notions (across the foundational theories of Whitehead and Russell's type theory, von Neumann set theory and Gödel-Bernays set theory).

Following Tarski, we treat the domain over which we reason as a set[20] and we may transform it on to itself with any function (transformation), which we can define in the language. We discover that with *some* transformations the objects of the domain remain the same, such as with the identity transformation, but other notions remain stable under some transformations and not others, such as the 'lesser than' relation. The notions (properties or relations between objects) which remain the same under *any* (recognised) transformation of the domain on to itself are then called 'invariant'. These are then identified as the *logical* notions, for Tarski. In this very precise sense, we say that the 'meaning' of these notions is invariant. It turns out that the invariant notions across Whitehead and Russell's type theory, von Neumann set theory and Gödel-Bernays set theory are: the logical connectives, negation, identity and the quantifiers.

Or rather, subsequently, we have worked out that the invariant notions Tarski found can be reduced to these. To be historically accurate: Tarski himself did not make this discovery directly. Instead, he found that there are only two invariant notions of *class*: the notion of a universal class, the notion of the empty class, (Tarski 1986 150). For Tarski, there are four invariant *binary relations*: "the universal relation which always holds between two objects, the empty relation which never holds, the identity relation which holds only between 'two' objects when they are identical, and its opposite the diversity relation." (Tarski 1986 150). Ternary, quaternary relations, and so on, also have a small, finite number of invariant notions under transformations, similar to the invariant notions over binary relations. The last notion which shows invariance is that of the cardinality of the domain, which reduces to the notion of quantifier under certain assumptions.

Tarski had the Whitehead and Russell logicist project very much in mind when he presented this material, so he went on to speculate whether this shows us that mathematics is really logic, and logical notions (so defined) are what tie

[20]This is in a loose sense of 'set', i.e., not tied to a particular set theory.

mathematics together. For Tarski, such a conclusion is too hasty. These results should be taken with a pinch of salt. I shall discuss *one* reason for this. There are others.

One reason is that the methodology adopted by Tarski will not determine for *every* notion whether it is invariant, or 'logical'. Tarski is quite frank about this, for, he asks the question: is the (set theoretic) membership relation a logical notion or not. It turns out that it *is* if we consider the membership relation as it is implicitly defined in Whitehead and Russell's type theory. But it is *not* if we consider the membership relation in von Neumann set theory, (Tarski 1986 152–3). As Shapiro puts it: "...on this [model theoretic] account, the logical-non-logical distinction would be an artefact of the logic [of the meta-theory]." (Shapiro 1991, 7). This is when we see ambiguity dis-ambiguated (if at all) by context, or meta-theory. Whether a relation is invariant depends on other meta-considerations that we use to show the invariance. The answers are not unique to the question: is this concept, symbolised in this way, invariant across theories? Nevertheless, we have, again, a cross-theory fixture: namely the invariant notions, and we hold them invariant relative to the meta-theory we used to demonstrate invariance. This is a type of partial re-enforcing of objectivity across theories.

Definition:

> *invariance objectivity* is when there are concepts, symbolised in a formal language, that are invariant under permutation of a domain on to itself. The invariance is relative to (at least) the meta-theory used to show the invariance.

There are other forms of crosschecking and fixtures in mathematics such as languages, formalisations, fix points and so on. In fact, mathematics is riddled with such. The more 'global' sort of objectivity we are after comes from taking all of these together.

I advance the thesis that

> It is the myriad collection of partial/ relative sorts of objectivity: objectivity preserving proof, crosscheck objectivity, invariance objectivity *and others* that *together* account for how it is that mathematics is objective in the face of the sort of pluralism canvassed here.

15.8 Analysis by Way of a Conclusion

For philosophers who are looking for a 'unified account' of mathematics, there are many mysteries left unexplained in existing attempts to provide such an account. Many of the mysteries concern mathematical practice. The pluralist gives honest answers.

For example, If we think that there is or should be a unified account, and we think that modern mathematics largely consists in proofs,[21] then we might ask why

[21] Both are widely held in the literature in philosophy of mathematics. And both are very suspect claims in light of pluralism.

there are so many proofs for some theorems. There are well over 100 proofs for pythagoras' theorem which are non-equivalent, except in the conclusion. They are not all 'suspect' proofs, made in some obscure part of mathematics. In other words, new proofs are not developed to assuage *doubt as to the truth*, or robustness, of the result. Put another way, they are not meant to be part of some *inductive* argument for Pythagoras' theorem. Each is a rigorous proof (up to the accepted standards of rigour at the time). Even in the case of theorems we judge less certain, *some* proofs are useful for assuaging doubt, but some are not; and yet, the institution of mathematics accepts them as valuable contributions to the field of mathematics. So what is the surplus information we gain from a proof, over the truth of the theorem proved?

The pluralist thinks of this question as indicative of a wider phenomenon. There is something unique and interesting about mathematics. Mathematics 'hangs together'; it seems to be objective and non-circular. It is not like a Popperian pseudo-science, but neither is very much of it checked against physical phenomena, and our material observations.[22] Moreover, the pluralist does not have the necessary philosophical convictions to attribute the hanging together to: absolute truth, consistency, embedding in a unique foundational theory or ontology. Ultimately, these are the wrong places to look according to the pluralist. The pluralist picture of mathematics is very different. The pluralist sees that mathematical theories can be applied to other mathematical theories, and individual theorems can serve in several theories, or infiltrate, quite disparate areas of mathematics. In fact, the pluralist claim is stronger than this. It is not that they *can* be so applied, it is that modern mathematics largely *consists in* such applications and infiltrations. See the chapter in this volume by Dzamonja and Kant. Moreover the wealth of crosschecking is exactly what re-enforces and leads to our producing several proofs of the same theorem; but more important, borrowing, crosschecking and fixtures is sufficient to warrant our confidence in the objectivity of mathematics in the *absence* of a unique foundation, and all that that entails philosophically.

In this paper I merely *introduced* the notion of 'fixtures', discussing two types of example, the ones from structuralism and the ones from Tarski's idea of using invariance to fix the idea of a 'logical' notion. Once we see some of these, others will suggest themselves to the reader when he revisits mathematical texts or articles. I propose the exercise of looking at the table of contents of a recent journal in logic or mathematics, and count how may articles are about limitative results, connections between theorems in different theories, in applications of one methodology to a 'foreign' area of mathematics and so on. There will be a significant percentage of such articles. It turns out that a lot of cutting edge contemporary mathematics is of this nature. Moreover, the borrowing and crosschecking is not all part of a unified outlook. Each borrowing or crosscheck depends on a context or a meta-theory.

[22]If we think that mathematics is more certain than observation, then *none* of it is 'checked' by observation.

Even by examining only a few of these sorts of borrowings and crosschecks (which pre-suppose 'fixtures'), and imagining that there are others, we learn two lessons.

One lesson is that mathematics does 'hang-together' and forms a distinctive discourse (distinctive in character, not in ontology).

The second lesson is that we do not need to rely on an ontology, a notion of absolute truth or on a unique theory to play the role of foundation, in order to justify pluralist mathematical practice, recognise its importance and give a firm and honest account of objectivity in mathematics.

Borrowing, crosschecks and fixtures supplant the philosophical triumvirate: ontology, knowledge and truth. Instead pluralists explain the triumvirate in other terms. The pluralist offers a rich account of the 'hanging together' of mathematics. No other discipline has developed a web of crosschecks as keenly as mathematics, and it is these crosschecks which allow us to loosen the stringent constraints of traditional foundationalism.

The pluralist diagnoses of why philosophy of mathematics took the course it did is that, within a Western tradition, philosophers take the triumvirate as primitive, or already understood, when this is not at all the case. In fact, they have misfired when they brought it to bear on the subject of mathematics.

Why would the pluralist say that traditional approaches in philosophy have misfired in the case of mathematics? It is partly a matter of (philosophical) temperament. Traditional Western philosophers tend towards a certain temperament (towards monism). Such philosophers think of crosschecking as *evidence for* the fact that there is a deep underlying truth or that there is an underlying mathematical ontology, or logic, or major foundational theory, or something to *explain* the miracle of application. As Tarski writes:

> The conclusion [that there are different answers to the question whether $\in$ is a logical notion] is interesting, it seems to me, because the two possible answers correspond to two different types of mind. A monistic conception of logic, set theory and mathematics, where the whole of mathematics would be a part of logic, appeals, I think, to a fundamental tendency of modern philosophers. [Tarski was giving this talk in 1936]. Mathematicians, on the other hand, would be disappointed to hear that mathematics, which they consider the highest discipline in the world, is a part of something so trivial as logic; and they therefore prefer a development of set theory in which set-theoretical notions are not logical notions. (Tarski 1986, 153).

Mathematicians today would be disappointed, not so much because logic is thought of as trivial, since it is not so judged today, but by the constraints on their creativity. Nevertheless, present day mathematicians do share Tarski's mathematician's concern about being held to a foundational standard, and that is because so much of the development of mathematics has no explicit roots in set theory or any other foundation. That is, mathematicians have, for the most part, quite disregarded the philosophical foundationalist aspirations. Instead, they rely on the crosschecks as confirmation of their results. And this is what foundations do under the more modern conception of 'foundation'. This is what allows mathematicians to borrow and to notice fixtures.

Moreover, each particular crosscheck is backed up by some sort of formal or informal proof. There are plenty of contexts where attempts at cross application do not work. It is not the case that everything in mathematics fits together in any way we choose, and it is the *failure* of borrowing which is also evidence for the objectivity and non-triviality of mathematics. This sort of objectivity is not grounded in an ontology.

What we *do* have evidence for is that there is a borrowing, crosschecking, cross-fertilisation and a *con*versation in mathematics; that the crosschecks are as rigorous and thorough as we choose. Sometimes we are slack, for example when we try to apply mathematics to physics, and find that we have to gerrymander the mathematics to fit the physical theory – see renormalisation.[23] In fact, mathematics is the discipline where the crosschecks are the most rigorous of any area of research. This accounts for the *phenomenology* of absolute truth and independence of mathematics. But phenomenology is not evidence for truth or independence. Our phenomenology sometimes misleads us.

Moreover, returning to the quotation from Tarski, the mathematician might well (depending on temperament, again) feel either unconcerned by metaphysical notions underlying his subject matter, or he might feel that raising such questions is alien to him, or he might adopt a schizophrenic attitude. All of these possible attitudes point to the variety of attitudes held by practicing mathematicians, and that there is such a variety testifies to the confusion, or lack of good and coherent *traditional* answer to the philosophical questions. The pluralist philosopher has a different temperament characterised by tolerance, honesty and modesty.

References

Brown, B., & Priest, G. (2004). Chunk and Permeate, A Paraconsistent Inference Strategy. Part I: The Infinitesimal Calculus. *Journal of Philosophical Logic., 33*(4), 379–388.

Hodges, W. (1997). *A shorter model theory*. Cambridge: Cambridge University Press.

Shapiro, S. (1991). *Foundations without foundationalism; A case for second-order logic* (Oxford logic guides) (Vol. 17). Oxford: Oxford University Press.

Shoenfield, J. R. (1967). (re-edited) (2000). *Mathematical logic*. Natick:: Association for Symbolic Logic, A.K. Peters.

Tarski, A. (1986). What are logical notions?. In J. Corcoran (Ed.), *History and Philosophy of Logic, 7*, 143–154.

Wright, C. (1992). *Truth and Objectivity*. Cambridge, MA: Harvard University Press.

[23]Renormalisation is a method of eliminating infinite quantities in certain calculations in electrodynamics. In electrodynamics, calculations involve finite quantities, this is plain from our physical conceptions. But the mathematical theory used includes infinite quantities, and they are the result of some straight experimental calculations. This is embarrassing, since it does not fit our physical conception. We therefore systematically (have a method to) eliminate such embarrassing quantities built into the *use* of the mathematical theory.

Chapter 16
From the Foundations of Mathematics to Mathematical Pluralism

Graham Priest

Abstract In this paper I will review the developments in the foundations of mathematics in the last 150 years in such a way as to show that they have delivered something of a rather different kind: mathematical pluralism.

16.1 Introduction

I suppose that an appropriate way to write an essay on the foundations of mathematics would be to start with a definition of the term, and then discuss the various theoretical enterprises that fall within its scope. However, I'm not sure that the term can be caught in a very illuminating definition. So I'm not going to do this.

What I want to do instead is to tell a story of how foundational studies developed in the last 150 years or so. I am not attempting to give an authoritative history of the area, however. To do justice to that enterprise would require a book (or two or three). What I want to do is put the development of the subject in a certain perspective—a perspective which shows how this development has now brought us to something of a rather different kind: mathematical pluralism. And what that is, I'll explain when we get there.

For the most part, what I have to say is well known. Where this is so, I shall just give some standard reference to the material at the end of each section.[1] When, towards the end of the essay, we move to material that is not so standard, I will give fuller references.

So, let me wind the clock back, and start to tell the story.

[1] A good general reference for the standard material is Hatcher (1982).

G. Priest (✉)
Department of Philosophy, CUNY Graduate Center, New York, NY, USA

Department of Philosophy, University of Melbourne, Melbourne, VIC, Australia

S. Centrone et al. (eds.), *Reflections on the Foundations of Mathematics*, Synthese Library 407, https://doi.org/10.1007/978-3-030-15655-8_16

16.2 A Century of Mathematical Rigor

The nineteenth century may be fairly thought of as, in a certain sense, the age of mathematical rigour. At the start of the century, many species in the genus of number were well known: natural numbers, rational numbers, real numbers, negative numbers, complex numbers, infinitesimals; but many aspects of them and their behaviour were not well understood. Equations could have imaginary roots; but what exactly is an imaginary number? Infinitesimals were essential to the computation of integrals and derivatives; but what were these 'ghosts of departed quantities', as Berkeley had put it?[2] The century was to clear up much of the obscurity.

Early in the century, the notion of a limit appeared in Cauchy's formulation of the calculus. Instead of considering what happens to a function when some infinitesimal change is made to an argument, one considers what happens when one makes a small finite change, and then sees what happens "in the limit", as that number approaches 0 (the limit being a number which may be approached as closely as one pleases, though never, perhaps, attained). Despite the fact that Cauchy possessed the notion of a limit, he mixed both infinitesimal and limit terminology. It was left to Weierstrass, later in the century, to replace all appeals to infinitesimals by appeals to limits. At this point, infinitesimals disappeared from the numerical menagerie.

Weierstrass also gave the first modern account of negative numbers, defining them as signed reals, that is, pairs whose first members are reals, and whose second members are "sign bits" ('$+$' or '$-$'), subject to suitably operations. A contemporary of Weierstrass, Tannery, gave the first modern account of rational numbers. He defined a rational number as an equivalence class of pairs of natural numbers, $\langle m, n\rangle$, where $n \neq 0$, under the equivalence relation, $\sim$, defined by:

$$\langle m, n\rangle \sim \langle r, s\rangle \text{ iff } m \cdot s = r \cdot n$$

Earlier in the century, Gauss and Argand had shown how to think of complex numbers of the form $x + iy$ as points on the two dimensional Euclidean plane—essentially as a pair of the form $\langle x, y\rangle$—with the arithmetic operations defined in an appropriate fashion.

A rigorous analysis of real numbers was provided in different ways by Dedekind, Weierstrass, and Cantor. Weierstrass' analysis was in terms of infinite decimal expansions; Cantor's was in terms of convergent infinite sequences of rationals. Dedekind's analysis was arguably the simplest. A *Dedekind section* is any partition of the rational numbers into two parts, $\langle L, R\rangle$, where for any $l \in L$ and $r \in R$, $l < r$. A real number can be thought of as a Dedekind section (or just its left-hand part).

[2] *The Analyst, or a Discourse Addressed to an Infidel Mathematician* (1734), §XXXV.

So this is how things stood by late in the century. Every kind of number in extant mathematics (with the exception of infinitesimals, which had been abolished) had been reduced to simple set-theoretic constructions out of, in the last instance, natural numbers.

What, then of the natural numbers themselves? Dedekind gave the first axiomatisation of these—essentially the now familiar Peano Axioms. This certainly helped to frame the question; but it did not answer it.[3]

16.3 Frege

Which brings us to Frege. Frege was able to draw on the preceding developments, but he also defined the natural numbers in purely set-theoretic terms. The natural number n was essentially the set of all n-membered sets (so that 0 is the set whose only member is the empty set, 1 is the set of all singletons, etc.). This might seem unacceptably circular, but Frege showed that circularity could be avoided, and indeed, how all the properties of numbers (as given by the Dedekind axioms) could be shown to follow from the appropriate definitions.

But 'follow from' how? The extant canons of logic—essentially a form of syllogistic—were not up to the job, as was pretty clear. Frege, then, had to develop a whole new canon of logic, his *Begriffsschrift*. Thus did Frege's work give birth to "classical logic".

Given Frege's constructions, all of the familiar numbers and their properties could now be shown to be sets of certain kinds. But what of sets themselves? Frege took these to be abstract (non-physical) objects satisfying what we would now think of as an unrestricted comprehension schema. Thus (in modern notation), any condition, $A(x)$, defines a set of objects $\{x : A(x)\}$. Because he was using second-order logic, Frege was able to define membership. Again in modern notation, $x \in y$ iff $\exists Z(y = \{z : Zz\} \wedge Zx)$.

Moreover, Frege took these set-theoretic principles themselves to be principles of pure logic. Hence all of arithmetic (that is, the theory of numbers) was a matter of pure logic—a view now called *logicism*. And this provided an answer to the question of how we may know the truths of arithmetic—or to be more precise, reduced it to the question of how we know the truths of logic. As to this, Frege assumed, in common with a well-worn tradition, that these were simply a priori.

Frege's achievement was spectacular. Unfortunately, as is well known, there was one small, but devastating, fly in the ointment, discovered by Russell. The naive comprehension principle was inconsistent. Merely take for $A(x)$ the condition that $x \notin x$, and we have the familiar Russell paradox. If B is the sentence $\{x : x \notin x\} \in \{x : x \notin x\}$ then $B \wedge \neg B$. Given the properties of classical logic, everything followed. A disaster.

[3]For the material in this section, see Priest (1998).

After the discovery of Russell's paradox, Frege tried valiantly to rescue his program, but unsuccessfully. The next developments of the *Zeitgeist* were to come from elsewhere.[4]

16.4 Russell

Namely, Russell—and his partner in logical crime, Whitehead. Russell was also a logicist, but a more ambitious one than Frege. For him, *all* mathematics, and not just arithmetic, was to be logic. In the first instance, this required reducing the other traditional part of mathematics—geometry—to logic, as well. This was relegated to Volume IV of the mammoth *Principia Mathematica*, which was never published.

But by this time, things were more complex than this. The work of Cantor on the infinite had generated some new kinds of numbers: transfinite ones. These were of two kinds, cardinals, measuring size, and ordinals, measuring order. Russell generalised Frege's definition of number to all cardinals: a cardinal number was *any* set containing all those sets between which there is a one-to-one correspondence. He generalised it further again to ordinals. An ordered set is *well-ordered* if every subset has a least member. An ordinal is any set containing all those well-ordered sets between which there is an order-isomorphism.

Of course, Russell still had to worry about his paradox, and others of a similar kind which, by that time, had multiplied. His solution was *type theory*. The precise details were complex and need not concern us here. Essentially, sets were to be thought of as arranged in a hierarchy of types, such that quantifiers could range over one type only. Given a condition with a variable of type i, $A(x_i)$, comprehension delivered a set $\{x_i : A(x_i)\}$; this set, however was not of type i, but of a higher type, and so it could not be substituted into the defining condition delivering Russell's paradox to produce contradiction.

Russell's construction faced a number of problems. For a start, it was hard to motivate the hierarchy of orders as a priori, and so as part of logic. Secondly, with his construction, Frege had been able to show that there were infinite sets (such as the set of natural numbers). The restrictions of type theory did not allow this proof. Russell therefore had to have an axiom to the effect that there was such a thing: the *Axiom of Infinity*. It was hard to see this as an a priori truth as well.[5]

On top of these, there were problems of a more technical nature. For a start, the hierarchy of types meant that the numbers were not unique: every type (at least, every type which was high enough) had its own set of numbers of each kind. This was, to say the least, ugly. Moreover, Cantor's work had delivered transfinite numbers of very large kinds. Type theory delivered only a small range of these.

[4]For the material in this section, see Zalta (2016).

[5]Earlier versions of type theory also required a somewhat problematic axiom called the *Axiom of Reducibility*. Subsequent simplifications of type theory showed how to avoid this.

Specifically, if $\beth_0 = \aleph_0$, $\beth_{n+1} = 2^{\beth_n}$, and $\beth_\omega = \bigcup_{n<\omega} \beth_n$, then type theory delivered only those cardinals less than $\beth_\omega$. Of course, one could just deny that there were cardinals greater than these, but *prima facie*, they certainly seemed coherent.

Finally, to add insult to injury, one could not even explain type theory without quantifying over all sets, and so violating type restrictions.

Russell fought gallantly against these problems—unsuccessfully.[6]

16.5 Zermelo

New developments arrived at the hands of Zermelo. He proposed simply to axiomatize set theory. He would enunciate axioms that were strong enough to deliver the gains of the nineteenth century foundational results, but not strong enough to run afoul of the paradoxes. His 1908 axiom system, strengthened a little by later thinkers, notably Fraenkel, appeared to do just this. The axioms were something of a motley, and so all hope of logicism seemed lost[7]; but, on the other hand, the system did not have the technical inadequacies of type theory.

The key to avoiding the paradoxes of set theory was to replace the naive comprehension schema with the *Aussonderung* principle. A condition, $A(x)$ was not guaranteed to define a set; but given any set, y, it defined the subset of y comprising those things satisfying $A(x)$. An immediate consequence of this was that there could be no set of all sets—or Russell's paradox would reappear. Indeed, all "very large" sets of this kind had to be junked, but with a bit of fiddling, the mathematics of the day did not seem to need these.

In particular, the Frege/Russell cardinals and ordinals were just such large sets. So to reduce number theory to set theory, a different definition had to be found. Zermelo himself suggested one. Later orthodoxy was to prefer a somewhat more elegant definition proposed by von Neumann. 0 is the empty set. $\alpha + 1$ is $\alpha \cup \{\alpha\}$, and given a set, X, of ordinals closed under successors, the ordinal which is the limit of these is $\bigcup X$. A cardinal was an ordinal such that there was no smaller ordinal that could be put in one-to-one correspondence with it.

Logicism had died. The fruits of nineteenth century reductionism had been preserved. The paradoxes had been avoided. The cost was eschewing all "large" sets; but this seemed to be a price worth paying.

The next developments came from a quite different direction.[8]

---

[6]Starting around the 1990s, there was a logicist revival of sorts, neo-logicism; but it never delivered the results hoped of it. For the material in this section, see Irvine (2015) and Tennant (2017).

[7]About 20 years later, in the work of von Neumann and Zermelo, a model of sorts was found: the cumulative hierarchy. This did provide more coherence for the axioms, but it did nothing to save logicism. On the contrary, it appeared to give set theory a distinctive non-logical subject.

[8]For the material in this section, see Hallett (2013).

16.6 Brouwer

In the first 20 years of the twentieth century, Brouwer rejected the idea that mathematical objects were abstract objects of a certain kind: he held them to be mental objects. Such an object exists, then, only when there is some mental procedure for constructing it (at least in principle). In mathematics, then, existence is constructibility. Brouwer took his inspiration from Kant. Mental constructions occur in time. Time, according to Kant, is a mental faculty which enforms sensations—or intuitions, as Kant called them. Hence, Brouwer's view came to be called *Intuitionism*. Intuitionism provides a quite different answer from that provided by logicism as to how we know the truths of mathematics: we know them in the way that we know the workings of our own mind (whatever that is).

Brouwer's metaphysical picture had immediate logical consequences. Given some condition, $A(x)$, we may have (at least at present) no construction of an object which can be shown to satisfy it; moreover, we may also have no way of showing that there is no such construction. In other words, both of $\exists x A(x)$ and $\neg\exists x A(x)$ may fail. Hence the Law of Excluded Middle fails. Nor is this the only standard principle of logic to fail. Suppose that we want to show that $\exists x A(x)$. We assume, for *reductio*, that $\neg\exists x A(x)$, and deduce a contradiction. This shows that $\neg\neg\exists x A(x)$; but this does not provide us with a way of constructing an object satisfying $A(x)$. Hence, it does now establish that $\exists x A(x)$. The Law of Double Negation (in one direction), then, also fails.

Brouwer did not believe in formalising logical inference: mental processes, he thought, could not be reduced to anything so algorithmic. But a decade or so later, intuitionist logic was formalised by Heyting and others. Unsurprisingly, it turned out to be a logic considerably weaker than "classical logic", rejecting, as it did, Excluded Middle, Double Negation, and other related principles.

Given the unacceptability of many classical forms of inference, Brouwer set about reworking the mathematics of his day. All proofs which did not meet intuitionistically acceptable standards had to be rejected. In some cases it was possible to find a proof of the same thing which was acceptable; but in many cases, not. Thus, for example, consider König's Lemma: every infinite tree with finite branching has at least one infinite branch. Such a branch may be thought of as a function, f, from the natural numbers to nodes of the branch, such that $f(0)$ is the root of the tree, and for all n, $f(n+1)$ is an immediate descendent of $f(n)$. We may construct f as follows. $f(0)$ has an infinite number of descendants by supposition. We then run down the branch preserving this property, thus defining an infinite branch. Suppose that $f(n)$ has an infinite number of descendants. Since it has only a finite number of immediate descendants, at least one of these must have an infinite number of descendants. Let $f(n+1)$ be one of these. The problem with this proof intuitionistically, is that we have, in general, no way of knowing which node or nodes these are, so we have no construction which defines the next value of the function. There is, then, no such function, since we have no way of constructing it.

Brouwer delivered ingenious constructions, which were intuitionistically valid, and which could do some of the things that classically valid constructions could do. Thus, in the case of König's Lemma, he established something called the *Fan Theorem*. However, it was impossible to prove everything which had a classical proof. Intuitionistic mathematics was therefore essentially revisionary. There are things which can be established classically which have no intuitionist proof.

This may make it sound as though intuitionist mathematics is a proper part of classical mathematics.[9] This, however, is not the case. True, not every proof that is classically valid is intuitionistically valid. But that means that there can be things which are inconsistent from a classical point of view, which are not so from an intuitionistic point of view. And this allows for the possibility that one may prove intuitionistically some things that are *not* valid in classical mathematics.

Take, for example, the theory of real numbers. Let U be the real numbers between 0 and 1; and think of these as functions from natural numbers to $\{1, 0\}$. Now consider a one-place function, F, from U to U. To construct F, we need to have a procedure which, given an input of F, f, defines its output, $F(f)$. And this means that for any n, we must have a way of defining $[F(f)](n)$. Since this must be an effective procedure, $[F(f)](n)$ must be determined by some "initial segment" of f—that is, $\{f(i) : i \leq m\}$ for some m. Hence, if f' agrees with f on this initial segment, $[F(f)](n)$ and $[F(f')](n)$ must be the same. It follows that $F(f)$ and $F(f')$ can be made as close as we please by making f and f' close enough. That is, all functions of the kind in question are continuous! This is simply false in classical real number theory: there are plenty of discontinuous functions.

Intuitionism is not, then, simply that sub-part of classical mathematics which can be obtained by constructive means: it is *sui generis*.

Ingenious though it was, though, intuitionist mathematics never really caught on in the general mathematical community. Mathematicians who did not accept Brouwer's philosophical leanings could see nothing wrong with the standard mathematics. Or perhaps more accurately, mathematicians were very much wedded to this mathematics, and so rejected Brouwer's philosophical leanings.[10]

16.7 Hilbert

The suspicion of classical reasoning was not restricted just to intuitionists, though. It was shared by Hilbert, who was as classical as they came. The discovery of Russell's paradox, and the apparently a priori principles that lead to it, was still something of a shock to the mathematical community; and Hilbert wanted a safeguard against

[9]Though one might also simply consider simply the constructive part of classical mathematics. See Bridges (2013).

[10]Further on all these things, see Iemhoff (2013).

things of this kind happening again. This inaugurated what was to become known as *Hilbert's Program.*

Hilbert wanted to *prove* that this could not happen. Of course, the proof had to be a mathematical one; and to prove anything mathematical about something, one has to have a precise fix on it. Hence, the first part of the program required such a fix on mathematics, or its various parts. This would be provided, Hilbert thought, by appropriate axiomatizations. Hilbert had already provided an axiom system for Euclidean geometry. So the next target for axiomatization was arithmetic. The axiomatization was to be based on classical logic—or at least the first-order part of Frege's logic, which Hilbert and his school cleaned up, giving the first really contemporary account of this.

Given the axiomatised arithmetic, this was then to be proved consistent. Of course given that the proof of consistency was to be a mathematical one, and the security of mathematical reasoning was exactly what was at issue in the project, there was an immediate issue. If our mathematical tools are themselves inconsistent, maybe they can prove their own consistency. Indeed, given that classical logic is being employed, if arithmetic is inconsistent, it can prove anything.

Hilbert's solution was to insist that the reasoning involved in a consistency proof be of a very simple and secure kind. He termed this *finitary*. Exactly what finitary reasoning was, was never defined exactly, as far as I am aware; but it certainly was even weaker than the constructive reasoning of intuitionists. The danger of contradiction, it seemed to Hilbert, arose only when the infinite reared essentially its enticing but dangerous head. Hence the reasoning of the consistency proof had to be something like simple finite combinatorial reasoning—most notably, symbol-manipulation.

This approach allowed a certain philosophical perspective. Take the standard language of arithmetic. Numerals are constituted by '0' followed by some number of occurrences of the successor symbol. Terms are composed from numerals recursively by applying the symbols for addition and multiplication. Equations are identities between terms. The Δ_0 fragment of the language is the closure of the equations under truth functions and bounded quantifiers (i.e., particular or universal quantifiers bounded by some particular number). The truth value of any Δ_0 statement can be determined in a finitary way. Terms can be reduced to numerals by the recursive definitions of addition and multiplication. Identities between numerals can be decided by counting occurrences of the successor symbol, and then truth functions do the rest, the bounded quantifiers being essentially finite conjunctions and disjunctions. So, according to Hilbert, we may take the Δ_0 statements to be the truly meaningful (contentful) part of arithmetic.

But what about the other statements? Given some axiom system for arithmetic, this will contain the finitary proofs of the true Δ_0 statements, but proofs will go well beyond this—notably, establishing statements with unbounded quantifiers. However, since the true Δ_0 statements are complete (that is, for any such statement, either it or its negation is true), the system is consistent iff it is a conservative extension of that fragment.[11] Thus, if the system is consistent, reasoning deploying statements not in the Δ_0 fragment can prove nothing new of this form. However, the statements might well have an instrumental value, in that using them can produce simpler and more expeditious proofs of Δ_0 statements. Hence, thought Hilbert, the non-Δ_0 sentences could be thought of as "ideal elements" of our reasoning—in much the same way that postulating an ideal "point at infinity" can do the same for proofs about finite points, or imaginary numbers can do the same for proofs about real numbers.[12]

The demise of Hilbert's Program is so well known that it hardly needs detailed telling. A young Gödel showed that any axiomatization of arithmetic of sufficient power—at least one that is consistent—must be incomplete. That is, there will be statements, A, such that neither A nor $\neg A$ is provable. Since (classically) one of these must be true, there was no complete axiomatization of arithmetic. Putting another nail in the coffin of the Program, Gödel also established that given such a system, there is a purely arithmetic statement which can naturally be thought of as expressing its consistency. However, this is not provable within the system—again if it is consistent. Since the system encodes all finitary reasoning (and much more), this seemed to show that a finitary proof of the consistency of even this incomplete system was impossible.[13]

16.8 Category Theory

So, by mid-century, this is how things stood: apart from some rearguard actions, the great foundational programs of the first part of the twentieth century were defunct. This, however, was not an end of the matter. New developments were to come from a quite new branch of mathematics: category theory.

It is common in mathematics to consider classes of structures of a certain kind: groups, topological spaces, etc. Important information about their common structure is delivered by the morphisms (structure preserving maps) between them. When the range of one morphism is the domain of another, such morphisms can be composed. If we write composition as $\bigcirc$, then the morphism $f \bigcirc g$ is a map which, when applied to an object x , delivers the object obtained by applying f to $g(x)$.

---

[11] If the system is not a conservative extension it proves the negation of some true Δ_0 sentence, and so the system is inconsistent. Conversely, if it is inconsistent, since it can prove everything, it is not a conservative extension.

[12] In these cases, the ideal elements are not statements, but objects. Hilbert discovered that quantifiers could be eliminated by the use of his ε-symbol. Thus, $\exists x A(x)$ is equivalent to $A(\varepsilon x A(x))$. One might—though I don't think Hilbert ever suggested this—take ε-terms themselves to signify ideal objects. In this way non-Δ_0 statements might be thought of as statements of Δ_0 form, but which concern these ideal objects (as well as, possibly, real ones).

[13] For the material in this section, see Zach (2013).

Starting in the late 1940s Eilenberg and MacLane generalised this way of looking at things, to deliver the notion of a category. The idea was taken up and developed substantially by later mathematicians, including Grothendieck and Lawvere.

A category is a bunch of objects, together with functions between them, thought of as morphisms, and often termed *arrows*, because of the way they are depicted diagrammatically. In fact, the objects may be dispensed with, since each may be identified with the identity function on it (which is a morphism). So the notion of a category may be axiomatised with a number of axioms concerning functional composition. A category is, then, any model of this axiom system (in the same way that a group, e.g., is any model of the axioms of the theory of groups). Hence, there is a category of all groups, all topological spaces, all sets, etc. The category of a particular kind of structures (e.g., sets) may justify further axioms concerning functional composition, in the same way that a consideration of Abelian groups justifies axioms additional to those of groups in general.

But foundationally there is now a problem. Since the consolidation of set theory in the early part of the century, it had been assumed that all mathematics could be formulated within set theory. One can, indeed, think naturally of a category as a set of a certain kind. But the problem is that categories such as those of all groups, all topological spaces, all sets—large categories, as they are called—are of the very large kind that had been excised from set theory by Zermelo in order to avoid Russell's paradox. It would seem, then, that set theory cannot provide any kind of foundation for category theory.

There are certain remedial measures one might essay. A large category is not a set, but one can think of it as a proper class, in the sense of NBG set theory (a weak form of second-order ZF)—proper classes being, in effect, sub-collections of sets which are not themselves members of anything. However, this is not generally good enough. For category theorists consider not only particular categories, but the category of functions between them. (Given two categories, the category of morphisms between them is called the *functor category*.) This is "too big" even to be a proper class.

One solution to these problems is to deploy the "Grothendieck hierarchy". This is the cumulative hierarchy, with levels, V_α, for every ordinal α, together with the assumption that there are arbitrarily large inaccessible cardinals. As is well known, if ϑ is an inaccessible cardinal, V_ϑ is a model of ZF set theory, so all the usual set theoretic operations can be performed within it. We may then think of the category of all sets, group, etc., as categories of objects in a V_ϑ. The categories themselves, their functor categories, etc., are not in V_ϑ, but are denizens of higher levels of the cumulative hierarchy. Category theory, then, must be thought of as "typically ambiguous", applying schematically to each V_ϑ.

The retrograde nature of this move is clear. The point of category theory is to chart commonalities of structure between *all* structures of a certain kind. The Grothendieck hierarchy explicitly reneges on this. One way to bring the point home is as follows. Suppose that we are considering a category of a certain kind, and we prove something of the form $\exists! x \forall y R(x, y)$. This might be some sort of representational theorem. Interpreting this at each level of the Grothendieck hierarchy, the uniqueness of the x in question is lost: all we have is one at every level.

An honest approach to category theory would seem, then, to take it to be a *sui generis* branch of mathematics. Some have even gone so far as to suggest that it should be taken as providing an adequate foundation for all mathematics, including set theory. The plausibility of this is delivered by the theory of *topoi*. Topoi are particularly powerful categories of a certain kind. (The category of all sets is one of them.) They can be characterised by adding further axioms concerning composition to the general axioms of category theory. All the standard constructions of set theory, at least all those which are involved in the reduction of the other normal parts of mathematics to set theory, can they be performed in a topos.

As a foundational strategy, the weakness of this move is evident. There are many topoi, and "standard mathematics" can be reconstructed in each one. We are, thus, back to the theoretical reduplication which plagued type theory.

I think it fair to say that what to make of all these matters is still *sub judice*. However, we are still not at the end of our story.[14]

16.9 Paraconsistency

We have so far met two formal logics in the foregoing: classical and intuitionistic. In both of these, the principle of *Explosion* is valid: $A, \neg A \vdash B$, for all A and B. The inference might be thought of as "vacuously valid" in these logics, since the premises can never hold in an interpretation. The principle is clearly counter-intuitive, though. Starting around the 1960s, the development of paraconsistent logic began, a *paraconsistent logic* being exactly one where Explosion is not valid. Using a paraconsistent logic we may therefore reason using inconsistent information in a perfectly sensible way. The information does not deliver triviality—that is, not everything can be established.

There are, in fact, many paraconsistent logics. Their key, semantically speaking, is to stop Explosion being vacuously valid, by including in the domain of reasoning, not only standard consistent situations, but also inconsistent ones. Thus, if p and q are distinct propositional parameters, there can be a situation where p and $\neg p$ hold, but q does not. This is not to suggest that these inconsistent situations may be actual. We reason, after all, about situations which are conjectural, hypothetical, etc. However, the view that some of these inconsistent situations are actual (that is, that what holds in them is actually true), is called *dialetheism*.[15]

The possibility of employing a paraconsistent logic opens up new possibilities in a number of the foundational matters which we have met.[16]

---

[14]For the material in this section, see Marquis (2014).

[15]On paraconsistency, see Priest et al. (2018b). On dialetheism, see Priest et al. (2018a).

[16]On inconsistent mathematics in general, see Mortensen (2017).

Thus, one of the possibilities that has been of much interest to paraconsistentists is set theory—and for obvious reasons: using a paraconsistent logic allows us to endorse the unrestricted comprehension schema. Contradictions such as Russell's paradox can be proved in the theory, but these are quarantined by the failure of Explosion. Moreover, it was proved that, with an appropriate paraconsistent logic, naive set theory (that is, set theory with unrestricted comprehension) is non-trivial.[17]

This raises the prospect of regenerating Frege's foundational project. Of course, having an unrestricted comprehension schema does not guarantee that this project can be carried through. The set-theoretic principles are strong, but the logic is much weaker than classical logic. Things other than Explosion need to be given up. Notably, the principle of *Contraction*, $A \to (A \to B) \vdash A \to B$, cannot be endorsed, because of Curry's Paradox.[18]

It was only relatively recently that Weber was able to show that much of Frege's program *can* be carried out in such a theory.[19] He showed that virtually all of the main results of cardinal and ordinal arithmetic can be proved in this set theory. Moreover, the theory can be used to prove the Axiom of Choice, as well as results that go beyond ZF set theory, such as the negation of the Continuum Hypothesis, and several large-cardinal principles. (Of course, in the context, this does not show that one cannot prove the negations of these as well.)

Weber's proofs have a couple of very distinctive elements. First, they use comprehension in a very strong form, namely:

$$\exists x \forall y (y \in x \leftrightarrow A)$$

where A may contain x itself. This provides the potential for having a fixed point, and so self-reference, built into the very characterisation of a set.[20] Next, Weber not only accommodates inconsistencies, but makes constructive use of them. Thus, a number of the results concerning cardinality, such as Cantor's Theorem, make use of sets of the form $\{x \in X : r \in r\}$, where r is the set of all sets which are not members of themselves, and so inconsistent.

Whether Weber's proof methods, and the various distinctions they require one to draw, are entirely unproblematic; and whether other aspects of set-theoretic reasonings (such as those required in model theory) can be obtained in naive set theory, are still questions for investigation.

Another of the foundational matters we have met, and with which paraconsistency engages, is that concerning Gödel's theorems. Gödel's first theorem is often glossed as saying that any "sufficiently strong" axiomatic theory of arithmetic is incomplete. In fact, what it shows is that it is either incomplete *or* inconsistent.

[17] See Brady (1989).

[18] With naive comprehension we can define a set, c, such that $x \in c \leftrightarrow (x \in x \to \bot)$. Contraction and *modus ponens* then quickly deliver a proof of $\bot$.

[19] See Weber (2010, 2012).

[20] Perhaps surprisingly, Brady's proof shows this strong form of comprehension to be non-trivial.

Of course, if inconsistency implies triviality, it is natural to ignore the second alternative. However, paraconsistency changes all that, since the theory may be complete, inconsistent, but non-trivial. (I note that there is nothing in the use of paraconsistent logic, as such, which problematises the proof of Gödel's theorem. The logic required of an arithmetic theory in order for it to hold is exceptionally minimal.)

Indeed, it is now known that there are paraconsistent axiomatic theories of arithmetic which contain all the sentences true in the standard model, and so which are complete (that is, for any sentence of the language, A, either A or $\neg A$ is in the theory). These theories are inconsistent, but non-trivial.[21]

What foundational significance this has depends, of course, on the plausibility of the claim that arithmetic might be inconsistent.

Implausible as this may seem, Gödel's theorem itself might be thought to lead in this direction. At the heart of Gödel's proof of his theorem, there is a paradox. Consider the sentence 'this sentence is not provable'. If it is provable, it is true; so it is not provable. Hence it is not provable. But we have just proved this; so it is.

Of course, this argument cannot be carried through in a consistent arithmetic, such as classical Peano Arithmetic, when proof is understood as proof in that system. This may be a matter of relief; or it may just show the inadequacy of the system to encode perfectly natural forms of reasoning.

Indeed, if one takes one of the axiomatic arithmetics containing all the truths of the standard model, there will be a formula of the language $Pr(x)$ which represents provability in the theory in the theory itself. It is then a simple matter to construct a sentence, G, in effect of the form $\neg Pr(\langle G \rangle)$ (where angle brackets indicate gödel coding), and establish that both $\neg Pr(\langle G \rangle)$ and $Pr(\langle G \rangle)$ hold in the theory.[22]

What of Gödel's second theorem? Since a paraconsistent theory of the kind we have just been talking about is inconsistent, one should not expect to prove consistency. But it is also non-trivial, i.e., some statements are not provable. This can be expressed by the sentence $\exists x \neg Pr(x)$, and this sentence can be proved in theories of the kind in question. Of course, this does not rule out the possibility that one may be able to prove $\neg \exists x \neg Pr(x)$ as well; and in what sense the proof of non-triviality is finitary may depend on other features of the arithmetic. To what extent these matters may be thought to help Hilbert's program is, then, still a moot point.[23]

---

[21] See Priest (2006, ch. 17).

[22] Additionally, one would expect that the schema $Pr(\langle A \rangle) \supset A$ would be provable in a theory of arithmetic in which $Pr(x)$ really did represent provability. In a consistent theory, it is not, as Löb's Theorem shows. However, the schema is provable in the above theories.

[23] A third foundational issue opened up by paraconsistency concerns category theory. Given that one can operate in a set theory with a universal set, it is possible to have a category of all sets, all groups, etc., where 'all' means *all*. The implications of this for the relationship between set theory and category theory are yet to be investigated.

16.10 Intuitionist and Paraconsistent Mathematics

We have now been rather swiftly through a story of the development of studies in the foundations of mathematics in the last 150 years. As we have seen, none of the foundational ideas we have looked at can claim to have met with uncontestable success. But looking back over developments, we can see that something else has emerged.

As we saw in Sect. 16.6, there are fields of intuitionist mathematics that are quite different from their classical counterparts.[24] Indeed, there are fields of intuitionist mathematics that have no natural classical counterpart.

Let me give just one example of this. This is the Kock-Lawvere theory of smooth infinitesimal analysis.[25] To motivate this, consider how one would compute the derivative of a function, $f(x)$, using infinitesimals. The derivative, $f'(x)$, is the slope of the function at x, given an infinitesimal displacement, i; so $f(x + i) - f(x) = if'(x)$. Now, as an example, take $f(x)$ to be x^3. Then $if'(x) = (x + i)^3 - x^3 = 3x^2i + 3xi^2 + i^3$. If we could divide by i, we would have $3x^2+3xi+i^2$. Setting i to 0 delivers the result—though how, then, did we divide by i?[26] If $i^2 = 0$, we have another route to the answer. For then it follows that for any infinitesimal, i, $if'(x) = 3x^2i$. We may not be able to divide by i, but suppose that $ai = bi$, for all i, implies that $a = b$ (this is the *Principle of Microcancellation*). $f'(x) = 3x^2$ then follows. This is exactly how the theory of smooth infinitesimal analysis proceeds.

Call a real number, i, a *nilsquare* if $i^2 = 0$. Of course, 0 itself is a nilsquare, but it may not be the only one! We may think of the nilsquares as infinitesimals. The theory of smooth infinitesimals takes functions to be linear on these. Given a function, f, there is a unique r such that, for every nilsquare, i $f(x+i)-f(x) = ri$. (In effect, r is the derivative of f at x.) This is the *Principle of Microaffineness*.[27]

Microaffineness implies that 0 is not the only nilsquare. For suppose that it is, then all we have is that $f(x + 0) - f(x) = r0$, and clearly this does not define a unique r. So:

[1] $\neg\forall i(i^2 = 0 \rightarrow i = 0)$

But now, why do we need intuitionist logic? Well, one might argue that 0 *is* the only nilsquare, which would make a mess of things. A typical piece of reasoning for this goes as follows. Suppose that i is a nilsquare and that $\neg i = 0$. Then i has an inverse, i^{-1}, such that $i \cdot i^{-1} = 1$. But then $i^2 \cdot i^{-1} = i$. Since $i^2 = 0$, it follows that $i = 0$.

---

[24] For a further account of some of these enterprises, see Dummett (2000, chs. 2, 3).

[25] On which, see Bell (2008).

[26] One answer to the conundrum is provided by non-standard analysis, an account of infinitesimals developed in the 1960s by Robinson. This deploys non-standard (classical) models of the theory of real numbers.

[27] Microcancellation follows. Take $f(x)$ to be xa. Then, taking x to be 0, Microaffineness implies that there is a unique r such that, for all i, $ai = ri$. So if $ai = bi$ for all i, $a = r = b$.

Hence, by *reductio*, we have shown that $\neg\neg i = 0$. If we were allowed to apply Double Negation, we could infer that $i = 0$—and so we would have a contradiction on our hands. But this move is not legitimate in intuitionist logic. We have just:

[**2**] $\forall i(i^2 = 0 \rightarrow \neg\neg i = 0)$

And we may hold both [1] and [2] together.

What we see, then, is that there are very distinctive fields of intuitionistic mathematics, quite different from the fields of classical mathematics. Moreover, one does not have to think that intuitionism is *philosophically* correct to recognise that these are interesting mathematical enterprises with their own integrity. They are perfectly good parts of pure mathematics. (Whether they have applications to areas outside of mathematics is a quite separate matter, and irrelevant to the present point.)

A similar point can be made with respect to paraconsistent mathematics. In Sect. 16.9 we saw that there were inconsistent mathematical theories relevant to various foundational enterprises: set theory and arithmetic. There are, however, many interesting inconsistent mathematical theories based on paraconsistent logics, which have no immediate application to foundational matters.[28] These include theories in linear algebra, geometry, and topology.

Let me give an example of one of these. This concerns boundaries. Take a simple topological space, say the one-dimensional real line. Divide it into two disjoint parts, left, L, and right, R. Now consider the point of division, p. Is p in L or R? Of course, the description under-determines an answer to the question. But when the example is fleshed out, considerations of symmetry might suggest that it is in both. Then, $p \in L$, and $p \in R$ so $p \notin L$—and symmetrically for R. So a description of the space might be that if $x < p$, x is (consistently) in L; if $p < x$, p is (consistently) in R; and p is both in and not in L, and in and not in R. Given an appropriate paraconsistent logic, the description is quite coherent.[29]

This might not seem particularly interesting, but the idea of inconsistent boundaries has interesting applications. One of these is to describe the geometry of "impossible pictures".[30] Consider the following picture:

The three-dimensional content of the picture is impossible. How should one describe it mathematically? Any mathematical characterisation will specify,

[28]See Mortensen (1995).

[29]Further on inconsistent boundaries, see Cotnoir and Weber (2015).

[30]For more on the following, see Mortensen (2010).

amongst other things, the orientations of the various faces. Now, consider the left-hand face, and in particular its lighter shaded part. This is 90° to the horizontal. Next, consider the top of the lower step on the right-hand side of the picture. This is 0° to the horizontal. Finally, consider the boundary between them (a vertical line on the diagram). This is on both planes. Hence it is at both 90° and 0° to the horizontal.[31] That's a contradiction, since it cannot be both; but that's exactly what makes the content of the picture impossible. Note that the characterisation of the content must deploy a paraconsistent logic, since it should not imply, e.g., that the top of the higher step is at 90° to the horizontal.

Hence, we see, again, that whatever one thinks about the truth of naive set theory, or inconsistent arithmetic, there are perfectly good mathematical structures based on a paraconsistent logic.

Both intuitionist logic and paraconsistent logic, then, provide perfectly coherent and interesting areas of mathematical investigation to which classical logic can be applied only with disaster.[32]

16.11 Mathematical Pluralism

What we have seen is that there are areas of relatively autonomous mathematical research. That is, there is a plurality of areas, such that there is no one of them to which all others can be reduced. This is mathematical pluralism. In truth, this pluralism was already clear in the case of category theory: attempts to reduce it to set theory or vice versa, were always straining at the seams. But intuitionist and paraconsistent mathematics have put the matter beyond doubt.[33] We have, in these cases, mathematical enterprises which are completely *sui generis*. (Note that

[31] And one can set things up in such a way that this does not imply that $90 = 0$.

[32] It seems to me that given any formal logic there could, at least in principle, be interesting mathematical theories based on this. However, intuitionistic logic and paraconsistent logic (and perhaps fuzzy logic; see Mordeson and Nair 2001) are the only logics for which this has so far really been shown.

[33] In fact, the matter is arguably the case even in classical set theory. We can investigate set theory in which the Axiom of Choice holds, and set theory in which the Axiom of Determinacy, which contradicts it, holds. In that case, however, a monist might claim that we are simply doing model theory, and investigating what holds in models of certain set-theoretic axiom systems. One might consider a similar claim in the cases mentioned in the text: we are just doing model-theory using classical logic, albeit of models of non-classical logics. But this suggestion seems lame. First, intuitionist real number theory and paraconsistent set theory are *not* done in this way. So the suggestion gets the mathematical phenomenology all wrong. Secondly, the insistence that the model theory be classical seems dogmatic. Investigations could proceed with intuitionist model theory, which would give quite different results. (Note that intuitionist model theory is well established, but there is as yet no such thing as paraconsistent model theory.) Thirdly, it is entirely unclear how to pursue this strategy in the case of category theory, simply because the foundational problem with category theory was precisely that its ambit appears to outstrip the classical models.

I am not maintaining that all these branches of mathematics are equally deep, rich, elegant, applicable, etc. That is a quite different matter. All I am claiming is that they are all equally legitimate pure mathematical structures.) What to make of the situation certainly raises interesting issues,[34] but that it now obtains cannot be gainsaid.[35]

Mathematical pluralism was certainly not an aim of work in the foundations of mathematics, but it has emerged none the less. In the eighteenth century, some mathematicians investigated Euclidean geometry, trying to prove that the Parallel Postulate was deducible from the other axioms. They did this by assuming its negation, and aiming for a contradiction, which, it turned out, was not coming. Their aim was not to produce non-Euclidean geometries; but by the nineteenth century, it became clear that this is exactly what they had done. In a similar way, the aim of research in the foundations of mathematics was not to establish mathematical pluralism—indeed, most researchers in the area took themselves to be mathematical monists. But that, it seems, is what, collectively, they have done. If one were Hegel, one would surely diagnose here a fasinating episode in the cunning of reason.[36]

References

Bell, J. L. (2008). *A primer of infinitesimal analysis* (2nd ed.). Cambridge: Cambridge University Press.

Brady, R. (1989). The non-triviality of dialectical set theory. In G. Priest, R. Routley, & J. Norman (Eds.), *Paraconsistent logic: Essays on the inconsistent* (pp. 437–71). München: Philosophia Verlag.

Bridges, D. (2013). Constructive mathematics. In E. Zalta (ed.), *Stanford encyclopedia of philosophy*. https://plato.stanford.edu/entries/mathematics-constructive/

Cotnoir, A., & Weber, Z. (2015). Inconsistent boundaries. *Synthese, 192*, 1267–94.

Dummett, M. (2000). *Elements of intuitionism* (2nd ed.). Oxford: Oxford University Press.

Hallett, M. (2013). Zermelo's axiomatization of set theory. In E. Zalta (Ed.), *Stanford encyclopedia of philosophy*. https://plato.stanford.edu/entries/zermelo-set-theory/

Hatcher, W. S. (1982). *The logical foundations of mathematics*. Oxford: Pergamon Press.

[34]Some discussion of it can be found in Priest (2013) and Shapiro (2014). Shapiro takes mathematical pluralism to entail logical pluralism. I am not inclined to follow him down that path. Given a mathematical structure based on a logic, L, reasoning in accord with L preserves truth-in-that-structures. This may not be truth *simpliciter* preservation.

[35]There are, as far as I can see, only two strategies for maintaining mathematical monism. One is the production of some kind of *ur*-mathematics to which all the kinds of mathematics we have met can be reduced. Maybe this could be some kind of foundational project for the twenty-first century; but nothing like this is even remotely on the horizon. The other strategy is simply to deny that the non-favoured kinds of mathematics really *are* mathematics. Now whether these theories are as deep, elegant, applicable, or whatever, as the favoured mathematics, might certainly be an issue. But it seems to me that denying that they are mathematics is the equivalent of the proverbial ostrich burying its head in the sand: these theories have clear mathematical interest.

[36]Many thanks go to Hartry Field, Arnie Koslow, and Zach Weber for very helpful comments on an earlier draft of this essay.

Iemhoff, R. (2013). Intuitionism in the philosophy of mathematics. In E. Zalta (Ed.), *Stanford encyclopedia of philosophy*. https://plato.stanford.edu/entries/intuitionism/
Irvine, A. (2015). *Principia mathematica*. In E. Zalta (Ed.), *Stanford encyclopedia of philosophy*. https://plato.stanford.edu/entries/principia-mathematica/
Marquis, J.-P. (2014). Category theory. In E. Zalta (Ed.), *Stanford encyclopedia of philosophy*. https://plato.stanford.edu/entries/category-theory/
Mordeson, J., & Nair, P. (2001). *Fuzzy mathematics: And introduction for engineers and scientists*. Heidelberg: Physica Verlag.
Mortensen, C. (1995). *Inconsistent mathematics*. Dordrecht: Kluwer.
Mortensen, C. (2010). *Inconsistent geometry*. London: College Publications.
Mortensen, C. (2017). Inconsistent mathematics. In E. Zalta (Ed.), *Stanford encyclopedia of philosophy*. https://plato.stanford.edu/entries/mathematics-inconsistent/
Priest, G. (1998). Number. In E. Craig (Ed.), *Encyclopedia of philosophy* (Vol. 7, pp. 47–54). London: Routledge.
Priest, G. (2006). *In contradiction* (2nd ed.). Oxford: Oxford University Press.
Priest, G. (2013). Mathematical pluralism. *Logic Journal of IGPL, 21*, 4–14.
Priest, G., Berto, F., & Weber, Z. (2018a). Diatheism. In E. Zalta (Ed.), *Stanford encyclopedia of philosophy*. https://plato.stanford.edu/entries/dialetheism/
Priest, G., Tanaka, K., & Weber, Z. (2018b). Paraconsistent logic. In E. Zalta (Ed.), *Stanford encyclopedia of philosophy*. https://plato.stanford.edu/entries/logic-paraconsistent/
Shapiro, S. (2014), *Varieties of logic*. Oxford: Oxford University Press.
Tennant, N. (2017). Logicism and neologicism. In E. Zalta (Ed.), *Stanford encyclopedia of philosophy*. https://plato.stanford.edu/entries/logicism/
Weber, Z. (2010). Transfinite numbers in paraconsistent set theory. *Review of Symbolic Logic, 3*, 71–92.
Weber, Z. (2012). Transfinite cardinals in paraconsistent set theory. *Review of Symbolic Logic, 5*, 269–93.
Zach, R. (2013). Hilbert's program. In E. Zalta (Ed.), *Stanford encyclopedia of philosophy*. https://plato.stanford.edu/entries/hilbert-program/
Zalta, E. (2016). Frege. In E. Zalta (Ed.), *Stanford encyclopedia of philosophy*, https://plato.stanford.edu/entries/frege/

Chapter 17
Does Mathematics Need Foundations?

Roy Wagner

Abstract This note opens with brief evaluations of classical foundationalist endeavors – those of Frege, Russell, Brouwer and Hilbert. From there we proceed to some pluralist approaches to foundations, focusing on Putnam and Wittgenstein, making a note of what enables their pluralism. Then, I bring up approaches that find foundations potentially harmful, as expressed by Rav and Lakatos. I conclude with a brief discussion of a late medieval Indian case study (Śaṅkara's and Nārāyaṇa's *Kriyākramakarī*) in order to show what an "unfounded" mathematics could look like. The general purpose is to re-evaluate the desiderata of foundational programs in mathematics.

17.1 Classical Foundations

A good starting point for discussing mathematical foundations is Frege. His concern is strongly ontological-epistemological, and rooted deep within post-Kantian philosophy. This concern includes, but is not limited to, what Penelope Maddy calls in this volume *metaphysical insight* (I take advantage of her useful terminology for foundational desiderata throughout this paper, and distinguish it with italics). As stated in the first paragraphs of the *Grundlagen* (1884), Frege wanted to clarify the concept of number in order to reduce the quantity of elementary mathematical concepts, explicate the principles of human reasoning, and find the underlying conceptual relations between mathematical statements. He believed that his logical definition of numbers would help achieve all this, while also settling the Kantian question about the analytic or synthetic nature of arithmetic.[1]

[1]It is important to recall that Frege's logical foundation of arithmetic does not extend to geometry. Frege believed geometry to be synthetic (that is, not reducible to logical analysis), as expressed

R. Wagner (✉)
Zurich, Switzerland
e-mail: roy.wagner@gess.ethz.ch

S. Centrone et al. (eds.), *Reflections on the Foundations of Mathematics*, Synthese Library 407, https://doi.org/10.1007/978-3-030-15655-8_17

The antinomies in Frege's work led Russell to scrutinize and reform the underlying logical apparatus (Russell 1903, Appendix B). But in order to maximize logical expressibility (in Maddy's terms, supplying a *generous arena*) while minimizing the risk of contradiction (a sort of *risk assessment* measured by the hierarchy of types), he had to sacrifice *metaphysical insight*. The axioms required for Russell's type theoretic logic are often considered rather contrived, not least by Russell himself, who described them as "the most arbitrary thing he and Whitehead had ever had to do, not really a theory but a stopgap" (Spencer Brown 1972, x).

Brouwer's motivations were rather different. He felt no need to explicate the concept of number, because, according to him, that concept was given to humans via the intuition of time (Brouwer 1981, ch. 1). But if one wanted to be faithful to this pre-linguistic intuition, then one's mathematical tools had to be scrutinized. Only those mental procedures that could be implemented within the framework opened up by the intuition of time (finite – but potentially infinite – constructions) were part of intuitionistic mathematics (in Maddy's terms, we would say he was looking to adhere to an *epistemic source*).

For Brouwer, language was a secondary representation, which was at risk of divergence from live meaning (Brouwer 1975[1919], 222–229). This mistrust of symbolic reasoning excluded from intuitionistic mathematics some of the procedures and constructs that Frege took for granted (such as the law of excluded middle applied to complete infinite collections). But the intuitionist approach did not completely reject classical mathematics as useful for whatever it is that it might be useful for. As Heyting put it, the subject of intuitionistic mathematics, namely "constructive mathematical thought, determines uniquely its premises and places it beside, not interior to, classical mathematics, which studies another subject, whatever subject that may be" (Heyting 1983 [1956], 69).

So far I have emphasized three different foundational purposes (which do not exhaust the various authors' motivations): Frege's conceptual explication (metaphysical insight), Brouwer's hygiene of procedure to preserve live, rather than linguistic, meaning (epistemic source), and Russell's attempts to exorcise contradictions (which came at the expense of both of the above). I now wish to point out how Hilbert's work would add some very different foundational desiderata.

Formalism can be described as the belief that mathematics is reducible to meaningless formal inscriptions subject to finitely verifiable rules, but Hilbert's version was more subtle than that. One should not confuse Hilbert's views on what mathematics *is* with the methodological issue of *how to decide* what should be allowed into mathematics. For the latter question, Hilbert's answer is clear: anything that is conservative (or consistent) with respect to finitary arithmetical knowledge. In order to verify this conservatism, one has to formalize mathematics, analyze finite sequences of mathematical symbols (propositions, proofs) by means of an *intuitive* finitary logic, and use this logic to verify consistency (Hilbert 1983[1926], 192).

---

in (1984[1874], 56) and (1971[1906], 9). Logic is therefore not the foundation of all mathematics according to Frege, but rather of arithmetic alone.

But when it comes to what a mathematical entity is (such as number or infinity), Hilbert remains vague. In the same paper, he says that mathematical signs for abstract concepts, such as infinity, "signify nothing" (Hilbert 1983[1926], 196), but also that infinity is "a concept of reason which transcends all experience and which completes the concrete as a totality", which is hardly nothing! (Hilbert 1983[1926], 201). On the other hand, he supposed finite numbers to have a clear intuitive or at least a practical meaning. Another indication that Hilbert did not view mathematical symbols as meaningless is his endorsement of the axiomatic method even for grounding empirical sciences such as physics (Corry 2004). So it seems that *metaphysical insight* and *epistemic source* did not interest Hilbert very much. He was aiming at something different.

Now Poincaré (1996 [1905–1906], 1034, 1039) and Wittgenstein (1976, 158–9) criticized logical and formal foundational programs for their circularity: one needs to be able to count in order to implement and verify the correctness of the symbolic manipulations that supposedly define and rule over numbers. While I believe that this criticism is justified in the case of Frege and Russell, I believe that it does not successfully apply to Hilbert. Indeed, Hilbert agrees with Kant (and Brouwer) that mathematics requires a non-logical given. But unlike Brouwer or Kant, Hilbert takes this given to be the realm of signs (rather than time or space), including "their properties, differences, sequences, and contiguities" (Hilbert 1983[1926], 196) – which practically already contains elementary arithmetic. His formal method sought to prove the consistency of formal systems using this intuitive knowledge, in order to certify the conservatism of more elaborate formal systems with respect to arithmetic. So there's no attempt to define number by number, only to use the intuitive aspects of numbers and symbols to investigate higher mathematics. This is, again, akin to what Maddy calls *risk assessment*.

Despite the blow suffered by Hilbert's program from Gödel's incompleteness theorem, Hilbert's program is, in a sense, the "winner" of the early twentieth century foundational debates. It allowed mathematicians to endorse any formalized theory whose consistency could be reduced (or at least compared) to that of some well understood system (ZF or its various extensions or competitors). For Hilbert's foundations to stand, we do need to take on faith the consistency of some potentially infinitary mathematics based on some symbolic intuition, but as long as the mathematical community has not found mathematical contradictions in these systems that could not be dissolved by further disambiguation, decoupling,[2] and monster-barring formalizations, this faith can be sustained.

But Hilbert's formal approach (which can be imposed on practically any mathematical theory, including model theoretic approaches to set theory, variants of category theory, etc.) did not only guarantee the plurality of mathematics (the *generous arena*, in Maddy's terminology). It did much more: it allowed to set aside questions of conceptual clarity (Hilbert's own conceptions were never

[2]By "decoupling" I mean breaking down a contradictory concept into several concepts such that each maintain some of its aspects in a non-contradictory manner.

quite clear!) and adherence to some pre-linguistic living intuitions. It exported to philosophy those questions that threatened mathematical consensus with dilemmas of *metaphysical insight* and *epistemic source* in exchange for a practical mechanism for obtaining mathematical consensus regarding different extensions of elementary arithmetic. The mathematical community could finally agree on how to agree, suspend judgment on whatever need not be agreed upon (e.g. which theory of infinity is "true"), and exile questions that cannot be expressed in Hilbert's framework (such as "what are numbers?") to the realm of obscure philosophy. Mathematicians could finally stop paying attention to philosophers and their kin.

In one sense, the above may be captured by what Maddy calls *shared standard* – a common framework against which to measure mathematical definitions and proofs – but this does not exhaust Hilbert's achievement. Being able to agree on what need not be agreed upon and casting it out of mathematics is perhaps *the* triumph of Hilbert's foundational program – a foundational desideratum that I add to Maddy's list and title: *bounding the scope of debate.*

But this achievement is not only about drawing a clear boundary around mathematics. It is also about making it extremely difficult for outsiders to cross the boundary and attempt to challenge it from within. The modern commitment to a higher level of formalization of mathematics has made sure that only highly trained mathematicians – those who have spent so much time in training and are so invested in the language of mathematics that they are not likely to risk its authority – could take a substantial part in the debate over mathematical values. This foundational desideratum may be titled: *keeping intruders out.* I believe that this has a lot more to do with the success of mathematical formalization than is usually acknowledged.

17.2 Pluralist Foundations

Given the above state of affairs, it is no surprise that while some philosophers continued to quibble in the polemical margins excluded by Hilbert's formalization, others took up a more unifying approach. Instead of being cornered into choosing a foundation for mathematics, they chose to explain that one needn't choose.

One prominent expression of this position in the twentieth century is due to Hilary Putnam. His *Mathematics without Foundation* (Putnam 1967) considered the set theoretic and modal approaches to the foundation of mathematics, and concluded that they offer complementary and reconcilable accounts: "Even if in some contexts the modal-logic picture is more helpful than the mathematical-objects picture, in other contexts the reverse is the case. ... Looking at things from the standpoint of many different 'equivalent descriptions,' considering what is suggested by all the pictures, is both a healthy antidote to foundationalism and of real heuristic value in the study of first-order scientific questions" (Putnam 1967, 19–20). Let a thousand flowers bloom!

What allowed Putnam to be so nonchalant about foundations is his belief in the reality of mathematics. Mathematical entities are so strongly entwined with the scientific project, that it makes no sense, according to Putnam, to think of them as less real than such entities as electrons or water. If we believe, with Putnam, that the scientific project is our most successful engagement with reality, we might as well endorse mathematical and physical entities (tangible and abstract alike) wholesale. Having such confidence in the reality of mathematics, we can show some lenience when it comes to its foundational description.[3]

Another pluralist approach was expressed by Wittgenstein. He was concerned that what was usually called "foundations of mathematics" had nothing to do with actually founding mathematics. Frege's or Russell's calculi were, for Wittgenstein, not the foundation of arithmetic, but new calculi, which could be related to other arithmetical counting and calculations techniques, so as to form the thick web of standards that mathematics is.

To understand this position we should first understand what makes a statement "mathematical". According to Wittgenstein, whether a statement is mathematical, is not determined by the objects that the statement supposedly refers to, but rather by its use (Wittgenstein 1976, 113–115). If we say that "2 and 3 are 5" or that "two apples and three apples are five apples", both statement can be empirical observations just as they can be mathematical statements, depending on how we act on them.

If these statements are used to make a practical prediction, then they are empirical. In such cases, if the prediction happens to be wrong, then it's simply wrong – just like a gut feeling or a weather forcast is sometimes right and sometimes wrong. In contrast, a mathematical use would set these statements as a standard against which things are described. In the latter case, "$2 + 3 = 5$" or "two apples and three apples are five apples" is our mathematical standard. When an empirical observation deviates from this standard, one is challenged to *explain* the discrepancy: perhaps we miscounted, or perhaps someone took an apple, or an apple decomposed before you were done counting, or one apple became two due to some quantum voodoo... What characterizes the use of a statement as mathematical in Wittgenstein's sense is that it assigns responsibility for bad predictions not to mathematics, but to our imperfect control of the situation or our mis-application of rules.

[3]Similar pluralist approaches are not unique to twentieth century formalized science. Here I will only mention one scholar who offered a pluralist solution to the problem of founding infinitesimals: Lazare Carnot. Carnot (1832, see also Schubring 2005 for an analysis) considered several approaches to infinitesimal analysis: the ancient method of exhaustion, actual infinitesimal indivisibles, Newton's fluxions and first and last ratios, intuitive notions of limits, and Lagrange's replacement of derivatives by power series expansions. To that Carnot added the formulation of infinitesimals as corrective variables, not necessarily small in value, which cancel out, if they eventually turn out equal to purely fixed numbers (the analogy here is to purely imaginary numbers, which must be zero if they are equal to real numbers). He considered all these systems worthy and potentially innovative. The choice between them, according to Carnot, should be pragmatic, rather than foundational (Carnot 1832, 120–121).

Now in this framework, Russell's or Frege's definitions of numbers are new techniques, which are subject to a standard that says that they must accord with our older forms of counting and calculating. If there's a mismatch between the old and new calculi, it must be explained – usually as an error in following the rules of either the old or new calculus. In cases of discrepancy between the old and new systems, we don't suspect that the systems do not match, but assume that there was an error in our adherence to the systems, even if we don't have a record of the process of calculating that would allow us to find the error (Wittgenstein 1976, Ch. 28). This holds even though in large enough calculations we are more likely than not to have inconsistent results, and are unlikely to even try using a logical calculus to do arithmetic. The situation is similar in the context of coupling everyday wordy mathematical proofs and Hilbert's formalized proofs. The standard is that every wordy proof is formalizable. If it's not, then there's a problem with the proof, not with the formalization standard.

Since calculating in Russell's or Frege's formalism is very inefficient, the real purpose of setting the mathematical standard that binds the new calculi with the old arithmetic is to draw some interesting analogies and project on arithmetic logical claims that are difficult or impossible to justify by looking at the old calculi alone (Wittgenstein 1976, 247). But it would be an error, according to Wittgenstein, to assume that logic grounds mathematics and follow logic-arithmetic analogies that are simply confusing or useless.

Of course, setting a standard is not something we can do arbitrarily. Imposing a new standard requires authority and/or some sort of motivating gain. If the standard requires users to waste a lot of effort on finding explanations for apparent discrepancies, the standard is less likely to be maintained. If discrepancies between classical arithmetic and the new calculi, or between wordy proofs and Hilbert's formal proofs were too big, that is, if people too often got into trouble trying to appeal to these standards – then they would likely be rejected as standards.[4]

So what enables mathematical standards is the fact that the available resources allow enough students to be trained to follow them more or less consensually without getting into too much trouble. As Wittgenstein put it, the facts on which our calculus is based are "the empirical facts that we generally remember mistakes in counting, don't twist the calculus around, and so on" and that "if six intelligent people add [I might add: 'small enough numbers'], they get the same result"

[4]Note that people would get into trouble trying to formalize proofs, because many proofs cannot be *actually* formalized with reasonable resources. Nevertheless, people do not actually run into trouble because they are not actually expected to fully formalize entire proofs. In practice, one may be required to present a semi-formalization of some specific sub-argument (replace some terms by a logical or set theoretical reconstruction, replace a qualitative statement or an implied construction by a more explicit formula, break down a derivation into sub-steps that are closer to known theorems, rules or axioms, etc.). This is something that mathematicians do manage to usually perform consensually to the extent required in actual mathematical deliberations and controversies. Still, there are people who think that formal proofs are not necessarily a very good standard against which to measure mathematics (e.g. Hersh 1997, Thurston 1994). For a more recent discussion of the formal/informal gap, see Rav (2007) and Tanswell (2015).

(Wittgenstein 1976, 291, 274). These facts are natural, psychological and social facts that have to do with human behavior, and they, according to Wittgenstein, form the grounds on which mathematical standards are rejected or endorsed. So, similarly to the case of Putnam, given Wittgenstein's kind of naturalistic foundations, pluralism with respect to mathematical languages or foundational systems is easy to tolerate.[5]

If we accept Wittgenstein's characterization of mathematics, then the pragmatic-naturalistic grounding of successful standards obviates the need for a foundational shared standard. Mathematics builds on many different practices, as well as on translation schemes between those practices, which people can be trained to follow consensually. Giving primacy to one standard over others by considering it a foundation helps fortify mathematical consensus (as this standard becomes the arbitrator when different standards conflict), but granting such primacy depends on already having a relatively robust system of translation schemes that tie different practices to each other without necessarily passing through the supposed foundation. Indeed, if this hadn't been the case, it would have been very difficult to choose one standard over others, and a dissensual mathematical culture would have emerged. So a foundation may strengthen what Maddy calls a *shared standard*, but cannot be given full credit for it.

Wittgenstein's pluralism went so far as to claim that an inconsistency does not invalidate a mathematical system, as long as the inconsistency doesn't lead us into problems (either because the results of the inconsistency are inconsequential, or because we know how to bypass the inconsistency even if we can't formalize how we do it – see Wittgenstein 1976, Chs. 21–22). A long history of inconsistent mathematics (infinitesimals from the seventeenth to the nineteenth century), the formal contradictions that still survive in theoretical physics, and contemporary paraconsistent mathematics (Mortensen 1995) all demonstrate that his position is not as unreasonable as it may sound to a philosopher or mathematician raised on Hilbertian formalism. That category theorists sometimes exceed the awkward bounds of Grothendieck Universes, despite their awareness of lurking contradictions (Ernst 2015, Krömer et al. 2009) is also witness to this fact.

This means that *risk assessment* as a foundational desideratum is perhaps better achieved through practical experience than through a foundational program. Moreover, I have already argued that the contribution of a foundation to a *shared standard* may be overrated, and what Maddy calls a *generous arena* is obviously obtained more fully in a pluralist framework than by any single foundation. Therefore, the main advantage of formal foundational programs (axiomatic, set- or category-theoretical) over pluralism might reduce to their ability to *bound the scope of debate* and *keep intruders out*, preventing philosophers and other critics from ruffling mathematics' feathers. Still, even though these are perhaps the most prominent

[5]For a discussion of why such rigorous standards are characteristic (perhaps even defining) of mathematics rather than of other sciences, I refer the reader to Wagner (2017, Ch. 3.3)

impacts of formal foundational systems, they are neglected in philosophical debates over foundational achievements and desiderata.

A more recent proponent of pluralism, Michele Friend (2014, see also this volume), has a more difficult task cut out for her, compared with Putnam or Wittgenstein. Her purpose is to convince philosophers and mathematicians that pluralism is tenable and makes sense even if we don't commit ourselves to Putnam's realism or Wittgenstein pragmatic naturalism. Since her account of pluralism has no independent ground to build on, it has to emerge from discussions within and against a motley of foundational positions (Friend 2014, part I). This makes it, I believe, an impressive and laudable attempt at bootstrapping a new mediating discourse from within a world of agonistic strands of discourse: "Pluralism is founded on the conviction that we do not have the necessary evidence to think that mathematics is one unified body of truths, or is reducible to one mathematical theory (foundation). The pluralist is simply agnostic on this issue, ... it is in the light of this agnostic attitude that pluralism is developed" (Friend 2014, 103).

Replacing the search for a single foundational standard by the construction of a mediating agnostic discourse between many different standards is a daunting project. However, proposing that proofs seek "to communicate something, to convey an idea, and to open a discussion" (Friend 2014, 214), this strand of pluralism has a chance of growing into something that will exceed being yet another scheme of translation between mathematical practices: it might actually be able to mediate between non-mathematicians and mathematicians. Given the current foundational success in *keeping intruders out*, this is a genuinely revisionary project.

17.3 Against Foundations

But pluralism is not our terminus. One may not only accept any or many foundations, one may reject the idea of foundations altogether. This rejection would not have to be based on the fact that no single foundation captures perfectly the entirety of mathematics. Rather, it would be based on the claim that foundations risk crippling mathematics. Yehuda Rav has addressed this issue in a string of papers (Rav 1999, 2007, 2008):

> Indeed, a strict axiomatization of analysis, or any field of mainstream mathematics, for that matter ... would be counter-productive, and essentially not feasible. The reason is simple, stemming from what can be called the transfer of technologies. ... Ideas and concepts that were developed independently in a particular area of mathematics are frequently used and applied in a different context, much to the enrichment of the latter. Encapsulating a major branch of mathematics, such as analysis, for example, within a rigid axiomatic framework, would block beforehand such unforeseeable developments... (Rav 2008, 134–135).

Trying to encapsulate the unformulated practice of mathematicians inside a foundational framework is considered here not only as arbitrary, but as actually harmful. The claim is that the *generous arena* of any given foundation or collection of foundations might not be generous enough when considered from the point of view of historical development and change. Rav does agree that foundations are crucial for the logician who proves theorems about formal theories (which Maddy calls *meta-mathematical corral*) – but these foundations need not constrain the innovative practitioner of such object theories, and the meta-mathematics that studies the object theories need not itself have a clear foundation (Rav 2008, 142).

Formal foundations, according to Rav, do not guide mathematicians. If anything, they are derived from the unformulated practice of mathematicians (Rav 2008, 125). Indeed, many successful mathematicians do an excellent job without being able to fully reconstruct the definitions and axioms of any foundational system, or recognize subtle violations of the axioms. Essential guidance is not really something that foundational programs provide – perhaps because they are too essentialist to capture something as historically changing and plural as mathematics.

Another anti-foundationalist voice in the history and philosophy of mathematics is Imre Lakatos, who stated in no uncertain terms that "knowledge has no foundations" (Lakatos 1976, 46). One of his arguments against foundations is linguistic: even a formal system is presented and couched in a natural language. But natural languages are unstable, contradictory and changing. So the meaning of formal rules is eventually susceptible to the uncertainties of natural languages (Lakatos 1976, 89–90). To that, one may object that while unexpected interpretations of formal rules *could*, in principle, come up, they *actually* do not, and consensus about which moves are allowed within a given formal system is effectively unthreatened at least for contemporary mathematics.

But while formalization can inhibit the emergence of unexpected interpretations of what is valid within a given system and make mathematics more consensual over larger spans of time, the polemical field of unexpected interpretations is not thereby extinguished – it is merely exported. In a formalized setting, interpretations relate to the translations between formalizations and applications or intuitive meanings, and disagreements emerge with respect to which phenomena or experiences are correctly represented by given formalizations and their consensual mathematical inferences. This expulsion of conceptual dialectics from mathematics to the non-mathematical realm of application or interpretation (which is a form of *keeping intruders out*) is considered by Lakatos a problematic turn of events (Lakatos 1976, 122).

Lakatos believed that committing to a specific foundational system without challenging it risks leading to degeneration. He believed that this is precisely what happened with the Euclidean and Newtonian postulates: they started as wild revolutionary ideas, and ended up becoming scientific Procrustean beds (Lakatos 1976, 48–49). For Lakatos, the dialectic of interpretations is not an unfortunate limitation of human language, but the productive engine of improvement. It is precisely because we can come up with new, unintended interpretations to old terms, that we can enrich and refine our knowledge (for an application of Lakatos's stance to set theory, see Tanswell 2017). Indeed, such interpretations force us to explore our proofs in new contexts and consider new forms of argument. In turn, these new arguments reshape our interpretations and concepts. This is the progressive dialectic of formal science.

According to Lakatos, rather than try to freeze this process by formalization, concepts should be made just as clear as required for the anticipated space of interpretations, and be revised when new interpretations arise. Lakatos would therefore not be content even with a plurality of foundations, as he was intent on encouraging dialectic changes of mathematical practices in an evolving world of interpretations. He would reject *bounding the scope of discussion* and *keeping intruders out* – those prominent achievements of formal foundational approaches – as desiderata.

17.4 Mathematics Without Foundations

One way to deal with all this is to accept foundational theories as long as they are not taken seriously *as* foundations – not so much a pluralism of foundations, but a pluralism with regards to partial translations between mathematical languages (formal and informal), none of which is granted primacy. But can we imagine a *highly evolved, proof oriented* mathematical culture that simply does not think in terms of foundations? I do not mean here a mathematical culture that is not certain about its foundations, or has imperfect foundations, or is polemical or pluralistic about its foundation (as was the case with the dominant strands of higher classical Greek, medieval Arabic and modern European mathematics). I mean a mathematics that simply doesn't practice foundations. Fortunately, we do not need to imagine such a culture. I believe it was really there, in Kerala (southwest India), in late medieval times, practiced by a group of mathematicians that hardly ever bothered with foundations (in the sense of trying to set a clear and unified logical ground, axioms or inference rules for mathematics).

Before I say something about this culture, I should make a disclaimer. Past mathematical cultures cannot replace our contemporary mathematics. Such cultures evolved in certain contexts and for certain functions that shaped them in ways that may not be relevant for other mathematical cultures. Other mathematical cultures could only expand our imaginations concerning what our mathematics could be. The relevance of this "could" has to be evaluated contextually for the inspirations it might (or might not) provide for future mathematics.

Turning back to our case study: Kerala mathematics in the fourtheenth to sixteenth century. This culture was the most advanced mathematical culture of its time throughout the world. Its breakthroughs include power series like approximations of trigonometric functions and the value of pi (surveys are available in Joseph 2009; Plofker 2009, ch. 7; Sarma et al. 2009). Here we are interested in one important text: Śaṅkara's Kriyākramakarī (Sarma 1975, see also Wagner 2018) – an encyclopedic commentary on Bhaskara II's twelfth century Līlāvatī, completed by his disciple Nārāyaṇa in the mid sixteenth century.

This work includes not only algorithms and general statements, but also detailed proofs (*upapatti* or *yukti*) in prose and in verse. Some historians of mathematics claimed that Indian mathematics had no interest in proofs (see Sarma et al. 2009,

267–70). This position has been refuted, as discussions of proofs and justification have been documented already in early classical Sanskrit mathematics (e.g. Keller 2012). Part of the problem is that the texts known to western scholars were mostly succinct summaries of algorithms in verse. But in India, these highly elliptic texts were accompanied by oral teaching and written commentaries, which included detailed justifications (for a list of sources see Sarma et al. 2009, xxix-xxxii, 294–96). Within this tradition, the most highly elaborate proofs that survive in Indian sources come from Kerala.

As observed by Srinivas (2005) and Raju (2007), Indian proofs did not start with axiomatic foundations. Indian proofs could rely on evidence elicited from observing a diagram or the use of analogies to physical situations. But while many Indian logical schools preferred observation over inference, this tendency does not reduce Indian proofs to inductive generalizations – a good proof had to give compelling reasons that would convince the community of specialists, not only show that a certain mathematical technique works by means of examples.

Moreover, the role of observation in proof does not make empirical reality the ground of mathematics, because mathematical astronomy included "fictional" non-observable and idealized entities that are viewed instrumentally, rather than realistically (Srinivas 2016, 11–12). These fictional entities can and do serve as starting points of mathematical proofs. This fits well with the fact that many Indian logical systems do not assume a bivalent truth value, which would equate fictional with false (Raju 2007).

One can find statements in the Kerala mathematical literature that might apear to be foundational, such as the claim that mathematical computations depend mostly on the rule of three, the Pythagorean theorem and arithmetical operations (Sarma et al. 2009, 30; a similar statement is made by Nīlakaṇṭha, another proponent of the same school, in the context of astronomical calculations, Sastri 1930, p.100). But this should not be taken seriously. There are obviously many other implicit and explicit mathematical tools in Indian proofs, and the Pythagorean Theorem (and, to an extent, also the rule of three) are themselves subject to justification within the same treatises.

If we follow the reasoning of proofs in the Kriyākramakarī "all the way down", we find several kinds of explicit grounding elements:

1. Arithmetic rules such as distributivity, associativity, etc.
2. Observation and cut-and-paste manipulations of two and three-dimensional geometric diagrams
3. Observation of simple physical models (such as ropes and beams)
4. Combinatorial common sense (for lack of better terminology)
5. The logic of inference (anumāna – this is used to explain the rule of three in terms of an underlying logical structure; see Hayashi 2000)
6. Analogies between different mathematical situations

Building on the above, one derives ever more complex algebraic and geometric statements, and ends up with the impressive achievements mentioned above.

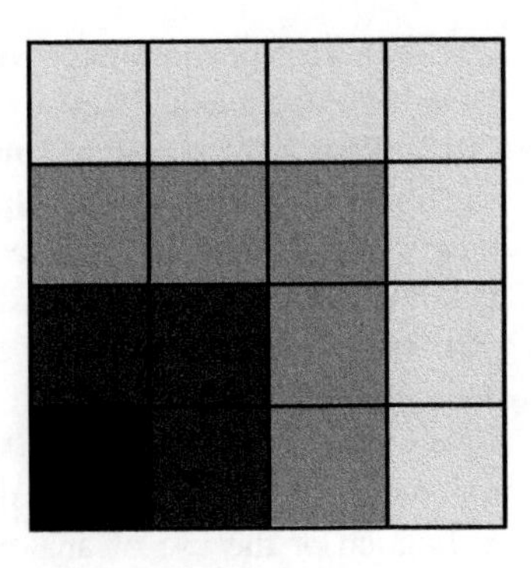

Diagram 17.1 The sum of consecutive odd numbers is a square number

So far, this is what we'd expect from an informal mathematical proof culture. What makes the Kriyākramakarī especially interesting is the fact that the relations between ground and conclusion are often very unstable and surprising. For example, we may expect to find proofs of the rule that we would express as $(a + b)(a - b) = a^2 - b^2$ to be derived from algebraic manipulations of the left hand product, or from a cut and paste manipulation of a diagram of a square divided in four parts – and the Kriyākramakarī indeed contains such proofs. It is, however, more surprising to find another proof at the end of the section on square roots, where this identity is derived from the fact that an arithmetic series of odd numbers starting from one sums to a square number (the latter is visually obvious from Diagram 17.1, and also proved arithmetically in the context of summing arithmetic series).

Given this fact, a difference of squares equals the tail of a finite arithmetic series of odd integers. Such a series can be viewed as the sum of two interlaced arithmetic series of consecutive integers. Then, by the general trick for summing arithmetic series (noting that the sum of two terms equally distant from the extremes of the series is constant), the interlaced arithmetic series sum to the required product of sum and difference (for example, $6^2 - 2^2 = (1 + 3 + 5 + 7 + 9 + 11) - (1 + 3) = 5 + 7 + 9 + 11 = (2 + 3) + (3 + 4) + (4 + 5) + (5 + 6) = (6 + 2)(6 - 2)$; Sarma 1975, 43–44). This proof is surprising because it is a rather difficult proof of a rather elementary fact, which depends on a method that is explained only later in the book.

Such alternation of geometric and algebraic methods is in fact prevalent in the Kriyākramakarī. A good example is the extended discussion of quadratic Diophantine problems of the following form: "find two numbers such that the sum and/or difference of their squares, together with some given perturbation, yield a square number" (e.g. find rational x and y such that $x^2 + y^2 + 1$ and $x^2 - y^2 + 1$ are squares of rational numbers). Each kind of problem is solved by at least one general rule with infinitely many instantiations. The methods of justification of these rules include:

1. Synthetic algebraic verification: show that plugging in the formulas given by the rule into the unknowns yields a result that can be algebraically rearranged as a square.

2. Algebraic analysis: show how the rule is derived from the conditions of the problem, heuristic algebraic models for the solution, and/or established algebraic identities.
3. Application of the numerical algorithm for root extraction to the sums and differences given by the rule; the fact that the algorithm terminates shows that the resulting expressions have rational roots (note that the *numerical* algorithms are applied here to general "variables")
4. Geometric cut-and-paste models showing that the relevant sums and differences are indeed squares.

These methods are often combined, but different emphasis on different methods is given in the proofs of different rules.[6]

One more kind of proof is available in the same context, which is, again, more surprising: modeling the algebraic problem by chords in a circle (Sarma 1975, 146). This method is used, for example, to show that for integer x and y, the squares of x and $\frac{x^2-y^2}{2y}$ sum to a square of a rational number. To prove this assertion consider Diagram 17.2. In this diagram, $y(2b + y) = x^2$, because chords in a circle intersect proportionally, and the diameter $2r$ equals $2(b + y)$. This means that $b = \frac{x^2-y^2}{2y}$. From the Pythagorean Theorem we have $r^2 = x^2 + b^2 = x^2 + \left(\frac{x^2-y^2}{2y}\right)^2$. But we also know that $r = b + y = \frac{x^2-y^2}{2y} + y$, which is rational. So the two squares sum to the square of a rational number.

This example is remarkable, because a rather simple algebraic fact, which can easily be explained algebraically, is proved here by means of a rather advanced geometrical reasoning, which, itself, is presented without proof and is related to mathematical material that shows up only later in the book.

In general, the multiplicity of methods and the variety of formulations within each method in the proofs of the Kriyākramakarī indicate that the author is interested

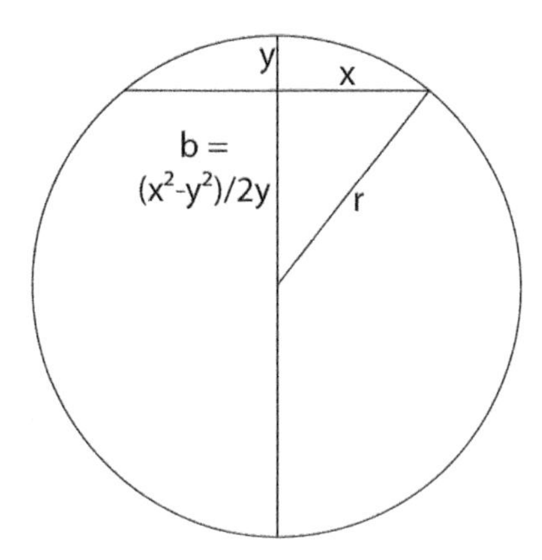

Diagram 17.2 Chords in a circle represent the solution of a Diophantine equation

[6]The "pulverization" method for solving such problems is treated in a separate section of the book.

in maximizing the understanding of *relations* between different mathematical methods and principles, rather than reducing mathematics to a unified foundation. The alternation of geometric and algebraic propositions that prove each other sometimes verges on circularity, where the two bodies of knowledge establish each other reciprocally.

17.5 Conclusion

On the one hand, this should make no sense from a foundational point of view. On the other hand, even in our foundationally oriented mathematical culture, similar Base-Case prevail. Induction-Step are consistently interested in the relations between different methods of proof. This often leads to attempts to prove a statement B from a statement A *and* vice-versa. From a foundational point of view, this makes sense where A and B are independent of the other axioms (e.g. the axiom of choice and the well-ordering principle). But such explorations take place also when A and B both follow from the given axioms, and there's no explicit logical control on the tools of the proof. In such situations A and B would be formally equivalent simply by virtue of being provable. When attempting to prove B from A after having proved A from B, it is often hard to tell if we do or do not "essentially" use the "main ideas" of the "independent" proof of B, or "really" derive B from A. Still, mathematicians have a strong interest in this kind of derivations.

The point I am trying to make with this case study is that a rich mathematical proof culture does not necessarily require foundations in order to thrive. Mathematics can in principle endorse Wittgenstein's holistic view and reject the ground/consequence division in favor of a system of correlation of practices, possibly containing circularities and manageable contradictions. It can further endorse Lakatos's view and expose mathematics to the sway of interpretive drifts by purposely not tying it to (a) formal system(s). Induction-Step modern research mathematics *should* adopt this approach is a different question, on which I remain here (with Friend) agnostic.

Shared standards and *risk assessment* can be achieved and have historically been achieved without committing to formal foundations. If they like, meta-mathematicians can keep their formalized *meta-mathematical corral* – this does not require them to impose it on the rest of us and restrict the *generous arena* of plural mathematics. True, the high level of consensus obtained from granting some formal foundation the power of final arbitration may suffer. Moreover, *bounding the scope of debate* and *keeping intruders out* would be more difficult to achieve without formal foundations. The latter, however, are perhaps not necessarily desirable. Once we accept that foundations are not strictly *necessary* for just *any* mathematical culture, then the debates over the reasons for adopting or rejecting foundational systems in *our* mathematical culture, and the role we assign to presumably foundational systems, become much more contingent and open ended.

References

Brouwer, L. E. J. (1975). *Collected works, Vol. 1: Philosophy and foundations of mathematics* (A. Heyting, Ed.). Amsterdam: North Holland Publishing Company.

Brouwer, L. E. (1981, January). *Brouwer's Cambridge lectures on intuitionism* (D. van Dalen, Ed.). New York: Cambridge University Press.

Carnot, L. (1832). *Reflexions on the metaphysical principles of the infinitesimal analysis*. Oxford: J.H. Parker.

Corry, L. 2004. *David Hilbert and the axiomatization of physics (1898–1918)*. Dordrecht: Kluwer.

Ernst, M. (2015). The prospects of unlimited category theory: Doing what remains to be done. *Review of Symbolic Logic, 8*(2), 306–327. https://doi.org/10.1017/S1755020314000495.

Frege, G. (1884). *Die Grundlagen der Arithmetik: eine logisch-mathematische untersuchung über den begriff der zahl*. Breslau: W. Koebner.

Frege, G. (1971). *On the foundations of geometry and formal theories of arithmetic* (E.W . Kluge, Ed.). New Haven: Yale University Press.

Frege, G. (1984). *Collected papers on mathematics, logic, and philosophy* (B. McGuinness, Ed.). Oxford: Blackwell.

Friend, M. (2014). *Pluralism in mathematics: A new position in philosophy of mathematics*. Dordrecht: Springer.

Hayashi, T. (2000). Govindasvamin's arithmetic rules cited in the Kriyakramakari of Sankara and Narayana. *Indian Journal of History of Science, 35*(3), 189–231.

Hersh, R. (1997). *What is mathematics, really?* Oxford: Oxford University Press.

Heyting, A. (1983). Disputation. In P. Benacerraf & H. Putnam (Eds.), *Philosophy of mathematics: Selected readings* (2nd ed., pp. 66–76). Cambridge: Cambridge University Press.

Hilbert, D. (1983). On the infinite. In P. Benacerraf & H. Putnam (Eds.), *Philosophy of mathematics: Selected readings* (2nd ed., pp. 66–76). Cambridge: Cambridge University Press.

Joseph, G. G. (2009). *A passage to infinity: Medieval Indian mathematics from Kerala and its impact*. New Delhi/Thousand Oaks: Sage.

Keller, A. (2012). Dispelling Mathematical Doubts. Assessing the mathematical correctness of algorithms in BhâSkara's commentary on the mathematical chapter of the Âryabhatîya. In K. Chemla (Ed.), *The history of mathematical proof in ancient traditions* (pp. 487–508). Cambridge: Cambridge University Press.

Krömer, R., McLarty, C., & Wright, M. (2009). Mini-workshop: Category theory and related fields: History and prospects. *Oberwolfach Reports, 6*(1), 463–492.

Lakatos, I. (1976). *Proofs and refutations: The logic of mathematical discovery*. Cambridge: Cambridge University Press.

Mortensen, C. E. 1995. *Inconsistent mathematics*. Dodrecht: Kluwer.

Plofker, K. (2009). *Mathematics in India*. Princeton: Princeton University Press.

Poincaré, H. (1996). Mathematics and logic. In W. B. Ewald (Ed.), *From Kant to Hilbert* (Vol. 2, pp. 1021–1070). Oxford: Oxford University Press.

Putnam, H. (1967). Mathematics without foundations. *The Journal of Philosophy, 64*(1), 5–22. https://doi.org/10.2307/2024603.

Raju, C. K. (2007). Cultural foundations of mathematics: The nature of mathematical proof and the transmission of the Calculus from India to Europe in the 16th c. In *Ce*. Delhi: Pearson Longman.

Rav, Y. (1999). Why do we prove theorems? *Philosophia Mathematica, 7*(1), 5–41. https://doi.org/10.1093/philmat/7.1.5.

Rav, Y. (2007). A critique of a formalist-mechanist version of the justification of arguments in mathematicians' proof practices. *Philosophia Mathematica, 15*(3), 291–320. https://doi.org/10.1093/philmat/nkm023.

Rav, Y. (2008). The axiomatic method in theory and in practice. *Logique et Analyse, 51*(202), 125–147.

Russell, B. (1903). *The principles of mathematics*. Cambridge: Cambridge University Press.

Sarma, K. V. 1975. *Līlāvatī of Bhāskarācārya with Kriyākramakarī of Śaṅkara and Nārāyaṇa*. Hoshiarpur: Vishveshvaranand Vedic Research Institute.

Sarma, K. V., Ramasubramanian, K., Srinivas, M. D., & Sriram, M. S. (2009). *Ganita-Yukti-Bhasa of Jyesthadeva*. Springer.

Sastri, S. (Ed.). (1930). *The Aryabhatiya of Aryabhatacarya, with the Bhasya of Nilakanthasomasutvan*. Trivandrum: Government Press.

Schubring, G. (2005). *Conflicts between generalization, rigor and intuition*. New York: Springer.

Spencer Brown, G. (1972). *Laws of form*. London: The Julian Press.

Srinivas, M. D. (2005). Proofs in Indian mathematics. In G. G. Emch, R. Sridharan, & M. D. Srinivas (Eds.), *Contributions to the history of Indian mathematics* (pp. 209–248). New Delhi: Hindustan Book Agency.

Srinivas, M. D. 2016. *On the nature of mathematics and scientific knowledge in Indian tradition*. http://iks.iitgn.ac.in/wp-content/uploads/2016/02/On-the-Nature-of-Mathematics-and-Scientific-Knowledge-in-Indian-Tradition-MD-Srinivas-2016.pdf.

Tanswell, F. (2015). A problem with the dependence of informal proofs on formal proofs. *Philosophia Mathematica, 23*(3), 295–310. https://doi.org/10.1093/philmat/nkv008.

Tanswell, F. S. (2017). Conceptual engineering for mathematical concepts. *Inquiry, 61*, 881–913. https://doi.org/10.1080/0020174X.2017.1385526.

Thurston, W. P. (1994). On proof and Progress in mathematics. *Bulletin of the American Mathematical Society, 30*(2), 161–177. https://doi.org/10.1090/S0273-0979-1994-00502-6.

Wagner, R. (2017). *Making and breaking mathematical sense: Histories and philosophies of mathematical practice*. Princeton: Princeton University Press.

Wagner, R. (2018). The Kriyākramakarī's integrative approach to mathematical knowledge. *History of Science in South Asia, 6*, 84–126. https://doi.org/10.18732/hssa.v6i0.23.

Wittgenstein, L. (1976). *Wittgenstein's lectures on the foundations of mathematics, Cambridge, 1939*. Chicago: University of Chicago Press.

Part V
Foundations in Mathematical Practice

Chapter 18
Foundations for the Working Mathematician, and for Their Computer

Nathan Bowler

Abstract We begin by discussing which aspects of the ZFC foundation for mathematics are useful for supporting mathematical practice. We focus in particular on those aspects which cater specifically to human styles of mathematical reasoning. We then consider some advantages of categorical or type theoretic foundations, and how these advantages rely on less human-friendly features of those systems. We close by raising the possibility that this latter type of foundation may become more appropriate as mathematical practice shifts towards a greater reliance on computers.

18.1 Disclaimer

I am not a philosopher but a working mathematician. Thus the aim of this paper is simply to present one possible perspective on foundational issues which appears to me from an outside perspective to be largely neglected by those working more directly on the foundations of mathematics. I am grateful to the anonymous referee for some helpful hints about how to give the paper a veneer of philosophical respectability.

18.2 What Use Are Foundations?

We all accept that good foundations are important, but why? To me, it's because they make my life as a working mathematician much easier. Foundations are sometimes informally motivated in less practical terms, by saying that they provide

---

N. Bowler (✉)
Universitat Hamburg, Hamburg, Germany
e-mail: Nathan.Bowler@uni-hamburg.de

S. Centrone et al. (eds.), *Reflections on the Foundations of Mathematics*,
Synthese Library 407, https://doi.org/10.1007/978-3-030-15655-8_18

an **Epistemic Source**[1] of basic knowledge on which further arguments can be based, that they enable **Metaphysical Insight** into the true nature of mathematical objects or that they give a **formal expression**[2] to our constructive mental activity. For now I want to set aside those ideas and focus on pragmatic advantages. So let's lower our eyes from the metaphysical sky to the earth, where foundations belong, and carefully examine how good foundations can support mathematical workmanship.

As a concrete example, if I want to use a theorem from group theory in my work then the requirements are minimal; I just have to check that some object I have built satisfies the definition of groups, and perhaps some further conditions from the theorem. I don't have to worry that group theorists might be relying on assumptions incompatible with the basic assumptions in my own discipline. Common foundations give us a shared language, across cultures and across disciplines. They give mathematics **portability**.

Common foundations also give us a **Generous Arena** of shared mathematical objects. If I hear about some exotic graph arising from an odd construction in functional analysis, I know that despite its origins it is one of the subjects of my study and could serve perfectly well as a counterexample to my own conjectures. That such things can be taken for granted is a sign of how reliable a mathematical infrastructure our current foundations provide.

Because modern foundations provide a powerful **Shared Standard**, I can have confidence in my results. By this I don't mean to claim that the currently accepted standards distil essential and indisputable truths about the world, such that any statements derived according to them must also stand beyond dispute. My claim is rather more banal: once I've proved something, I don't have to worry that someone else will prove the opposite. I can also have confidence in other people's results. When I prove something which other people have shown to be false, I always know that I can go back over my proof and I'll find a mistake somewhere. In this way I am insulated from inconsistency.

It isn't that there's some mathematical guarantee that I'll never run into an inconsistency; Gödel pulverised any such hopes.[3] It's more that if there is an inconsistency then the chances of me being the one to stumble on it are tiny. And

[1]In this paper, capitalised terms in bold are taken from Maddy's classification of various aspects of what it is to be a foundation in Maddy (2017). Basic familiarity with that classification is helpful but hopefully not essential to follow this paper.

[2]Uncapitalised terms in bold are names I have chosen for other things we might hope for in a foundation, not discussed by Maddy. **Formal expression** seems to me to be a serious omission, given its importance for intuitionism. Intuitionists consider mathematics to be the result of the constructive mental activity of humans rather than the discovery of fundamental principles claimed to exist in an objective reality; accordingly, the aim of a foundation is to give a formal expression to that activity.

[3]More precisely, he showed that if a formal system is expressive enough to encode first order arithmetic then it can only prove its own consistency if it is itself inconsistent. Since the mathematics I care about builds on arithmetic and since I don't want to work in an inconsistent system, this leaves no hope of a proof, of the kind Hilbert wanted, of the consistency of that mathematics from within a simpler and more evident subsystem.

even if I did, it wouldn't be just my problem. Mathematicians around the world, including dedicated teams of foundational specialists, would jump on the problem at once and would quickly come up with a patch. This means that I can leave **Risk Assessment** regarding possible inconsistencies to the experts. Note that the benefit for the working mathematician of a good foundation is not so much that it enables **Risk Assessment** as that it insulates us from the need to worry about it.[4]

In contrast to the practical approach taken in the last few paragraphs it was once common to claim that, in giving an axiomatisation for mathematics, the point is to gather some (perhaps self-evident) axioms and principles to serve as an **Epistemic Source**, on the basis of which unshakeable knowledge can be built up. This might be plausible if the axiomatisation were the obvious one: extensionality plus universal comprehension. Those really are self-evident principles. The problem is that they lead to the Russell paradox. Self-evident claims are unfortunately not always true. Our current foundations are too obviously shaped by a trial-and-error attempt to cover as much ground as possible whilst avoiding paradox for such a view of them to be sustainable.

Another suggestion is that our foundations give **formal expression** to our intuitive human reasoning. That doesn't fit with my experience. Forming my intuitions into a proof formal enough to be published feels like nailing jelly to a wall. Instead our foundations give a **Shared Standard** determining which sorts of reasoning will count as mathematics. In return for the price of fitting my own arguments to this 'mathematics transfer protocol', I have the benefit of belonging to a community whose other members follow the same rules, and on this basis rests my confidence that their results will not contradict my own.

Most dubious of all is the idea that our foundations give **Metaphysical Insight** into the true nature of mathematical objects, including such deep discoveries as the fact that the number 3 is the union of the number 1 and the ordered pair (0,1). This and similar examples show that the way mathematical objects are implemented in the **Generous Arena** provided by our foundation has no metaphysical significance.

Even if these less practical goals could be fulfilled by some alternative foundation, it would not be widely adopted if it did not also serve mathematicians by letting them talk to each other, easily use each other's results and have confidence in those results. It is these pragmatic considerations which are really essential for any foundation of mathematics.

The main lesson I want to draw from all of this is that mathematical foundations are not immutable, and that changes in foundations can be justified on purely pragmatic grounds.

[4]More bluntly put, most of what falls under **Risk Assessment** is irrelevant for the working mathematician and only good from the point of view of foundational specialists. The same goes for the role of a foundation as a **Meta-mathematical Corall**.

18.3 Why ZFC Suits Me

Having briefly considered in general terms what foundations are good for, let's now focus on the particular advantages of the most widely used current foundation, ZFC. In this section, I hope to illustrate that this formalism is particularly well adapted to the quirks of its human users.

Firstly, it is simple, self-contained and memorable enough that we are able to learn it. Learning to work with this formalism isn't completely trivial, but it can be done within a single course of lectures. That is enough to allow this formalism to be absorbed and used by people all around the world. We saw in the last section that this universality is essential. So since the users are humans the learnability is also essential.

Part of what makes ZFC so learnable is that it is easy to picture using the cumulative hierarchy. This isn't just a nice extra feature that allows set theory lecturers to draw pretty pictures on the board. It is vital. To see this, we must pause to reflect on how much of mathematics relies on leveraging the particular toolkit of intuitions and skills we have inherited from our ancestors.

The human brain is not particularly well suited to abstract reasoning or calculation. Concerning our calculation abilities, as Dehaene points out in 'The Number Sense' (Dehaene 1997), the brain 'can accurately code only sets whose numerosity does not exceed 3, and it tends to confuse numbers as they get larger and closer'. No doubt your own unhappy memories confirm his observation that 'we have ... a hard time remembering the small number of equations that make up the multiplication table'. As for abstract reasoning, not even a tenth of people can solve an elementary logical problem called the Wason selection task (Wason 1977):

> You are shown a set of four cards placed on a table, each of which has a number on one side and a colored patch on the other side. The visible faces of the cards show 3, 8, red and brown. Which card(s) must you turn over in order to test the truth of the proposition that if a card shows an even number on one face, then its opposite face is red?

We are able to get around these limitations to some extent by translating arithmetical or logical problems into terms that our brains are adapted to deal with. Dehaene reports that the brain 'tends to associate the range of numerical quantities with a spatial map, thus legitimizing the metaphor of a mental number line oriented in space' and that there is 'an automatic association between numbers and space'. Similarly, performance on the Wason selection task increases greatly when it is presented in the context of a social norm such as 'if a person is drinking beer then he must be over 20 years old' (Cosmides and Tooby 1992).

Our brains have been honed over millennia to have strong and reliable intuitions for certain highly specific tasks. Much of modern mathematics is built around methods for exploiting these intuitions in formal contexts. Dehaene provides an excellent example:

> the English mathematician John Wallis, in 1685, made a unique gift to the mathematical community when he introduced a concrete representation of complex numbers – he first saw that they could be envisioned as a plane where the "real" numbers dwelled along a

horizontal axis. To function in an intuitive mode, our brain needs images – and as far as number theory is concerned, evolution has endowed us with an intuitive picture only of positive integers.

Similarly, the true power of a subject like topology is not that it allows us to see the relation between donuts and coffee cups, which after all is intuitively clear, but that it allows us to leverage the same intuitions of shape and space in reasoning about formal objects such as spectra of rings or infinite products of discrete spaces. A similar exploitation of spatial intuitions lies behind the power of algebraic geometry. As a further example, our excellence at strategising and developing tactics to outwit competitors is used in the many flavours of game theory and is reflected by the use of algorithms all over mathematics. If there were a way to get formal arguments out of the ability to recognise faces, it would be the highlight of modern mathematics.

Returning to the question of the benefits of ZFC set theory, it is clear that it relies on this kind of **intuition-leveraging**. Humans are adapted to build complex tools, which means building up objects from component parts, themselves built from yet simpler components. This is precisely the intuition exploited by the idea of the cumulative hierarchy, and it allows us to visualise and understand many of the axioms of ZF. Our intuitions about decision-making, which underlie the axiom of choice, are rather different but no less strong.

This notion of **intuition-leveraging** is reminiscent of that of intrinsic justification; we might say that an axiom is *intrinsically justified* if our intuitions tell us that the axiom is true of the objects of discourse. There is however a key difference. To leverage our intuitions it is only necessary that the objects of discourse behave sufficiently like other objects, ones for which we do have an intuitive grasp, that our intuitions for the latter objects are helpful in working with the former ones. Thus our intuitions about shape and space tell us, on the face of it, nothing about spectra of rings, since these spaces don't resemble the spaces in our experience. But by pretending that they do we can exploit our intuitions to help us work more easily with such spectra. Similarly, although one might dispute that our intuitions about decision making still apply when infinitely many choices are to be made simultaneously, there is no question that these intuitions help us to work with the axiom of choice.

An independent advantage of ZFC for human users is its **clean ontology**. Mathematical objects like the set of natural numbers simply exist and have certain properties. We are so used to this way of doing mathematics that it might seem inevitable, but as I shall discuss in the next section it is perfectly possible to imagine a foundation for mathematics without this feature. For now it will suffice to focus on two aspects of this ontology: the way it is achieved by means of a neat distinction between syntax and semantics, and the extravagantly wasteful nature of the collection of objects this foundation provides.

The ZFC foundation has two clearly distinguished parts. First there is a formal account of what counts as a mathematical statement or a mathematical proof; this is provided by first-order logic. Then there is the particular theory to which such reasoning may be applied, namely the axioms of ZFC. This latter part of the foundation is what provides the ontology; it determines that the fundamental objects of mathematical discourse are sets and gives us flexible ways to build and work with these sets.

This allows us to maintain in our[5] heads the distinction between arguments, which are simply *about* mathematical objects and belong to the first part of the formalism, and the objects themselves, which belong to the second. The details of how the arguments go can be detached from the properties which they establish. Thus to use the results of other mathematicians I don't need to keep track of any information about the syntactic details of their arguments.

Of course this distinction is not always as clean as we would like. It is perfectly possible to treat the arguments themselves as mathematical objects, with fruitful if often unsettling consequences. One of these is that first-order logic is incapable of pinning down the properties of even such a familiar object as the set of natural numbers (Gödel again). On the other hand the second part of the formalism is, strictly speaking, not the collection of sets but only a first-order theory of sets. As just noted, there is no hope that such a theory could completely pin down the universe of sets.

Nevertheless, enough properties of sets are determined by the theory that we don't run into problems by treating it as describing a fixed and platonic realm. This makes working in ZFC very comfortable for human users. Even for statements which cannot be proved nor disproved, such as the continuum hypothesis, ZFC panders to our assumption that there must be some truth of the matter through its incorporation of the law of the excluded middle: CH must be either true or false, we just don't know which.[6]

The second aspect of the ontology which is tailored to human ways of thinking is its profligacy. Almost all mathematical reasoning is about objects which appear[7] in the cumulative hierarchy just after the natural numbers, at most within the next dozen layers. So what is all the other junk there for? Well, it is built with constructions which we occasionally need to make use of. Indeed, this is one of the main uses of the ZF formalism in practice: the extremely flexible set-building notation.

---

[5]The word 'our' here refers to working mathematicians. As will shortly be explained, the distinction in our heads looks rather shaky when examined closely. Those with more philosophical sophistication, and we ourselves in our more reflective moments, know that there are thorny issues to be resolved here. One possible point of view is that a modern approach to foundations essentially involves identification of sets, numbers etc. with syntactical objects, a view which would have deep consequences. My own view is that such an identification is not forced on us, but explaining the reasons for this would go far beyond the scope of this paper. In any case, the point here is that, whether such a distinction is ultimately tenable or not, working as if it were is a useful practice which allows the mathematician to focus on mathematics rather than philosophy.

[6]As in the last paragraph but one, no claim about the extent to which LEM actually holds for CH is intended here, nor do I intend to disparage the sophisticated points of view on the matter held by modern set theorists. The claim is just that LEM is a useful working hypothesis which allows nonspecialists to ignore the philosophical issues and get on with their work.

[7]Or rather can be implemented; see the next two paragraphs.

There is a sense in which the flexible constructions provided by ZFC play a more fundamental role than the specific sets constructed with them. The identification of natural numbers or other mathematical objects with particular sets is, after all, rather arbitrary. Why should we suppose, following von Neumann, that the elements of the number 2 are 0 and 1, rather than using Dedekind's definition and taking the only element of 2 to be 1? Indeed, where did we get the idea that natural numbers should have elements at all? Worse still, we are told that the elements of the rational number 2 are different again, meaning that the natural number 2 is not a rational number.

A better way to put the matter would not be to say that natural numbers *are* sets, but that they can be *implemented* as sets, by using the powerful set-building constructions of ZFC. A canonical implementation is a helpful part of the **Shared Standard**, in the same way that it is helpful if everyone drives on the same side of the road, but the choice of which implementation to use is fairly arbitrary. The same goes for ordered pairs, for rational, real or complex numbers, and so on to the higher edifices of mathematical study.

Occasionally we run into mathematical objects which we need the full strength of ZFC to implement or to reason about. It would feel very unnatural to normally work in a weaker formalism, and to only work with a stronger formalism capable of supporting those implementations when we need them. Since we naturally think in terms of a fixed realm of objects, it only makes sense that if it's fine to use those extra constructions when we need them then the objects they construct must be sitting there the whole time.

Indeed, this fits with our day-to-day experience as human investigators of the world. We confine our attention to a small part of the universe, unworried by the fact that there is a great multitude of objects with which we do not concern ourselves. Why should our relation to the universe of mathematics be otherwise?

In pointing out the ways in which ZFC is tailored to our human oddities, I'm not trying to claim that we should strive to overcome these oddities and replace ZFC with a better foundation, one which isn't held back by human nature. My point is that ZFC is a great foundation for us precisely because it is tailored to work with our human nature by **leveraging** our **intuitions**, by giving us a **clean ontology** to work with, and by providing an **Arena** for the implementation of mathematical objects which is **Generous** to the point of wastefulness.

18.4 Categories and Type Theory: Wonderfully Alien

Category theory and type theory have some compelling advantages over set theory, but a closer look shows that they come at the cost of making the foundation unfriendly for human users. Indeed, some of their most attractive features are so closely tied to utterly unintuitive ways of thinking that even if we did use such a foundation we would have trouble taking full advantage of its power.

As a warm-up before we get to these pricklier features, let's consider a familiar but less compelling argument for the advantages of category theoretic over set theoretic foundations. When a group theorist studies a particular group, they don't care what its elements actually are, but only how they combine under multiplication and inversion. For example, if they're studying the simple group of order 60, its elements might be even permutations of the set $\{1, 2, 3, 4, 5\}$, or rotations of a regular dodecahedron in $\mathbb{R}^3$, or pairs of 2×2 matrices over the field of integers modulo 5; these things live at very different levels in the cumulative hierarchy, but that doesn't matter.[8] The idea is that by stripping away the unnecessary substructural detail a category theoretic approach can provide **Essential Guidance** about the nature of the structure itself.

This argument loses a little of its force when we recall that mathematical objects are usually implemented as sets, rather than fundamentally being sets. A typical group theorist does not care which particular implementation of the simple group of order 60 they are using. With this in mind, there are two different ways in which we could interpret the claim that a categorical foundation provides **Essential Guidance**.

The first is to note that it provides a unified account of just what it means to have an implementation for many kinds of mathematical object. Indeed, the language of category theory is very useful for isolating and expressing exactly what we require from such implementations. On this view, category theoretic ideas can serve as useful organising principles but they do not play a foundational role. Instead, they help us understand what it means to implement objects in an existing foundation.[9] Furthermore, as Maddy notes (Maddy 2017), although category theoretic ideas are well suited to dealing in this way with the objects of algebraic topology or algebraic geometry, they don't work so well for the objects of other fields such as analysis or combinatorics.

A second and more extreme interpretation is that category theory allows us to dispense entirely with the dirty business of implementing mathematical objects and to work in an implementation-agnostic foundation. But that would be to throw out the baby with the bathwater, since we would lose the *that* along with the *how*. We would lose all the benefits of a **clean ontology**. So this is no good if we want a human-friendly foundation. If we give up on the idea that our foundation should be human-friendly, then the benefits which open up are far more powerful than **Essential Guidance**, though also considerably more alien.

As we move to a discussion of these new advantages, it will be helpful to keep in mind the dichotomy between guidance for implementation within an existing foundation and the attempt to build an implementation-agnostic foundation. The same dichotomy will shortly appear again, but expressed in terms of interpretation of syntax rather than implementation of objects.

---

[8]Thanks to Peter Johnstone for this lovely example.

[9]A more nuanced approach, championed by McLarty (2004), is to maintain a set theoretic foundation, but use a first order theory of functions and their composition rather than of membership. The fact that the basic theory here is categorical allows a smoother implementation of many mathematical objects. This foundation enjoys many of the benefits discussed in the last section, though it involves a tradeoff of **intuition-leveraging** against **Essential Guidance**. For a fuller discussion, see Sect. 18.5.

Since you and I are humans, let's work our way up to these unintuitive advantages slowly, starting with a concrete experience familiar to all mathematicians. No doubt it has happened to you, after proving some statement, that you realised that you could prove something more general with exactly the same argument. What precisely does this mean? Well, the argument is something which can be written on a piece of paper, a bit of syntax. In order to see this argument as proving your original result, you employ an interpretation of that syntax. But with a slightly broader interpretation of that same syntax, perhaps just relaxing restrictions on how some variables in the main definitions can be interpreted, you get the more general result too.

Sometimes we modify the interpretation of the syntax more drastically. A familiar case of this is in arguments by duality. There are examples of duality all over mathematics: in projective geometry, the theory of planar graphs, matroid theory and category theory. In each of these cases the duality has both a semantic and a syntactic component. Semantically, a construction is given which assigns to each of the objects of study a dual object. This means that, for any statement about the objects, we can consider what it would mean when applied instead to the dual objects. Thus for each concept we obtain a dual concept and for each theorem we obtain a dual theorem. This transformation can be made entirely syntactic; typically it consists of adding the prefix 'co' in front of certain words and removing it from others. In other words, when we argue by duality this can be seen as employing the original argument but with an alternate interpretation of its syntax.

Using duality in this way can be extremely powerful, and it is about as far as this sort of trick is usually pushed. But we can go a lot further. One natural option is to exploit the extra flexibility of interpretation available for a more stripped-down syntactic framework. Typically the arguments we produce are expressed (or rather expressible) in the default syntactic framework of modern mathematics, which is first-order logic. But if your argument does not use the full strength of first-order logic then you can wring a lot more out of it.

If, for example, you didn't use the principle of the excluded middle at any point then you can interpret your argument in the internal language of any topos.[10] So you may be able to generalise your result to a version in which all objects have a continuously varying structure parameterised by your favourite topological space or an action of your favourite group, to give two of the most banal examples.

The trick here was that if the argument can be expressed with a more stripped-down syntax then it can be transferred to a richer variety of contexts. Category theory has some clever tools to allow very fine-grained exploitation of this 'argument lifting' idea. There is a large hierarchy of levels of syntactic complication, and for each level a corresponding categorical structure in which syntax at that level

[10] A topos is a highly structured kind of category. The prototypical example is the category of sheaves over a topological space. There are many other examples, such as categories of actions of a fixed group. By means of the internal language, first-order statements in the language of sets can be given a meaning within the topos. See for example Johnstone (2002) or MacLane and Moerdijk (1992).

can be interpreted, with simpler syntax corresponding to more flexible structures.[11] This ranges down from the interpretation of higher-order intuitionistic logic in topoi through the interpretation of algebraic theories in categories with finite products.

We can get a better grasp of how this works by examining that last case in more detail for a particular algebraic theory, namely the theory of groups. Recall that a *group object* in a category with finite products consists of an object G together with maps $e\colon 1 \to G$ and $i\colon G \to G$ and $m\colon G \times G \to G$ satisfying the usual rules for the identity, inverses and multiplication of a group. Many straightforward facts about groups, such as the uniqueness of inverses, also hold for group objects. Using the trick of argument lifting it is clear why this must be the case. The arguments for these elementary properties of groups are simple enough that they can be expressed in algebraic logic and so they can be interpreted in any category with finite products.

Of course for more advanced statements within group theory more complex arguments are needed, which cannot be expressed within algebraic logic. In standard presentations of group theory no attempt is made to clearly demarcate which arguments can be expressed in this way, just as no attempt is made to demarcate which arguments avoid the law of the excluded middle and so can be lifted to any topos. Many arguments will fall at some level of the hierarchy in between these two, and so the natural categorical setting for interpreting them will be categories with more structure than just finite products but less structure than is required for topoi.

It would be very helpful if each argument were flagged with the information of exactly how much syntactic complication is needed to express it, since this would also make clear the proper context to which it can be lifted. But it is clear that such a presentation of group theory would be painful for a human learner; the extra information would have to be learned by rote. On the other hand, storing and using information of this type is a task for which computers are ideally suited.

You might recognise the following argument schema if you have ever taught undergraduate students:

Lemma 4.1 *If P then Q.*

Proof Suppose not for a contradiction. Then we have P but $\neg Q$. Since P, [derivation of Q from P] and so Q. This contradicts our assumption that $\neg Q$.※ □

This argument is needlessly complicated, with a gratuitous use of the principle of the excluded middle. I usually discourage my students from making arguments of this kind on purely aesthetic grounds, but we have seen above that there is a potential pragmatic reason. By following the discipline of using only the syntactic forms and logical principles which are strictly needed in an argument it is possible to increase the number of contexts to which it can be lifted.

The suite of tools for **argument lifting** provided by category theory appears again in a thinly disguised form in type theoretic foundations for mathematics. Corresponding to the various kinds of structure which might be imposed on a

[11] For a discussion of a decent representative sample of such levels, see Johnstone (2002).

category we have the various kinds of 'type constructor'. For example, finite products in category theory correspond to product types in type theory. Arguments correspond to terms which can be built in the type theory, and an argument being expressible at a particular level of the hierarchy means that its corresponding term is present in the type theory with the appropriate type constructors for that level.

Seen this way, the benefits of the **modularity** of type theoretic foundations become clear. By mixing and matching different type constructors, we can express each argument in a formalism carefully tuned to match the logical and syntactic strength it needs. This is strongly reminiscent of the modular and flexible nature of programming languages, where the ability to choose a language with structures and libraries suited to the task in hand is a great advantage.

There is a further advantage to keeping track of the strength of the formalism required for each particular mathematical argument; it allows **decidability-tracking**. If the formalism is simple enough, then aspects of it may be decidable. For example, it may be decidable whether a given term inhabits a given type or whether two such terms are judgmentally equal. More strongly, it may sometimes be decidable, for a given type, whether or not there is any term inhabiting it. This is an advantage from a computational rather than a human perspective; human mathematicians are unlikely to follow fixed algorithms to decide whether mathematical claims are true, but it is convenient to save work by implementing such algorithms on computers and therefore useful to keep track of where such algorithms are likely to be helpful.

We've already seen a couple of ways in which the tools provided by categorical or type theoretic foundations are better suited to the grasp of computers than of humans, but this becomes even clearer when we consider how incompatible they are with two of the aspects of ZFC which make it so comfortable for us to use: the straightforward ontology and the clean separation of syntax and semantics.

To see how the ontology falls apart, let's return to the example of the natural numbers and think about how their theory should be treated if we take seriously the discipline of keeping track of the weakest appropriate level of interpretation for each argument. Already in the very weak context of a category with finite products[12] it is possible to define a notion of 'natural numbers object'. Just as for group objects, many of the elementary properties of the natural numbers also hold for natural numbers objects, because the arguments proving them can be interpreted in any category with finite products. So number theory, like group theory, can be stratified according to how flexibly the arguments can be interpreted.

So what are the most elementary arguments involving natural numbers *about*? From our current perspective, the natural answer is that they are about the things for which we can interpret them, that is, natural numbers objects in categories with finite products. But this has the worrying implication that more advanced arguments involving natural numbers are about a more restricted class of things, because we

[12]In fact there is even a weaker notion of natural numbers object which can be defined in any category with a terminal object.

need more categorical structure in order to interpret them. This rather uncomfortable situation is very different from the **clean ontology** given by ZFC, in which all these arguments are about a single object.

There is a compromise open to us if we want to preserve the **clean ontology** within a categorical foundation. We could say that number theoretic arguments are about the actual natural numbers but happen to have a syntactic form which also allows them to be interpreted in a wider context. But notice that this means taking an external view of those other sites of interpretation. We must see them as objects already instantiated on the basis of some other foundation. So this more comfortable approach does not allow us to exploit **argument lifting** in the way that we structure the foundation.

To see what it would mean to take an interpretation-agnostic view of these structures, and to take seriously the idea of building **argument lifting** into the foundation, we must return to type theory. In place of the notion of a natural numbers object we find there the notion of a natural numbers type. Some terms witnessing elementary properties of the natural numbers type can already be built in the presence of relatively few and simple type constructors. Those corresponding to more advanced arguments, however, can only be formed in the presence of stronger type constructors.

So how should we think about this? Working in such a modular foundation, can all these arguments still be thought of as being about the same object? There is a part of me that, perhaps trying to hang on to a unitary ontology, generates the following story: the natural numbers are indeed a single object, which in some contexts is rather blurry, because the type theoretic environment is not strong enough to determine it fully. Stronger type constructors cast a sharper light on it, and so when they are present its properties become clearer and crisper.

As you pause to smile at this clumsy attempt to preserve some semblance of an ontology, please note that this is only a concern for me because I am human. My ontological worries are strictly irrelevant to what it means to work successfully within such a foundation. A computer, unburdened by them, could happily produce and manipulate terms at various levels of complexity, never losing track of just which constructors are used to produce which terms.

Keeping track of such rote facts is just the sort of thing at which we humans struggle. Imagine the inconvenience if, to make use of other people's results, you had to pay attention not only to the details of what they proved but also to the syntactic form of their arguments and the logical principles used. We saw in the last section that ZFC avoids such difficulties by means of a clean separation between syntax and semantics. In category theoretic and type theoretic foundations the dividing wall between them has crumbled.

In category theory the syntax breaks through and invades the semantics; free categories with a particular structure tend to encode precisely the syntax which may naturally be interpreted in categories with that structure. Thus interpretation of that syntactic structure can be seen purely semantically as a structure preserving functor from that free category to the category in which the interpretation takes place. Interpretations of type theory can be seen in precisely this light, since the arrows in these free categories often correspond precisely to equivalence classes of terms under convertibility. So in a type theoretic foundation the conquest of semantics by syntax is complete.

Another way in which syntax and semantics merge in type theory is the exploitation of the Curry-Howard correspondence to treat propositions and mathematical objects on an equal footing. For a human, whilst it is interesting to note the similarities of structure between propositional and set theoretic constructions, identifying the two can be a little disorienting. But for a computer system the simplification entailed by this identification is a great boon.

I have argued that some of the best features of type theoretic foundations are also the features which make them least comfortable for human users. A weak form of **Essential Guidance** is compatible with a human-friendly foundation, but to gain the advantages of **argument lifting** and **decidability tracking** it would be necessary to jettison the **clean ontology** and **intuition-leveraging** that we are used to. Similarly, the **modularity** available within a type theoretic foundation is in conflict with the aim of having a single **Generous Arena**. We could make, at best, clumsy use of such a foundation and we would not be able to exploit its full power. A computer system, on the other hand, could work very efficiently in that context, storing and exploiting metadata about the syntactic strength of each argument and using computational approaches guided by the varying levels of decidability available at each level of the modular foundational structure.

18.5 Shifting Foundations

Modern mathematics relies ever more on the use of computers, not only to perform brute-force searches and complex calculations but also to encode and verify the mathematical arguments themselves. By the 1990s, computers were already used to prove hypergeometric identities, and to derive new ones (Petkovsek et al. 1996). They could also resolve open problems in other fields, such as the Robbins Problem (Bernd 1998). Nowadays it is totally normal to rely on computer systems to solve basic problems in algebra and calculus, and theorem provers are sophisticated enough to tackle much more complex problems such as the Boolean Pythagorean Triples Problem (Marijn et al. 2016).

Automated theorem provers are now usually integrated into interactive proof assistants, which makes the balance of contributions from the human and the computer harder to estimate. But following the trend from the last paragraph, it is reasonable to expect that over time more and more of the work will be done by computers and less and less by humans. If, as I have argued, our current foundation is optimised for human use but is not so well suited for computer systems, this may necessitate a shift from set theoretic towards type theoretic foundations. Foundations are after all not inscribed in stone; the main criteria for a good foundation are pragmatic.

The main area in which this shift is beginning to take place is in proof checking and verification systems, for which the main contenders sensibly use a type theoretic architecture. However, translating existing mathematics into a form which can be processed by such a system remains somewhat arduous. It will be many years yet before it is reasonable to expect mathematicians to submit a formal verification along with each of their papers, and the process of settling on a standard for doing this is likely to be fraught.

If there is a shift in foundations of the kind I envision, it is unlikely to be orderly. There will be a number of competing standards with different practical advantages, and the competition between them will take a long time to resolve. Indeed, there is no reason to expect it to resolve at all. Just as practical considerations may drive a shift to new foundations, they may also drive the mathematical community away from reliance on a single foundation.

What, after all, is the practical reason why we currently require a fixed and unified foundation? It is that otherwise we would regularly need to translate arguments back and forth between foundations, and this would not always even be possible. We would have to keep careful track of the logical strength of the arguments used in order to know which results could be transferred to which foundational settings. We just saw, however, that a key advantage of type theoretic foundations is that they do allow us to keep track of logical strength in this way. Furthermore, since type theoretic arguments have a clear and constrained syntax, automated tools for translating arguments from one foundational setting to another in a way which preserves information about their strength are likely to be achievable.

This brings out a further advantage of type theoretic foundations which we overlooked in the last section, because we were still supposing a single unified foundation. But the same features we discussed there also support **automated translation** between foundations. For Maddy (2017), a **Generous Arena** is important because it allows import and export of objects and methods between different mathematical fields:

> What's needed is a single arena where all the various structures studied in all the various branches can coexist side-by-side, where their interrelations can be studied, shared fundamentals isolated and exploited, effective methods exported and imported from one to another, and so on.

Towards the end of her discussion of categorical foundations, after reiterating the point in this quotation, Maddy writes 'I leave it to the reader's conscience to decide whether **Generous Arena** ... should be jettisoned'. But perhaps this is not so much a moral issue as a practical one. **Automated translation** gives an alternative and more flexible way to achieve the same goals. This goes far beyond the **modularity** of type theoretic foundations discussed in the last section, in the same way that the flexibility provided by alternative programming languages goes beyond that provided by modularity of individual languages. It involves **automated tracking of logical strength** which similarly goes far beyond the **decidability tracking** mentioned in the last section, and is likely to be of more use to working mathematicians than the kinds of **Risk Assessment** provided by traditional foundations.

Since our current reasons for using a single unchanging foundation do not apply to foundations based in computer systems, we may well see a continual competition between constantly developing and improving foundational standards, as is currently the case for programming languages. The choice of foundation used for a particular mathematical enterprise would then be driven by pragmatic considerations. The dozen or so decades surrounding the start of the third millennium may turn out to be the only period in history in which mathematics has a unified foundation.

As an example, different foundations may be better suited to pure and applied mathematics. This is particularly apparent for quantum mechanics, a field whose philosophical foundations remain under-clarified. According to the Kochen-Specker theorem (1967) there is no way to assign actual values to quantum mechanical observables in such a way that they have definite intrinsic values at any given time, independent of the device used to measure them. Thus the mathematics of quantum mechanics does not straightforwardly fit with the traditional kinds of ontology used in physics. Because of this there are a number of competing *interpretations*, most famously the Copenhagen Interpretation and the Many-Worlds Interpretation. These give indistinguishable predictions but take very different stances on such fundamental questions as whether there is a single universe (Copenhagen) or many other parallel worlds (Many-Worlds). But these worries, like the ontological worries about type theory in the last section, only arise for human users.

Since the ontology of quantum mechanics is so vexed, there is no clear advantage to formalising it in a foundation with a **clean ontology**, so in this case foundations optimised in other directions may be the best choice. For example, there has been a surge of interest in categorical formulations of quantum mechanics in recent decades. Once serious contender is the *categorical quantum mechanics* of Abramsky and Coecke. They give a survey of their ideas in Abramsky and Coecke (2009), where the key benefits they mention (apart from new perspectives on old ideas) are 'identifying the fundamental mathematical structures at work' (this is **Essential Guidance**) and 'an effective calculational formalism' with 'automated software tool-support' (this is linked to **decidability tracking**).

Returning to pure mathematics, what are the implications of these ideas for those working on type theoretic foundations? Since such foundations are better suited to use by computer systems, it would be natural to develop them principally with computer systems and not human users in mind. This means that there is no need to try to build in some of the aspects which human users would require, such as a **clean ontology**, **intuition-leveraging**, or a single **Generous Arena**. Instead, the aspects making the foundation more usable for humans could be constructed at the level of a user interface.

We can get some fragmentary initial ideas about what such a user interface might involve by considering three very different research projects. The first is the development, by Ganesalingam and Gowers (2017), of 'a program that solves elementary mathematical problems, mostly but not exclusively in metric space theory, and presents the solutions in a form that is hard to distinguish from solutions that human mathematicians might write'. The problems which are solved by this program are far from the cutting edge of what is achievable by modern automated theorem provers, being the sort of exercises I might set to my students to check that they have understood the definitions. But the proofs produced by this program are almost indistinguishable from the answers that good students might write. As such, they are far more human-readable than the usual output of a theorem prover. A tool like this would be a useful final processing step in the interactions of computer-based mathematical systems with humans.

Perhaps closer to the foundational issues we have examined in this paper is the work of McLarty (2004). He outlines a set-theoretic axiomatisation ETCS, due to Lawvere, which is expressed in terms of functions and their composition rather then of set membership. ETCS is considerably weaker than ZFC. McLarty explains that '[results that] mathematicians normally use are all theorems of ETCS. Ones dealing with $\aleph_\omega$ or higher cardinals require extensions of ETCS'. He presents a suitable extension ECTS+R which is inter-interpretable with ZFC.

In the light of our earlier considerations, this suggested foundation is lacking on a number of fronts. Considered from the human point of view, if we want to sometimes allow ourselves the indulgence of the stronger foundation ETCS+R then we must accept the ontological consequences. If we are to act as if we are dealing with straightforwardly existing mathematical objects, then they must have the properties implied by that stronger foundation even when we argue about them in ways expressible in the weaker one. Similarly the objects provided by the stronger foundation don't disappear when we use the weaker one. So if we want a **clean ontology** then we must take ETCS+R, not just ETCS with the option of R, as our foundation. But in doing so we give up on the **intuition-leveraging** power of ZFC, since the axiom R has much less intuitive traction than the corresponding ZFC axiom of replacement.[13] Even taking into account the gain in **Essential Guidance**, this sacrifice is so great that ETCS+R is less suited for human users than ZFC.

From the computer perspective, however, McLarty's foundation does not go nearly far enough. In terms of **modularity** it is extremely coarse-grained, allowing little more than the bolting on of replacement or large cardinal axioms to the already (by categorical standards) powerful system ETCS. Accordingly it provides no resources for **argument lifting** or **decidability tracking**, let alone **automated tracking of logical strength**.

McLarty's proposal seems to fall between two stools, but perhaps this could be turned to its advantage if we consider its possibilities as a user interface for a computer-based foundation. In this role it wouldn't matter that it lacks the resources mentioned in the previous paragraph, since they would be the responsibility of the back end. The similarity to the established foundation ZFC would make it fairly easy for humans to learn, and the **Essential Guidance** it provides would play a

[13]the other axioms of ETCS also arguably have less intuitive bite than their ZFC counterparts, but this is particularly clear in the case of R.

practical role, by ensuring that human users phrase their mathematics in terms which lack the kind of implementation-dependent baggage which would make it hard to translate into the type-theoretic foundation used by the computer. Of course, ETCS was not designed with this role in mind, but it at least gives a rough idea of the kind of foundation which could be designed, not for humans or computers, but for the interface between them.

The third project was similarly never intended for the role of a user interface, but it serves to illustrate what might be involved in building a bridge between the kind of ontology-agnostic approach to quantum mechanics advocated above and ontology-bound human users. It is the *topos foundation for theories of physics*, due to Döring and Isham, as summarised in Döring and Isham (2011). They begin by constructing a topos which is designed in such a way that its internal language appears to have a bearing on a particular physical system, coded as a Hilbert space. Within this topos, they find an analogue of the classical state space. Then they show that the Kochen-Specker theorem is equivalent to the assertion that there is no map from the terminal object to this *state object*. Bearing in mind that in the category of sets maps from the terminal object to a set correspond to elements of that set, this can be seen as saying that there are no elements, in the classical sense, of the state space.

Having thus made precise the impossibility of fitting quantum systems into a classical ontology, they carefully investigate to what extent our usual ways of speaking about the world can be recovered within this topos. They outline a formal language in which propositions about a system are represented by sub-objects of its state object. Then truth values can be assigned to propositions with the aid of a truth object. In this way many of the familiar ways of talking about physical systems can be given an interpretation. Much care is needed in setting up this framework in such a way that the claims of the formal language are not of the kind which the Kochen-Specker theorem would render meaningless.

The key benefit of this work is not that it makes quantum systems technically easier to work with (in fact it does the opposite). It is that it gives a method for taking our standard ways of talking about the world, built to work with a **clean ontology**, and fitting them, where possible, to the mathematics of quantum systems. Where this is not possible, it gives a precise account of this impossibility. If my earlier suggestion of ignoring issues of ontology in the foundations of quantum mechanics is to be followed, then some theory which can play these roles will be an essential part of any user interface.

None of these three examples comes close to showing what a full user interface between computer and human foundations for mathematics could look like; each just gives a tantalising glimpse of a part of the problem, together with a disconcerting insight into the difficulty of even these small parts of the task. The question of how to design a good interface to allow a human to seamlessly interact with a mathematical system which, under the hood, is based on a type theoretic foundation, is an important and neglected practical problem.

Acknowledgements Only the ideas in the last section of this paper are original. The ideas in the first two sections emerged from discussions with Thomas Forster and Adrian Mathias. In particular, I picked up the idea of **portability** as a key benefit of good foundations from Mathias and the idea that mathematical objects are implemented within foundations rather than found there from Forster. Unfortunately neither they nor I were able to find suitable citations for these ideas in their published work.

The **argument lifting** idea in the third section is completely standard, and can be found at least implicitly in much of the category theoretic literature. I picked it up mostly from Peter Johnstone, through his teaching and his book (Johnstone 2002).

I am grateful to all three of them for helpful, if conflicting, critical comments on an earlier version of this paper, as well as to the anonymous referee for explaining how I could make it more presentable to philosophers.

References

Abramsky, S., & Coecke, B. (2009). Categorical quantum mechanics. In K. Engesser, D. M. Gabbay, & D. Lehmann (Eds.), *Handbook of quantum logic and quantum structures* (pp. 261–323). Amsterdam: Elsevier.

Cosmides, L., & Tooby, J. (1992). Cognitive adaptations for social exchange. In: J. H. Barkow, L. Cosmides, & J. Tooby (Eds.), *The adapted mind: Evolutionary psychology and the generation of culture*. New York: Oxford University Press.

Dahn, B. I. (1998). Robbins algebras are boolean: A revision of McCune's computer-generated solution of Robbins problem. *Journal of Algebra, 208*(2), 526–532.

Dehaene, S. (1997). *The number sense*. London: Penguin.

Döring, A., & Isham, C. (2011). *"What is a thing?" Topos theory in the foundations of physics* (pp. 753–937). Berlin/Heidelberg: Springer.

Ganesalingam, M., & Gowers, W. T. (2017). A fully automatic theorem prover with human-style output. *Journal of Automated Reasoning, 58*(2), 253–291.

Heule, M. J. H., Kullmann, O., & Marek, V. W. (2016). Solving and verifying the boolean pythagorean triples problem via cube-and-conquer. In N. Creignou & D. Le Berre (Eds.), *Theory and applications of satisfiability testing – SAT 2016* (pp. 228–245). Cham: Springer International Publishing.

Johnstone, P. T. (2002). *Sketches of an elephant: A topos theory compendium*. Oxford logic guides. New York: Oxford University Press.

Kochen, S., & Specker, E. P. (1967). The problem of hidden variables in quantum mechanics. *Journal of Mathematics and Mechanics, 17*(1), 59–87.

MacLane, S., & Moerdijk, I. (1992). *Sheaves in geometry and logic: A first introduction to topos theory* (Universitext). Berlin: Springer.

Maddy, P. (2017). Set-theoretic foundations. In Caiceds et al. (Eds.), Foundations of Mathematics. Providence: American Mathematical Society.

McLarty, C. (2004). Exploring categorical structuralism. *Philosophia Mathematica, 12*(1), 37–53.

Petkovsek, M., Wilf, H., & Zeilberger, D. (1996). *A=B*. Wellesley: Peters.

Wason, P. C. (1977). Self-contradictions. In P. N. Johnson-Laird & P. C. Wason (Eds.), *Thinking: Readings in cognitive science*. Cambridge: Cambridge University Press.

Chapter 19
How to Frame a Mathematician

Modelling the Cognitive Background of Proofs

Bernhard Fisseni, Deniz Sarikaya, Martin Schmitt, and Bernhard Schröder

Abstract Frames are a concept in knowledge representation that explains how the receiver, using background information, completes the information conveyed by the sender. This concept is used in different disciplines, most notably in cognitive linguistics and artificial intelligence. This paper argues that frames can serve as the basis for describing mathematical proofs. The usefulness of the concept is illustrated by giving a partial formalisation of proof frames, specifically focusing on induction proofs, and relevant parts of the mathematical theory within which the proofs are conducted; for the latter, we look at natural numbers and trees specifically.

19.1 Introduction

Frames are a concept in knowledge representation that explains how the receiver, using background information, completes the information conveyed by the sender. This concept is used in different disciplines, most notably in cognitive linguistics and artificial intelligence. This paper argues that frames can serve as the basis for describing mathematical proofs. The usefulness of the concept is illustrated by

B. Fisseni
Faculty of the Humanities, University of Duisburg-Essen, Essen, Germany

Leibniz-Institut für Deutsche Sprache, Digitale Sprachwissenschaft, Mannheim, Germany

D. Sarikaya
Institute of Philosophy, University of Hamburg, Hamburg, Germany

M. Schmitt
Center for Information and Language Processing, Ludwig-Maximilians-Universität München, Munich, Germany

B. Schröder (✉)
Institute for German Studies, Linguistics, University of Duisburg-Essen, Essen, Germany
e-mail: bernhard.schroeder@uni-due.de

S. Centrone et al. (eds.), *Reflections on the Foundations of Mathematics*,
Synthese Library 407, https://doi.org/10.1007/978-3-030-15655-8_19

giving a partial formalisation of proof frames, specifically focusing on induction proofs, and relevant parts of the mathematical theory within which the proofs are conducted; for the latter, we look at natural numbers and trees specifically.

19.2 Motivation

Why would one want to use frames for modelling mathematical proofs? We have three arguments which also connect our concept to related work in different areas of formal mathematics and philosophy.

First, using frames is not as unheard of as it may seem: The schemata and tactics underlying semi-automatic provers such as Coq[1] und Isabelle[2] can be seen as frames. These schemata also contain patterns of proofs that are partially filled in by the mathematician conducting the proof, and are partially completed by heuristics implemented in the provers. For instance, in the excerpt from an interactive session with Coq presented in Fig. 19.1, Coq's tactic `induction` generates two subgoals

```
Coq < Lemma induction_test : forall n:nat, n = n -> n <= n.
1 subgoal

============================
forall n : nat, n = n -> n <= n


Coq < intros n H.
1 subgoal

n : nat
H : n = n
============================
n <= n


Coq < induction n.
2 subgoals

H : 0 = 0
============================
0 <= 0

subgoal 2 is:  S n <= S n
```

Fig. 19.1 Example run for an induction on natural numbers in Coq, taken from the Coq reference manual (Coq Development Team 2017, section 8.5.2)

[1]https://coq.inria.fr/

[2]https://isabelle.in.tum.de/

from a proof about natural numbers, which correspond to the *base case* and the *induction step* as discussed in the remainder of this article.

Secondly, for comprehending and checking proofs, mathematicians at least to some extent complete the informal proofs in mathematical texts using their expertise and the proof schemata they have acquired in their mathematical life. A common task for students in BA theses is to enrich a given published (expert) proof with information that completes the proof schemata and hence shows that they have comprehended the proof. Even the task of *finding* a proof appears to be closely related to frames. There are different proof techniques which might be triggered, for instance by the domain of discourse, as in the example of Turán's Theorem discussed at length in Sect. 19.6.5: Firstly, we see another typical technique besides induction, namely an *extremal argument*, secondly, we do not use standard induction but *strong induction*. This shows that both frames can be active at the same time and can even interact, as will be discussed later.

A typical remark made by experienced mathematicians is that beginners must present and be presented more explicit proofs, but that for experts a reduction in detail is even beneficial in optimising the effort needed to process a proof. This can be seen as support for the assumption that the proof schemata are cognitively 'real' and, at least in the beginning, explicitly acquired.

Thirdly, in the philosophy of mathematics, it is also discussed how to deal with gaps. Azzouni for instance introduces the *derivation-indicator view* (see Azzouni 2004, 2009), arguing that while in the daily life a mathematician deals with proofs, those proofs are actually indicative of underlying derivations. Carl and Koepke (2010) argue that this view might be deeply linked to our first motivating point. There is quite an extensive literature discussing the connection between informal and formal proofs and especially debating Azzouni's views, including Rav (2007) and Tanswell (2015) or in this Volume Carl: (Chap. 14).

With frames we present a semantic abstraction layer which is situated between the textual layer and a full formal derivation. This layer is more on the cognitive side compared to Proof Representation Structures (Cramer et al. 2009; Cramer 2013) and can provide strategies to fill gaps on the way from the text towards a formal representation. We surmise that these frames represent what mathematicians use for processing proofs, but they might also be usable for guiding proof assistants or provers. Our main argument in this article is that this level of representation can help to explain why only some aspects of proofs are made explicit in actual natural language proofs, and others can be left implicit, supplied by the implicit knowledge about the proof structure. In the terminology of frames, we say that the slots of these frames can be filled with contextually available or inferable information. For instance, the induction variable must be identified and a base case must be supplied, even if it is 'evidently trivial' and therefore need not be explicitly stated. This view can also provide an account of similarity between proofs: If the frame instances one constructs during the reading of a proof are sufficiently similar (e.g., unifiable, ignoring nasty details such as variable names), the proofs can be considered the same, and will ultimately lead to the same derivation as well. Thus text can be seen as consisting of frame indicators, and frames as derivation indicators.

On the way from a textual representation to a frame representation, two kinds of completing information can be distinguished: On the one hand, slots provided by frames must be filled with adequate values, such as identifying the induction variable. On the other hand, we also must fill gaps in the proof, as in the case of the omitted induction base case.

In a way, this links to half-automated provers, which also need less information than fully formalised proof checkers. However, while in the former the restrictions are imposed by the power of 'tactical' components and run-time considerations, we consider the human mathematician to be the measure of explicitness for textbook proofs, and provers to be the measure of explicitness for formal proofs.

19.3 Frames

In this section, we first give a high-level overview of the frame concept and will then illustrate the structure of frames below referring to FrameNet's semantic annotation of verbs.

The concept of frame in artificial intelligence is generally traced back to a report by Minsky (1974) which introduces frames as a general "data-structure for representing a stereotyped situation, like being in a certain kind of living room, or going to a child's birthday party" (p. 1), organised in a hierarchy (called *frame-system* by Minsky). 'Situation' should not be taken too seriously, as frames can be used to model concepts in the widest sense, and Minsky illustrates the use of frames with a multitude of examples from vision to story understanding.[3] Each frame is characterised by a set of features, which can be (sub-)frames again. Features, also called *slots*, can have default values and can carry constraints; *slots* are *filled* by concrete values. The concept has later been elaborated to formal systems such as the Frame Representation Language (FRL, cf. Roberts and Goldstein 1977a,b), which, among others, allows to attach a procedure to a slot that will calculate the value of a feature from other values when needed. In the world of mathematics, we can imagine that when talking about a circle in Euclidean geometry, we need the centre and the radius to describe the circle; we can calculate its diameter, if needed.[4] This helps reduce redundancy of the representation. Also, FRL allows to define that values are constrained, so that we could specify that radii must be positive, or buyer

[3]Schank and Abelson (1977) develop the related concept of scripts, which adds a temporal dimension. Though we do not model temporal progression explicitly, our use of frames also resembles scripts, if one reads the constituents of a proof frame as a plan for linear text organisation.

[4]This shows a certain kind of freedom we have in defining frames; we could also define the circle based on the centre and the diameter, of course.

and seller be different.[5] A default assumption connected to the mathematical frame *function* might be that the limit of a function is called 'ε' and that it is small but greater than zero. Conversely, one can assume that the name ε within the area of analysis denotes a limit with the respective properties, and the use of ε contributes to evoking the greater frame of limiting. Giving a different name to a limit or a different meaning to ε would however, strictly speaking, not be excluded.

An important linguistic theory using frames is FrameNet.[6] A development of Fillmore's (1968) Case Grammar, FrameNet develops a hierarchical model of (mainly) verb semantics where the features are the participants and their semantic roles. Verbs evoke frames which then provide certain roles that can be taken on by entities in the discourse universe. Roles are generally divided into core and non-core roles; this division captures the salience and optionality of the roles within a frame.[7] For example, *Commerce_buy* has the core roles *Buyer* and *Goods*, while *Seller*, *Money* and *Means* (e.g., cash vs. check) and many others are non-core roles.

Hyponym verbs either inherit roles from their parents or provide *perspectives* on a more general frame by making roles more or less optional. For instance, *Commerce_goods-transfer* inherits from *Transfer* and thus has more specific roles (e.g., *Buyer* instead of *Donor*), but also new ones (*Money*, i.e., the price). *Commerce_buy* and *Commerce_sell* are perspectives on *Commerce_goods-transfer*, but differ in that the role of *Seller* is not a core role for *buy*. Some of the frames are lexical, e.g., the ones tied to *buy* and *sell*; the more general ones are often non-lexical.

The differences in role assignment thus capture semantic (and pragmatic) differences between verbs. Consider a simple example. Explicitly realised frame elements are annotated in the sentence:

(1) a. [$_{\text{Buyer}}$ John] bought [$_{\text{Goods}}$ a beautiful medieval book] [$_{\text{Time}}$ yesterday].
b. [$_{\text{Seller}}$ Peter] sold [$_{\text{Goods}}$ a beautiful medieval book] to [$_{\text{Buyer}}$ John] for [$_{\text{Money}}$ twenty Euros].

This can be represented in a feature-value matrix, which also illustrates subframes (in the TIME field). The exclamation mark indicates core roles, and $[\![\mathit{expression}]\!]$ is the semantics of the *expression*. We use *point-in-time*, *person*, *money*, *purpose* as labels (for types) that constrain the potential fillers. To spare the reader the enumeration of slots that have not been realised explicitly, we abbreviate the frame, indicating this with the ellipsis dots. From now on, we will always assume that a description in this way is generally partial and ellipsis dots are not needed.

[5]Our feature structures will employ subtyping and inheritance for this purpose.

[6]See, e.g., Ruppenhofer et al. (2006) and the project's website at https://framenet.icsi.berkeley.edu

[7]Even core roles can be omitted sometimes, as in *John finally sold his car*.

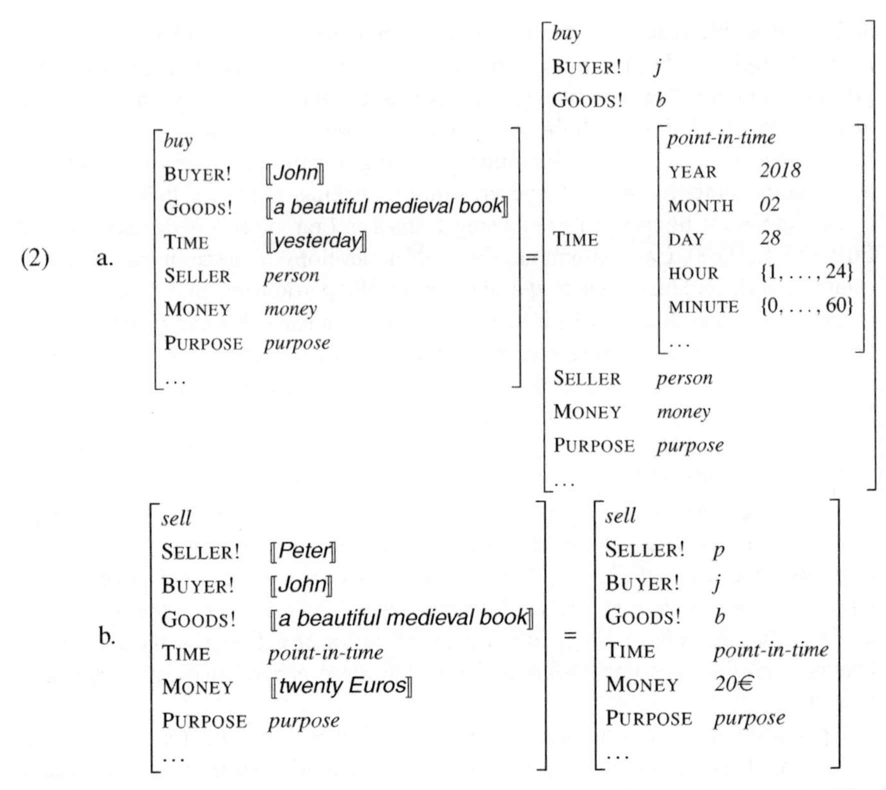

While both sentences evoke the similar frames, the *Seller* slot need not be filled in the first case. However, it is present and could be filled, just as the *Time* slot specifying when the transfer occurs is present (but usually non-core) with most verbs, and can be filled with more or less specific values.

When considering mathematical texts, the ideal of a fully formal proof does not easily allow for non-core roles remaining unexpressed without having default values. This has to do with a key difference between the scope of our and FrameNet's approach: While FrameNet is interested in the grammatical realisation of single sentences, we are concerned with the semantic modelling of a complete proof. Moreover, the specific pragmatics of proofs in mathematics demand that a certain semantic completeness of understanding can be achieved when working through it. An example of non-core constituents of a proof may therefore be metacomments about the proof strategy, which we will leave aside in this paper.

The important points about frames we will use in the following are: Frames are used to define concept hierarchies, where each concept has features (slots) which can have default values and can be complex (sub-frames). When instantiating frames, their values can be given explicitly, they can be supplied by default values (or relations), and the filling of slots can be guided by constraints from mathematical convention, or from the context of the proof.

In a full cognitive system, we also need rules to relate the frames and to supply default conclusions or inferences and we need belief revision. To some extent, these can also be represented as frames, but other formalisations may be more elegant; nothing hinges on this assumption for the rest of the article.

19.4 Frames as Feature Structures

Frame representations can be seen as feature structure descriptions in the sense of Carpenter (1992), and this is the formalism we will be using later to model our frames as well.

The formalism assumes that a feature structure is a labelled graph whose labels indicate *features* (i.e., from one node, a label can occur only once), and whose terminal nodes and whose subgraphs respectively represent *values*. Those values are atomic in the case of terminal nodes and in the case of subgraphs they are complex, i.e., they have features and values themselves. Nodes are assumed to be typed and the types are assumed to be organised hierarchically. Features are introduced at a certain level in the hierarchy and are inherited by descendants.

These graphs are described by feature-value matrices as introduced in the preceding section. In the feature structure descriptions, the boxes indicate structure sharing; where the shared structures can sensibly be represented by mathematical variables, we abuse notation by boxing letters etc. An example of this is given with $\boxed{s}$ and $\boxed{x}$ in Fig. 19.4 presented in the next section.

19.5 Induction as a Frame

Induction is a mathematical proof technique that can be used to prove statements about a set of inductively definable structures. A formal induction proof consists of at least two parts: in the so-called *base case*, the statement is proved for a least element of the set (which shall be called M for brevity; it must be partially ordered and have at least one minimum); in the so-called *induction step*, an implication is proved: under the assumption that the statement holds for some $x \in M$, it is proved for a value building on x, e.g., the successor of x. The assumption is called *induction hypothesis*. The term *induction variable* denotes the identifier used for the elements of M throughout the proof (i.e., x above). Although most induction proofs only have one base case and one induction step, in general, it is possible to have more of each. There are virtually countless variations. It might be possible to have two induction steps, for even and uneven numbers, we might take a step from n to $2n$ and $2n+1$ as often used in computational complexity, or even without a base case. The so called *strong* induction (sometimes also called *weak* induction) shows that a statement for k is true if it holds (i.e., can be shown to be true) for all smaller cases. Since the empty universal quantifier is true, this gives us the base case immediately.

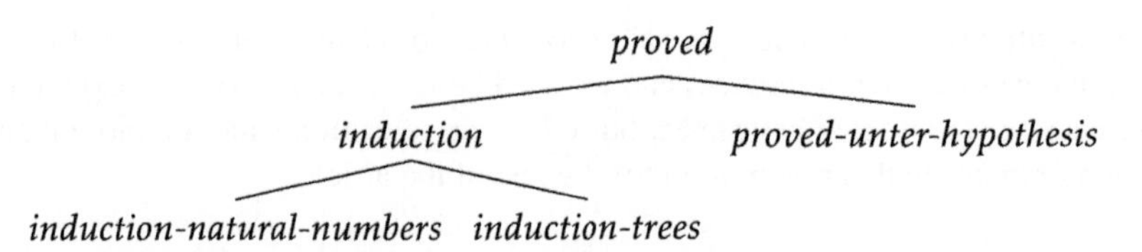

Fig. 19.2 Inheritance hierarchy of frames in the *proved* family

$$\begin{bmatrix} \textsc{Assertion} & \textit{sentence} \\ \textsc{Proof} & \textit{list(proof-step} \vee \textit{assumption} \vee \textit{definition} \vee \textit{goal)} \end{bmatrix}$$

Fig. 19.3 Proved frame

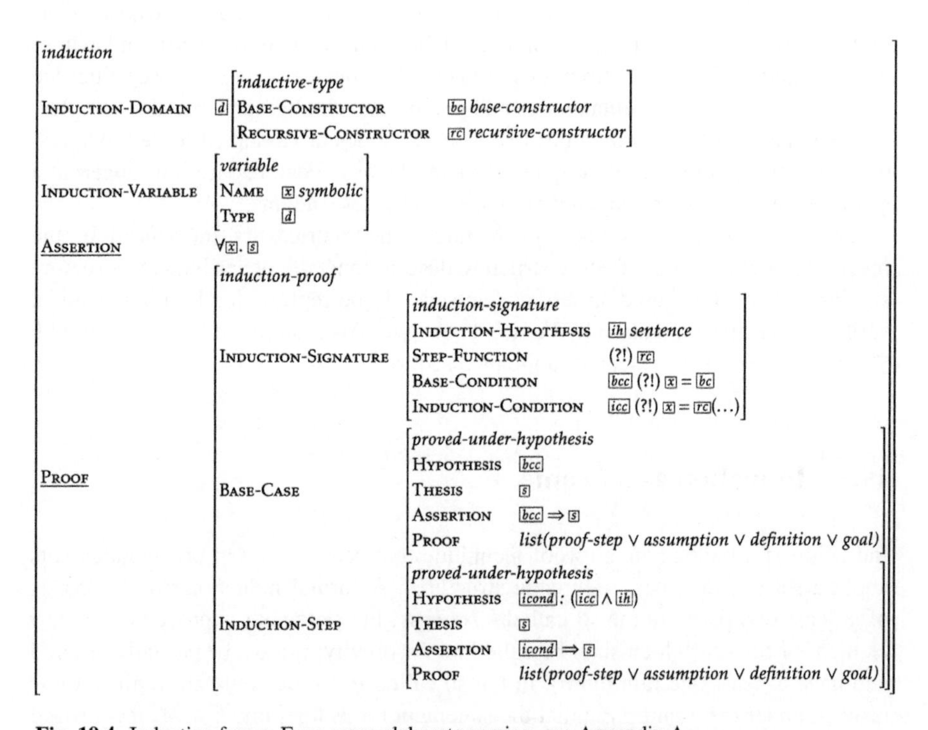

Fig. 19.4 Induction frame. For a more elaborate version, see Appendix A

We annotated a collection of proofs from different areas of mathematics, which the examples in this paper are taken from. Based on the parts of a formal induction proof and on our annotation experience, we propose the following frame structure as a formalisation of relevant aspects of the mental representation of a natural language proof by induction as shown in Figs. 19.2–19.4.

As indicated in Sects. 19.3 and 19.4, we assume that the structures can be organised in a hierarchy, where more specific structures are derived from more general structures and inherit some of their features, as shown in Fig. 19.2.

First, we define a frame for a general proof. This structure – called *proved* – consists of an assertion of the to-be-proved statement and the actual proof (cf. Fig. 19.3).

An *induction* is a special type of proof. Therefore, the induction frame is also a *proved* frame and thus inherits the features ASSERTION and PROOF (underlined in Fig. 19.4). On the same level, we introduce the new feature INDUCTION-VARIABLE, which captures the induction variable as described above in this section.

The particular structure of induction proofs is captured by the PROOF feature as follows. Both the BASE-CASE and the INDUCTION-STEP contain a condition which only holds during the respective part of the proof. These conditions constrain the induction variable to be the least element of M in the base case and to be greater than the least element during the induction step. These conditions are rarely stated explicitly but often play an integral role in the proof.

The BASE-CASE and INDUCTION-STEP features constitute the actual proof as they contain an instance of the *proved* frame. Additionally, the STEP-FUNCTION feature captures the specific successor function of the induction step.[8]

Finally, the feature INDUCTION-HYPOTHESIS references the special assumption described in the beginning of this section.

The following types or, equivalently, constraints are used in Fig. 19.4: *sentence* is the type of sentential semantic object, which can be either a natural language *expression* or a mathematical *formula*. *function* is the type of a semantic object for a mathematical function. *symbolic* is not a semantic, but a syntactic, non-sentential object such as a variable name. Furthermore, we defined *proved* and *induction* as types for the respective frames described in this section.

19.5.1 Constraints

Our feature structure formalism allows us to formulate constraints and thus capture additional aspects of knowledge about induction proofs that often remain implicit. An example is the value of the ASSERTION feature in a INDUCTION-STEP: It is normally a more specific version of the toplevel ASSERTION value of the induction proof. We describe this connection in Fig. 19.4.

19.5.2 Default and Inferred Values

In the case of natural numbers being the subject of an induction proof, some of the features have default values. Table 19.1 gives a summary.

[8] If there is more than one induction step, each of them can have a different successor function. As several base cases and induction steps are possible, the BASE-CASES and INDUCTION-STEPS features should be considered list-valued in principle, even if we use a simplified singleton notation (with features BASE-CASE and INDUCTION-STEP, but see Appendix A for a more complete version of the frame) in most examples.

Table 19.1 Default values for the case of natural numbers. s is the sentence that is to be proved; as explained in the text,"(?!) *value*" indicates default values; we used "min(ℕ)" because the natural numbers can start with 0 or 1

INDUCTION-VARIABLE	(?!) $\begin{bmatrix} \text{NAME} & n \\ \text{TYPE} & \mathbb{N} \end{bmatrix}$
BASE-CONDITION	(?!) $n = \min(\mathbb{N})$
STEP-FUNCTION	(?!) $\lambda n.\, n + 1$
INDUCTION-CONDITION	(?!) $n = n' + 1$
INDUCTION-HYPOTHESIS	(?!) $s[n'/n]$

Fig. 19.5 A term and a graph describing the same binary tree

```
(Node 2
(Node 1 EmptyTree EmptyTree)
(Node 3 EmptyTree EmptyTree))
```

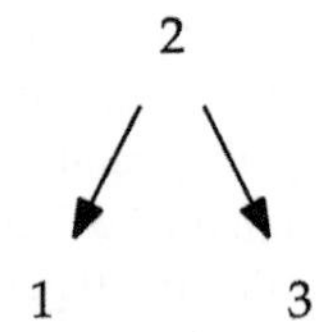

Table 19.2 Default values for the algebraic data type Tree as defined in the text

INDUCTION-VARIABLE	(?!) $\begin{bmatrix} \text{NAME} & t \\ \text{TYPE} & \textit{Tree} \end{bmatrix}$
BASE-CONDITION	(?!) $t =$ EmptyTree
STEP-FUNCTION	(?!) Node
INDUCTION-CONDITION	(?!) $t =$ Node $x\ t_1\ t_2$
INDUCTION-HYPOTHESIS	(?!) $s[t_1/t] \wedge s[t_2/t]$

In Sect. 19.3 we already mentioned that besides default values and explicitly given values, we can also have values that are calculated and more generally: inferred. We do not specifically mark explicitly given values, but indicate "(?!) default values" and "(⇝) inferred values" as illustrated in this sentence.

In general, any inductively defined structure naturally induces default values for a proof by induction. Consider, e.g., the following definition of an algebraic data type for binary trees in the functional programming language Haskell (see Marlow 2010):

```
data Tree a = EmptyTree
| Node a (Tree a) (Tree a)
```

This piece of code defines a term representation for binary trees. An example is shown in Fig. 19.5. An overview of the default values that naturally arise from this definition is given in Table 19.2.

19.6 Example Proofs

In this section we apply the induction frame presented above and show how mathematicians implicitly use frames in constructing or understanding proof texts. We start with a relatively straightforward representation of an induction proof and then continue with a bare-bones proof where knowledge of frame elements is needed to complete the gaps and make sense of the overall structure.

19.6.1 A Proposition from Linear Algebra

We will show how the frame concept can be applied to a proof text and in how far semantic text explication which relates textual information to parts of the induction frame is needed for the adequate interpretation of the text.

Let us consider the proof of proposition 4.4.6 in Kowalski (2016, 92f). We received two different versions of the same proof by asking a master student from physics and a PhD student from a logic-related programme, to reformulate the proof in order to explain the central proof ideas to first semester students. They wrote their version independently of each other.

19.6.2 First Reformulation

The first version is shown in (3). It addresses only the induction part of the proof in Kowalski (2016, 92f) which is not worked out there.[9]

(3) a. Before we start with the proof, we show that the induction scheme is easy to generalize. In the classical form the induction step requires that one derive a statement for $n + 1$ out of a statement depending on n.

$$n \to n + 1 \tag{3.1}$$

But there we throw away the fact that the logic of induction is a chain of proved statements, hence consists of all statements before n. Thus, we can use the modified version of the induction step:

$$\forall i \leq n \to n + 1 \tag{3.2}$$

b. The statement $t_1 = 0$ was already shown.

c. The remaining step would be to show $t_1 = \cdots = t_i = 0 \implies t_{i+1} = 0$ Therefore we apply $f^{k-(i+1)}$ on

[9]In this and in the following proofs the subdivision by small Latin letters has been added by the authors of this paper. References to equations have been renumbered according to the scheme used in the rest of this article.

$$t_1 v + \cdots + t_{i+1} f^i(v) + \cdots + t_{k-1} f^{k-2}(v) = 0, \quad (3.3)$$

which leads to

$$t_1 f^{k-(i+1)}(v) + \cdots + t_{i+1} f^{k-1}(v) = 0 \quad (3.4)$$

higher order terms drop out. Finally $t_{i+1} = 0$ follows directly from eq. 4 when we use $t_1 = \ldots t_i = 0$ and $f^{k-1}(v) \neq 0$.

The proposition in Kowalski (2016, 92f) the proof in (3) refers to is:

(4) PROPOSITION 4.4.6. Let V be a finite-dimensional **K**-vector space and let f be a nilpotent endomorphism of V. Let $n = \dim(V)$. Then $f^n = 0$. More precisely, for any vector $v \neq 0$ in V, and $k \geq 0$ such that $f^k(v) = 0$ but $f^{k-1}(v) \neq 0$,[10] the vectors

$$(v, f(v), \ldots, f^{k-1}(v))$$

are linearly independent.

(3) proves the linear independence of $(v, f(v), \ldots, f^{k-1}(v))$ by actually showing that $t_1 = \ldots = t_k = 0$ follows from $t_1 v + t_2 f(v) + \ldots + t_k f^{k-1}(v) = 0$. The proof fragment builds on the assumption that $f_k = 0$ and $f^{k-1} \neq 0$ and on the proof for the base case $t_1 = 0$ in Kowalski (2016, 93).

The proof text consists of an introducing comment (3-a) and the derivation argument itself comprising (3-b) and (3-c). In (3-b) and (3-c), there is no explicit mention of the induction scheme or of one of the slots of the induction frame (cf. Fig. 19.4). We can, however, find some indicators of an induction scheme, especially the proof goal $t_1 = \cdots = t_i = 0 \implies t_{i+1} = 0$, which is reminiscent of the step from an induction hypothesis to the thesis of the induction step, although being non-typical. The expert reader, of course, will expect that the universal claim ("*for any vector...*") will require an inductive proof.

The induction frame is referenced in the comment part (3-a). This part has two main functions: It sets the frame, and it modifies the standard scheme of the induction step, i.e., it overrides a default setting of the relation between the INDUCTION-HYPOTHESIS and the THESIS of the STEP-STATEMENT in instantiating a strong induction i.e., here an inference from the *conjunction of all* instances of a predicate $P[j]$ for $1 \leq j \leq i$, and not just from P[i] itself, to $P[i+1]$. Equipped with this background, the reader can (hypothetically) identify the whole second part of the text (3-b) and (3-c) as an induction with $t_1 = \cdots = t_i = 0$ as INDUCTION-HYPOTHESIS and $t_{i+1} = 0$ as the THESIS of the INDUCTION-STEP. The ASSERTION of the whole frame, which is not stated explicitly, can be reconstructed as $\forall i (t_i = 0)$. From that the INDUCTION-VARIABLE has to be inferred as i with type $\mathbb{N}$. The STEP-FUNCTION is the default one. The BASE-CONSTRUCTOR (and subsequently also the BASE-CASE and the INDUCTION-CASE) chooses its value from the respective default alternative.

[10]The original text reads: $f^{k-1}(v) = 0$. The misprint was corrected here in (4).

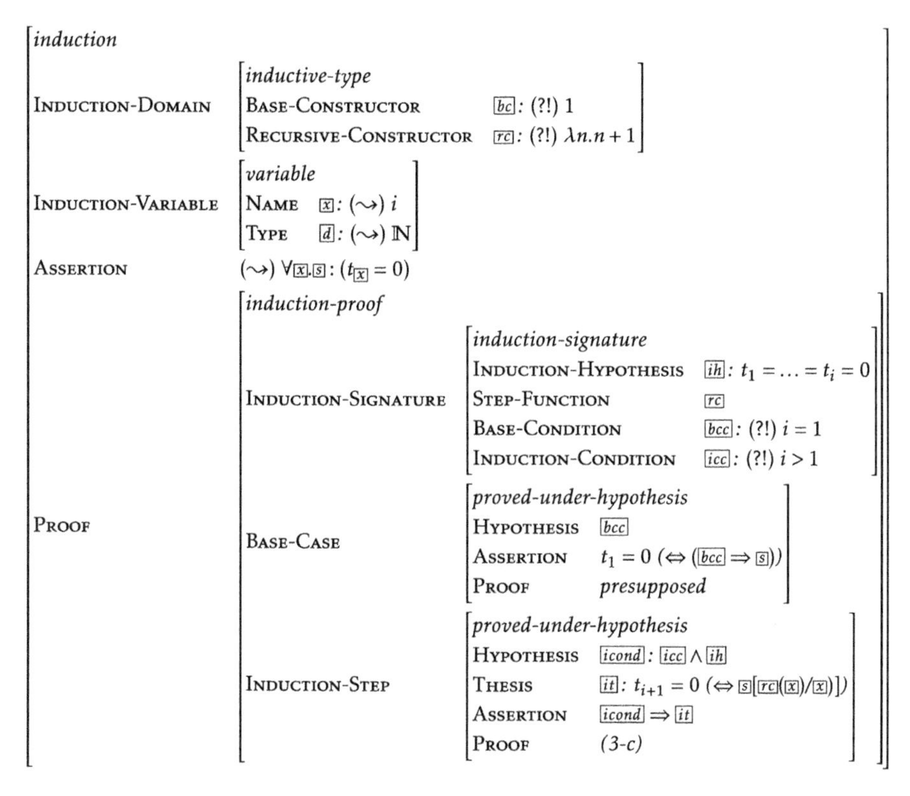

Fig. 19.6 Induction frame as instantiated by (3), (?!) ... are default values, and (⇝) ... are inferred values, other values are given in the text, but not necessarily marked as slot values

(3-b) points then to $t_1 = 0$ as the ASSERTION of the BASE-CASE. The BASE-CASE slot of the INDUCTION-SIGNATURE has therefore to be filled with $i = 1$. That confirms that the BASE-CONSTRUCTOR (and subsequently also the BASE-CONDITION and the INDUCTION-CONDITION) chooses its value from the respective default alternative. Figure 19.6 sums up how the frame slots are instantiated by the proof in (3).

This short sketch illustrates that in order to capture the whole argument of (3-a) a reader has to fill in missing slots of the induction frame by some inferences from the information given in the proof text, i.e., the reader interprets the text under the hypothesis that the relevant information for filling the frame slots is provided in the text.

19.6.3 Second Reformulation

Let us now consider a second proof by a PhD student in a logic-related programme of Philosophy who finished a Master's degree in Mathematics.

(5) a. ***Prop. (4.4.6)*** *Let V be a finite-dimensional* $\mathbf{K}$*-vector space and let* f *be a nilpotent endomorphism of V. Let* $n = \dim(V)$. *Then* $f^n = 0$. *More precisely, for any vector* $v \neq 0$ *in* V, *and* $k \geq 0$ *such that* $f^k(v) = 0$ *but* $f^{k-1}(v) \neq 0$, *the vectors*

$$(v, f(v), \dots, f^{k-1}(v))$$

are linearly independent.

b. **Proof** Let V be a finite-dimensional $\mathbf{K}$-vector space and let f be a nilpotent endomorphism of V. Let $n = \dim(V)$.

c. If it is shown that for any vector $v \neq 0$ in V and $k \geq 0$ such that $f^k(v) = 0$ and $f^{k-1}(v) \neq 0$ the vectors $(v, f(v), \dots, f^{k-1}(v))$ are linearly independent, then $k \leq n$, because $(v, f(v), \dots, f^{k-1}(v))$ are k linearly independent vectors and there are at most n linearly independent vectors in V. Since f is nilpotent for any $v \in V$ there is a k such that $f^k(v) = 0$. Then $f^m(v) = 0$ for any $m \geq k$. So, $f^n(v) = 0$ for all $v \in V$, because $k \leq n$, which implies $f^n = 0$. Thus, the second claim of the proposition implies the first.

d. We now show the second claim. Let $v \neq 0$ be a vector in V and $k \geq 0$ such that $f^k = 0$ and $f^{k-1} \neq 0$. To show that $(v, f(v), \dots, f^{k-1}(v)$ are linearly independent, let $t_0, t_1, \dots, t_{k-1}$ be elements of K such that

$$t_0 v + t_1 f(v) + \dots + t_{k-1} f^{k-1}(v) = 0. \quad (5.5)$$

We will show by induction on $i \leq k-1$ that $t_i = 0$ for all $i \leq k-1$, which shows that $(v, f(v), \dots, f^{k-1}(v))$ are linearly independent.

e. **Base Case:** We consider t_0, and show $t_0 = 0$. We apply f^{k-1} to equation (5.5):

$$f^{k-1}(t_0 v + t_1 f(v) + \dots + t_{k-1} f^{k-1}(v)) = f^{k-1}(0)$$

$$t_0 f^{k-1}(v) + t_1 f^k(v) + \dots + t_{k-1} f^{2k-2}(v) = 0 \quad \text{(Linearity)}$$

$$t_0 f^{k-1}(v) = 0 \quad (f^k(v), f^{k+1}(v), \dots, f^{2k-2}(v) \text{ are } 0)$$

$$t_0 = 0 \quad (f^{k-1}(v) \neq 0).$$

f. **Induction Step:** Now assume that $t_0, \dots t_i = 0$ for an arbitrary, fixed $i < k-1$. We show that $t_{i+1} = 0$. We substitute $t_0 = 0, \dots t_i = 0$ in (5.5) and apply f^{k-i-2}:

$$f^{k-i-2}(t_{i+1} f^{i+1}(v) + \dots + t_{k-1} f^{k-1}(v)) = f^{k-i-2}(0)$$

$$t_{i+1} f^{k-i-2+i+1}(v) + \dots + t_{k-1} f^{k-i-2+k-1}(v) = 0 \quad \text{(Linearity)}$$

$$t_{i+1} f^{k-1}(v) + \dots + t_{k-1} f^{2k-i-3}(v) = 0$$

$$t_{i+1} f^{k-1}(v) = 0 \qquad (f^k(v), f^{k+1}(v), \ldots, f^{2k-i-3}(v) \text{ are } 0)$$

$$t_{i+1} = 0 \qquad (f^{k-1}(v) \neq 0).$$

This ends the induction and the proof.

In contrast to the first version, one sees that this version contains more details and is a little longer. The induction scheme is mentioned and the slots are not only mentioned but even marked by the usage of bold print. The proof text consists of an introduction of the given facts (5-b). Note the indirect discourse we can find typically in mathematical texts. The author then shows in (5-c) that the second part of the proposition is indeed a strengthening of the first.

In (5-d) she states the method of the proof and makes the induction variable, i.e., i explicit, she also unravels the definition of *linear independent*. In (5-e) we can follow the base case. We note the typical explication of proofsteps in the right column on longer parts of calculations. Similarly to this the induction step is explained in (5-f). We note that the induction assumption is not stated explicitly as an induction assumption. Asking why the author made the decision, she said that adding the "*Now assume that*" was sufficient for her to make explicit that an assumption is used here.

This amplification of the original proof illustrates that the frames are (in some sense) cognitively real. While we can assume that the author of this second reformulation was able to comprehend the original proof before reformulating it, she chose to make the frame and the slots explicit, without any explicit knowledge of our frame model as presented above. We asked her to make the proof comprehensible for a first semester student. We assume that her changes are caused by the assumption that first year students need the explicit information.

19.6.4 The Original Proof

The original proof in Kowalski (2016, 93) – cited here as (6) – is even more reduced. It just mentions the induction frame in (6-c) after giving an example for the step from the base case to the immediate successor in (6-b). That i is to be the INDUCTION-VARIABLE with $i = 1$ as the BASE-CONDITION can be inferred from the step from t_1 to t_2 in (6-b) and the universal quantification in (6-c). Only the ASSERTION is stated explicitly in (6-c). The whole proof as specified in (3-c) has to be reconstructed from the hint that in the step from $i = 1$ to $i = 2$ f^{k-2} has to be applied instead of f^{k-1}, which leads to the general term f^{k-i+1} for the step from i to $i + 1$ (Fig. 19.7).

(6) a. Proof. First, the second statement is indeed more precise than the first: let $k \geq 1$ be such that $f^k = 0$ but $f^{k-1} \neq 0$; there exists $v \neq 0$ such that $f^{k-1}(v) \neq 0$, and we obtain $k \leq n$ by applying the second result to this vector v. We now prove the second claim. Assume therefore that $v \neq 0$ and

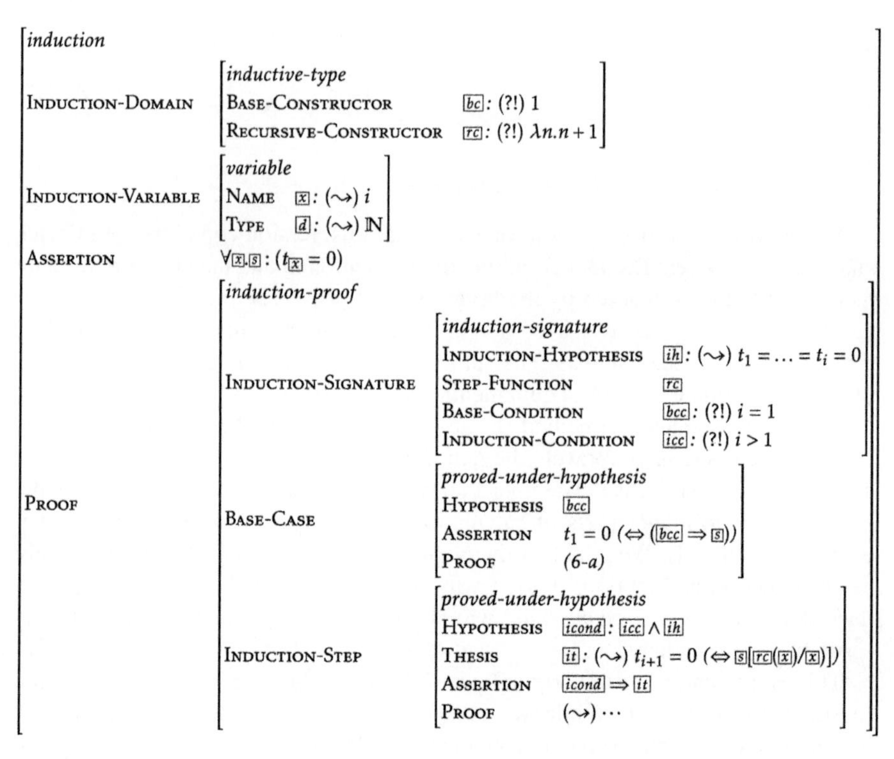

Fig. 19.7 Induction frame as instantiated by (6), default and inferred values are marked as in Fig. 19.6

that $f^k(v) = 0$ but $f^{k-1}(v) \neq 0$. Let $t_0, \ldots, t_{k-1}$ be elements of **K** such that

$$t_1 v + \cdots + t_{k-1} f^{k-1[sic!]}(v) = 0.$$

Apply f^{k-1} to this relation; since $f^k(v) = \ldots = f^{2k-2[sic!]}(v) = 0$, we get

$$t_1 f^{k-1}(v) = t_1 f^{k-1}(v) + t_2 f^k(v) + \cdots + t_{k-1} f^{2k-2[sic!]}(v) = 0,$$

and therefore $t_1 f^{k-1}(v) = 0$. Since $f^{k-1}(v)$ was assumed to be non-zero, it follows that $t_1 = 0$.

b. Now repeating this argument, but applying f^{k-2} to the linear relation (and using the fact that $t_1 = 0$), we get $t_2 = 0$.

c. Then similarly we derive by induction that $t_i = 0$ for all i, proving the linear independence stated.

Kowalski (2016, 93); in the first equation, the exponent $k - 1$ has to be replaced by $k - 2$; in the line below and the second equation, $2k - 2$ by $2k - 3$.

19.6.5 Turán's Theorem

So far we have shown how the concept of frames can be applied to rather basic proofs, now we turn to a more advanced example. The proof (7), taken from Diestel (2006, 165f), has a transparent structure, but superficially does not follow the induction frame in a linear order. It is therefore well-suited to explain how frame slots are filled in reading the text, and it also contains some gaps to be filled with knowledge of the frame.

(7) **Theorem 7.1.1.** (Turán 1941)
For all integers r, n with $r > 1$, every graph $G \not\supseteq K^r$ with n vertices and $\mathrm{ex}(n, K^r)$ *edges*[11] *is a $T^{r-1}(n)$.*
Proof.

a. We apply induction on n. For $n \leq r-1$ we have $G = K^n = T^{r-1}(n)$ as claimed. For the induction step, let now $n \geq r$.
Since G is edge-maximal without a K^r subgraph, G has a subgraph $K = K^{r-1}$.

b. By the induction hypothesis, $G - K$ has at most $t_{r-1}(n-r+1)$ edges, and each vertex of $G - K$ has at most $r-2$ neighbours in K.

c. Hence

$$\|G\| \leq t_{r-1}(n-r+1) + (n-r+1)(r-2) + \binom{r-1}{2} = t_{r-1}(n); \qquad (7.6)$$

the equality on the right follows by inspection of the Turán graph $T^{r-1}(n)$ (Fig. (7-c)).

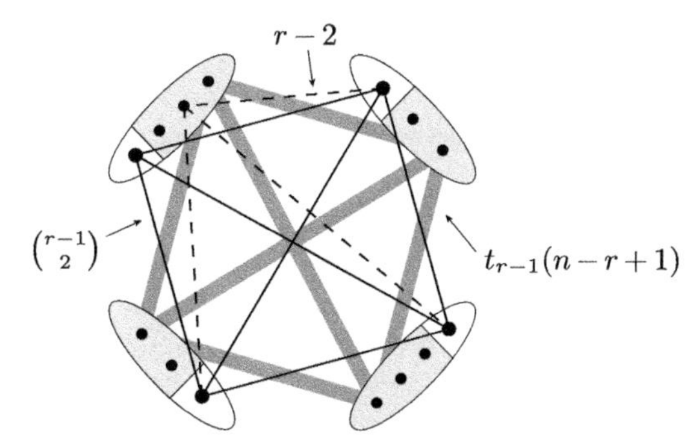

Fig. 7.1.3 The equation from (1) for $r = 5$ and $n = 14$

Since G is extremal for K^r (and $T^{r-1}(n) \not\supseteq K^r$), we have equality in (7.6). Thus, every vertex of $G - K$ has exactly $r-2$ neighbours in K – just like the vertices $x_1, \ldots, x_{r-1}$ of K itself. For $i = 1, \ldots, r-1$ let

$$V_i := \{v \in V(G) \mid vx_i \notin E(G)\}$$

[11] $\mathrm{ex}(n, H)$ means the number of edges of a graph $G \not\supseteq H$ that is extremal for n and H, i.e., the maximal amount of edges a graph on n vertices could have, without having a subgraph (isomorphic to) H.

> be the set of all vertices of G whose $r - 2$ neighbours in K are precisely the vertices other than x_i. Since $K^r \nsubseteq G$, each of the sets V_i is independent, and they partition $V(G)$. Hence, G is $(r-1)$-partite. As $T^{r-1}(n)$ is the unique $(r-1)$-partite graph with n vertices and the maximum number of edges, our claim that $G = T^{r-1}(n)$ follows from the assumed extremality of G. □

Diestel (2006, 165f)

The proof is about graphs, hence it is in principle possible to use induction on the number of vertices, i.e., induction on the natural numbers. In Graph Theory it is mostly common to use strong induction, since an operation on the vertex set of the graph might remove more than one vertex. Hence we need the induction hypothesis for all smaller numbers of vertices. Here the induction step for graphs G with n vertices presupposes by the induction hypothesis (7-b) that the theorem has been proved for graphs $G - K^{r-1}$ with $n - r + 1$ vertices.

Beside this we can see a second proof technique in the proof of Turán's Theorem, namely an *extremal argument*. Turán's Theorem is actually often considered as the starting point of a whole new subfield of *extremal combinatorics* or *extremal graph theory*. This technique reoccurs in different proofs in a manner that the mathematical community decided that it constitutes this field. We can only scratch the surface of the formalisation of this frame here, but we would argue that it corresponds to another frame. There needs something to be extremal to constitute an extremal argument. *Extremal* can mean *minimal* or *maximal*. In this case we have considered a maximal structure without any further specific properties. We can consider different graph invariants. In this case it was the number of edges, but we can in principle take any invariant, like the lengths of the shortest cycle in the graph, i.e., its girth. To summarise, one slot of the frame is filled by the extremal structure, another by the kind of extremality (*minimal* or *maximal*), and a third slot would contain an equivalence relation on the respective structures (e.g., the number of edges). An in-depth analysis of different extremal arguments would of course yield to more features of the frame.

We can infer even more from this passage. Knowing that this is one of the first results in *extremal combinatorics*, it offers the possibility to consider maximal counter examples. Actually, within one field proofs are often designed closely based on few basic *techniques*. A list of those techniques might be found for instance in a handbook article such as Bollobás (1995) for extremal combinatorics. We surmise that these techniques constitute a frame hierarchy, every technique a frame of its own, and proofs are generally primarily designed as interactions of these frames, similar to what we have shown in the discussion of Turán's proof in this section.

19.7 Conclusion

In Sect. 19.6 we have given examples of four different proof texts, two textbook proofs and two reformulations for one of them. All of them show that frame knowledge is needed to reconstruct the derivation from the text and that such knowledge is applied when elaborating a proof, respectively. We treat the examples in a way suggesting that the relevant inferences are driven by expectations generated by the slot structure of the frame and default values. In this respect mathematical texts and their interpretation do not seem to differ much from other sorts of descriptive texts.

A The Full Induction Frame

There are strong relationships between induction and case distinctions. The frame in Fig. 19.8 gives an idea how this can be described in our approach. It also accommodates the case that the induction domain has multiple base constructors and recursive constructors.

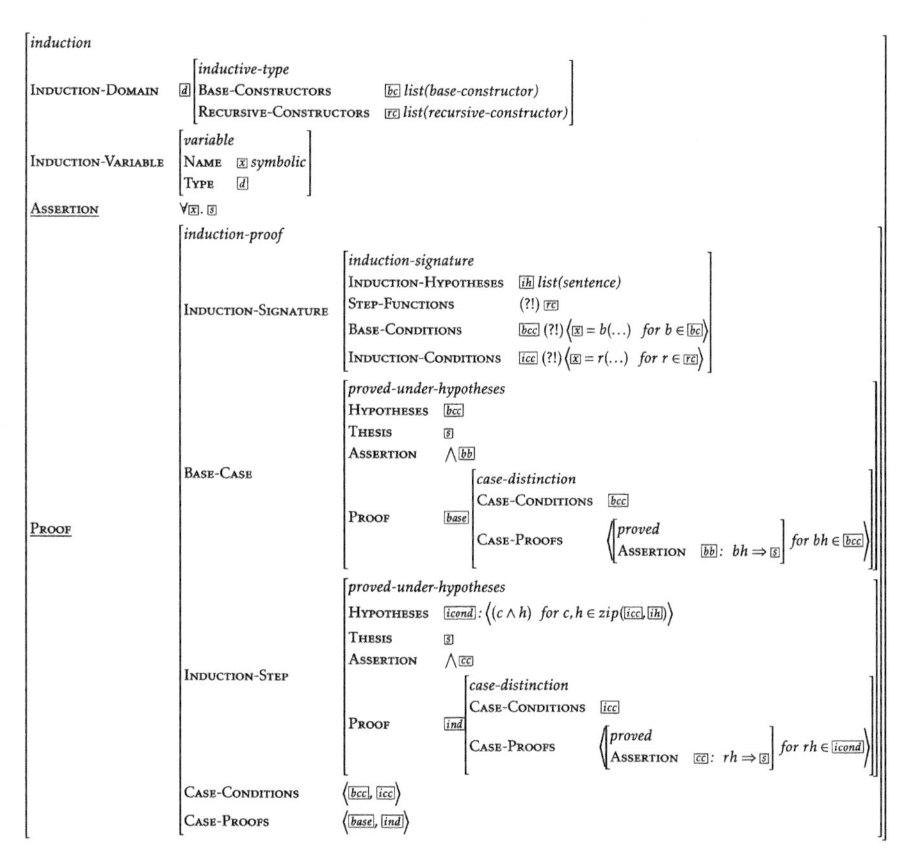

Fig. 19.8 A fuller Induction Frame which also illustrates the connection to case distinctions

References[12]

Azzouni, J. (2004). The derivation-indicator view of mathematical practice. *Philosophia Mathematica, 12*(2), 81–106.

Azzouni, J. (2009). Why do informal proofs conform to formal norms? *Foundations of Science, 14*(1–2), 9–26.

Bollobás, B. (1995). Extremal graph theory. In R. L. Graham, M. Grotschel, & L. Lovasz (Eds.), *Handbook of combinatorics* (Vol. 2, pp. 1231–1292). Amsterdam/New York: Elsevier.

Carl, M., & Koepke, P. (2010). Interpreting Naproche—An algorithmic approach to the derivation-indicator view. In *Proceedings of the International Symposium on Mathematical Practice and Cognition* (pp. 7–10).

Carpenter, B. (1992). *The logic of typed feature structures* (Cambridge tracts in theoretical computer science). Cambridge: Cambridge University Press.

Coq Development Team. (2017). *The Coq proof assistant: Reference manual, version 8.7.2*. INRIA. https://coq.inria.fr/refman/index.html

Cramer, M. (2013). *Proof-checking mathematical texts in controlled natural language*. Ph.D. thesis, Rheinische Friedrich-Wilhelms-Universität Bonn.

Cramer, M., et al. (2009). The Naproche project: Controlled natural language proof checking of mathematical texts. In N. E. Fuchs (Ed.), *Controlled natural language* (Lecture notes in computer science, Vol. 5972, pp. 170–186). Heidelberg: Springer.

Diestel, R. (2006). *Graph theory*. Heidelberg: Springer.

Fillmore, C. (1968). The case for case. In E. Bach & R. T. Harms (Eds.), *Universals in linguistic theory* (pp. 1–88). London: Holt, Rinehart, and Winston.

Kowalski, E. (2016). *Linear algebra* (Lecture notes). ETH Zurich, published at https://people.math.ethz.ch/~kowalski/script-la.pdf.

Marlow, S. (Ed.). (2010). *Haskell 2000 language report*. Haskell 2010 Committee. https://www.haskell.org/onlinereport/haskell2010/.

Minsky, M. (1974). *A framework for representing knowledge*. Technical report. Cambridge, MA: MIT.

Rav, Y. (2007). A critique of a formalist-mechanist version of the justification of arguments in mathematicians' proof practices. *Philosophia Mathematica, 15*(3), 291–320. https://doi.org/10.1093/philmat/nkm023

Roberts, R. B., & Ira, P. Goldstein (1977a). *The FRL manual*. Technical report. Cambridge, MA: MIT.

Roberts, R. B., & Ira, P. Goldstein (1977b). *The FRL primer*. Technical report. Cambridge, MA: MIT.

Ruppenhofer, J., et al. (2006). *FrameNet II: Extended theory and practice*. Distributed with the frameNet data. Berkeley: International Computer Science Institute.

Schank, R. C., & Abelson, R. P. (1977). *Scripts, plans, goals and understanding: An inquiry into human knowledge structures*. Hillsdale: L. Erlbaum.

Tanswell, F. (2015). A problem with the dependence of informal proofs on formal proofs. *Philosophia Mathematica, 23*(3), 295–310. https://doi.org/10.1093/philmat/nkv008

Túrán, P. (1941). Egy gráfelméleti szélsőértékfeladatról – Eine Extremalaufgabe aus der Graphentheorie. *Matematikai és Fizikai Lapok, 48*, 436–452.

[12]Web sites were checked 2017-12-15, 12:09.

Chapter 20
Formalising Mathematics in Simple Type Theory

Lawrence C. Paulson

Abstract Despite the considerable interest in new dependent type theories, simple type theory (which dates from 1940) is sufficient to formalise serious topics in mathematics. This point is seen by examining formal proofs of a theorem about stereographic projections. A formalisation using the HOL Light proof assistant is contrasted with one using Isabelle/HOL. Harrison's technique for formalising Euclidean spaces is contrasted with an approach using Isabelle/HOL's axiomatic type classes. However, every formal system can be outgrown, and mathematics should be formalised with a view that it will eventually migrate to a new formalism.

20.1 Introduction

Let's begin with Dana Scott:

> No matter how much wishful thinking we do, the theory of types is here to stay. There is *no other way* to make sense of the foundations of mathematics. Russell (with the help of Ramsey) had the right idea, and Curry and Quine are very lucky that their unmotivated formalistic systems are not inconsistent.[1] (Scott 1993, p. 413)

The foundations of mathematics is commonly understood as referring to philosophical conceptions such as logicism (mathematics reduced to logic), formalism (mathematics as "a combinatorial game played with the primitive symbols") (von Neumann 1944, p. 62), Platonism ("mathematics describes a non-sensual reality, which exists independently ... of the human mind") (Gödel 1995, p. 323) and intuitionism (mathematics as "a production of the human mind") (Heyting 1944, p. 52). Some of these conceptions, such as logicism and formalism, naturally

[1] Italics in original.

L. C. Paulson (✉)
Computer Laboratory, University of Cambridge, Cambridge, England
e-mail: lp15@cam.ac.uk

S. Centrone et al. (eds.), *Reflections on the Foundations of Mathematics*,
Synthese Library 407, https://doi.org/10.1007/978-3-030-15655-8_20

lend themselves to the idea of doing mathematics in a formal deductive system. Whitehead and Russell's magnum opus, *Principia Mathematica* (Whitehead and Russell 1927), is the quintessential example of this. Other conceptions are hostile to formalisation. However, a tremendous amount of mathematics has been formalised in recent years, and this work is largely indifferent to those philosophical debates.

This article is chiefly concerned with the great body of analysis and topology formalised by John Harrison, using higher-order logic as implemented in his HOL Light proof assistant (Harrison 1996). The original motive for this work was to verify implementations of computer arithmetic, such as the calculation of the exponential function (Harrison 2000), prompted by the 1994 floating-point division bug that forced Intel to recall millions of Pentium chips at a cost of $475 million (Nicely 2011). Another great body of mathematics was formalised by Georges Gonthier using Coq: the four colour theorem (Gonthier 2008), and later, the odd order theorem (Gonthier et al. 2013). Here the motive was to increase confidence in the proofs: the first four colour proof involved thousands of cases checked by a computer program, while the proof of the odd order theorem originally appeared as a 255-page journal article. Finally there was the Flyspeck project, to formalise Thomas Hales's proof of the Kepler conjecture, another gigantic case analysis; this formalisation task was carried out by many collaborators using HOL Light and Isabelle/HOL, so again higher-order logic.

Higher-order logic is based on the work of Church (1940), which can be seen as a simplified version of the type theory of Whitehead and Russell. But while they were exponents of logicism, today's HOL Light and Isabelle/HOL users clearly aren't, or at least, keep their views secret.

Superficially, Coq users are indeed exponents of intuitionism: they regularly refer to constructive proofs and stress their rejection of the excluded middle. However, this sort of discussion is not always convincing. For example, the abstract announcing the Coq proof of the odd order theorem declares "the formalized proof is constructive" (Gonthier et al. 2013, p. 163). This theorem states that every finite group of odd order is solvable, and therefore a constructive proof should provide, for a given group G of odd order, evidence that G is solvable. However, the solvability of a finite group can be checked in finite time, so no evidence is required. So does the constructive nature of the proof embody anything significant? It turns out that some results in the theory of group modules could only be proved in double-negation form (Gonthier et al. 2013, p. 174).

Analysis changes everything. Constructive analysis looks utterly different from classical analysis. As formulated by Bishop (Bishop and Bridges 1985), we may not assume that a real number x satisfies $x < 0 \vee x = 0 \vee x > 0$, and $x \neq 0$ does not guarantee that $xy = 1$ for some real y. In their Coquelicot analysis library, Boldo et al. assume these classical principles, while resisting the temptation to embrace classical logic in full (Boldo et al. 2015, §3.2).

The sort of constructivism just described therefore seems to lack an overarching philosophical basis or justification. In contrast, Martin-Löf's type theory was intended from the start to support Bishop-style constructive analysis (Martin-Löf 1975); this formal calculus directly embodies Heyting's intuitionistic interpretation of the logical constants (Martin-Löf 1996). It is implemented as the Agda (Bove et al. 2009) programming language and proof assistant.

It's worth remarking that the very idea of fixing a formalism as the *foundation* of intuitionistic mathematics represents a sharp deviation from its original conception. As Heyting wrote,

> The intuitionistic mathematician ... uses language, both natural and formalised, only for communicating thoughts, i.e., to get others or himself to follow his own mathematical ideas. Such a linguistic accompaniment is not a representation of mathematics; still less is it mathematics itself. (Heyting 1944, p. 52–3)

Constructive logic is well supported on the computer. However, the choice of proof assistant is frequently dictated by other considerations, including institutional expectations, the availability of local expertise and the need for specific libraries. The popularity of Coq in France is no reason to imagine that intuitionism is the dominant philosophy there.

Someone wishing to formalise mathematics today has three main options:

- Higher-order logic (also known as simple type theory), where types are built inductively from certain base types, and variables have fixed types. Generalising this system through polymorphism adds considerable additional expressiveness.
- Dependent type theories, where types are parameterised by terms, embodying the propositions-as-types principle. This approach was first realised in NG de Bruijn's AUTOMATH (de Bruijn 1980). Such systems are frequently but not necessarily constructive: AUTOMATH was mainly used to formalise classical mathematics.
- Set theories can be extremely expressive. The Mizar system has demonstrated that set theory can be a foundation for mathematics in practice as well as in theory (Bancerek and Rudnicki 2002). Recent work by Zhan (2017) confirms this point independently, with a high degree of automation.

All three options have merits. While this paper focuses on higher-order logic, I make no claim that this formalism is the best foundation for mathematics. It is certainly less expressive than the other two. And a mathematician can burst free of any formalism as quickly as you can say "the category of all sets". I would prefer to see a situation where formalised mathematics could be made portable: where proofs could be migrated from one formal system to another through a translation process that respects the structure of the proof.

20.2 Higher-Order Logic on the Computer

A succinct way to describe higher-order logic is as a predicate calculus with simple types, including functions and sets, the latter seen as truth-valued functions.

Logical types evolved rapidly during the twentieth century. For Whitehead and Russell, types were a device to forestall the paradoxes, in particular by enforcing the distinction between sets and individuals. But they had no notation for types and never wrote them in formulas. They even proved (the modern equivalent of) $V \in V$, concealing the type symbols that prevent Russell's paradox here (Feferman 2004). Their omission of type symbols, which they termed *typical ambiguity*, was a precursor to today's polymorphism. It seems that they preferred to keep types out of sight.

Church (1940) provided a type notation including a type ι of individuals and a separate type o of truth values, with which one could express sets of individuals (having type $o\iota$), sets of sets of individuals (type $o(o\iota)$) etc., analogously to the cumulative hierarchy of sets, but only to finite levels. Church assigned all individuals the same type.

Other people wanted to give types a much more prominent role. The mathematician NG de Bruijn devoted much of his later career, starting in the 1960s, to developing type theories for mathematics:

> I believe that thinking in terms of *types* and *typed sets* is much more natural than appealing to untyped set theory.... In our mathematical culture we have learned to keep things apart. If we have a rational number and a set of points in the Euclidean plane, we cannot even imagine what it means to form the intersection. The idea that both might have been coded in ZF with a coding so crazy that the intersection is *not empty* seems to be ridiculous. If we think of a set of objects, we usually think of collecting things of a certain type, and set-theoretical operations are to be carried out inside that type. Some types might be considered as subtypes of some other types, but in other cases two different types have nothing to do with each other. That does not mean that their intersection is empty, but that it would be insane to even *talk* about the intersection. (de Bruijn 1995, p. 31)[2]

De Bruijn also made the case for polymorphism:

> Is there the drawback that working with typed sets is much less economic then with untyped ones? If things have been said for sets of apples, and if these same things hold, *mutatis mutandis*, for sets of pears, does one have to repeat all what had been said before? No. One just takes a type variable, ξ say, and expresses all those generalities for sets of things of type ξ. Later one can apply all this by means of a single instantiation, replacing ξ either by *apple* or by *pear*. (de Bruijn 1995, p. 31)

His work included the first computer implementations of dependent type theories. However, his view that apples and pears should have different types, using type variables to prevent duplication, is universally accepted even with simple type theory.

[2] Italics in original.

20.2.1 Why Simple Type Theory?

What is the point of choosing simple type theory when powerful dependent type theories exist? One reason is that so much can be done with so little. HOL Light "sets a very exacting standard of correctness" and "compared with other HOL systems, ... uses a much simpler logical core."[3] Thanks to this simplicity, fully verified implementations now appear to be in reach (Kumar et al. 2016). Isabelle/HOL's logical core is larger, but nevertheless, concepts such as quotient constructions (Kaliszyk and Urban 2011), inductive and coinductive definitions (Blanchette et al. 2014; Paulson 1997), recursion, pattern-matching and termination checking (Krauss 2010) are derived from Church's original HOL axioms; with dependent type theories, such features are generally provided by extending the calculus itself (Giménez 1995).

The other reason concerns automation. Derivations in formal calculi are extremely long. Whitehead and Russell needed hundreds of pages to prove $1 + 1 = 2$ (Whitehead and Russell 1927, p. 360).[4] Proof assistants must be capable of performing lengthy deductions automatically. As a general rule, more expressive formalisms are more difficult to automate, and some technical features of constructive type theories interfere with automation. It is striking to consider the extent to which the Ssreflect proof language and library has superseded the standard Coq libraries. Gonthier and Mahboubi write

> Small-scale reflection is a formal proof methodology based on the pervasive use of computation with symbolic representations. ... The statements of many top-level lemmas, and of most proof subgoals, explicitly contain symbolic representations; translation between logical and symbolic representations is performed under the explicit, fine-grained control of the proof script. The efficiency of small-scale reflection hinges on the fact that fixing a particular symbolic representation strongly directs the behaviour of a theorem-prover. (Gonthier and Mahboubi 2010, p. 96)

So Ssreflect's remarkable success (Gonthier 2008; Gonthier et al. 2013) appears to involve sacrificing a large degree of mathematical abstraction. The Coquelicot analysis library similarly shies away from Coq's full type system:

> The Coq system comes with an axiomatization of standard real numbers and a library of theorems on real analysis. Unfortunately, ... the definitions of integrals and derivatives are based on dependent types, which make them especially cumbersome to use in practice." (Boldo et al. 2015, p. 41)

In the sequel, we should be concerned with two questions:

- whether simple type theory is sufficient for doing significant mathematics, and
- whether we can avoid getting locked into *any* one formalism.

---

[3]http://www.cl.cam.ac.uk/~jrh13/hol-light/

[4]In fact the relevant proposition, $*54 \cdot 43$, is a statement about sets. Many of the propositions laboriously worked out here are elementary identities that are trivial to prove with modern automation.

The latter, because it would be absurd to claim that any one formalism is all that we could ever need.

20.2.2 Simple Type Theory

Higher-order logic as implemented in proof assistants such as HOL Light (Harrison 1996) and Isabelle/HOL (Nipkow et al. 2002) borrows the syntax of types in the programming language ML (Paulson 1996). It provides

- *atomic types*, in particular `bool`, the type of truth values, and `nat`, the type of natural numbers.
- *function types*, denoted by $\tau_1 \Rightarrow \tau_2$.
- *compound types*, such as τ `list` for lists whose elements have type τ, similarly τ `set` for typed sets. (Note the postfix notation.)
- *type variables*, denoted by `'a`, `'b` etc. They give rise to polymorphic types like `'a` $\Rightarrow$ `'a`, the type of the identity function.

Implicit in Church, and as noted above by de Bruijn, type variables and polymorphism must be included in the formalism implemented by a proof assistant. For already when we consider elementary operations such as the union of two sets, the type of the sets' elements is clearly a parameter and we obviously expect to have a single definition of union. Polymorphism makes that possible.

The terms of higher-order logic are precisely those of the typed λ-calculus: identifiers (which could be variables or constants), λ-abstractions and function applications. On this foundation a full predicate calculus is built, including equality. Note that while first-order logic regards terms and formulas as distinct syntactic categories, higher-order logic distinguishes between terms and formulas only in that the latter have type `bool`.

Overloading is the idea of using type information to disambiguate expressions. In a mathematical text, the expression $u \times v$ could stand for any number of things: $A \times B$ might be the Cartesian product of two sets, $G \times H$ the direct product of two groups and $m \times n$ the arithmetic product of two natural numbers. Most proof assistants make it possible to assign an operator such as $\times$ multiple meanings, according to the types of its operands. In view of the huge ambiguity found in mathematical notations—consider for example xy, $f(x)$, $f(X)$, df/dx, x^2y, $\sin^2 y$—the possibility of overloading is a strong argument in favour of a typed formalism.

20.2.3 Higher-Order Logic as a Basis for Mathematics

The formal deductive systems in HOL Light and Isabelle/HOL closely follow Church (1940). However, no significant applications can be tackled from this primitive starting point. It is first necessary to develop, at least, elementary theories of the natural numbers and lists (finite sequences). General principles of recursive/inductive definition of types, functions and sets are derived, by elaborate constructions, from the axioms. Even in the minimalistic HOL Light, this requires more than 10,000 lines of machine proofs; it requires much more in Isabelle, deriving exceptionally powerful recursion principles (Blanchette et al. 2014). This foundation is already sufficient for studying many problems in functional programming and hardware verification, even without negative integers.

To formalise analysis requires immensely more effort. It is necessary to develop the real numbers (as Cauchy sequences for example), but that is just the beginning. Basic topology including limits, continuity, derivatives, power series and the familiar transcendental functions must also be formalised. And all that is barely a foundation for university-level mathematics. In addition to the sheer bulk of material that must be absorbed, there is the question of duplication. The process of formalisation gives rise to several number systems: natural numbers, integers, rationals, reals and complex numbers. This results in great duplication, with laws such as $x + 0 = x$ existing in five distinct forms. Overloading, by itself, doesn't solve this problem.

The need to reason about n-dimensional spaces threatens to introduce infinite duplication. Simple type theory does not allow dependent types, and yet the parameter n (the dimension) is surely a natural number. The theory of Euclidean spaces concerns $\mathbb{R}^n$ for any n, and it might appear that such theorems cannot even be stated in higher-order logic. John Harrison found an ingenious solution (Harrison 2005): to represent the dimension by a type of the required cardinality. It is easy to define types in higher-order logic having any specified finite number of elements. Then $\mathbb{R}^n$ can be represented by the type $n \rightarrow$ real, where the dimension n is a type. Through polymorphism, n can be a variable, and the existence of sum and product operations on types even allow basic arithmetic to be performed on dimensions. It must be admitted that things start to get ugly at this point. Other drawbacks include the need to write $\mathbb{R}$ as $\mathbb{R}^1$ in order to access topological results in the one-dimensional case. Nevertheless, this technique is flexible enough to support the rapidly expanding HOL Light multivariate analysis library, which at the moment covers much complex analysis and algebraic topology, including the Cauchy integral theorem, the prime number theorem, the Riemann mapping theorem, the Jordan curve theorem and much more. It is remarkable what can be accomplished with such a simple foundation.

It's important to recognise that John Harrison's approach is not the only one. An obvious alternative is to use polymorphism and explicit constraints (in the form of sets or predicates) to identify domains of interest. Harrison rejects this because

> it seems disappointing that the type system then makes little useful contribution, for example in automatically ensuring that one does not take dot products of vectors of different lengths or wire together words of different sizes. All the interesting work is done by set constraints, just as if we were using an untyped system like set theory. (Harrison 2005, p. 115)

Isabelle/HOL provides a solution to this dilemma through an extension to higher-order logic: *axiomatic type classes* (Wenzel 1997). This builds on the idea of polymorphism, which in its elementary form is merely a mechanism for type schemes: a definition or theorem involving type variables stands for all possible instances where types are substituted for the type variables. Polymorphism can be refined by introducing classes of types, allowing a type variable to be constrained by one or more type classes, and allowing a type to be substituted for a type variable only if it belongs to the appropriate classes. A type class is defined by specifying a suite of operations together with laws that they must satisfy, for example, a partial ordering with the operation $\leq$ satisfying reflexivity, antisymmetry and transitivity or a ring with the operations 0, 1, $+$, $\times$ satisfying the usual axioms. The type class mechanism can express a wide variety of constraints using types themselves, addressing Harrison's objection quoted above. Type classes can also be extended and combined with great flexibility to create specification hierarchies: partial orderings, but also linear and well-founded orderings; rings, but also groups, integral domains, fields, as well as linearly ordered fields, et cetera. Type classes work equally well at specifying concepts from analysis such as topological spaces of various kinds, metric spaces and Euclidean spaces (Hölzl et al. 2013).

Type classes also address the issue of duplication of laws such as $x + 0 = x$. That property is an axiom for the type class of groups, which is inherited by rings, fields, etc. As a new type is introduced (for example, the rationals), operations can be defined and proved to satisfy the axioms of some type class; that being done, the new type will be accepted as a member of that type class (for example, fields). This step can be repeated for other type classes (for example, linear orderings). At this point, it is possible to forget the explicit definitions (for example, addition of rational numbers) and refer to axioms of the type classes, such as $x + 0 = x$. Type classes also allow operators such as $+$ to be overloaded in a principled manner, because all of those definitions satisfy similar properties. Recall that overloading means assigning an operator multiple meanings, but when this is done through type classes, the multiple meanings will enjoy the same axiomatic properties, and a single type class axiom can replace many theorems (Paulson 2004).

An intriguing aspect of type classes is the possibility of recursive definitions over the structure of types. For example, the lexicographic ordering $\leq$ on type τ `list` is defined in terms of $\leq$ on type τ. But this introduces the question of circular definitions. More generally, it becomes clear that the introduction of type classes goes well beyond the naïve semantic foundation of simple type theory as a notation for a fragment of set theory. Recently, Kunčar and Popescu (2015) have published an analysis including sufficient conditions for overloaded constant definitions to be sound, along with a new semantics for higher-order logic with type classes. It remains much simpler than the semantics of any dependent type theory.

20.2.4 A Personal Perspective

In the spring of 1977, as a mathematics undergraduate at Caltech, I had the immense privilege of attending a lecture series on AUTOMATH given by de Bruijn himself, and of meeting him privately to discuss it. I studied much of the AUTOMATH literature, including Jutting's famous thesis (Jutting 1977) on the formalisation of Landau's *Foundations of Analysis*.

In the early 1980s, I encountered Martin-Löf's type theory through the group at Chalmers University in Sweden. Again I was impressed with the possibilities of this theory, and devoted much of my early career to it. I worked on the derivation of well-founded recursion in Martin-Löf's type theory (Paulson 1986a), and created Isabelle originally as an implementation of this theory (Paulson 1986b). Traces of this are still evident in everything from Isabelle's representation of syntax to the rules for Π, Σ and $+$ constructions in Isabelle/ZF. The logic CTT (constructive type theory) is still distributed with Isabelle,[5] including an automatic type checker and simplifier.

My personal disenchantment with dependent type theories coincides with the decision to shift from extensional to intensional equality (Nordström et al. 1990). This meant for example that $0+n = n$ and $n+0 = n$ would henceforth be regarded as fundamentally different assertions, one an identity holding by definition and the other a mere equality proved by induction. Of course I was personally upset to see several years of work, along with Constable's Nuprl project (Constable et al. 1986), suddenly put beyond the pale. But I also had the feeling that this decision had been imposed on the community rather than arising from a rational discussion. And I see the entire homotopy type theory effort as an attempt to make equality reasonable again.

20.3 Example: Stereographic Projections

An example will serve to demonstrate how mathematics can be formalised using the techniques described in Sect. 20.2.3 above. We shall compare two formalisations of a theorem: the HOL Light original and the new version after translation to Isabelle/HOL using type classes.

The theorem concerns stereographic projections, including the well-known special case of mapping a punctured[6] sphere onto a plane (Fig. 20.1). In fact, it holds under rather general conditions. In the two-dimensional case, a punctured circle is flattened onto a line. The line or plane is infinite, and points close to the puncture are mapped "out towards infinity". The theorem holds in higher dimensions with the sphere generalised to the surface of an n-dimensional convex bounded set and the plane generalised to an affine set of dimension $n-1$. The mappings are continuous bijections between the two sets: the sets are *homeomorphic*.

[5] http://isabelle.in.tum.de

[6] *Punctured* means that one point is removed.

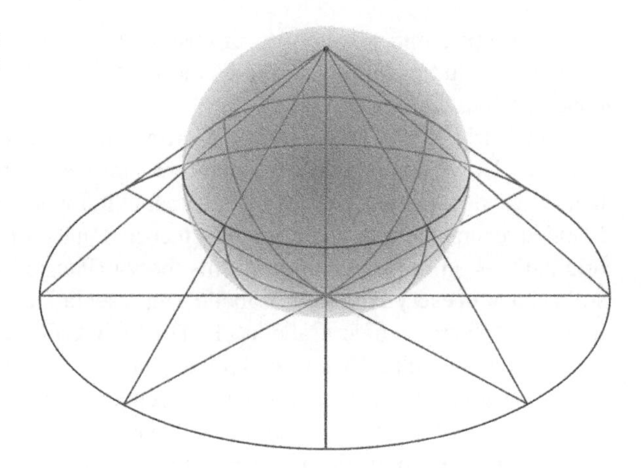

Fig. 20.1 3D illustration of a stereographic projection from the north pole onto a plane below the sphere (Original: Mark.Howison at English Wikipedia This version: CheChe • CC BY-SA 4.0)

The theorem we shall examine is the generalisation of the case for the sphere to the case for a bounded convex set. The proof of this theorem is formalised in HOL Light[7] as shown in Fig. 20.2. At 51 lines, it is rather short for such proofs, which can be thousands of lines long.

The HOL Light proof begins with the statement of the desired theorem. We see logical syntax coded as ASCII characters: `!` $= \forall$ and `/\` $= \wedge$. Moreover, the `DELETE` operator refers to the removal of a set element ($S - \{a\}$). Words such as `convex` and `bounded` denote predicates defined elsewhere. Infix syntax is available, as in the symbol `homeomorphic`. We see John Harrison's representation of $\mathbb{R}^n$ in the type `real^N->bool` and in particular, `!s:real^N->bool` abbreviates "for all $s \subseteq \mathbb{R}^N$". Note that the constraint on the dimensions is expressed through the concept of affine dimension rather than some constraint on `M` and `N`. This statement is legible enough, and yet the notation leaves much to be desired, for example in the necessity to write `&1` (the ampersand converting the natural number 1 into an integer).

```
!s:real^N->bool t:real^M->bool a.
     convex s /\ bounded s /\ a IN relative_frontier s /\
     affine t /\ aff_dim s = aff_dim t + &1
     ==> (relative_frontier s DELETE a) homeomorphic t
```

We have to admit that the proof itself is unintelligible. Even a HOL Light user can only spot small clues in the proof text, such as the case analysis on whether the set s is empty or not, which we see in the first line, or the references to previous lemmas. If we look carefully, we might notice intermediate statements being proved, such as

```
~(t:real^N->bool = {})
```

or

[7] File https://github.com/jrh13/hol-light/blob/master/Multivariate/paths.ml

```
let HOMEOMORPHIC_PUNCTURED_SPHERE_AFFINE_GEN = prove
 (`!s:real^N->bool t:real^M->bool a.
       convex s /\ bounded s /\ a IN relative_frontier s /\
       affine t /\ aff_dim s = aff_dim t + &1
       ==> (relative_frontier s DELETE a) homeomorphic t`,
  REPEAT GEN_TAC THEN ASM_CASES_TAC `s:real^N->bool = {}` THEN
  ASM_SIMP_TAC[AFF_DIM_EMPTY; AFF_DIM_GE; INT_ARITH
   `--(&1):int <= s ==> ~(--(&1) = s + &1)`] THEN
  MP_TAC(ISPECL [`(:real^N)`; `aff_dim(s:real^N->bool)`]
    CHOOSE_AFFINE_SUBSET) THEN REWRITE_TAC[SUBSET_UNIV] THEN
  REWRITE_TAC[AFF_DIM_GE; AFF_DIM_LE_UNIV; AFF_DIM_UNIV; AFFINE_UNIV] THEN
  DISCH_THEN(X_CHOOSE_THEN `t:real^N->bool` STRIP_ASSUME_TAC) THEN
  SUBGOAL_THEN `~(t:real^N->bool = {})` MP_TAC THENL
   [ASM_MESON_TAC[AFF_DIM_EQ_MINUS1]; ALL_TAC] THEN
  GEN_REWRITE_TAC LAND_CONV [GSYM MEMBER_NOT_EMPTY] THEN
  DISCH_THEN(X_CHOOSE_TAC `z:real^N`) THEN STRIP_TAC THEN
  MP_TAC(ISPECL
   [`s:real^N->bool`; `ball(z:real^N,&1) INTER t`]
       HOMEOMORPHIC_RELATIVE_FRONTIERS_CONVEX_BOUNDED_SETS) THEN
  MP_TAC(ISPECL [`t:real^N->bool`; `ball(z:real^N,&1)`]
       (ONCE_REWRITE_RULE[INTER_COMM] AFF_DIM_CONVEX_INTER_OPEN)) THEN
  MP_TAC(ISPECL [`ball(z:real^N,&1)`; `t:real^N->bool`]
       RELATIVE_FRONTIER_CONVEX_INTER_AFFINE) THEN
  ASM_SIMP_TAC[CONVEX_INTER; BOUNDED_INTER; BOUNDED_BALL; CONVEX_BALL;
              AFFINE_IMP_CONVEX; INTERIOR_OPEN; OPEN_BALL;
              FRONTIER_BALL; REAL_LT_01] THEN
  SUBGOAL_THEN `~(ball(z:real^N,&1) INTER t = {})` ASSUME_TAC THENL
   [REWRITE_TAC[GSYM MEMBER_NOT_EMPTY; IN_INTER] THEN
    EXISTS_TAC `z:real^N` THEN ASM_REWRITE_TAC[CENTRE_IN_BALL; REAL_LT_01];
    ASM_REWRITE_TAC[] THEN REPEAT(DISCH_THEN SUBST1_TAC) THEN SIMP_TAC[]] THEN
  REWRITE_TAC[homeomorphic; LEFT_IMP_EXISTS_THM] THEN
  MAP_EVERY X_GEN_TAC [`h:real^N->real^N`; `k:real^N->real^N`] THEN
  STRIP_TAC THEN REWRITE_TAC[GSYM homeomorphic] THEN
  TRANS_TAC HOMEOMORPHIC_TRANS
    `(sphere(z,&1) INTER t) DELETE (h:real^N->real^N) a` THEN
  CONJ_TAC THENL
   [REWRITE_TAC[homeomorphic] THEN
    MAP_EVERY EXISTS_TAC [`h:real^N->real^N`; `k:real^N->real^N`] THEN
    FIRST_X_ASSUM(MP_TAC o GEN_REWRITE_RULE I [HOMEOMORPHISM]) THEN
    REWRITE_TAC[HOMEOMORPHISM] THEN STRIP_TAC THEN REPEAT CONJ_TAC THENL
     [ASM_MESON_TAC[CONTINUOUS_ON_SUBSET; DELETE_SUBSET];
      ASM SET_TAC[];
      ASM_MESON_TAC[CONTINUOUS_ON_SUBSET; DELETE_SUBSET];
      ASM SET_TAC[];
      ASM SET_TAC[];
      ASM SET_TAC[]];
    MATCH_MP_TAC HOMEOMORPHIC_PUNCTURED_AFFINE_SPHERE_AFFINE THEN
    ASM_REWRITE_TAC[REAL_LT_01; GSYM IN_INTER] THEN
    FIRST_X_ASSUM(MP_TAC o GEN_REWRITE_RULE I [HOMEOMORPHISM]) THEN
    ASM SET_TAC[]]);;
```

Fig. 20.2 HOL Light code for the stereographic projection theorem

```
~(ball(z:real^N,&1) INTER t = {})
```

though in the latter case it is unclear what z is. The formal proof consists of program code, written in a general-purpose programming language (OCaml) equipped with a library of proof procedures and supporting functions, for that is what HOL Light is. A HOL Light proof is constructed by calling proof procedures at the OCaml command line, but one could type in any desired OCaml code. Users sometimes write such code in order to extend the functionality of HOL Light. Even if their code is incorrect,[8] they cannot cause HOL Light to generate false theorems. All LCF-style proof assistants employ a similar kernel architecture.

In recent years, I have been embarked on a project to translate the most fundamental results of the HOL Light multivariate analysis library into Isabelle. The original motivation was to obtain the Cauchy integral theorem (Harrison 2007), which is the gateway to the prime number theorem (Harrison 2009) among many other results. I was in a unique position to carry out this work as a developer of both Isabelle and HOL. The HOL family of provers descends from my early work on LCF (Paulson 1987), and in particular the proof tactic language, which is perfectly preserved in HOL Light. The 51 lines of HOL Light presented above are among the several tens of thousands that I have translated to Isabelle/HOL. Figure 20.3 presents my version of the HOL Light proof above, as shown in a running Isabelle session. Proof documents can also be typeset with the help of LaTeX, but here we have colour to distinguish the various syntactic elements of the proof: keywords, local variables, global variables, constants, etc.

The theorem statement resembles the HOL Light one but uses the Isabelle **fixes/assumes/shows** keywords to declare the premises and conclusion. (It is typical Isabelle usage to minimise the use of explicit logical connectives in theorem statements.) Harrison's construction `real^N` isn't used here; instead the variable a is declared to belong to some arbitrary Euclidean space. An advantage of this approach is that types such as *real* and *complex* can be proved to be Euclidean spaces despite not having the explicit form `real^N`.

The proof is written in the Isar structured language, and much of it is legible. An affine set U is somehow obtained, with the same dimension as S, which we note to be nonempty, therefore obtaining some element $z \in U$. Then we obtain a homeomorphism between *rel_frontier S* and *sphere z 1* $\cap$ *U*, using a previous result.[9] Then an element is removed from both sides, yielding a new homeomorphism, which is chained with the homeomorphism theorem for the sphere to yield the final result. And thus we get an idea how the special case for a punctured sphere intersected with an affine set can be generalised to the present result.

The Isar proof language (Wenzel 2007), inspired by that of the Mizar system (Trybulec 1993), encourages the explicit statement of intermediate results and obtained quantities. The notation also benefits from Isabelle's use of mathematical symbols, and a further benefit of type classes is that a number like 1 belongs to all numeric types without explicit conversion between them.

[8]Malicious code is another matter. In HOL Light, one can use OCaml's `String.set` primitive to replace T (true) by F. Given the variety of loopholes in programming languages and systems, not to mention notational trickery, we must be content with defences against mere incompetence.

[9]Because the HOL Light libraries were ported en masse, corresponding theorems generally have similar names and forms.

```
proposition homeomorphic_punctured_sphere_affine_gen:
  fixes a :: "'a :: euclidean_space"
  assumes "convex S" "bounded S" and a: "a ∈ rel_frontier S"
      and "affine T" and affS: "aff_dim S = aff_dim T + 1"
    shows "rel_frontier S - {a} homeomorphic T"
proof -
  obtain U :: "'a set" where "affine U" "convex U" and affdS: "aff_dim U = aff_dim S"
    using choose_affine_subset [OF affine_UNIV aff_dim_geq]
    by (meson aff_dim_affine_hull affine_affine_hull affine_imp_convex)
  have "S ≠ {}" using assms by auto
  then obtain z where "z ∈ U"
    by (metis aff_dim_negative_iff equals0I affdS)
  then have bne: "ball z 1 ∩ U ≠ {}" by force
  then have [simp]: "aff_dim(ball z 1 ∩ U) = aff_dim U"
    using aff_dim_convex_Int_open [OF ‹convex U› open_ball]
    by (fastforce simp add: Int_commute)
  have "rel_frontier S homeomorphic rel_frontier (ball z 1 ∩ U)"
    by (rule homeomorphic_rel_frontiers_convex_bounded_sets)
       (auto simp: ‹affine U› affine_imp_convex convex_Int affdS assms)
  also have "... = sphere z 1 ∩ U"
    using convex_affine_rel_frontier_Int [of "ball z 1" U]
    by (simp add: ‹affine U› bne)
  finally have "rel_frontier S homeomorphic sphere z 1 ∩ U" .
  then obtain h k where him: "h ` rel_frontier S = sphere z 1 ∩ U"
                    and kim: "k ` (sphere z 1 ∩ U) = rel_frontier S"
                    and hcon: "continuous_on (rel_frontier S) h"
                    and kcon: "continuous_on (sphere z 1 ∩ U) k"
                    and kh:  "⋀x. x ∈ rel_frontier S ⟹ k(h(x)) = x"
                    and hk:  "⋀y. y ∈ sphere z 1 ∩ U ⟹ h(k(y)) = y"
    unfolding homeomorphic_def homeomorphism_def by auto
  have "rel_frontier S - {a} homeomorphic (sphere z 1 ∩ U) - {h a}"
  proof (rule homeomorphicI)
    show h: "h ` (rel_frontier S - {a}) = sphere z 1 ∩ U - {h a}"
      using him a kh by auto metis
    show "k ` (sphere z 1 ∩ U - {h a}) = rel_frontier S - {a}"
      by (force simp: h [symmetric] image_comp o_def kh)
  qed (auto intro: continuous_on_subset hcon kcon simp: kh hk)
  also have "... homeomorphic T"
    by (rule homeomorphic_punctured_affine_sphere_affine)
       (use a him in ‹auto simp: affS affdS ‹affine T› ‹affine U› ‹z ∈ U››)
  finally show ?thesis .
qed
```

Fig. 20.3 The stereographic projection theorem in Isabelle/HOL

20.4 Discussion and Conclusions

The HOL Light and Isabelle proofs illustrate how mathematical reasoning is done in simple type theory, and what it looks like. The Isabelle proof demonstrates that simple type theory can deliver a degree of legibility, though the syntax is a far cry from normal mathematics. The greater expressiveness of dependent type theories has not given them any advantage in the domain of analysis: the leading development (Boldo et al. 2015) is not constructive and downgrades the role of dependent types.

As I have remarked elsewhere (Paulson 2018), every formal calculus is ultimately a prison. It will do some things well, other things badly and many other things not at all. Mathematics write their proofs using a combination of prose and beautiful but highly ambiguous notations, such as $\partial^2 z/\partial x^2 = y^2/x^2 + e^x y \cos y$. Formal proofs are code and look like it, even if they are allowed to contain special symbols and Greek letters. The various interpretations of anomalous expressions such as $x/0$ are also foundational, and each formalism must adopt a clear position when one might prefer a little ambiguity. (Both HOL Light and Isabelle define $x/0 = 0$, which some people find shocking.) Develop our proof tools as we may, such issues will never go away. But if past generations of mathematicians could get used to REDUCE and FORTRAN, they can get used to this.

The importance of legibility can hardly be overstated. A legible proof is more likely to convince a sceptical mathematician: somebody who doesn't trust a complex software system, especially if it says $x/0 = 0$. While much research has gone into the verification of proof procedures (Kumar et al. 2016; Schlichtkrull 2016), all such work requires trusting similar software. But a mathematician may believe a specific formal proof if it can be inspected directly, breaking this vicious cycle. Ideally, the mathematician would then gain the confidence to construct new formal proofs, possibly reusing parts of other proofs. Legibility is crucial for this.

These examples, and the great proof translation effort from which they were taken, have much to say about the process of porting mathematics from one system to another. Users of one system frequently envy the libraries of a rival system. There has been much progress on translating proofs automatically (Kaliszyk and Krauss 2013; Obua and Skalberg 2006), but such techniques are seldom used. Automatic translation typically works via a proof kernel that has been modified to generate a trace, so it starts with an extremely low-level proof. Such an approach can never deliver legible proofs, only a set of mechanically verified assertions. Manual translation, while immensely more laborious, yields real proofs and allows the statements of theorems to be generalised to take advantage of Isabelle/HOL's type classes.

All existing proof translation techniques work by emulating one calculus within another at the level of primitive inferences. Could proofs instead be translated at the level of a mathematical argument? I was able to port many proofs that I did not understand: despite the huge differences between the two proof languages, it was usually possible to guess what had to be proved from the HOL Light text, along with many key reasoning steps. Isabelle's automation was generally able to fill the gaps. This suggests that in the future, if we start with structured proofs, they could be translated to similarly structured proofs for a new system. If the new system supports strong automation (and it must!), the translation process could be driven by the text alone, even if the old system was no longer available. The main difficulty would be to translate statements from the old system so that they look natural in the new one.

The huge labour involved in creating a library of formalised mathematics is not in vain if the library can easily be moved on. The question "is simple type theory the right foundation for mathematics?" then becomes irrelevant. Let's give Gödel the last word (italics his):

> Thus we are led to conclude that, although everything mathematical is formalisable, it is nevertheless impossible to formalise all of mathematics in a *single* formal system, a fact that intuitionism has asserted all along. (Gödel 1986, p. 389)

Acknowledgements Dedicated to Michael J C Gordon FRS, 1948–2017. The development of HOL and Isabelle has been supported by numerous EPSRC grants. The ERC project ALEXANDRIA supports continued work on the topic of this paper. Many thanks to Jeremy Avigad, Johannes Hölzl, Neel Krishnaswami, Andrew Pitts, Andrei Popescu and the anonymous referee for their comments.

References

Bancerek, G., & Rudnicki, P. (2002). A compendium of continuous lattices in Mizar. *Journal of Automated Reasoning, 29*(3–4), 189–224.
Benacerraf, P., & Putnam, H. (Eds.). (1983). *Philosophy of mathematics: Selected readings* (2nd ed.). Cambridge: Cambridge University Press.
Bishop, E., & Bridges, D. (1985). *Constructive analysis*. Berlin: Springer.
Blanchette, J. C., Hölzl, J., Lochbihler, A., Panny, L., Popescu, A., & Traytel, D. (2014). Truly modular (co)datatypes forIsabelle/HOL. In G. Klein & R. Gamboa (Eds.), *Interactive Theorem Proving—5th International Conference, ITP 2014* (LNCS, Vol. 8558, pp. 93–110). Springer.
Blazy, S., Paulin-Mohring, C., & Pichardie, D. (Eds.). (2013). *Interactive Theorem Proving—4th International Conference* (LNCS, Vol. 7998). Springer.
Boldo, S., Lelay, C., & Melquiond, G. (2015). Coquelicot: A user-friendly library of real analysis for Coq. *Mathematics in Computer Science, 9*(1), 41–62.
Bove, A., Dybjer, P., & Norell, U. (2009). A brief overview of Agda—A functional language with dependent types. In S. Berghofer, T. Nipkow, C. Urban, & M. Wenzel (Eds.), *TPHOLs* (LNCS, Vol. 5674, pp. 73–78). Springer.
Church, A. (1940). A formulation of the simple theory of types. *Journal of Symbolic Logic, 5*, 56–68.
Constable, R. L. et al. (1986). *Implementing mathematics with the Nuprl proof development system*. Englewood Cliffs: Prentice-Hall.
de Bruijn, N. G. (1980). A survey of the project AUTOMATH. In J. Seldin & J. Hindley (Eds.), *To H.B. Curry: Essays in combinatory logic, lambda calculus and formalism* (pp. 579–606). London: Academic Press.
de Bruijn, N. G. (1995). On the roles of types in mathematics. In P. de Groote (Ed.), *The Curry-Howard isomorphism* (pp. 27–54). Louvain-la-Neuve: Academia.
Feferman, S. (2004). Typical ambiguity: Trying to have your cake and eat it too. In G. Link (Ed.), *100 years of Russell's paradox* (pp. 131–151). Berlin/Boston: Walter de Gruyter.
Giménez, E. (1995). Codifying guarded definitions with recursive schemes. In P. Dybjer, B. Nordström, & J. Smith (Eds.), *Types for Proofs and Programs: International Workshop TYPES'94* (pp. 39–59). Springer.
Gödel, K. (1986). Review of Carnap 1934: The antinomies and the incompleteness of mathematics. In S. Feferman (Ed.), *Kurt Gödel: Collected works* (Vol. I, p. 389). New York: Oxford University Press.

Gödel, K. (1995). Some basic theorems on the foundations of mathematics and their implications. In S. Feferman (Ed.), *Kurt Gödel: Collected works* (Vol. III, pp. 304–323). New York: Oxford University Press. Originally published in 1951.

Gonthier, G. (2008). The four colour theorem: Engineering of a formal proof. In D. Kapur (Ed.), *Computer mathematics* (LNCS, Vol. 5081, pp. 333–333). Berlin/Heidelberg: Springer.

Gonthier, G. & Mahboubi, A. (2010). An introduction to small scale reflection in Coq. *Journal of Formalized Reasoning, 3*(2).

Gonthier, G., Asperti, A., Avigad, J., Bertot, Y., Cohen, C., Garillot, F., Le Roux, S., Mahboubi, A., O'Connor, R., Ould Biha, S., Pasca, I., Rideau, L., Solovyev, A., Tassi, E., & Théry, L. (2013). A machine-checked proof of the odd order theorem. In Blazy et al. (2013) (pp. 163–179).

Harrison, J. (1996). HOL Light: A tutorial introduction. In M. K. Srivas & A. J. Camilleri (Eds.), *Formal Methods in Computer-Aided Design: FMCAD'96* (LNCS, Vol. 1166, pp. 265–269). Springer.

Harrison, J. (2000). Floating point verification in HOL light: The exponential function. *Formal Methods in System Design, 16*, 271–305.

Harrison, J. (2005). A HOL theory of Euclidean space. In J. Hurd & T. Melham (Eds.), *Theorem proving in higher order logics: TPHOLs 2005* (LNCS, Vol. 3603, pp. 114–129). Springer.

Harrison, J. (2007). Formalizing basic complex analysis. In R. Matuszewski & A. Zalewska (Eds.), *From insight to proof: Festschrift in honour of Andrzej Trybulec* (Studies in logic, grammar and rhetoric, Vol. 10(23), pp. 151–165). University of Białystok.

Harrison, J. (2009). Formalizing an analytic proof of the prime number theorem. *Journal of Automated Reasoning, 43*(3), 243–261.

Heyting, A. (1944). The intuitionist foundations of mathematics. In Benacerraf and Putnam (1983) (pp. 52–61). First published in 1944.

Hölzl, J., Immler, F., & Huffman, B. (2013). Type classes and filters for mathematical analysis in Isabelle/HOL. In Blazy et al. (2013) (pp. 279–294).

Jutting, L. (1977). *Checking Landau's "Grundlagen" in the AUTOMATH system*. PhD thesis, Eindhoven University of Technology.

Kaliszyk, C., & Urban, C. (2011). Quotients revisited for Isabelle/HOL. In W. C. Chu, W. E. Wong, M. J. Palakal, & C.-C. Hung (Eds.), *SAC'11: Proceedings of the 2011 ACM Symposium on Applied Computing* (pp. 1639–1644). ACM.

Kaliszyk, C., & Krauss, A. (2013). Scalable LCF-style proof translation. In Blazy et al. (2013) (pp. 51–66).

Krauss, A. (2010). Partial and nested recursive function definitions in higher-order logic. *Journal of Automated Reasoning, 44*(4), 303–336.

Kumar, R., Arthan, R., Myreen, M. O., & Owens, S. (2016). Self-formalisation of higher-order logic: Semantics, soundness, and a verified implementation. *Journal of Automated Reasoning, 56*(3), 221–259.

Kunčar, O., & Popescu, A. (2015). A consistent foundation for Isabelle/HOL. In C. Urban & X. Zhang (Eds.), *Interactive Theorem Proving—6th International Conference, ITP 2015* (LNCS, Vol. 9236, pp. 234–252). Springer.

Martin-Löf, P. (1975). An intuitionistic theory of types: Predicative part. In H. Rose & J. Shepherdson (Eds.), *Logic Colloquium'73* (Studies in logic and the foundations of mathematics, Vol. 80, pp. 73–118). North-Holland.

Martin-Löf, P. (1996). On the meanings of the logical constants and the justifications of the logical laws on the meanings of the logical constants and the justifications of the logical laws. *Nordic Journal of Philosophical Logic, 1*(1), 11–60.

Nicely, T. R. (2011). Pentium FDIV flaw. FAQ page online at http://www.trnicely.net/pentbug/pentbug.html

Nipkow, T., Paulson, L. C., & Wenzel, M. (2002). *Isabelle/HOL: A proof assistant for higher-order logic*. Springer. Online at http://isabelle.in.tum.de/dist/Isabelle/doc/tutorial.pdf

Nordström, B., Petersson, K., & Smith, J. (1990). *Programming in Martin-Löf's type theory. An introduction*. New York: Oxford University Press.

Obua, S. & Skalberg, S. (2006). Importing HOL into Isabelle/HOL. In U. Furbach & N. Shankar (Eds.), *Automated Reasoning: Third International Joint Conference, IJCAR 2006. Proceedings*, Seattle, 17–20 Aug 2006 (LNAI, Vol. 4130, pp. 298–302). Springer.
Paulson, L. C. (1986a). Constructing recursion operators in intuitionistic type theory. *Journal of Symbolic Computation, 2*, 325–355.
Paulson, L. C. (1986b). Natural deduction as higher-order resolution. *Journal of Logic Programming, 3*, 237–258.
Paulson, L. C. (1987). *Logic and computation: Interactive proof with Cambridge LCF*. Cambridge/New York: Cambridge University Press.
Paulson, L. C. (1996). *ML for the working programmer* (2nd ed.). Cambridge: Cambridge University Press.
Paulson, L. C. (1997). Mechanizing coinduction and corecursion in higher-order logic. *Journal of Logic and Computation, 7*(2), 175–204.
Paulson, L. C. (2004). Organizing numerical theories using axiomatic type classes. *Journal of Automated Reasoning, 33*(1), 29–49.
Paulson, L. C. (2018). Computational logic: Its origins and applications. *Proceedings of the Royal Society of London A: Mathematical, Physical and Engineering Sciences, 474*(2210). https://doi.org/10.1098/rspa.2017.0872
Schlichtkrull, A. (2016). Formalization of the resolution calculus for first-order logic. In J. C. Blanchette & S. Merz (Eds.), *Interactive Theorem Proving: 7th International Conference, ITP 2016. Proceedings*, Nancy, 22–25 Aug 2016 (LNCS, Vol. 9807, pp. 341–357). Springer.
Scott, D. S. (1993). A type-theoretical alternative to ISWIM, CUCH, OWHY. *Theoretical Computer Science, 121*, 411–440. Annotated version of the 1969 manuscript.
Trybulec, A. (1993). Some features of the Mizar language. http://mizar.org/project/trybulec93.pdf/
von Neumann, J. (1944). The formalist foundations of mathematics. In Benacerraf and Putnam (1983) (pp. 61–65). First published in 1944.
Wenzel, M. (1997). Type classes and overloading in higher-order logic. In E. L. Gunter & A. Felty (Eds.), *Theorem Proving in Higher Order Logics: TPHOLs'97* (LNCS, Vol. 1275, pp. 307–322). Springer.
Wenzel, M. (2007). Isabelle/Isar—A generic framework for human-readable proof documents. *Studies in Logic, Grammar, and Rhetoric, 10*(23), 277–297. From Insight to Proof—Festschrift in Honour of Andrzej Trybulec.
Whitehead, A. N., & Russell, B. (1962). *Principia mathematica*. Cambridge: Cambridge University Press. Paperback edition to *56, abridged from the 2nd edition (1927).
Zhan, B. (2017). Formalization of the fundamental group in untyped set theory using auto2. In M. Ayala-Rincón & C. A. Muñoz (Eds.), *Interactive Theorem Proving—8th International Conference, ITP 2017* (pp. 514–530). Springer.

Chapter 21
Dynamics in Foundations: What Does It Mean in the Practice of Mathematics?

Giovanni Sambin

Abstract The search for a synthesis between formalism and constructivism, and meditation on Gödel incompleteness, leads in a natural way to conceive mathematics as dynamic and plural, that is the result of a human achievement, rather than static and unique, that is given truth. This foundational attitude, called dynamic constructivism, has been adopted in the actual development of topology and revealed some deep structures that had remained hidden under other views. After motivations for and a brief introduction to dynamic constructivism, an overview is given of the changes it induces in the practice of mathematics and in its foundation, and of the new results it allows to obtain.

I believe that reflection on foundations, that is philosophy of mathematics, should not be separated from the development of mathematics itself, otherwise its purpose would be vacuous, of pure confirmation of what already exists. For mathematicians in particular the study of foundations should lead to greater awareness, and should also serve to modify and improve the mathematics they produce. Foundational research should go beyond the search for consistency or the repetition of the controversy between Brouwer and Hilbert, albeit in updated forms.

The search for a synthesis between formalism and constructivism, and a meditation on Gödel incompleteness open to discussion, rather than a dogmatic denial of its meaning as for example Bourbaki did, leads in a natural way to conceive mathematics as dynamic and plural, that is the result of a human achievement, rather than static, complete, unique, that is, a given absolute truth. In other

Some of the content of this paper was anticipated in my talks at Continuity, Computability, Constructivity (CCC 2017), Nancy (France), 29 June 2017 and at Incontro di Logica Matematica, AILA, Padova (Italy), 25 August 2017. I thank my wife Silvia for support and advice. I thank the editors, in particular my corresponding editor Deniz Sarikaya, for patience and empathy.

G. Sambin (✉)
Department of Mathematics, University of Padova, Padova, Italy
e-mail: sambin@math.unipd.it

S. Centrone et al. (eds.), *Reflections on the Foundations of Mathematics*,
Synthese Library 407, https://doi.org/10.1007/978-3-030-15655-8_21

words, it becomes possible to see the domain of mathematics as a product of a Darwinian process of evolution, as is the case with all the other domains of science. The common vision of mathematics as a manifestation of truth is modified and mathematics is transformed into a theory of abstractions and the information they manage, produced by ourselves through a dynamic process starting from reality. The effectiveness and objectivity of mathematics are not given, but conquered through the dynamic interaction between abstraction and application to reality. I call dynamic constructivism this foundational attitude.

For over 25 years (since 1991 with Sambin (1991) in theory, and since the mid-1990s also in the practice of mathematics (Sambin 2002, 2008, 2011, 2012, 2015, 2017)) I have engaged myself in developing a dynamic vision of mathematics at the same time putting it to the test of practice in the actual development of topology.

It is very interesting to confirm that on the basis of dynamic constructivism one can really develop topology. But even more interesting, and largely unexpected, is the discovery that assuming profoundly such a change of attitude towards foundations leads also to many new results. The concrete development of topology (see Sambin (2020, to appear) of which this chapter is an anticipation) has revealed some deep structures underlying topology which in theory were already visible in the 1930s of the twentieth century but were not seen in practice because of the dazzling light of the accepted view.

After a brief explanation of the motivations and a brief introduction to dynamic constructivism and the Minimalist Foundation that corresponds to it, the purpose of this chapter is to illustrate what changes it induces in the practice of mathematics and what novelties it allows to obtain in mathematics (symmetry and duality in topology and in general a deeper connection between logic and topology, continuity as a commutative square, mathematisation of existential statements, embedding of topology with points into that without points, algebrisation of topology,. . .) and in its foundation (a foundational system with two levels of abstraction linked by the forget-restore principle, conservativity of ideal aspects over real mathematics,. . .).

So, in hindsight, one can see all the signs of a new Kuhnian paradigm also for mathematics.

21.1 Towards a New Paradigm in Mathematics: Dynamic Constructivism

The purpose of this section is to introduce the general motivations and characteristics of the new attitude towards foundations that I have called dynamic constructivism. It seems to me that the best way to motivate and illustrate dynamic constructivism is to start with a brief overview of the historical development of the debate on foundations[1] and a description of the current situation, as I see it.

[1]The chapters by Priest and by Wagner, as well as the general introduction of this volume, contain analogous overviews.

More than a philosophical examination, my intention is to attempt to acquire awareness on the part of mathematicians. Finally, I illustrate schematically some methodological principles showing what my proposal means in practice.

21.1.1 The Problem of Foundations, Hilbert Program, Enriques Criterion and Gödel Theorems

The problem of foundations is the attempt to give answers to some questions such as: What is mathematics? What is its meaning? Where does it come from? The problem of foundations has been seriously posed only twice in the history of mankind. The first time it was faced in ancient Greece by many scholars, including Pythagoras, Zeno, Plato, and finally Euclid, who expresses all of mathematics in geometric language.

The discovery of non-Euclidean geometry, the birth of abstract algebra, the need of a rigorous formulation of analysis in the nineteenth century, strained intuition and made the problem of foundations for the second time pressing and inevitable. The question was: is it possible to still keep everything together in one system and give it a meaning? A positive answer was seen in set theory, later called naive, developed mainly by Cantor, with contributions by Frege, Dedekind and Peano.

This theory worked well, apparently, but, as history teaches us, it soon gave rise to some paradoxes: among the first, that of Burali-Forti in 1896 and among the best known, that of Russell in 1901.

The paradoxes were not a completely negative experience, because they forced a rethinking, but they shook the entire community of mathematicians and logicians because it was clear that such a problem could not remain ignored. The first thirty years of the twentieth century are commonly known as the period of crisis of foundations.

It is well known there were three lines of thought that developed then: formalism (chiefly by Hilbert), constructivism (chiefly by Brouwer), and logicism (chiefly by Russell). I will briefly discuss formalism and constructivism, which are the two currents of thought that played a fundamental role in the formulation of my new approach.

What goes by the name of Hilbert's program was to prove the coherence of an axiomatised version of set theory, with the only purpose of avoiding contradictions and thus demonstrating its consistency in a purely formal way, by studying it meta-mathematically. ZFC, namely Zermelo-Fraenkel plus the axiom of Choice, is an axiomatic set theory that tries to avoid the paradoxes, and retain the power of Cantor's naive theory of sets. Hilbert's request was to demonstrate the consistency of ZFC in an irrefutable way, that is, using only the tools of what he calls finitary, or real mathematics.

Brouwer argued that the solution provided by Hilbert was not sufficient because it removed the ground under the feet of mathematics, leaving it devoid of meaning. Federigo Enriques, a contemporary of Hilbert and Brouwer, but closer to the latter, had a less extreme position. While expressing gracefully in a beautiful old style

Italian, he does not spare a few lashes to Hilbert: "Hilbert seeks for a logical demonstration [of consistency], but we do not quite see in what sense the views of the illustrious geometer should be understood." (Enriques 1906, page 134, footnote).

And again in what I call *Enriques' criterion*: "If then you would not lose yourself in a dream devoid of sense, you should not forget the *supreme condition of positivity*, by means of which the critical *judgement must affirm or deny*, in the last analysis, *facts* either particular or general." (Enriques 1906, page 11, my emphasis).

ZFC does not meet Enriques' criterion: it does not talk about facts, no concrete interpretation of it has been given. Moreover, we do not even know whether it is consistent, in the sense that we do not have a proof and probably we will never have one. In fact, the request that all mathematics can be done within ZFC means that within mathematics it is not possible to prove the consistency of set theory, since ZFC does not prove Con(ZFC) by Gödel's second incompleteness theorem.

For some mysterious reasons, the weakest theory from a theoretical point of view became the most common. What happened? After Gödel theorems, Bourbaki's attitude, which is basically a denial of the problem, became common. Can't we prove consistency? We declare it is not a problem! We are not interested in the problem of foundations, we consider it solved once and for all. Logic and the problem of foundations are not part of mainstream mathematics, those who still deal with it are not first class mathematicians.

An explicit expression of this is given by Dieudonné, who in Dieudonné (1970) says: "On foundations we believe in the reality of mathematics, but of course when philosophers attack us with their paradoxes we rush to hide behind formalism and say: 'Mathematics is just a combination of meaningless symbols' and then we bring out Chaps. 1 and 2 on set theory. Finally we are left in peace to go back to our mathematics and do it as we have always done, with the feeling each mathematician has that he is working with something real. This sensation is probably an illusion, but is very convenient. That is Bourbaki's attitude towards foundations." The choice of words (attack/hide, meaningless/real, feeling/illusion) is particularly illuminating.

We will see later on how the bourbakists were wrong and how overcoming this rigid and irreverent position towards the problem of foundations can be very fruitful, even within mathematics.

In any case, the problem of the meaning of mathematics and of what one is talking about when one talks of mathematical entities remains open. Trust in a theory for which there is no proof of consistency is more like an act of faith than a criterion provided with some kind of scientific content. When bourbakists declare that the problem of foundations does not concern them is the same as looking at a tree without grasping the forest behind it.

As for me, I believe that this attitude cannot be the basis of a healthy foundation of mathematics. ZFC, which was supposed to be the solution, has become part of the problem which, in addition, hinders any advancement towards more solid and convincing innovations.

21.1.2 Towards a Synthesis

Analysing the formalist approach by Hilbert and the constructivist one by Brouwer, we can somehow perceive the traces of a dynamic similar to that between Hegelian thesis and antithesis which, however, precisely because of its dynamic/dialectic nature, requires an overcoming, a synthesis.

The classical approach proposed by Hilbert program, that declares mathematics to be self-sufficient, that is, capable of justifying its own meaning by virtue of consistency of its formalised version such as ZFC, is destined to confront dialectically with the other approach, that of Brouwer which, on the contrary, requires each statement of mathematics to have a computational meaning, that is, a meaning provided operatively by applications to concrete entities.

The dialectic, that is, the dynamic interplay that has seen the opposition between formalists and constructivists, has generated in the course of the twentieth century gradually more detailed positions, without however triggering a real overcoming.

The abstractness of the formalist position as opposed to the sometimes slavish concreteness of the constructivist one can, in fact, trigger a mechanism for synthesis that brings together the positive of the two positions, and elevates them generating new conceptions and new perspectives.

Bishop's book (Bishop 1967) showed that constructive mathematics, in particular analysis, does not depend on Brouwer's subjective views and is much simpler than previously expected. After the book Bishop (1967), constructive mathematics has become a rich and vibrant field of research. However, it has not been successful among mathematicians because they fear that too large a part of classical mathematics should be amputated, and they perceive motivations as still unclear and partly subjective.

One can see that after Bishop many other fields of mathematics (geometry, algebra, measure theory,...) can be developed constructively and some entirely new branches have been created (I have contributed myself with formal topology (Sambin 1987)).

These advancements in constructive mathematics, unimaginable a hundred and even fifty years ago, should make both parties more relaxed: formalists no longer have reasons to fear a revolution that amputates mathematics, constructivists no longer need to take refuge in motivations foreign to mathematics, because they are aware of their own potential.

So today times are ripe and the moment is right to start a synthesis, which already manifests itself, in my opinion, in terms of a dynamic and minimalist approach.

One problem of the formalist treatment, via ZFC, is the surrender of our intuition in the face of an alleged superior authority. For example, ZFC proves the existence of a well ordering of real numbers, but nobody is able today, and probably also in the future, to exhibit it. That well ordering exists only in a platonic world. So an argument based on acts of faith, such as that of the truth of ZFC, produces a piece of concrete information, such as that on real numbers, that cannot be obtained with a proof made entirely of concrete arguments (that is, which can only be obtained by passing through the full power of ZFC).

In the technical terms of logic, this is a situation of non-conservativity: a higher, abstract system proves some statements about a lower, more concrete one, and therefore with higher priority, that cannot be obtained in the lower system itself.

The reason why non-conservativity is commonly accepted lies, in my opinion, in the fact that one assumes bivalence, that is, the classical notion of truth, and completeness of theories, and these two are postulated because one presumes one can achieve security and stability through immobility.[2] Thus one falls into purely ideal knowledge.

The tragic side of this choice is that one prefers to destroy one's intuition rather than to give up, or modify, ideology. I believe we should abandon this kind of vision, especially in a global world that needs to accept a plurality of views, and therefore needs more soft, flexible, adaptable views, that base their strength on internal awareness rather than external authority or force.

One problem of the constructivist treatment is that, fearing the dangers of the formalist escape towards pure linguistic expressions without content, it has sought a solution by banning any ideal aspect from mathematics and restricting only to computational content (numerical meaning). This position is crystal clear in Bishop, for example.

I believe this has really sacrificed mathematics. Mathematics is not just computation, it is made up of many other ingredients: spatial intuition, mainly, but also formal deduction, axiomatic method, etc.

These are the shortcomings we should try to adjust. The task is not pure presentation, it also requires some conceptual change. In fact, in order to be able to make a synthesis, we must expand both visions, by distinguishing total information, related to computation and therefore to real mathematics, from partial information, related to ideal aspects.

Hilbert's vision implicitly included real mathematics in constructive mathematics and asked for consistency of the theory of ideal aspects. Then, in a bit crude words, a synthesis should be sought by modifying this vision, on the one hand by identifying real mathematics with the constructive one in the sense of effective (which also includes a treatment of infinity), and on the other hand by strengthening, after Gödel's lesson, the request for consistency of ideal mathematics into a request for its conservativity over real mathematics.

The attempt to make a synthesis between formalism and constructivism requires abandoning the idea of an absolute and complete truth and all acts of faith in an authority external to us, and instead seeking a perspective that allows conservativity of ideal entities with respect to real mathematics. Thus we are guided in a natural way towards a dynamic vision.

An open and profound reflection on Gödel incompleteness theorems, in contrast to the dogmatic and superficial denial of its meaning by Bourbaki, naturally leads to similar conclusions. Meditation on Gödel theorems leads us to the conclusion that truth and mathematics cannot be reduced to a system chosen once and for all, but

[2] Assuming classical logic and completeness of the more concrete system, conservativity of the higher one coincides with its consistency.

there is always something left out to be conquered, step by step. Every time we try to block truth and mathematics in a system, due to the phenomenon of incompleteness we can observe that something is left out. But then, again, we might as well conceive mathematics in a dynamic way.

Therefore the main challenge, and my proposal, is to move from a view of mathematics as static, complete, unique, transcendent, that is, an absolute given truth, to one in which mathematics is seen as a dynamic, partial, plural, evolutionary human achievement. This is the paradigm shift that I propose.

In other words, we should move from a view of mathematics external to us, with some religious flavour, to a totally human vision, based on evolution. Bishop basically says that mathematics has a special status, but he does not explain why. His objection to ZFC is due to its lack of numerical meaning, and not because he is reluctant to concede mathematics a transcendent nature (see page 1 of Bishop 1967).

Evolution is now commonly accepted in science, except in the domain of mathematics. It becomes possible to see also the domain of mathematics as the product of a Darwinian evolutionary process, as with all other domains of science. The difference is "only" that the domain of mathematics, its object of study, that is mathematical entities, are produced by ourselves human beings, rather than by nature (one could say: produced by nature, through us natural human beings).

I wrote about this idea in detail in Sambin (1991) (in Italian). In Sambin (2002), I express it this way: "It is now generally believed by biologists and by neuroscientists that nothing beyond biology and evolution is theoretically necessary to explain human beings, that is, their bodies *and* their minds. [...] No metaphysics, no ghosts, no homunculi. This is a big intellectual change, which has lead to a new global cultural attitude. My general claim is simply that the same holds for logic and mathematics. That is, a naturalistic, evolutionary attitude is fully sufficient, and [...] actually convenient for explaining not only the human body and mind, but also all human intellectual products, including logic and mathematics, which are just the most exact of these. That is, mathematics is a product of our minds and so to explain it we require no more than what is needed by biologists to explain the mind."

In a nutshell, mathematics does not come from heaven, but is built entirely by humans in a dynamic process that selects and stabilizes what works, and drops what does not work, just as in Darwinian natural evolution.

A dynamic vision, which includes different qualities of information, is the key to solving the problems left unresolved by other approaches to foundations:

1. to constructivists, it provides access to ideal entities, as partial information, or non-numerical information, without having to modify their own conception of computational mathematics,
2. to formalists, it allows not to loose meaning, by making it clear that ideal entities are not a truth above our heads, that we must accept even at the cost of sacrificing our intuition, but rather some fictitious entities that we introduce to simplify the treatment of real mathematics, but without modifying it, let alone contradict it.

I have deeply assumed such a change of foundational attitude in my practical work of mathematical research (in constructive topology), even before being able to communicate it adequately. Unexpectedly, in a certain sense, this has also brought with it many new results and changes in the practice of mathematics.

The purpose of the present paper is to illustrate these novelties in practice, both in methodology (a foundation which is open, minimalist, flexible, intentionally incomplete and with two levels of abstraction,...), and in results (a deeper connection between logic and topology, a mathematization of existential statements, an embedding of topology with points into topology without points, algebrization of topology,...).

Not long ago, with hindsight, that is after having implemented it without being aware of it, I realized that these changes could be seen as the practical signs of a paradigm shift, in the sense of Kuhn (1962), even for mathematics.

21.1.3 A Plural and Dynamic View of Mathematics

Now I spend a few words to explain how it is possible to conceive mathematics as obtained dynamically, not only the historical process of growth of knowledge, but also the actual construction of its entities. What is mathematics? Let us look at facts and rephrase the question in these terms: what is a mathematical abstraction?

One starts from reality, one extracts some features considered relevant, one gets something that is more sharply defined and easier to manipulate in place of the things from which it comes, which are more complex and articulated, and finally the result of such manipulation is applied to reality itself.

It is much simpler and more efficient to manipulate symbols than things. For example, if one has to count large quantities of goods, it is better to do it with the help of simpler tokens in place of goods themselves, so that one can easily manipulate them, rather than counting goods one by one in reality.

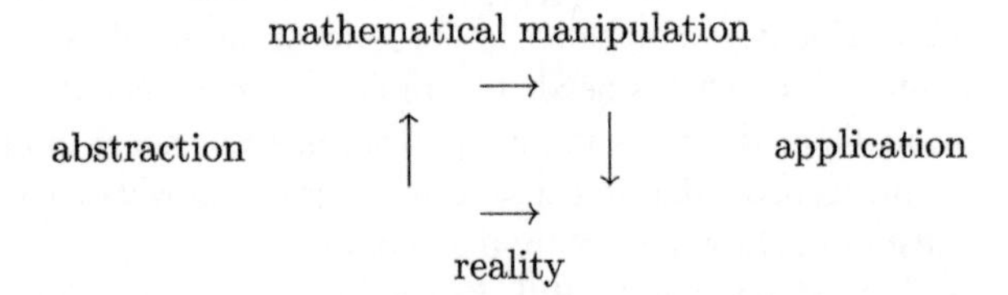

When one reads it as a progressive process of abstraction, mathematics is well explained through a diagram like that: one abstracts, one manipulates, and then one applies. If it works, that is, if the diagram is commutative, the abstraction is accepted and one goes on. If it does not work, one adjusts it or looks for a better one. This is somehow the beginning of everything.

Every culture has its own mathematics. It is useful to man for survival, a sort of continuation of natural evolution. The main idea is that mathematics is an evolutionary advantage that all cultures have developed. In my opinion, mathematics is the exploration and use of abstract notions and structures produced by our thought and that can be useful in a completely reliable way to understand the world.

It should not be difficult to agree with this perspective. However, if one takes it seriously, it has many clear consequences to which not many people come.

The first is that nothing is given. Every notion is the result of human abstraction starting from reality, which includes previously constructed mathematics.

The second is that, since abstractions are the result of a human process, there is no reason why there should be only one way of obtaining them. Each different foundation expresses a different criterion with which to abstract, a different style of abstraction. If mathematics is the creation and study of abstract concepts, a foundation is a choice of how to abstract them from reality.

The third is that application is an essential and integral part of mathematics, otherwise the dynamic process would not even begin. Application is part of mathematics, it is not something separate, as in the traditional division between pure mathematics and applied mathematics. The isolation of mathematics from its applications means that it focuses on the manipulation of entities, and then in fact forgets the importance of the process of abstraction with which such entities and concepts are created, that is the process by which we arrive at definitions.

The fourth consequence is that the question "what are mathematical entities?" is replaced by: why and how do we build them? how do we communicate them? to which domains can we fruitfully apply them? It is always a matter of referring to us, and not to something given externally to us. We are the measure of things in the sense that we humans equip ourselves with the tools we believe to be useful to improve our lives.

Remember that the object of study of all other sciences today has an evolutionary explanation. Why should mathematics be the *only* exception?

The fifth consequence is that objectivity of mathematics is a result, and not something given. Objectivity is a goal, which is reached through a laborious dynamic process, and not a place to start from, set down by others.

21.1.4 Some Principles of Dynamic Constructivism

To enter immediately into the heart of what we have developed so far, let's move on and see how this new vision is put into practice. In order to obtain some meta-mathematical results (such as normalization, extraction of programs, computational or realizability interpretation, etc.) one needs a formal system to be specified in all detail. The Minimalist Foundation (MF), introduced in Maietti and Sambin (2005) and Maietti (2009), seems the best one can do to express the ideas of dynamic constructivism in a formal system. We here will focus on conceptual novelties of MF that are made possible by the dynamic view.

Adopting dynamic constructivism in practice means, from a formal point of view, developing mathematics using MF as a foundation. From an informal point of view, it means following these four principles:

1. Pluralism in foundations and hence in mathematics[3] We have already said that there are many different ways to abstract. One could say that the choice

[3]See the chapters by Friend, Priest, and Wagner in this volume for other views on pluralism.

of a foundation (or foundational system) represents one of these ways, which corresponds to the choice of which kind of information is considered relevant and which is not, and therefore what should be maintained or left out, in the process of abstraction.

Every different style with which abstractions are made, that is every different foundation, yields a different way of doing mathematics. First comes the style of abstraction, that is the foundation, and then the resulting mathematics. There is no single foundation nor a privileged one. Each foundation produces a different kind of mathematics and should be respected.

For example, constructive mathematics does not coincide with the constructivization of classical mathematics, that is to say with classical mathematics developed according to stricter principles of proof; because at least half of the work, and actually the most exciting half, consists in finding appropriate definitions, corresponding to the different way of abstracting.

Pluralism is an immediate consequence. But pluralism includes, in my opinion, also finding a common root for seemingly incompatible points of view or approaches. MF is compatible with the most relevant foundations for both constructive and classical mathematics, predicative and impredicative (constructive type theory, calculus of construction, internal theory of toposes, axiomatic set theory, both classical and constructive, Feferman's explicit mathematics) in the sense that each of them can be obtained as an extension of MF.

2. Accept open concepts and incomplete theories Since the construction of mathematics is a never-ending process and – as we said – nothing is given a priori, in advance, one should avoid whatever blocks the process, as it happens when assuming a fixed universe of all sets, or of all subsets, or of all propositions.

So many of our notions, through which we describe reality, are open, intrinsically incomplete. It is not necessary to assume that there is a universe of all sets, or equivalently that sets exist independently of us humans. Simply, whenever we run into an aggregate of objects meeting the conditions we have established to be called a set, we will call that aggregate a "set".

Contrary to a common vision, this is not a limitation but the source of a more relaxed vision and a deeper understanding. The consistency of the minimalist foundation MF becomes a theorem with every rigour in the sense specified later, contrary to the case of axiomatic set theory ZFC.

3. Preserve all conceptual distinctions and all fragments of information There is no reason why all aspects of our abstract thinking should be reducible to a single notion, such as that of set. What about the notion of operation, for example? Or about the notion of proposition: why should it be the same as a set?

When mathematics ceases to be conceived as existing in itself, as the gift offered to us by some transcendental assumption, all its results (not only theorems or solutions to problems, but also definitions, intuitions, conceptual distinctions, etc.) are the result of the human struggle for knowledge, and thus all of them become valuable and must be kept and respected, as they are, without reducing everything to a single notion, like that of set.

As a result, many more primitive notions than usual are maintained so that, alongside the notion of set, we will have the notion of collection and proposition, even in their form under assumptions (which produce the notions of operation, subset, relation, function, etc.). Moreover, we adopt a logic, such as intuitionistic (or constructive) logic, which is more refined in maintaining conceptual distinctions, such as that between φ and $\neg\neg\varphi$, or that between $\exists x\varphi$ and $\neg\forall x\neg\varphi$.

In other words, on the one hand we want to keep under control the validity of every assumption of foundational character and, on the other, keep all information, of any kind, even partial, unless it can be restored automatically.

Being respectful of all notions becomes being careful not to forget or fictitiously introduce any information. A high resolution of concepts (such as a high resolution image) is obtained – perhaps paradoxically for some – through a minimalist choice of assumptions. The examples in Sect. 21.1.5 should remove all doubts about this.

Dealing also with partial information means going beyond the position of those constructivists, as Bishop and Martin-Löf, who, in order to eliminate any ideal aspect and keep information under control, have self-limited themselves to the case in which one has complete information, as in computation.

In particular, Martin-Löf constructive type theory has total control over information, and in particular it is intensional. This is fine for meta-mathematical properties and for formalization in a proof-assistant, but it is not so good for doing mathematics, which is extensional (instructions defining an operation do not appear in the concept of equality, the proof that $P(a)$ is true is not relevant to conclude that a belongs to the subset corresponding to P,...).

To actually use type theory in the practice of mathematics, in particular to get the usual extensional concept of subset and of relation, we need to relax total control and we "forget" some piece of information. As I hint further, a way to do this is to abandon the propositions-as-sets principle, that is at the base of Martin-Löf type theory (and that can be read as a rigorous expression of the so-called BHK interpretation of intuitionistic logic), and introduce the concept of proposition as primitive, and independent from the concept of set.

At that point, an autonomous justification of intuitionistic logic is needed as an alternative to BHK. We believe we can explain logical constants by taking the difference between levels of abstraction seriously: each logical constant is the reflection at the (more abstract) level of object language of a link between assertions at the (more concrete) meta-level. But this is another story, which is not going to be told here, see Sambin et al. (2000).

4. Keep all the different levels of abstraction It is a fact that communication can have different levels of reference, which in mathematics is expressed through the fundamental distinction between language and metalanguage, but also different levels of abstraction. In the Minimalist Foundation the intensional aspects coexist with the extensional ones, and this is achieved by formulating MF with two levels of abstraction. The extensional level is as close as possible to the practice of constructive mathematics, while the intensional level is closer to the formalization of mathematics, for example in a proof assistant.

In addition, the distinction between the ideal and real aspects of mathematics can be kept within MF, through the set/collection and operation/function distinctions (see below).

In conclusion, it can be shown that the Minimalist Foundation satisfies all the properties that a foundation of a dynamic view of mathematics should have in order to be completely satisfactory. This means it is:

Applicable, meaningful This means that it fulfils Enriques' criterion: applications are the facts we talk about. In rigorous terms, "facts" consist of the computational content, as expressed by a *realizability interpretation*, see Maietti and Maschio (2014).

Reliable We trust in its consistency because we have a proof of it, and not by acts of faith or emotions. This follows from the realizability interpretation, that is from the fact that Enriques' criterion is satisfied.

Consistency also follows from the existence of a normalization procedure for derivations. Normalization and realizability are proved at the intensional level, but then their consequences propagate to the extensional level, since the two levels of MF are linked according to the forget-restore principle, as proved in Maietti (2009).

Consistency of MF is immediately reducible to that of a fragment of Martin-Löf type theory, for which several proofs are available in the literature.

Universal It allows one to express and preserve all significant conceptual distinctions. Every notion has a clear and specific meaning. That is why it is minimalist in assumptions, because it aims to be maximal in conceptual distinctions.

It is universal also in the sense that all foundations can be expressed as extensions of MF, adding appropriate assumptions.

These properties make it a very natural framework for pluralism, because they allow to express all possible foundations within the same picture and therefore to compare them more easily. In particular, it is the best possible environment for "reverse mathematics", as it provides a rigorous common basis for analyzing the various other foundational hypotheses.

Precise It can be formalized with the precision required by the *formalization of mathematics* in a proof assistant (see Sacerdoti Coen and Tassi 2011).

All this is possible! The Minimalist Foundation MF fully agrees with the perspective of dynamic constructivism.

Some positive features are the following. Having many levels of abstraction means being able to choose the best one for the problem in question or for different purposes. For example, we can maintain our spatial intuitions even when we treat computation. No basic principle is assumed that can "lose" information and no information is assumed to be given or restored for free.

Mathematics becomes a theory (in fact, the various possible theories) for the management of abstractions and the information they transmit. Since in reality we often have different qualities of information, even partial or less sharp information should be preserved.

Having all the conceptual distinctions, namely keeping all the ingredients of mathematics, means keeping our intuitions even when it comes to computations (or programs). And it also means being able to express all the nuances of specific concrete problems.

21.1.5 Minimalism in Practice

In practice, forgetting the intensional level that is used only to obtain meta-mathematical results and formalization in a proof assistant, the ideas behind MF are simple, even for mathematicians with no concern for foundations, that is, even without specifying the formal system. Even for them a concept of proposition, of generated or non-generated set and judgments of this kind, but dependent on an argument (set-indexed family of sets, set-indexed family of elements, etc.) are clear and intuitive, on an informal level. Nothing else is needed.

The real and profound difference that I propose, with respect to the common vision, is that mathematical entities do not exist in themselves, but it is us who build them. Moreover, none of the common assumptions are made on these entities, for the reasons we now see.

Following the general principle that suggests keeping all conceptual distinctions, we wish to distinguish a positive notion of existence, an effective notion of set and of function, from the corresponding notions, but negative or not effective.

1. A positive notion of existence Following the general principle, we wish to distinguish a positive (constructive) notion of existence $\exists x\varphi$, which allows us, at least ideally, to build a witness d that certifies existence, that is, for which $\varphi(d)$ is valid, from a negative (classical) $\neg\forall x\neg\varphi$, in which one proves that the denial of existence leads to an absurdity. This means that we cannot accept as universally valid the law of the excluded middle, that says that for any proposition φ, either φ is true or its negation $\neg\varphi$ is true, even if we do not know which of these:

LEM (*Law of Excluded Middle*)) $\varphi \vee \neg\varphi$ true for all propositions φ.

In fact, assuming LEM, the equivalence between $\exists x\varphi$ and $\neg\forall x\neg\varphi$ is derivable (this is a generalization of the de Morgan law $\varphi \vee \psi \leftrightarrow \neg(\neg\varphi \;\&\; \neg\psi)$).

I would like to stress and clarify right away that, in our approach, where there is no absolute and predetermined concept of truth, the question at stake is not to believe in the truth of LEM or not, but to decide whether we are interested in maintaining or not a distinction between two different types of information ($\exists$ and $\neg\forall\neg$) that the assumption of LEM nullifies.

That is, LEM destroys a different quality of information, such as that between $\exists x\varphi$ and $\neg\forall x\neg\varphi$. So much so that many classical mathematicians do not even perceive the difference. The matter is deciding whether this suits us or not. So, in reverse, those who are not willing to give up LEM show that they do not care, or do not consider it relevant, to keep the distinction between $\exists x\varphi$ and $\neg\forall x\neg\varphi$.

We are interested in maintaining a positive understanding of the existential quantifier, which cannot be derived or borrowed from anything other. Therefore we cannot assume validity of LEM, and this means that we must adopt constructive (or intuitionistic) logic.

2. An effective notion of set The relation of the distinction between generated or non-generated set, with the axiom of the power set is conceptually analogous to the one just seen between positive and negative exists, with the principle LEM. The axiom of the power set says that:

PSA (*Power Set Axiom*) if X s a set, then $\mathcal{P}X$ is a set,

where $\mathcal{P}X$ is formed by all subsets of X. As we now see, assuming this axiom means not considering the distinction between two different notions, which we call here set and collection, within the notion of set of classical mathematics.[4]

The notion of set has for us an effective meaning; that is, we call set the situation in which we have a finite number of definite rules to construct all elements. The formal way of expressing that all the elements are obtained by iterated use of the rules is to assume that a principle of proof by induction holds. Our sets are therefore inductively generated.

The notion of element x of a set X is included in the previous explanation: x is an element of X, written $x \in X$ as usual, if x is obtained by applying (a finite number of times) the rules defining X.

Therefore the question: what is a set? has for us a practical answer. To give a set means to exhibit the finite number of rules by which we obtain all its elements. We are the ones who build sets, and so it is sufficient to remember the rules with which we (or others) have built them. This view is hardly compatible with an axiomatic approach, in which information is not retained of what the specific rules for building a set are.

Typical examples of sets in our sense are finite sets, the set $\mathbb{N}$ of natural numbers, lists of elements over a set, or the product of sets, etc.

A collection is an aggregate of objects without exhaustive rules to build them, that is, without principles of induction. So, while a set is effective, real, completely communicable and stable over time (as determined univocally by a finite number of rules), a collection is an ideal, open concept, which theoretically could change.

A typical example of collection is given by $\mathcal{P}X$, all subsets of a set X. In fact, a constructive notion says that a subset of a set X is a proposition with an argument $x \in X$. And that two subsets are equal if for each x of X they are equivalent as propositions. Then it is easy to show, with an argument by diagonalization, that there cannot be a finite number of rules with which all subsets of X can be built. So $\mathcal{P}X$ is definable constructively only as a collection.

[4]Note that it is not a mere matter of words; if the classical meaning of the word set is considered by now fixed, our sets might be called proper, or small, or effective sets.

Assuming the validity of PSA, as in ZFC, means not being interested in the distinction of sets from collections, that is, sets that are actually constructible from those that are not. Conversely, rejection of PSA allows us to preserve an effective notion of set, and keep it distinct from that of collection.

3. An effective notion of function The situation is quite similar for a third conceptual distinction, even if it is seldom made and preserved.

We call *operation* a family of elements indexed on a set $g(x) \in S\ (x \in X)$. This means that we know how to obtain an explicit description of the value $g(x)$, when we are provided with an explicit description of the argument x. So what we call operation here coincides with the common notion of function in constructivism, for example in Bishop and Martin-Löf.

We wish to distinguish this notion from that of function, in the sense of functional relation, which is typical of classical mathematics. A (binary) relation $R(x, a)$ from X to S is a proposition that depends on two arguments $x \in X$ and $a \in S$ (instead of only one argument, as in the notion of subset). And two relations are said to be equal if for all x and a they give equivalent propositions. Then here a function is nothing but a total and singlevalued relation, that is a relation that satisfies $(\forall x \in X)(\exists! a \in S) R(x, a)$. This is the usual set-theoretic definition.

Informally, we know that on every argument x the value of the function exists and is unique, but we do not know how to obtain it explicitly. So, in a sense, a function can be thought of as an operation built by someone else, but who does not give us the instructions with which she built it. In this case we have no chance to restore them (for example, if the other person is an opponent in a two-handed game and the operation is her strategy in choosing the moves, from our point of view that is only a function).

A link between the two notions is present in any case, in the sense that every operation $g(x) \in S\ (x \in X)$ gives rise to a function $\hat{g}(x, a) \equiv (g(x) = a)$, the graph of g. But we do not accept validity of the converse, that is, that each function is the graph of an operation, because this would be like expecting to be able to explicitly describe the values of g, even when only their existence and uniqueness are known.

As we will see, this is a simple way to get a mathematical characterization of Brouwer's choice sequences. Therefore we cannot accept without restriction the validity of the so-called axiom of unique choice (for operations):

AC! (*Axiom of Unique Choice*): $(\forall x \in X)(\exists! a \in S) R(x, a) \to (\exists g \in OP(X, S))(\forall x \in X) R(x, g(x))$.

where R is a relation and $g \in OP(X, S)$ is an operation from X to S. In fact, this would mean that every function is the graph of an operation, which is precisely what we wish to avoid. Of course, a fortiori we cannot accept the axiom of choice AC in the form $(\forall x \in X)(\exists a \in S) R(x, a) \to (\exists g \in OP(X, S))(\forall x \in X) R(x, g(x))$.

Given that many consider AC! a tautology, or (in the constructive context) many consider AC! valid simply by the constructive interpretation of quantifiers, it is convenient to explain the way in which its validity can be kept under control. The issue is to distinguish a weak, although positive, concept of existence $\exists$ from a concept of strong existence, such as that expressed by Σ in Martin-Löf type theory.

We say that $\exists x\varphi(x)$ is true when we have a guarantee that a witness d that makes $\varphi(d)$ true can eventually be found, at least ideally, even if no operation is available to provide it (as instead it happens for Σ). An example of such existence $\exists$ (besides the one appearing in the game above) could be given by the situation in which our wedding ring is lost at a party in the open air by a friend: we know with positive certainty that there is a wedding ring in our friend's garden, but we have no effective procedure that allows us to find it.

In order to maintain this further distinction (between $\exists$ and Σ), another principle, which is typical of Martin-Löf type theory, must be kept under control, namely the so-called propositions-as-sets, which identifies sets with propositions (and therefore $\exists$ with Σ). In fact, if in the formulation of $\mathtt{AC!}$ above we replace $\exists$ with Σ, then, as it is well known, it turns out to be derivable.

In conclusion, to preserve the distinction between operations and functions we must keep propositions distinct from sets, and this is the main departure from Martin-Löf type theory.

21.2 New Mathematics Arising from Dynamic Constructivism

The most personal and profound reason why I chose to undertake the challenge of writing a book Sambin (2020, to appear) and I faced all the relative effort is to show, to myself and others, that one can "rebuild the universe" of mathematical thought according to the new criteria that I have developed over many years of research activity and that I have seen possible thanks to the synthesis between formalism and constructivism of which I spoke above (in Sect. 21.1.2).

I wanted to show that one can build an environment in which I could always feel at ease, even in the technically most difficult steps, because I have experienced that even to those it was possible to give an explanation and a clear explicit motivation. I hope readers too will feel comfortable within this new approach.

So my motivation is foundational (or philosophical) as much as mathematical. It is my belief that, on the one hand, the study of a (constructive) foundation of mathematics necessarily goes together with putting it into practice developing mathematics *actually*, and on the other hand, the development of mathematics should always also include awareness of the foundation in which this is possible.

My main foundational need is to be able to have a mathematics that admits an explanation without any kind of "transcendental" assumption, that is devoid of any form of platonism, in which for every entity one can tell how and why it was built or introduced by us human beings. This means showing with facts that there is a mathematics that explicitly maintains a clear, direct or indirect link with the concrete reality of which it speaks abstractly.

In my opinion, mathematics is an arsenal of conceptual tools, very well thought out and developed, with which we organise our knowledge of reality. Constructivism means, ultimately, full awareness of what notions and principles are used, or what abstractions and idealizations apply, when we look – mathematically – at reality.

What I pursue, in the final analysis, is a mathematics in which one can see how dynamic constructivism, described above, is put into practice.

So, instead of bringing further theoretical arguments in favour of dynamic constructivism that little change mathematics itself, I intend to show by facts that this foundation, in addition to standing up theoretically, works very well in practice.

The purpose of the book I already mentioned several times and that here I am presenting in preview is to show, with definitions and theorems, that doing mathematics in a radically new way, both conceptually and technically, is totally possible and reasonable. The result is a very "clean" mathematics, in which all ingredients coexist and interact with each other and everything maintains a clear intuition.

To my knowledge, the development of topology I introduce and discuss in Sambin (2020, to appear) is the first and only systematic version (as distinct from meta-mathematical results that only demonstrate the possibility) since the time of Hermann Weyl (1918) of abstract mathematics developed in a strict and explicit predicative manner.

Following a tradition for mathematical textbooks, the foundation is left at an informal level, that is to say, the concepts that are used are illustrated only in words, just as before the advent of formal systems and computers. Although it is not necessary to know the formal system for the Minimalist Foundation to be able to read it, the whole book can be formally expressed within MF. Moreover, it can be formalized in a proof assistant (as indeed it was done on Matita (Asperti et al. 2011)). Even this alone would be very important conceptually.

If the deepest motivation of what I propose is foundational, why then do we speak primarily of topology? The reasons are two, of different nature.

1. Topology is the discipline that deals with the connections between finite and infinite. This makes it fundamental in classical mathematics, and for us even inescapable. In fact, as we will see, it allows us to introduce constructively (in our sense) ideal entities, which are determined only by an infinite amount of information (such as real numbers, or better streams and choice sequences), as ideal points over an effective structure, and therefore always maintaining the distinction between what is infinite/ideal and what is finite/effective/real (as required in Sect. 21.1.2).

A fascinating outcome is that the study of foundations is inextricably linked to topology in a very concrete sense. So topology is not a topic like many others in mathematics: we would have gone through it anyway.

2. In practical terms, topology is the most difficult subject to develop constructively (in our sense), and so if we succeed with that, we succeed also with all the rest of mathematics.

Now, even without seeing formal details of the foundational system MF that corresponds to dynamic constructivism, I would like to make it clear immediately that the answer to questions such as: can we really develop mathematics, topology in particular, on the basis of apparently such a weak foundation? and how much of classical mathematics can be done? is that not only can mathematics be done, but that one can do a lot of it. And the part that remains excluded seems to be not so useful for applications. In any case, even when it is not possible to do something on the basis of MF, it is interesting to examine what additional principles would be needed to obtain it.

As the book shows with facts, much of topology can be done only on the basis of MF. The only addition to MF that must be done throughout the book is a principle of generation by induction and co-induction which is needed to generate positive topologies starting from an axiom-set.

Developing topology on the basis of MF means making explicit and keeping much more pieces of information than usual on stage. These are kept hidden by "strong" foundations because one expects to be able to rebuild them without actually being able to do so, and this means, as a matter of fact, they are not really considered interesting.

Such additional information makes computational interpretation and computer implementation possible at all, or much easier. All of mathematics we do over MF offers no obstacle to its implementation in a proof assistant.

However, there is no reason to fear one has to change jobs: it is mathematics as always, its practice does not change much. The reader will never doubt that it is still clearly mathematics, and not computer code.

On the contrary, since MF enjoys the advantage of being compatible with any other foundation, readers do not need to learn a new foundation. They can keep their own vision, and all what they read remains immediately comprehensible and valid. In other words, using MF makes it easier to increase that part of mathematics that is common to all foundations (contrary to the very unpleasant fact, either of not being able to transfer a result from one foundation to another, or of being able to do so only at the cost of modifying the proof).

Similarly, this new way of doing mathematics acts as an intermediate between the usual way and the formalization required by a proof assistant, and remains the same regardless of the various possible implementations. The advantage over implementations from classical mathematics directly to computer code is that the information needed for implementation is maintained (and not abstracted, as usual) and is included in the mathematics itself. This allows an easier accumulation of knowledge on the formalization of mathematics in a proof assistant.

In addition, introducing this new way of doing mathematics makes it possible for mathematicians, and not just computer scientists, to work towards formalization in a computer. This is why we introduce the extensional level of MF and extensional notions by way of propositions (subset, relation, etc.). Lack of usual extensional notions is, in my opinion, the main reason why mathematics on the basis of Martin-Löf type theory was not much developed so far.

If what can be done of topology to someone seems little or trivial, or reached through a too long path, just think that it is all effective, implementable. If we keep in mind that what we are doing is effective topology, then it becomes clear that we can do a lot, indeed it is surprising that we can do virtually everything.

Summing up, MF works very well: it allows us to go beyond Bishop, both in mathematics (topology, axiomatic method, algebra,...), and in foundations (formalization, implementation).

To get all this a price has been paid: the main task was to start everything anew from the beginning and find the (constructively) correct basic topological definitions. This is very different from reformulating constructively the definitions of classical mathematics.

It is a widespread belief that there is only one mathematics, the classical one, and that the purpose of constructivism is to develop as much as possible of it while avoiding "strong" but constructively unreliable principles, such as the principle of the excluded middle or impredicative definitions. This is the opinion of Bishop (1967), for example.

With this perspective, the enthusiasm of discovery is only in the hands of classical mathematicians; what remain to constructivists is only the "moral" satisfaction of a convincing proof. It is then no wonder if, under this perspective, constructive mathematics is considered sufficiently interesting only by mathematicians with particular sensitivities. In fact, what can be the appeal of finding only new proofs, following more stringent criteria, of well-known facts about well-known concepts?

However, things are not like that at all. A dynamic and pluralist vision leads to a different conclusion. If a foundation is the choice of the criteria with which one abstracts from reality to obtain mathematical entities, then a different style of abstraction, that is a different foundation, gives rise to different entities, which classical mathematicians neglect de facto, and therefore with them a different mathematics, with different problems.

In my opinion, the most fascinating part of research in constructive mathematics is that it reveals new structures and new ways of mathematical thought. But to create new mathematics, the foundation must be used in practice, and not only studied meta-mathematically, beginning with definitions, without any dependence on or deference towards the consolidated classical or impredicative tradition. One starts from scratch, there is no need for ZFC.

This was done for the Minimalist Foundation. The reward of this patient work is the surprising and intriguing discovery that the use *in practice* of a different foundation, much weaker than previous ones, has fostered the emergence of a new mathematics, new concepts and new structures that had remained entirely unnoticed, almost hidden by too strong assumptions such as PSA, LEM and AC.

Thus it turns out that the additional information required and retained by MF to maintain certain conceptual distinctions (see Sect. 21.1.5) has its own clear "logical" structure, which lies under the usual notions and results, and which actually illuminates usual topology, leading to results that are new also for the usual approach (classical or impredicative).

I insist on concrete development because the most interesting thing is that in order to carry out what is classically done by applying "strong" principles, we are forced by practice to invent new concepts and new tools.

A typical example is given by overlap, a notion that one must invent to be able to speak smoothly and constructively of the inhabited intersection of two subsets $D, E \subseteq X$. Formally, one puts:

$$D \between E \equiv (\exists x \in X)(x \epsilon D \ \& \ x \epsilon E).$$

It is useless in classical mathematics, since there it is equivalent to $D \cap E \neq \emptyset$. But, as we will see, the use of $\between$ in practice pulls out many structures that previously had remained hidden.

The use of a foundation that preserves a greater distinction between concepts is analogous to the use of a fishing net with narrower meshes: it takes much more to the surface. It is also like using a scalpel instead of a razor. The use of new tools or means of exploration (telescope, microscope, off-road vehicle, laser, computer, soft climbing shoes,...) has always led to new, previously unimaginable discoveries. Similarly, the use of the new MF foundation has allowed us to bring out some new, very clear and solid facts, as we are now going to see.

The production of a new mathematics is an important confirmation of the validity of the foundational setting. Of this we speak from now on.

21.2.1 Duality, Symmetry and Other Structures at the Base of Topology

According to the common definition, a topology ΩX on a set of points X is a subcollection $\Omega X \subseteq \mathcal{P}X$. Usually, one just says ΩX is a given subset of $\mathcal{P}X$. Recall that a better management of information means we distinguish sets from collections, and hence do not assume Power Set Axiom PSA. Moreover, constructively we should explain what it means that $\Omega X \subseteq \mathcal{P}X$ is given.

The best we can do is to start from a base for open subsets, that is, a family of subsets ext (a) $\subseteq X$ $(a \in S)$ indexed on a second set S, called basic neighbourhoods. And then obtain ΩX by closing under (set-indexed) unions over these.

Equivalently, one can reduce to $(X, \Vdash, S)$, where $\Vdash$ is the relation defined by $x \Vdash a \equiv x \epsilon \text{ext}(a)$. So the basic structure is two sets linked by a relation; one set of points, one set of indexes, or observables, and a relation $\Vdash$ linking the two. I have called it a basic pair.

The presence of the second set S is the source of several discoveries. Actually, topology can be seen as the result of the dynamics between two sets X and S, through the relation $\Vdash$. In this setting, one can define basic notions of topology, like open, closed, cover, positivity, continuity,..., and see that they have a deep structural justification, based on symmetry and logical duality. This was not visible before, since using PSA the topology ΩX itself is considered to be a set, and hence the presence of S is not necessary.

The interior of a subset is usually defined to contain all points which are in some basic neighbourhood that is all contained in the subset. Similarly, the closure of a subset is all points which, whenever they are in a neighbourhood, that neighbourhood overlaps the subset. Then in a basic pair interior and closure of a subset $D \subseteq X$ are defined by:

$$x \in \operatorname{int} D \equiv \exists a(x \Vdash a \ \& \ \operatorname{ext} a \subseteq D), \qquad x \in \operatorname{cl} D \equiv \forall a(x \Vdash a \rightarrow \operatorname{ext} a \between D).$$

Since $E \between D$ is the logical dual of $E \subseteq D \equiv (\forall x \in X)(x \in E \rightarrow x \in D)$, the first discovery is that the formulas defining int and cl are exactly the logical dual of each other, obtained one from another by swapping $\forall$, $\exists$ and $\rightarrow$, $\&$. This had not been seen assuming classical logic, because one "simplifies prematurely" using excluded third and double negation.

By analysing definitions, one can see that int and cl are obtained by composing more elementary operators between $\mathcal{P}X$ and $\mathcal{P}S$. We define $\Diamond, \Box$, ext, rest, by putting:

$$\begin{array}{ll} a \in \Diamond D \equiv \operatorname{ext} a \between D & \qquad x \in \operatorname{ext} U \equiv \Diamond x \between U \\ a \in \Box D \equiv \operatorname{ext} a \subseteq D & \qquad x \in \operatorname{rest} U \equiv \Diamond x \subseteq U \end{array}$$

where $\Diamond x \equiv \{a \in S : x \Vdash a\}$. Since we need it later anyway, we recall here that ext, rest, $\Diamond, \Box$ are nothing but the four possible images of a subset along a relation $r : X \rightarrow S$ when $r \equiv \Vdash$. The use of $\between$ to express an existential quantification, besides $\subseteq$ to express a universal one, makes definitions clear and compact:

$$\begin{array}{llll} \text{existential image} & a \in r\,D \equiv r^{-}a \between D & \text{existential anti-image} & x \in r^{-}U \equiv r\,x \between U \\ \text{universal image} & a \in r^{-*}\,D \equiv r^{-}a \subseteq D & \text{universal anti-image} & x \in r^{*}U \equiv r\,x \subseteq U \end{array} \qquad (21.1)$$

where $rx \equiv \{a \in S : xra\}$ for all $x \in X$, and $r^{-}a \equiv \{x \in X : xra\}$ for all $a \in S$. So $\Diamond$, ext are existential and $\Box$, rest are universal. Then by definitions it is

$$\operatorname{int} = \operatorname{ext} \Box, \qquad \operatorname{cl} = \operatorname{rest} \Diamond,$$

that is, int and cl are nothing but a mathematical representation of quantifiers of the form $\exists\forall$ and $\forall\exists$, respectively. This is evident if one uses relativized quantifiers, since then $x \in \operatorname{int} D \equiv (\exists a \in \Diamond x)(\forall x \in \operatorname{ext} a)(x \in D)$ and $x \in \operatorname{cl} D \equiv (\forall a \in \Diamond x)(\exists x \in \operatorname{ext} a)(x \in D)$, so that the formal difference between int and cl is just the swap between $\exists$ and $\forall$.

With no conditions on $\Vdash$, the requirements on the structure $\mathcal{X} = (X, \Vdash, S)$ are perfectly symmetric. So $\mathcal{X}^{sym} \equiv (S, \Vdash^{-}, X)$, where $a \Vdash^{-} x \equiv x \Vdash a$, is also a basic pair. Then a second discovery is that we can define also other operators $\mathcal{J}, \mathcal{A}$ on the side of S, so-called formal side, which are symmetric of int, cl, respectively:

$$\mathcal{J} = \Diamond \operatorname{rest}, \quad \mathcal{A} = \Box \operatorname{ext}.$$

In fact, $\Diamond$ is symmetric of ext, and $\Box$ of rest, that is, $\Diamond$ coincides with ext^{sym}, the operator ext in the symmetric basic pair $\mathcal{X}^{sym}$, and similarly for rest.

As we now see, this is the beginning of pointfree topology. $\mathcal{A}$ was already known. The novelty here is the introduction of the operator $\mathcal{J} \equiv \Diamond\ \mathsf{rest}$, dual of $\mathcal{A}$, and the definition of $U \subseteq S$ as formal closed if $U = \mathcal{J}U$.

Since $\mathsf{ext} \dashv \Box$ and $\Diamond \dashv \mathsf{rest}$ are adjunctions, the operators $\mathsf{int}, \mathcal{J}$ are reductions (contractive, monotone, idempotent) while $\mathsf{cl}, \mathcal{A}$ are saturations (expansive, monotone, idempotent). Moreover, open subsets of X, that is fixed point for int, are all and only those of the form $\mathsf{ext}\,U$, for some $U \subseteq S$. They form a complete lattice which is isomorphic to fixed points for $\mathcal{A}$, which are hence called formal open subsets. By duality, something fully analogous holds for closed subsets. The situation is summarized in the following nice drawing, reminding some from the Middle Ages:

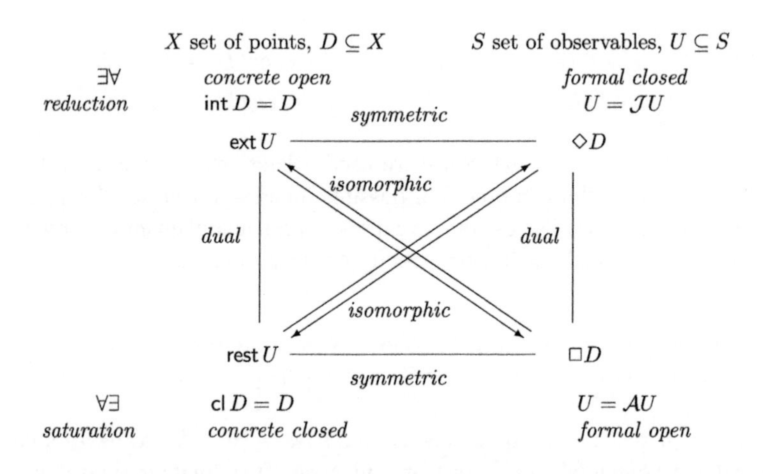

After lecturing about these structures for over twenty years in the five continents, to the best of my knowledge I can say that they were never published before. It is an example of results which were possible in the 1930s, but one had to wait for a logician in the 1990s to discover them, so sixty years later than what could happen, simply because he looked for something more informative than the classical foundation and did not stop! This is the real reason of these satisfactory discoveries; a long delay due only to a different mindset. They were buried under ideology, namely excess of assumptions.

In fact, assuming classical logic with LEM, the positive information of the form $\forall\exists$ which characterizes cl is identified with, and hence replaced by, the combination $\neg\exists\forall\neg$, that is absence or negation of an information of the form $\exists\forall$, which characterizes int. Thus closed subsets are uniquely determined by open subsets by way of negation, or complement, and closure is explicitly definable by $\mathsf{cl} \equiv -\,\mathsf{int}\,-$ (complement of interior of complement). In conclusion, the strict logical duality between int and cl is reduced to complementation, and hence becomes invisible. Even without LEM, closure can be uniquely determined by interior through an impredicative definition. As a consequence, closed subsets and the operator cl are anyway by and large ignored.

In our approach, closure and interior are mathematical notions representing two combinations of quantifiers, $\forall\exists$ and $\exists\forall$ respectively, which cannot be reduced to each other by negation. So they must be treated independently.

Assuming PSA makes the second set S useless, and hence symmetry cannot emerge. In fact, ΩX itself is a set, and then $(X, \Omega X)$ amounts to the basic pair $(X, \epsilon, \Omega X)$ with $\Vdash = \epsilon$. So the definition of $\mathcal{J}$ also cannot emerge (as it happened, as a matter of facts).

This is sufficient, in my opinion, to disprove the claim that the classical paradigm is just "absolute truth".

The discovery of duality and symmetry manifests a clear logical structure underlying the notions of open and of closed subsets. It is then natural to investigate whether the same or similar structure underly also other notions or results of topology. It has been exciting to make in this way some other discoveries, which were previously made de facto impossible due to the excessive "power" of ZFC.

The next discovery is that in the framework of basic pairs one can characterize continuity in a simple structural way. In fact, the presence of bases S, T allows one to see that a function $f : X \to Y$ is continuous from $(X, \Vdash, S)$ into $(Y, \Vdash', T)$ if and only if there is a relation s between S and T such that the square is commutative, namely $\Vdash' \circ f = s \circ \Vdash$. Then by symmetry it is natural to consider a relation r also between X and Y, and continuity becomes $\Vdash' \circ r = s \circ \Vdash$. Formally, one can easily prove the equivalence of all the following conditions:

1. r is continuous, that is $r\,x \between \mathrm{ext}\, b \to \exists a(x \Vdash a \;\&\; \mathrm{ext}\, a \subseteq r^{-}\mathrm{ext}\, b)$,
2. r^{-} is open,
3. $r^{-}\mathrm{ext}\, b = \mathrm{ext}\,(s_r{}^{-}b)$ for all $b \in T$, where $a\, s_r\, b \equiv \mathrm{ext}\, a \subseteq r^{-}\mathrm{ext}\, b$,
4. there exists a relation $s : S \to T$ such that $\Vdash' \circ r = s \circ \Vdash$.

In other words, continuity is the same as commutativity of a square of relations between sets:

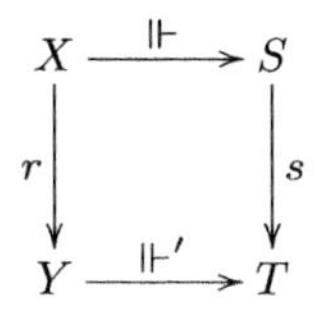

So, when continuity gets free from its bond with functions and when bases are explicit, a fully structural characterisation of it becomes visible. After looking for a reference with no success, I believe this also is a new discovery. It had not be seen classically, since one gets rid of bases "too early" on the assumption they can be reconstructed from ΩX.

Passing from functions to relations gives of course greater generality (a fact which widens the range of applications), but what is interesting here is that it makes the underlying logical structure more explicit. Restricting to functions somehow hides logic, since for every function f the equation $f^- = f^*$ holds, that is, existential and universal anti-images coincide. Finally, as we see below, with relations one can finally give mathematical evidence to the old claim that the pointfree approach is more general than traditional pointwise topology.

I cheated a little, when I said that a basic pair is sufficient to build a topological space. In fact, traditionally $(X, \Omega X)$ is a topological space if open subsets ΩX are closed under unions and finite intersections. Up to here, $\Vdash$ in a basic pair $(X, \Vdash, S)$ is *any* relation. If open subsets ΩX are defined as those of the form $\mathrm{ext}\, U \equiv \cup_{b \epsilon U} \mathrm{ext}\ b$ for some $U \subseteq S$, then by its definition ΩX is closed under unions.

Then we recall that ΩX is closed also under finite intersections if and only if $\mathrm{ext}\,(a) \subseteq X$ $(a \in S)$ is a base for a topology. In our framework this amounts to a condition on $\Vdash$ which, putting $c \,\epsilon\, a \downarrow b \equiv \mathrm{ext}\, c \subseteq \mathrm{ext}\, a\ \&\ \mathrm{ext}\, c \subseteq \mathrm{ext}\, b$, can be expressed by two simple equations:

B1: $\mathrm{ext}\, a \cap \mathrm{ext}\, b = \mathrm{ext}\,(a \downarrow b)$ every two neighbourhoods have a common refinement,
B2: $\mathrm{ext}\, S = X$ every point has a neighbourhood.

Validity of B1 and B2 together is called *convergence*. A *concrete space* is a basic pair $(X, \Vdash, S)$ satisfying convergence.

Since all the rest is a simple application of logic, the discovery here is that convergence is the only mathematical postulate to obtain topological spaces.

We now see how this idea is extended to morphisms. The trick is to introduce a notion of point "in the sense of topology", that is, which can be characterized in terms of open and of closed subsets. So we say that a subset of points $D \subseteq X$ is *convergent* if:

D is strongly inhabited: $D \between \mathrm{ext}\, a$ for some a,
D is filtering: $D \between \mathrm{ext}\, a\ \&\ D \between \mathrm{ext}\, b \rightarrow D \between \mathrm{ext}\,(a \downarrow b)$ for all a, b,

and that a subset $\alpha \subseteq S$ is an *ideal point of* $\mathcal{X}$ if:

α is inhabited,
α is filtering: $D \between \mathrm{ext}\, a\ \&\ D \between \mathrm{ext}\, b \rightarrow D \between \mathrm{ext}\,(a \downarrow b)$ for all a, b,
α is formal closed.

A consequence of the dynamics between X and S is that $\Diamond$, rest form a bijection between convergent subsets and ideal points in $\mathcal{X}$. Thus the notion of convergent subset is a properly constructive substitute of the notion of irreducible closed subset, see Johnstone (1982).

Then the following are equivalent expressions of convergence of $\mathcal{X}$:

B1 and B2 hold, so $\mathcal{X}$ is a concrete space,
every singleton in X is convergent, written $\mathcal{P}_1 X \subseteq Conv(\mathcal{X})$,
every $\Diamond x$ is an ideal point of $\mathcal{X}$.

A commutative square (r, s) is said to be convergent if it satisfies any of the following equivalent conditions:

r maps convergent subsets into convergent subsets,
s maps ideal points into ideal points,
r^{-} respects finite intersections,
$r^{-}\text{ext}\,(b \downarrow_{\mathcal{Y}} c) = \text{ext}\,(s^{-}b \downarrow_{\mathcal{X}} s^{-}c)$, for all $b, c \in T$, and $r^{-}\text{ext}\,T = \text{ext}\,S$.

The idea is that such an r is a "function in a topological sense", rather than a set-theoretic one (which means it maps singletons into singletons).

So we have two categories. **BP** is the category of basic pairs and relation-pairs (r, s), that is commutative squares, up to a suitable notion of equality (which also turns out to have a purely structural nature, namely equality of diagonals). It is essentially only applied logic, in the sense that all its notions are just an expression of certain combinations of quantifiers. The category **CSpa** of concrete spaces is obtained as a subcategory of **BP** by requiring convergence, both on objects and on morphisms.

All proofs are very elementary, once definitions and statements are clear. With classical "eyes", it is very hard to see these little facts; and in fact they had not be seen for over sixty years. Building on them, one can develop topology with points in a constructive and very structural way, showing that logic and topology are much closer to each other than one would expect before. This is, in my opinion, of great foundational interest.

21.2.2 Pointfree Topology Is Real

The category of concrete spaces provides a satisfactory constructive version of topological spaces. However, the book Sambin (2020, to appear) is mostly on the pointfree approach to topology. Why do we need pointfree topology?

Adopting an effective notion of set, that is distinguishing sets from collections, means that in most examples of topological spaces points form a collection, and not a set. However, it is a matter of facts that in all such examples there is a family of basic neighbourhoods indexed on a set S. That is, effectivity lies on the base, which means on the side of observables, and not on the side of points. So, since our aim is to maximize effectivity, we start from the effective part, namely the pointfree side, make it as rich as possible, and introduce points and spaces with points as ideal notions, using them as "late" as possible. The deep motivation for the pointfree approach is as simple as that.

The general method to introduce and justify all pointfree definitions is *to axiomatise the structures induced on the formal side*, that is, deposit as much as possible of the categories of basic pairs and of concrete spaces on their formal side, and then introduce pointfree notions as an axiomatization of such structures. As we now see in practice, this method is very efficient. Owing to the discovery of symmetry and of logical duality, it produces pointfree notions which are definitely richer than one would previously expect.

Axiomatizing the structure deposited by a basic pair $\mathcal{X} = (X, \Vdash, S)$ on S produces the notion of *basic topology*, namely a triple $(S, \mathcal{A}, \mathcal{J})$ satisfying:

$\mathcal{A}$ saturation: $U \subseteq \mathcal{A}V \leftrightarrow \mathcal{A}U \subseteq \mathcal{A}V$,
$\mathcal{J}$ reduction: $\mathcal{J}U \subseteq V \leftrightarrow \mathcal{J}U \subseteq \mathcal{J}V$,
$\mathcal{A}/\mathcal{J}$ compatibility: $\mathcal{A}U \between \mathcal{J}V \leftrightarrow U \between \mathcal{J}V$.

Breaking the symmetry, when $\mathcal{X}$ is a concrete space the trace on S is validity of the equation $\mathcal{A}U \cap \mathcal{A}V = \mathcal{A}(U \downarrow V)$, where $U \downarrow V \equiv \cup_{a \in U, b \in V} a \downarrow b$ and $a \downarrow b \equiv \mathcal{A}\{a\} \cap \mathcal{A}\{b\}$; we call it pointfree convergence. A *positive topology* is a basic topology satisfying (pointfree) convergence.

The saturation $\mathcal{A}$ was part of the original definition of formal topology in Sambin (1987). Moreover, pairs $(S, \mathcal{A})$ with $\mathcal{A}$ satisfying convergence can be seen as a predicative presentation of locales (Sambin 1987; Battilotti and Sambin 2006). If they are equipped with a predicate Pos as required in Sambin (1987), they correspond to open locales. The novelty of positive topologies, both for constructive and classical mathematics, is the presence of the reduction $\mathcal{J}$ besides $\mathcal{A}$, and the condition of compatibility to connect them. This makes them substantially more general than formal topologies and locales.

Perhaps expressing the content of compatibility on the usual side of points might help to appreciate its novelty. By the symmetry of basic pairs, the operators int and cl on X also give rise to a basic topology, that is, int is a reduction, cl is a saturation, and their link is compatibility $\text{cl}\, D \between \text{int}\, E \leftrightarrow D \between \text{int}\, E$ for all $D, E \subseteq X$. Compatibility here means (as one can check with simple logical manipulations) that all points in $\text{cl}\, D$ are also limit points for D in the topology of all open subsets.[5]

Although $\mathcal{J}$ is new, from the point of view of the underlying logical structure the introduction of $\mathcal{A}$ and $\mathcal{J}$ are equally natural. $\mathcal{A}$ is a way to express quantifications of the form $\forall\exists$, that is $\Box\text{ext}$, and $\mathcal{J}$ of the form $\exists\forall$, that is $\Diamond$ rest. The interaction between the saturation $\mathcal{A}$ and the reduction $\mathcal{J}$ through compatibility is a trace of the logical duality between $\forall\exists$ and $\exists\forall$. One can only guess that $\mathcal{J}$ went unobserved because to express compatibility quickly and perspicuously one needs overlap, and incredibly a specific notation for it, such as the sign $\between$ adopted here, had not been systematically used before.[6]

The presence of $\mathcal{J}$ considerably strengthens the expressive power of pointfree topology. In fact, it provides a primitive treatment of closed subsets in pointfree terms by defining $U \subseteq S$ to be *formal closed* when $\mathcal{J}U = U$.

The fact that $\mathcal{J}$ was lacking in the constructive tradition is particularly surprising. Assuming classical logic, in every basic pair the equation $\mathcal{J} = -\mathcal{A}-$ holds, exactly

[5] Beware that limit points in this sense require an impredicative definition, and thus we cannot use them to define closure in terms of interior.

[6] The notation $\between$ is now rapidly spreading in the community of constructive mathematicians.

for the same reasons as $\mathrm{cl} = -\mathrm{int}-$ above. So in every structure $(S, \mathcal{A})$ one can define $\mathcal{J}$ by putting $\mathcal{J} \equiv -\mathcal{A}-$ and treat formal closed subsets thus. But this choice is not viable constructively, because it does not correspond to the underlying logical structure and negation (and hence complementation) behaves very differently. So, if constructivists only use open subsets, they will prevent themselves from using closed subsets at all, while this is something which classical mathematicians can do, in their way. For constructivists, competing with classical mathematicians by adopting their notions and treating them in a constructive foundation is like fighting while keeping one hand on the back! With this handicap (restrict to classical definitions and treat them constructively), they know even before the start that they will loose the competition.

That is why we *need* to have an independent notion of closed subsets, both in pointwise and in pointfree topology. I believe there is no valid reason for constructivists to refrain from their introduction. Much of the debate on constructivism should be revised at the light of this perspective.

After basic pairs and basic topologies, the independent treatment of the notions of open and closed is systematically kept in all subsequent definitions, beginning with that of their morphisms. We take the properties of s in a relation-pair (r, s) as defining conditions and say that a continuous relation from $(S, \mathcal{A}, \mathcal{J})$ into $(S', \mathcal{A}', \mathcal{J}')$ is a relation s from S into S' such that s preserves formal closed and s^{-*} preserves formal open subsets, that is, if $U = \mathcal{J}U$ and $V = \mathcal{A}V$, then $sU = \mathcal{J}'sU$ and $s^{-*}V = \mathcal{A}'s^{-*}V$, respectively. These conditions are an immediate consequence of the underlying combinations of quantifiers. As positive topologies are obtained from basic topologies by adding convergence, their morphisms are obtained by adding only its preservation. If $(r, s) : \mathcal{X} \rightarrow \mathcal{Y}$ is a convergent relation-pair, then s satisfies $\mathcal{A}s^{-}(b \downarrow c) = \mathcal{A}(s^{-}b \downarrow s^{-}c)$ and $\mathcal{A}s^{-}T = \mathcal{A}S$; we call formal map any continuous relation satisfying these two conditions. With these morphisms, basic topologies and positive topologies become categories, called **BTop** and **PTop**, respectively.

Convergence aside, all definitions, including those of arrows, seem to flow almost by themselves out of purely logical considerations. I have chosen the name Basic Picture for the theory of basic pairs and of basic topologies because it can be described as the study of pure logical dynamics between sets, when they are connected by relations. In this sense, it is nothing but applied logic.

At the same time, this study is carried on by means of definitions which have a strong topological meaning. In a certain sense, topological notions can be viewed as a quick way to deal with different combination of quantifiers. So the basic picture is also a generalization of topology.

Topology in the stricter sense, either with points as in concrete spaces or pointfree as in positive topologies, is obtained as a special case simply by adding at will a single mathematical assumption, the module of convergence.

So the basic picture provides elegant, purely logical structures underneath topology, which means a safe foundation to topology and a deeper understanding of familiar notions. In a few words, logic and topology are much closer (constructively) than one would imagine before (classically).

All these structures were left in the dark by the classical perspective since there they collapse to trivial properties of complements, which can be understood as purely mathematical. They remain visible only with the finer distinctions imposed by constructive logic; actually, this could be a good reason for some classical mathematicians to learn it.

An interesting novelty, which was not expected at the beginning, is that our "weak" framework finally offers a precise mathematical expression and proof to the expectation, which is well present since the time of Grothendieck, of a notion of space generalizing that of topological space. Indeed, we achieve this with positive topologies.

The well-known link between pointfree and standard topology (with points) is a categorical adjunction between topological spaces **Top** and locales **Loc**. Such adjunction can be considerably strengthened and becomes an embedding, provided that topological spaces are replaced with concrete spaces and locales with positive topologies. That is, we can prove that the category of concrete spaces **CSpa** embeds into that of positive topologies **PTop**. By what precedes, it is clear that the embedding functor maps objects and morphisms of **CSpa** into the structures they induce on their formal side.

This gives a precise statement, and a rigorous proof in terms of categories, to the original and deepest motivation of pointfree topology, namely the claim that it is more general than traditional topology with points. Contrary to a common belief, this is an example showing that constructive mathematics can provide results which are new also for classical mathematicians. It is not known whether such an embedding is possible without exploiting the presence of the new notion of positivity relation $\ltimes$.

One of the specific merits of formal topology is the introduction of effective methods (generation by induction) in topology, see Sambin (1987) and in particular Coquand et al. (2003). This highlights the deepest and most interesting role of topology, in my opinion, that is the interplay between effective/discrete and ideal/continuous. With the presence of $\mathcal{J}$ we see, as a further novelty, that also coinductive methods appear on the stage.

In order to study generation of $\mathcal{A}$ and $\mathcal{J}$ on a set S, it is better to look at them equivalently as relations. We call cover the relation $\lhd$ (between elements and subsets) defined by $a \lhd U \equiv a \,\epsilon\, \mathcal{A}U$; dually, the positivity relation $\ltimes$ is defined by $a \ltimes U \equiv a \,\epsilon\, \mathcal{J}U$. The operator $\mathcal{A}$ is a saturation if and only if $\lhd$ is reflexive ($a \,\epsilon\, U \rightarrow a \lhd U$) and transitive ($a \lhd U \;\&\; U \lhd V \rightarrow a \lhd V$, where $U \lhd V \equiv (\forall a \,\epsilon\, U)(a \lhd V)$). Similarly for $\mathcal{J}$ and $\ltimes$. In this language, compatibility is expressed by $a \lhd U \;\&\; a \ltimes V \rightarrow U \ltimes V$, where $U \ltimes V \equiv (\exists b \,\epsilon\, U) b \ltimes V$.

In the same way as cover $\lhd$ is generated by induction (Coquand et al. 2003), now dually positivity $\ltimes$ is generated by coinduction (Martin-Löf and Sambin 2020, To appear). To do this, we start from an axiom-set, that is, any set-indexed family of subsets $C(a, i) \subseteq S$ ($a \in S, i \in I$),[7] which we think of as axioms saying that

[7]The case in which I depends on a, as in Coquand et al. (2003), is only apparently more general.

$C(a, i)$ covers a, for every $i \in I$. The induction and coinduction rules have the form:

$\lhd$-generation: for all $a \in S, i \in I$

$$\text{refl}\ \frac{a \epsilon U}{a \lhd U} \qquad \text{trax}\ \frac{i \in I \quad C(a,i) \lhd U}{a \lhd U} \tag{21.2}$$

$\lhd$-induction: for all $a \in S$, $W \subseteq S$

$$\frac{a \lhd U \quad U \subseteq W \quad \begin{array}{c}[i \in I, C(b,i) \subseteq W] \\ | \\ b \epsilon W\end{array}}{a \epsilon W} \tag{21.3}$$

$\ltimes$-cogeneration: for all $a \in S, i \in I$

$$\text{corefl}\ \frac{a \ltimes U}{a \epsilon U} \qquad \text{cotrax}\ \frac{i \in I \quad a \ltimes U}{C(a,i) \ltimes U} \tag{21.4}$$

$\ltimes$-coinduction: for all $a \in S$, $W \subseteq S$

$$\frac{a \epsilon W \quad W \subseteq U \quad \begin{array}{c}[b \epsilon W, i \in I] \\ | \\ C(b,i) \between W\end{array}}{a \ltimes U} \tag{21.5}$$

One can see that $\lhd$ is the least cover such that $a \lhd C(a, i)$, or equivalently $C(a, i) \subseteq \mathcal{A}U \to a \lhd U$, holds for all $a \in S, i \in I$. Dually, $\ltimes$ is indeed the greatest positivity such that $a \ltimes U \to C(a, i) \between \mathcal{J}U$. Moreover, $\ltimes$ is compatible with $\lhd$, so that $(S, \lhd, \ltimes)$ is a basic topology. We refer to Coquand et al. (2003) for technical details to be imposed on the axiom-set to obtain that the resulting cover is convergent, so that $(S, \lhd, \ltimes)$ is a positive topology.

To make the above result meaningful, we need to postulate, as anticipated, that indeed (21.2), (21.3) and (21.4), (21.5) define two relations between elements and subsets; this is the principle called `ICAS`, Induction and Coinduction on Axiom-Sets.

The above theorem of generation is a central result for a development of topology which aims to be as effective as possible. Reflecting on the fact that one starts from *any* axiom-set, its generality should be evident. All common example of spaces, as real numbers, Baire space, Zariski topology, Scott domains, etc., fall under its range and thus can be defined in an effective way.

We have a look only at the example of Baire space, since it is the most illuminating on the interaction between real and ideal aspects. We need to spend a few words on notation. We write $\mathbb{N}^*$ for the set of lists of natural numbers. We write $k * l$ for the concatenation of k, l and extend this notation to sets by putting $k * \mathbb{N} \equiv \{k * n : n \in \mathbb{N}\}$. We define two operations on lists such that $k = tail(k) * head(k)$ holds for all $k \in \mathbb{N}^*$.

We write $\lhd_{\mathbb{N}}$ and $\ltimes_{\mathbb{N}}$ for the relations generated inductively by the rules in the first line and coinductively by those in the second line, respectively:

$$refl. \ \frac{k \epsilon U}{k \lhd_{\mathbb{N}} U} \qquad \zeta \ \frac{k \lhd_{\mathbb{N}} U}{k * a \lhd_{\mathbb{N}} U} \qquad \digamma \ \frac{k * S \lhd_{\mathbb{N}} U}{k \lhd_{\mathbb{N}} U}$$

$$corefl. \ \frac{k \ltimes_{\mathbb{N}} U}{k \epsilon U} \qquad \frac{k * a \ltimes_{\mathbb{N}} U}{k \ltimes_{\mathbb{N}} U} \qquad \frac{k \ltimes_{\mathbb{N}} U}{k * S \ltimes_{\mathbb{N}} U}$$

The rules ζ and $\digamma$ are exactly Brouwer's rules of inductive generation of bars (see for instance Martino and Giaretta 1981) in the present notation. One can check formally that they are equivalent to an axiom-set. In fact, we choose I to contain only two elements, $\flat$ and $\sharp$, and put $C(k, \flat) \equiv \{tail(k)\}$, $C(k, \sharp) \equiv k * S$. Then the general scheme (21.2) for $\flat \in I$ gives

$$\frac{tail(k) \lhd U}{k \lhd U} \quad \text{for all } k \in S^*,$$

which clearly is equivalent to ζ for all $k \in S^*$. For $\sharp \in I$ we get exactly $\digamma$. Then by the general result $(\mathbb{N}^*, \lhd_{\mathbb{N}}, \ltimes_{\mathbb{N}})$ is a positive topology, called Baire positive topology. The same procedure works also after replacing $\mathbb{N}$ with any set; if we chose a set only with 0, 1 as elements, we obtain Cantor positive topology.

By applying (21.5) above with $W \equiv U$ one derives that U is formal closed (that is $a \epsilon U \leftrightarrow a \ltimes U$) if and only if U splits the axioms, that is $b \epsilon U \ \& \ i \in I \rightarrow C(b, i) \between U$ holds for all $b \in S, i \in I$. So, in the case of Baire positive topology, $U \subseteq \mathbb{N}^*$ is formal closed if and only if

$$k \epsilon U \rightarrow k * \mathbb{N} \between U \quad \text{and} \quad k \epsilon U \rightarrow tail(k) \epsilon U. \tag{21.6}$$

This shows that an inhabited formal closed subset of Baire positive topology is, apart from the property of decidability, exactly the same as a spread in Brouwer's intuitionism. This fact confirms, in my opinion, the interest of both notions.

We have tried so far to maximize effectivity, and we have seen that this means developing the pointfree approach to topology. It is now the right time to follow what we proposed in Sect. 21.1.2 and treat also ideal aspects. These are expressed by re-introducing points through the notion of ideal point over a positive topology. Thus ideal notions are introduced locally, case by case, and not once and for all, as with actual infinite sets of ZFC.

The general method to obtain pointfree definitions here works particularly well. In fact, we define ideal points of a positive topology $\mathcal{S} = (S, \lhd, \ltimes)$ to be those subsets of S which satisfy the same conditions met by subsets $\Diamond x$ for $x \in X$ in a concrete space $\mathcal{X}$. But this task has already been performed, through the definition of ideal point of $\mathcal{X}$ (page 478). This can be expressed in the language of $\mathcal{S}$ by noting that $D \between \text{ext}\, a$ iff $a \in \Diamond D$. So we say that a subset $\alpha \subseteq S$ is an *ideal point of* $\mathcal{S}$ if:

α is inhabited: $\alpha \between S$,
α is filtering: $a \in \alpha \;\&\; b \in \alpha \rightarrow a \downarrow b \between \alpha$ for all a, b,
α is formal closed: $a \in \alpha \leftrightarrow a \ltimes \alpha$.

To help intuition, which says that α is the subset of its own approximations, we write $\alpha \Vdash a$ for $a \in \alpha$. Then α is filtering if $\alpha \Vdash a \;\&\; \alpha \Vdash b \rightarrow (\exists c \in a \downarrow b)\alpha \Vdash c$. And α is formal closed if and only if α enters $\ltimes$, that is $\alpha \Vdash a \;\&\; \alpha \subseteq U \rightarrow a \ltimes U$. From this, by compatibility, it follows that α splits the cover $\lhd$, that is $\alpha \Vdash a \;\&\; a \lhd U \rightarrow \alpha \between U$.

Ideal points of $\mathcal{S}$ form a subcollection $IPt(\mathcal{S})$ of $\mathcal{P}S$, which in general is not a set. This is not a problem, because there role is only to please our spatial intuition, as we now show. For every positive topology $\mathcal{S}$, we define $Ip(\mathcal{S}) \equiv (IPt(\mathcal{S}), \ni, \mathcal{S})$ (where $\alpha \ni a \equiv a \in \alpha$) and call it the *ideal space* associated with $\mathcal{S}$. As in a concrete space, we can equip $IPt(\mathcal{S})$ with a topology. Open subcollections are of the form $\text{ext}\,(U)$, which contains α if $\alpha \between U$, and closed ones of the form $\text{rest}\,(U)$, which contains α if $\alpha \subseteq U$.

Ip is extended to morphisms of **PTop** by putting $Ip(s) \equiv (s_\exists, s)$ for every formal map $s : \mathcal{S} \rightarrow \mathcal{T}$. Here $s_\exists$ is just a notation for the operation $\alpha \mapsto s\alpha$, where $s\alpha$ is the existential image of α along s. One can show that the conditions defining $s : \mathcal{S} \rightarrow \mathcal{T}$ as a formal map contain what is sufficient to make $s_\exists : IPt(\mathcal{S}) \rightarrow IPt(\mathcal{T})$ a continuous mapping.

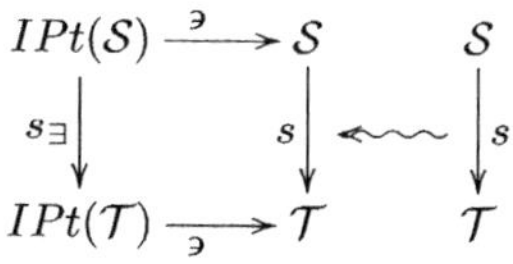

$$
\begin{array}{ccccc}
IPt(\mathcal{S}) & \xrightarrow{\ni} & \mathcal{S} & & \mathcal{S} \\
{\scriptstyle s_\exists}\downarrow & & {\scriptstyle s}\downarrow & \leftsquigarrow & \downarrow{\scriptstyle s} \\
IPt(\mathcal{T}) & \xrightarrow[\ni]{} & \mathcal{T} & & \mathcal{T}
\end{array}
$$

The category of ideal spaces **ISpa** is defined as the "image" of **PTop** under Ip. Conversely, one can define a functor Up : **ISpa** $\rightarrow$ **PTop** which forgets all what Ip added. Thus **ISpa** is isomorphic to **PTop** in a trivial way. However, passing from **PTop** to **ISpa** is not at all useless. The reason is that ideal spaces and ideal maps are a decoration of effective structures with some redundant information which is very useful to support our spatial intuition.

The case of an ideal point of Baire positive topology on $\mathbb{N}^*$ is particularly illuminating and relevant. In fact, as we now see, it provides a rigorous mathematical formulation of Brouwer's notion of choice sequence.

From a purely geometric point of view, it is clear what a choice sequence α is. One can interpret nodes in $\mathbb{N}^*$ as sequences of choices of elements in $\mathbb{N}$ which say how to proceed upwards starting from nil, and which halt after a finite number of choices. Choice sequences are the same, except that they do not halt. So one can see very well that choice sequences are obtained from the notion of finite list $k \in \mathbb{N}^*$ by extrapolation to the infinite.

Brouwer proposed to clarify this intuitive notion by first defining spreads, as recalled above, and then explaining what it means for a choice sequence to be in a spread. I claim that his idea is perfectly well expressed by saying that a choice sequence in the spread U is nothing but and ideal point $\alpha \subseteq U$. In fact, α itself is a spread, since it is inhabited and formal closed. Since it is inhabited, by (21.6) it contains nil. Again by (21.6), if $\alpha \Vdash k \equiv k \epsilon \alpha$ then also $k * \mathbb{N} \between \alpha$, which means that any node in α has at least one successor. But since α is filtering, there is at most one such successor. In other words, a choice sequence for us is the same as a convergent spread, that is an ideal point. An argument confirming this identification is that one can prove that ideal points (of Baire positive topology) coincide with functions (not operations!) from $\mathbb{N}$ to $\mathbb{N}$.

Our notion of choice sequence is ideal (apart from its name, ideal point) since we require no condition of effectivity on the subset α. So choice sequences are an intrinsically infinitary, or ideal notion, while basic neighbourhoods, that is finite sequences, are real.

The definitions of cover and positivity induced by $Ip(\mathcal{S})$ on S are $a \lhd_{Ip(\mathcal{S})} U \equiv \forall\alpha(\alpha \Vdash a \rightarrow \alpha \between U)$ and $a \ltimes_{Ip(\mathcal{S})} U \equiv \exists\alpha(\alpha \Vdash a \ \& \ \alpha \subseteq U)$, respectively. Such definitions are impredicative, since they contain quantifications on collections, and thus they are acceptable only as ideal notions .

One can easily see that, by our definition of ideal point, the original cover $\lhd_{\mathcal{S}}$ on S is always contained in the ideal cover $\lhd_{Ip(\mathcal{S})}$ induced by $Ip(\mathcal{S})$, that is $a \lhd_{\mathcal{S}} U \rightarrow \forall\alpha(\alpha \Vdash a \rightarrow \alpha \between U)$. Dually, $\exists\alpha(\alpha \Vdash a \ \& \ \alpha \subseteq U) \rightarrow a \ltimes_{Ip(\mathcal{S})} U$ holds for every $\mathcal{S}$.

Nothing guarantees that the original $\lhd_{\mathcal{S}}$ on S coincides with $\lhd_{Ip(\mathcal{S})}$ induced on S by $Ip(\mathcal{S})$. This is the reason why in the ideal space corresponding to $\mathcal{S}$ it is convenient to keep $\mathcal{S}$ itself, rather than the naked set S. On the contrary, the equality of the two covers is a very interesting property. We say that $\mathcal{S}$ is spatial if $\lhd_{\mathcal{S}} = \lhd_{Ip(\mathcal{S})}$; this simply follows the definition of spatial locale. Instead, new is the dual notion of $\mathcal{S}$ reducible, which holds if $\ltimes_{\mathcal{S}} = \ltimes_{Ip(\mathcal{S})}$, see Ciraulo and Sambin (2019).

Spatiality and reducibility of $\mathcal{S}$ express important properties of $\mathcal{S}$, which are particularly interesting in the case of Baire positive topology. As first noticed in Fourman and Grayson (1982), spatiality of Baire positive topology is just an equivalent formulation (in our terms) of Bar Induction:

$$\texttt{BI} \qquad \forall\alpha(k \epsilon \alpha \rightarrow U \between \alpha) \rightarrow k \lhd U$$

The dual to spatiality, reducibility, should be valid in Baire positive topology. In fact, it is says that every spread is inhabited by a choice sequence:

$$\texttt{SH} \qquad k \ltimes U \rightarrow \exists\alpha(k \epsilon \alpha \ \& \ \alpha \subseteq U)$$

and this looks somehow implicit in Brouwer's definition of choice sequence.

Both BI and SH are perfectly precise and clear mathematical statements. They are intuitively obvious, one can even think of them as a way to clarify the intuitive notion of choice sequence. But from the perspective of MF and positive topology they look as unprovable. Since we keep ideal and real mathematics well distinct, a way out of this difficulty seems available to us. In fact, most probably we cannot prove BI and SH, but if we could prove meta-mathematically that such ideal principles are conservative over real, pointfree topology, then we could continue to use them without affecting effectivity.[8] Trying to prove conservativity of BI and SH over Baire positive topology is work in progress.

This expression of the notion of choice sequence is very fragile, since it depends on foundational choices. Assuming axiom of unique choice AC! amounts to killing the distinction between lawlike sequences (that is, sequences which follow a law) and choice sequence, and every sequence is lawlike (as in Bishop). Absence of AC!, as in MF, shows that Brouwer and Bishop were talking about two different notions, and thus can reconcile them. Thus we can reconcile spatial intuition with computational interpretation, in the sense that they can to live together without ignoring each other, if we pass from one flat system (of which we can ask only consistency) to a system with two levels, the real (positive topologies, pointfree) and the ideal (ideal spaces, pointwise), so that having two languages and theories, we can speak about conservativity of one with respect to the other. I do not mean here to identify literally Hilbert's real, or finitary mathematics with pointfree topology, but rather suggest that both are self-justified and used to justify ideal mathematics.

21.2.3 The Dark Side of the Moon: Mathematizing Existentials

Adopting classical logic, mathematical notions which are defined by a formula containing an existential quantifier can be reduced to other notions by negation. The first example is $D \between E \equiv \exists x(x \in D \ \& \ x \in E)$, which is classically equivalent to $\neg\forall x\neg(x \in D \ \& \ x \in E)$, and hence $D \between E$ is definable as $D \cap E \neq \emptyset$. Similarly, the quantification $\exists\forall$ defining $\text{int} = \text{ext}\,\Box$ is equivalent to $\neg\forall\exists\neg$, so that int is classically definable as $-\ \text{cl}\,-$. By symmetry, $\mathcal{J}$ is classically definable by putting $\mathcal{J} \equiv -\mathcal{A}-$. This is clearly not possible constructively since $\exists$ has a positive character which cannot be expressed by way of negation.

The relation $\between$ and the operator $\mathcal{J}$ are new mathematical notions introduced in order to express such existential quantifiers. They seem to be the right tools to gain access to and explore a whole new territory of mathematics, the *mathematization of existentials*, that is, a positive mathematical treatment of statements involving the existential quantifier. This had remained invisible to a classical perspective, because

[8]Note that Kleene's construction proving that $\text{BI} + \text{AC!} + \text{CT} \vdash \bot$, which is commonly understood as showing that effectivity is incompatible with choice sequences and Bar Induction, is not a problem here because of the absence of AC!.

of the excessive strength of its negation. I like to call it *the dark side of the moon*. It looks really paradoxical to me that $\between$ and $\mathcal{J}$ had been ignored for so long also in the constructive tradition.

Using $\between$ and $\mathcal{J}$, which express $\exists$ and $\exists\forall$ in an abstract form, one can see that every notion of traditional topology is accompanied by its existential dual, which in classical mathematics was definable by negation. So formal closed subsets are dual to open ones and basic topologies besides $\mathcal{A}$ contain its dual $\mathcal{J}$, which means that cover $\lhd$ is accompanied by positivity $\ltimes$, arrows preserve the universal $\mathcal{A}$ as well as existential $\mathcal{J}$, generation by induction is accompanied by that by coinduction, etc.

Working with $\between$ and $\ltimes$ in the practice of research demands a sort of internal revolution; in authentic research, in my opinion, one should always be ready for this. One aim of the book Sambin (2020, to appear) is to show that the mathematics of existentials is reasonable and clean. It actually often allows to express better, and sometimes understand better, also some aspects of classical mathematics (a little example is the clean description (21.1) of images of subsets exploiting $\between$ besides $\subseteq$).

After realizing that using $\between$ is worthwhile and after getting familiar with it, it becomes a natural step further to express $\between$ in algebraic terms. The standard algebraic counterparts of the structure of subsets of a set is given by complete Boolean algebras (cBa) and complete Heyting algebras (cHa)[9] in the classical and constructive tradition, respectively. They axiomatize properties of operations of subsets with respect to inclusion, which is a universal notion. They are thus here enriched with a new primitive $\asymp$, a binary relation which corresponds to the existential notion $\between$ of overlap between subsets, in the same way as the common notion of partial order $\leq$ corresponds to inclusion $\subseteq$. One then adds some natural axioms on $\asymp$ and its connections with the operations of a cHa:

symmetry if $p \asymp q$ then $q \asymp p$
preservation of meet: if $p \asymp q$ then $p \asymp p \wedge q$
splitting of join: $q \asymp \bigvee_{i \in I} p_i$ iff $q \asymp p_i$ for some $i \in I$,
density: $\forall r(p \asymp r \to q \asymp r) \to p \leq q$,

This new structure has been called an *overlap algebra*.

Density is a strong assumption, which is not algebraic in some strict technical sense. It can be used to show that, reasoning classically, an overlap algebra is exactly the same as a cBa, where $\asymp$ is defined by $p \asymp q \equiv p \wedge q \neq 0$. One can guess that this explains why overlap algebras did not appear in the classical tradition. However, the same result does not hold reasoning constructively over a cHa. Once again, one can see that reasoning classically on a classical definition allows one to obtain something which is not possible by reasoning constructively over the same definition purged of classical aspects. Rather than saying no to some classical arguments on negation, a

[9]We do not consider here some fine distinctions between the three notions of cHa, frames or locales.

constructivist should work harder and find positive mathematical notions to express constructively the illusory strength supplied by classical logic!

The situation is summarized by a table exhibiting the lack of expressiveness in the established constructive tradition:

	Classically	Intuitionistically	Our new notion
Power of a set	cBa = overlap algebra	cHa	Overlap algebra = cHa + $\gtrless$

By mimicking the classical definition and only blocking the classical arguments when dealing with it, before fully appreciating the richness of the original notion in the classical framework, one runs the risk of undermining its expressive power. An overlap algebra is a *strictly* positive notion, which one cannot define by negation without using classical arguments. The presence of $\gtrless$ permits a positive definition of atoms, from which it follows easily that every atomic overlap algebra coincides, up to isomorphisms, with the power of a set. However, the notion of overlap algebra is much more general, see Ciraulo and Sambin (2010).

By exploiting $\between$ once again, one can find a suitable notion of arrows to turn overlap algebras into a category **OA**. We first consider the case of two overlap algebras $\mathcal{P}X$ and $\mathcal{P}S$ for some sets X and S. Using overlap, one can find an abstract characterization of a relation between X and S in terms of operators on subsets, that is operations between $\mathcal{P}X$ and $\mathcal{P}S$. The crucial trick is to define two operations $F : \mathcal{P}X \to \mathcal{P}S$ and $F' : \mathcal{P}S \to \mathcal{P}X$ to be symmetric, written $F \cdot|\cdot F'$, when $FD \between U$ iff $D \between F'U$, for all $D \subseteq X$ and $U \subseteq S$. This notion has a clear existential flavour, and is put aside the familiar notion of adjoint operators $F \dashv G$. Then one can prove that any four operators $F, G' : \mathcal{P}X \to \mathcal{P}S$ and $F', G : \mathcal{P}S \to \mathcal{P}X$ satisfy $F \dashv G$, $F' \dashv G'$ and $F \cdot|\cdot F'$ if and only if there is a relation r from X to S such that $F = r$, $F' = r^-$, $G = r^*$ and $G' = r^{-*}$, where r, r^-, r^{-*}, r^* are images of subsets along r, as defined in (21.1).

This says that relations between X and S are characterized in a way which extends immediately to all overlap algebras. Therefore the notion of morphism between two arbitrary overlap algebras $\mathcal{P}$, $\mathcal{Q}$ is a quadruple of operations $f, f^{-*} : \mathcal{P} \to \mathcal{Q}$ and $f^-, f^* : \mathcal{Q} \to \mathcal{P}$ which form a symmetric pair of adjunctions, that is, such that $f \dashv f^*$, $f^- \dashv f^{-*}$ and $f \cdot|\cdot f^-$.

It is evident, I believe, that without overlap it is very difficult to see this characterization of relations. As a matter of facts, to my knowledge it was not discovered, perhaps even in the classical framework.

The increase of expressive power provided by the new primitive $\gtrless$ is sufficient to formulate most of the basic picture and positive topology in algebraic terms. With overlap algebras the distinction between sets and collections is lost, and so we revert to the usual terminology with points.

The definition of basic topology in algebraic terms, in the terminology of the side of points, becomes an overlap algebra $\mathcal{P}$ equipped with a closure operator C (characterized by $p \leq Cq \leftrightarrow Cp \leq Cq$) and an interior operator I (characterized by $Ip \leq q \leftrightarrow Ip \leq Iq$) which satisfy compatibility: if $Ip \gtrless Cq$ then $Ip \gtrless q$. This is a simple notion, over which one can easily develop an algebraic version of constructive topology. By contraposition, previous absence of $\between$, and hence of compatibility, says that a proper constructive algebraic treatment of topology was missing.

When opens are expressed by a reduction, convergence is easily expressed in algebraic terms by requiring them to be closed under meets: $Ip \wedge Iq \leq I(p \wedge q)$. To express a situation which is valid on topological spaces, we can add a condition of C-I-density: $\forall r(p \between Ir \to q \between Ir) \to p \leq Cq$.

One thus reaches an algebraic version of the basic picture, and hence of topology. So the soundness of our (set theoretic) definitions gets confirmed by the fact that they extend seamlessly to the more general framework of overlap algebras.

My claim is that an algebraic version of open subsets, which is a cHa classically and also intuitionistically, should be an overlap algebra with a reduction I to express open subsets (or a saturation if we are on the formal side). A topology is again a cHa classically, because one defines closed subsets by negation. Intuitionistically it is also cHa, but there one doesn't have a proper notion of closed. I believe the correct notion here is that of overlap algebra equipped with both a reduction I and a saturation C, with suitable axioms. So the table above is extended to:

	Classically	Intuitionistically	Our new notion
Power of a set	cBa	cHa/locale	Overlap algebra
Open subsets	cHa	cHa/locale	Overlap algebra + I
Topology	cHa	cHa/locale	Overlap algebra + I + C

Here again one can see that adopting only the classical definition, constructivists deprive themselves of some tools which they can use to express positively some notions which the classicist treats by way of negation.

21.3 What Can We Achieve? Benefits of a Paradigm Shift

The current situation is really funny. In all the sciences of nature and life, from astronomy to biology, the evolutionary vision is accepted as a matter of fact (except for small minorities). Mathematics, the science on which all others should be based, has instead remained the only one in which a truth that transcends time and humanity is necessary. One accepts that the man descends from the ape, that the bat is similar to the mouse, but it is believed that sets have always existed, and wait only for us to describe them. Is not this a profound inconsistency that needs to be fixed?

Times are ripe to bring a dynamic-evolutionary point of view also within mathematics. I think I have shown with facts that this different conceptual attitude is not only possible, but that indeed it brings with it several advantages, both in theory and in practice.

It can also have a profound impact on the working mathematician. If completeness is impossible by Gödel's theorems, we can give up immediately and accept without any regret the incompleteness of our knowledge. If truth and so also mathematics are not given, and therefore they are neither complete nor bivalent, then it is no longer correct that all that is not false must be true, and all that is consistent exists, and this means that we are not compelled to consider true or existing even statements or entities devoid of meaning.

Without referring to any absolute truth, mathematics is freed from all fundamentalism with a religious flavour and, paradoxically, becomes better founded. And we can feel free from the oppression of a truth hanging over our heads, and relieved of the nightmare of an external authority above us that tells us what is to be considered true, significant or correct, and to which we must be subject even if we do not understand well what it is. Then we can see that mathematics is our creation and management of abstract concepts, and of the actual information that we express with them.

All this on the one hand relaxes us and makes us free, but on the other at the same time makes us responsible. As no mathematics exists before us, we should always be aware of the abstractions by which we pass from reality to mathematical notions. And we must accept, as for all human creations, a plurality of the principles with which we do this, that is a plurality of foundations, and then finally of mathematics to which they give rise. Mathematics indeed cannot be a single one, no mathematics is the absolute truth.

In conclusion, we become more relaxed, free and creative, but also more responsible. In one word, adults (as opposed to Bourbaki).

The cultural attitude of dynamic constructivism is expressed in practice by the Minimalist Foundation MF. MF leaves room for everyone, in the sense that it admits all other foundations as an extension, and basically does not ask for any act of faith. It is therefore ideal as the foundation of an evolutionary, dynamic mathematics, open to different conceptions. Expressing one's position on the basis of a common foundation/language may be helpful in dissolving some communication problems. Assuming the distinction we made here between real mathematics and ideal mathematics, it can be said that, for classicists, mathematics is only the ideal one, while for (orthodox) constructivists mathematics is only the real one. It will be difficult for them to understand each other without clarifying this. In our framework we can treat both, and clarify how they are connected.

Adopting a new foundation like MF really means wearing a pair of new glasses with which to look at reality. Not only it works very well in the practice of mathematics, but it also allows one to see many things that nobody had seen before. A surprise, in general, is that the additional information that one has to keep to operate in this minimalist way is not pure "code" for the computer, but has its own clear logical structure.

In a nutshell, the discovery is that logic and topology are much more deeply linked than before, and that logic becomes useful also in the process of giving definitions. This increased role of logic has a general explanation. In fact, if one does not believe in a truth established outside us, then the least we can ask is that what we are talking about has a clear logical structure.

The turning point is when creativity is freed from the cage of the absolute and of completeness, and gives rise to several new definitions. What is exciting, indeed very appealing, is that in this way many mathematical innovations have emerged, to achieve which, while keeping the old attitude, the (almost) 25 years it took me to develop them would not have been sufficient. All the technical innovations presented in Sect. 21.2 have been induced by dynamic constructivism.

Some results have been found constructively that seem to me of interest to any non-biased mathematician, even classic. And some new facts have been found which had escaped also the "orthodox" constructivists. All this is due to greater attention on both types of information: abstract and computational.

Constructivism therefore does not mean to remake constructively what classicists do. A constructive foundation is not handmaid of the classical one, so much so that one discovers structures that the classical foundation had not seen, even if they were elementary.

Being constructivist does not mean having to give up something in advance. There is no a priori amputation of classical mathematics. Before saying that something is out of reach, one has to try to obtain it.

Independence of thought from the classical vision manifests itself in practice in the freedom to accept or leave, modify, innovate classical definitions. In particular, we see that an abstract predicative mathematics, that is not only computation, is entirely possible.

Due to the methodological choice of compatibility of the minimalist foundation with all others, no ideological commitment is required to appreciate the mathematical innovations it induces. In practice, the new structures of the Basic Picture and Positive Topology remain new and intelligible, and may be of interest, even for mathematicians with different foundational principles, or without active interest in foundational questions.

In addition to novelties, it must be remembered that everything that is done over MF automatically is consistent, indeed, automatically has a concrete meaning, as there is a realizability interpretation. So Enriques' criterion is satisfied.

I think I have shown that one can do mathematics after Cantor, that is, with sets and their generality, without having to submit to the platonist-formalist vision. The common vision today is not necessary, it is only *one* of the possible visions, so much so that it had not seen many things.

Mathematics, even the most advanced, can be done over a foundation which we not only know to be consistent, but of which we also have a specific, computational interpretation. This shows that acts of faith, like that in ZFC, are not necessary. On the contrary, it becomes clear that they are the fruit of a cultural attitude that should now become obsolete, since it no longer has any reason to be.

In the common vision, the role of foundations is to provide safe grounds to the activity of mathematicians, that all in all is taken as a matter of fact. Something similar happens also for the recent approach called Univalent Foundation UF or homotopy interpretation of type theory. UF says nothing about the conception of mathematics, which apparently is to be accepted as given, just like in the classic paradigm; in particular, it keeps silent on pluralism. Only this attitude can explain why UF uses an advanced mathematical discipline, such as homotopy theory, as a foundation of mathematics. A foundation that starts from the top. Apart from these conceptual differences, there seem to be some convergence between MF and some outcomes of UF, and this should be investigated.

Instead, we say that the activity of doing mathematics must go together (not in each mathematician, of course, but as a principle) with the study of its foundations. We put back mathematics together with its foundations. Perhaps it is precisely the dynamic interaction between the study of foundations and the development of mathematics that is the most specific feature of my work.

The fact that someone succeeds in doing mathematics without any dogma, proves that this is possible. I believe that this simple observation should be for a true scientist reason enough to try to do the same, inasmuch as I believe it is a constituent part of science to do without a dogma, whenever this is possible.

Many will continue to think that mathematics is classical mathematics. But now they will have fewer arguments: constructivism amputates mathematics (false), the classical vision is in any case a necessary preliminary to constructivism (false), constructivists will never find anything really new (false), etc.

After all, what I propose is not a "constructive" mathematics, as opposed to classical mathematics, but simply a mathematics which is well-founded and dynamic. However trivial, this is, incredibly, a new paradigm in the sense of Kuhn: to organize the way we interpret the world, and not to describe a platonic universe above our heads.

References

Asperti, A., Maietti, M. E., Sacerdoti Coen, C., Sambin, G., & Valentini, S. (2011). Formalization of formal topology by means of the interactive theorem prover Matita. In J. H. Davenport, W. M. Farmer, J. Urban, & F. Rabe (Eds.), *Intelligent computer mathematics* (Lecture notes in artificial intelligence, Vol. 6824, pp. 278–280). Berlin/Heidelberg: Springer.

Battilotti, G., & Sambin, G. (2006). Pretopologies and a uniform presentation of sup-lattices, quantales and frames. *Annals of Pure and Applied Logic, 137*, 30–61.

Bishop, E. (1967). *Foundations of constructive analysis*. Toronto: McGraw-Hill.

Ciraulo, F., & Sambin, G. (2010). The overlap algebra of regular opens. *Journal of Pure and Applied Algebra, 214*, 1988–1995.

Ciraulo, F., & Sambin, G. (2019). Reducibility, a constructive dual of spatiality. *Journal of Logic and Analysis, 11*, 1–26.

Coquand, T., Sambin, G., Smith, J., & Valentini, S. (2003). Inductively generated formal topologies. *Annals of Pure and Applied Logic, 104*, 71–106.

Dieudonné, J. (1970). The work of Nicholas Bourbaki. *American Mathematical Monthly, 77*, 134–145.

Enriques, F. (1906). Problemi della scienza, Zanichelli. English translation: Problems of Science (introd. J. Royce; trans. K. Royce). Chicago: Open Court, 1914.

Fourman, M. P., & Grayson, R. J. (1982). Formal spaces. In A. S. Troelstra & D. van Dalen (Eds.), The L.E.J. Brouwer centenary symposium (Studies in logic and the foundations of mathematics, pp. 107–122). Amsterdam: North-Holland.

Johnstone, P. T. (1982). *Stone spaces*. Cambridge/New York: Cambridge University Press.

Kuhn, T. S. (1962). *The structure of scientific revolutions* (2nd enl. ed., 1970). Chicago: University of Chicago Press.

Maietti, M. E. (2009). A minimalist two-level foundation for constructive mathematics. *Annals of Pure and Applied Logic, 160*, 319–354.

Maietti, M. E., & Maschio, S. (2014). An extensional Kleene realizability model for the minimalist foundation. In *20th International Conference on Types for Proofs and Programs, TYPES 2014* (pp. 162–186).

Maietti, M. E., & Sambin, G. (2005). Toward a minimalist foundation for constructive mathematics. In L. Crosilla & P. Schuster (Eds.), *From sets and types to topology and analysis. Towards practicable foundations for constructive mathematics* (Oxford logic guides, Vol. 48, pp. 91–114). Oxford/New York: Oxford University Press.

Martin-Löf, P., & Sambin, G. (2020, to appear). Generating positivity by coinduction. In Sambin (2020, to appear).

Martino, E., & Giaretta, P. (1981). Brouwer, Dummett, and the bar theorem. In S. Bernini (Ed.), Atti del Congresso Nazionale di Logica, Montecatini Terme, 1–5 ottobre 1979 (pp. 541–558). Naples: Bibliopolis.

Sacerdoti Coen, C., & Tassi, E. (2011). Formalizing overlap algebras in Matita. *Mathematical Structures in Computer Science, 21*, 1–31.

Sambin, G. (1987). *Intuitionistic formal spaces – a first communication*. In D. Skordev (Ed.), *Mathematical logic and its applications* (Vol. 305, pp. 187–204). New York: Plenum Press.

Sambin, G. (1991). Per una dinamica nei fondamenti. In G. Corsi & G. Sambin (Eds.), Nuovi problemi della logica e della filosofia della scienza (pp. 163–210). Bologna: CLUEB.

Sambin, G. (2002). *Steps towards a dynamic constructivism*. In P. Gärdenfors, J. Wolenski, & K. Kijania-Placek (Eds.), *In the scope of logic, methodology and philosophy of science* (Volume One of the XI International Congress of Logic, Methodology and Philosophy of Science, Cracow, Aug 1999, pp. 261–284). Amsterdam: Elsevier.

Sambin, G. (2008). *Two applications of dynamic constructivism: Brouwer's continuity principle and choice sequences in formal topology*. In M. van Atten, P. Boldini, M. Bourdeau, & G. Heinzmann (Eds.), *One hundred years of intuitionism (1907–2007)* (The Cerisy Conference, pp. 301–315). Basel: Birkhäuser.

Sambin, G. (2011). A minimalist foundation at work. In D. DeVidi, M. Hallett, & P. Clark (Eds.), *Logic, mathematics, philosophy, vintage enthusiasms. Essays in honour of John L. Bell* (The Western Ontario series in philosophy of science, Vol. 75, pp. 69–96). Dordrecht: Springer.

Sambin, G. (2012). Real and ideal in constructive mathematics. In P. Dybjer, S. Lindström, E. Palmgren, & G. Sundholm (Eds.), *Epistemology versus ontology, essays on the philosophy and foundations of mathematics in honour of Per Martin-Löf* (Logic, epistemology and the unity of science, Vol. 27, pp. 69–85). Dordrecht: Springer.

Sambin, G. (2015). Matematica costruttiva. In H. Hosni, G. Lolli, & C. Toffalori (Eds.), Le direzioni della ricerca logica in Italia (Edizioni della Normale, pp. 255–282). Springer.

Sambin, G. (2017). C for constructivism. Beyond clichés. *Lett. Mat. Int., 5*, 1–5. https://doi.org/10.1007/s40329-017-0169-1

Sambin, G. (2020, to appear). *Positive topology and the basic picture. New structures emerging from constructive mathematics* (Oxford Logic Guides). Oxford/New York: Oxford University Press.

Sambin, G., Battilotti, G., & Faggian, C. (2000). Basic logic: Reflection, symmetry, visibility. *The Journal of Symbolic Logic, 65*, 979–1013.

Weyl, H. (1918). Das Kontinuum. Kritische Untersuchungen über die Grundlagen der Analysis, Veit, Leipzig. English trans.: Pollard, S., & Bole, T. (1987). *The continuum. A critical examination of the foundation of analysis*. Philadelphia: Th. Jefferson University Press.

Printed by Printforce, the Netherlands